AF358814

Zebrafish Models for Development and Disease 3.0

Zebrafish Models for Development and Disease 3.0

Editors

James A. Marrs
Swapnalee Sarmah

Basel • Beijing • Wuhan • Barcelona • Belgrade • Novi Sad • Cluj • Manchester

Editors
James A. Marrs
Indiana University-Indianapolis
Indianapolis, IN
USA

Swapnalee Sarmah
Indiana University-Indianapolis
Indianapolis, IN
USA

Editorial Office
MDPI AG
Grosspeteranlage 5
4052 Basel, Switzerland

This is a reprint of articles from the Special Issue published online in the open access journal *Biomedicines* (ISSN 2227-9059) (available at: https://www.mdpi.com/journal/biomedicines/special_issues/zebrafish_3).

For citation purposes, cite each article independently as indicated on the article page online and as indicated below:

Lastname, A.A.; Lastname, B.B. Article Title. *Journal Name* **Year**, *Volume Number*, Page Range.

ISBN 978-3-7258-1593-7 (Hbk)
ISBN 978-3-7258-1594-4 (PDF)
doi.org/10.3390/books978-3-7258-1594-4

Contents

James A. Marrs and Swapnalee Sarmah
Another Swim in the Extensive Pool of Zebrafish Research
Reprinted from: *Biomedicines* **2024**, *12*, 546, doi:10.3390/biomedicines12030546 . . . . **1**

Avital Baniel, Limor Ziv, Zohar Ben-Moshe, Ofer Sarig, Janan Mohamad, Alon Peled, et al.
Cutaneous and Developmental Effects of *CARD14* Overexpression in Zebrafish
Reprinted from: *Biomedicines* **2022**, *10*, 3192, doi:10.3390/biomedicines10123192 . . . . **4**

Tanya Scerbina and Robert Gerlai
Acute Administration of Ethanol and of a D1-Receptor Antagonist Affects the Behavior and
Neurochemistry of Adult Zebrafish
Reprinted from: *Biomedicines* **2022**, *10*, 2878, doi:10.3390/biomedicines10112878 . . . . **14**

**Bridgette E. Drummond, Brooke E. Chambers, Hannah M. Wesselman, Shannon Gibson,
Liana Arceri, Marisa N. Ulrich, et al.**
osr1 Maintains Renal Progenitors and Regulates Podocyte Development by Promoting *wnt2ba*
via the Antagonism of *hand2*
Reprinted from: *Biomedicines* **2022**, *10*, 2868, doi:10.3390/biomedicines10112868 . . . . **35**

Yasong Zhao, Xiaohui Li, Guili Song, Qing Li, Huawei Yan and Zongbin Cui
Comparative Transcriptome Analysis Provides Novel Molecular Events for the Differentiation
and Maturation of Hepatocytes during the Liver Development of Zebrafish
Reprinted from: *Biomedicines* **2022**, *10*, 2264, doi:10.3390/biomedicines10092264 . . . . **56**

**Sara Bozzer, Luca De Maso, Maria Cristina Grimaldi, Sara Capolla, Michele Dal Bo,
Giuseppe Toffoli and Paolo Macor**
Zebrafish: A Useful Animal Model for the Characterization of Drug-Loaded Polymeric NPs
Reprinted from: *Biomedicines* **2022**, *10*, 2252, doi:10.3390/biomedicines10092252 . . . . **74**

Jeng-Wei Lu, Liang-In Lin, Yuxi Sun, Dong Liu and Zhiyuan Gong
Effect of Lipopolysaccharides on Liver Tumor Metastasis of *twist1a/kras*^{V12} Double Transgenic
Zebrafish
Reprinted from: *Biomedicines* **2022**, *10*, 95, doi:10.3390/biomedicines10010095 . . . . **87**

**Chiara Tobia, Daniela Coltrini, Roberto Ronca, Alessandra Loda, Jessica Guerra,
Elisa Scalvini, et al.**
An Orthotopic Model of Uveal Melanoma in Zebrafish Embryo: A Novel Platform for Drug
Evaluation
Reprinted from: *Biomedicines* **2021**, *9*, 1873, doi:10.3390/biomedicines9121873 . . . . **105**

Xiang Chen, Yuwen Huang, Pan Gao, Yuexia Lv, Danna Jia, Kui Sun, et al.
Knockout of *mafba* Causes Inner-Ear Developmental Defects in Zebrafish via the Impairment of
Proliferation and Differentiation of Ionocyte Progenitor Cells
Reprinted from: *Biomedicines* **2021**, *9*, 1699, doi:10.3390/biomedicines9111699 . . . . **118**

Xixi Li, Guili Song, Yasong Zhao, Jing Ren, Qing Li and Zongbin Cui
Functions of *SMC2* in the Development of Zebrafish Liver
Reprinted from: *Biomedicines* **2021**, *9*, 1240, doi:10.3390/biomedicines9091240 . . . . **139**

Jeng-Wei Lu, Yuxi Sun, Pei-Shi Angelina Fong, Liang-In Lin, Dong Liu and Zhiyuan Gong
Lipopolysaccharides Enhance Epithelial Hyperplasia and Tubular Adenoma in
Intestine-Specific Expression of *kras*^{V12} in Transgenic Zebrafish
Reprinted from: *Biomedicines* **2021**, *9*, 974, doi:10.3390/biomedicines9080974 . . . . **152**

Yang-Wen Hsieh, Yi-Wen Tsai, Hsin-Hung Lai, Chi-Yu Lai, Chiu-Ya Lin and Guor Mour Her
Depletion of Alpha-Melanocyte-Stimulating Hormone Induces Insatiable Appetite and Gains
in Energy Reserves and Body Weight in Zebrafish
Reprinted from: *Biomedicines* **2021**, *9*, 941, doi:10.3390/biomedicines9080941 . . . . **173**

**James Hentig, Kaylee Cloghessy, Manuela Lahne, Yoo Jin Jung, Rebecca A. Petersen,
Ann C. Morris and David R. Hyde**
Zebrafish Blunt-Force TBI Induces Heterogenous Injury Pathologies That Mimic Human TBI
and Responds with Sonic Hedgehog-Dependent Cell Proliferation across the Neuroaxis
Reprinted from: *Biomedicines* **2021**, *9*, 861, doi:10.3390/biomedicines9080861 . . . . **191**

Priyadharshini Manikandan, Swapnalee Sarmah and James A. Marrs
Ethanol Effects on Early Developmental Stages Studied Using the Zebrafish
Reprinted from: *Biomedicines* **2022**, *10*, 2555, doi:10.3390/biomedicines10102555 . . . . **225**

biomedicines

Editorial

Another Swim in the Extensive Pool of Zebrafish Research

James A. Marrs * and Swapnalee Sarmah

Department of Biology, Indiana University Indianapolis, 723 West Michigan Street, Indianapolis, IN 46202, USA; swapnalee.sarmah@yahoo.com
* Correspondence: jmarrs@iu.edu; Tel.: +1-317-278-0031

Citation: Marrs, J.A.; Sarmah, S. Another Swim in the Extensive Pool of Zebrafish Research. *Biomedicines* 2024, 12, 546.
https://doi.org/10.3390/biomedicines12030546

Received: 15 February 2024
Revised: 22 February 2024
Accepted: 26 February 2024
Published: 29 February 2024

1. Neural Systems

The zebrafish has gained utility in modeling biomedical phenomena for discovery research. Over 55,000 entries are found by searching the term "zebrafish" on PubMed. The creativity of the research community adds new approaches and targets for zebrafish models. Experiments using genome-wide methods have helped identify its mutations, transcription and chromatin structure. Genetic models are being applied to understand gene function and human disease. Experimental manipulations are powerful and reveal new mechanisms and biological knowledge stemming from zebrafish. This third volume of "Zebrafish Models for Development and Disease" adds new knowledge to the growing body of literature. An introduction highlighting a few of the papers in this Special Issue is presented below.

Traumatic brain injury is a widespread problem. Trauma in general is poorly understood and understudied. Growing attention is being paid to traumatic brain injury to support patients and to understand neural trauma [1,2]. David Hyde's laboratory developed a consistent and reproducible model using adult zebrafish [3]. Their model produces a variety of injury phenotypes, from brain tissue disruptions to systemic effects, like edema and inflammation. The zebrafish is a regenerative model that allows investigators to evaluate these responses. Sonic hedgehog signaling was shown to regulate neural proliferation and regeneration responses following injury.

Alcohol consumption produces widespread societal problems. Progress is being made to identify its causation and treatment, including genetic interactions [4,5]. Animal models help us understand alcohol's effects on the brain, and zebrafish have been used to great effect in Robert Gerlai's laboratory. In their contribution to this Special Issue, adult zebrafish were treated with D1-dopamine-receptor antagonists and alcohol to determine whether these effects interact [6]. Experiments compared inbred AB and heterogenous SFWT genetic strains. They found synergy between D1-dopamine-receptor antagonists and alcohol for shoaling behavior but found additive effects on exploratory behavior. They next tested the effects of D1-dopamine-receptor antagonists and alcohol treatments on neurotransmitter levels and found interactions between dopamine and DOPAC (metabolite of dopamine) levels. Only alcohol affected the serotonin levels, showing specificity in their model. Behavior paradigms are needed to extend zebrafish models' utility in neuroscience, and the groundwork in this study is important and necessary.

2. Tumor Biology

Zebrafish are versatile for producing animal models for various conditions, including tumor progression [7]. Sara Rezzola and colleagues developed a zebrafish platform to evaluate drug efficacy on human and mouse uveal melanoma cells [5]. They transplanted uveal melanoma tumor xenografts into the zebrafish eye near the developing choroid vasculature. These cells grew and invaded the eye tissues. Using a luciferase-expressing cell line, they quantified the tumor response to chemotherapeutic drugs. This is an exciting addition to the assay arsenal.

Inflammation contributes to tumor progression; this was studied in the zebrafish in Zhiyuan Gong's laboratory using a transgenic *krasV12*-expressing oncogene in the intestine and using lipopolysaccharide or dextran sulfate sodium inflammatory compounds as treatment [7]. Treatment with the inflammatory compounds in the transgenic *krasV12*-expressing fish increased neutrophils and macrophages in the intestine. The synergy of the oncogene and inflammation produced more hyperplasia and tumorogenesis, including specific cellular proliferation, apoptosis and signaling responses. Tumor progression changed the cellular composition and morphology in the intestine. The authors plan to use this model to investigate tumor initiation mechanisms and test antitumor drugs.

3. Liver Biology

Exploiting zebrafish as models for liver biology represents an opportunity to analyze, validate and consolidate our understanding derived from other models, like rodents [8]. Zebrafish are outstanding developmental models, and the study conducted in Zongbin Cui's laboratory used the fish system to characterize the transcriptome changes during liver development [8]. Hepatocytes were sorted using flow cytometry at three stages. The authors identified genes whose transcription changed over these times. Gene ontogeny (GO) was used to categorize the types of activities that changed during development. Initial changes (60 to 72 h postfertilization; hpf) were seen in the cell cycle, DNA replication, DNA repair, RNA processing and transcription regulation. Later (72 to 96 hpf), the Kyoto Encyclopedia of Genes and Genomes (KEGG) was used to categorize the changes that occurred in the hepatocytes, including changes in the cell cycle, RNA degradation, ubiquitin-mediated proteolysis, signaling pathways, basal transcription factors and glycan degradation. These pathways included similar activities as those seen at the earlier stages. In addition, the metabolic pathways were upregulated, including nucleic acid bases, energy carriers, amino acids, ABC transporters and p53, which participates in many pathways. This study will provide a guide for future studies of liver development, stem cell biology and liver disease processes.

4. Kidney Biology

The kidney is an organ that highlights simplicity in the zebrafish system; it only has one fused glomerulus and two nephrons in the embryo [9], but still has the powers of genetics and regenerative capacity found in zebrafish [10]. Genetic defects, like polycystic kidney disease, are modeled in zebrafish [11], as are other ciliopathies [12]. Indeed, zebrafish are useful models to study kidney development and disease [13,14].

In Dr. Rebecca Wingert's laboratory, Drummond et al. identified the *osr1* gene mutation that affects kidney development [15]. This gene normally promotes the podocyte lineage, which are the cells forming the glomerulus filter. Osr1 is a zinc finger transcription factor, and the authors show that Osr1 promotes the expression of the paracrine signaling ligand Wnt2ba. Indeed, they show that Wnt2ba expression can partially rescue *osr1* mutant fish, allowing for renal podocyte development. They illustrate that the Osr1 and Hand2 transcription factors antagonize each other, regulating the podocyte progenitor pathway. This is an exceptional study showing the power of zebrafish models to dissect complex developmental biology pathways.

5. Conclusions

Overall, this Special Issue illustrates the spectrum of ways that zebrafish can be applied in biomedical research. New and creative approaches continue to uncover knowledge about development and disease. New technologies, particularly genome-wide methods, are helping us to understand biological and disease processes. The future is bright, and there are extraordinary possibilities for growth in the zebrafish research ecosystem.

Author Contributions: Writing—original draft preparation, J.A.M.; writing—review and editing, J.A.M. and S.S. All authors have read and agreed to the published version of the manuscript.

Funding: This work was supported by NIH/NIAAA 1 R21 AA026711.

Data Availability Statement: Not applicable.

References

1. Blennow, K.; Brody, D.L.; Kochanek, P.M.; Levin, H.; McKee, A.; Ribbers, G.M.; Yaffe, K.; Zetterberg, H. Traumatic brain injuries. *Nat. Rev. Dis. Primers* **2016**, *2*, 16084. [CrossRef] [PubMed]
2. Maas, A.I.R.; Menon, D.K.; Manley, G.T.; Abrams, M.; Akerlund, C.; Andelic, N.; Aries, M.; Bashford, T.; Bell, M.J.; Bodien, Y.G.; et al. Traumatic brain injury: Progress and challenges in prevention, clinical care, and research. *Lancet Neurol.* **2022**, *21*, 1004–1060. [CrossRef] [PubMed]
3. Hentig, J.; Cloghessy, K.; Lahne, M.; Jung, Y.J.; Petersen, R.A.; Morris, A.C.; Hyde, D.R. Zebrafish Blunt-Force TBI Induces Heterogenous Injury Pathologies That Mimic Human TBI and Responds with Sonic Hedgehog-Dependent Cell Proliferation across the Neuroaxis. *Biomedicines* **2021**, *9*, 861. [CrossRef] [PubMed]
4. Agrawal, A.; Brislin, S.J.; Bucholz, K.K.; Dick, D.; Hart, R.P.; Johnson, E.C.; Meyers, J.; Salvatore, J.; Slesinger, P.; Collaborators, C.; et al. The Collaborative Study on the Genetics of Alcoholism: Overview. *Genes Brain Behav.* **2023**, *22*, e12864. [CrossRef] [PubMed]
5. Johnson, E.C.; Salvatore, J.E.; Lai, D.; Merikangas, A.K.; Nurnberger, J.I.; Tischfield, J.A.; Xuei, X.; Kamarajan, C.; Wetherill, L.; Collaborators, C.; et al. The collaborative study on the genetics of alcoholism: Genetics. *Genes Brain Behav.* **2023**, *22*, e12856. [CrossRef] [PubMed]
6. Scerbina, T.; Gerlai, R. Acute Administration of Ethanol and of a D1-Receptor Antagonist Affects the Behavior and Neurochemistry of Adult Zebrafish. *Biomedicines* **2022**, *10*, 2878. [CrossRef] [PubMed]
7. Mione, M.C.; Trede, N.S. The zebrafish as a model for cancer. *Dis. Model. Mech.* **2010**, *3*, 517–523. [CrossRef] [PubMed]
8. Goessling, W.; Sadler, K.C. Zebrafish: An important tool for liver disease research. *Gastroenterology* **2015**, *149*, 1361–1377. [CrossRef]
9. Poureetezadi, S.J.; Wingert, R.A. Little fish, big catch: Zebrafish as a model for kidney disease. *Kidney Int.* **2016**, *89*, 1204–1210. [CrossRef] [PubMed]
10. Gemberling, M.; Bailey, T.J.; Hyde, D.R.; Poss, K.D. The zebrafish as a model for complex tissue regeneration. *Trends Genet.* **2013**, *29*, 611–620. [CrossRef] [PubMed]
11. Chen, Z.; Drummond, I.A. Polycystin-2, mechanosensing, and left-right asymmetry in autosomal dominant polycystic kidney disease. *Kidney Int.* **2023**, *104*, 638–640. [CrossRef]
12. Delvallee, C.; Dollfus, H. Retinal Degeneration Animal Models in Bardet-Biedl Syndrome and Related Ciliopathies. *Cold Spring Harb. Perspect. Med.* **2023**, *13*, a041303. [CrossRef] [PubMed]
13. Chambers, B.E.; Weaver, N.E.; Lara, C.M.; Nguyen, T.K.; Wingert, R.A. (Zebra)fishing for nephrogenesis genes. *Tissue Barriers* **2023**, 2219605. [CrossRef]
14. Hukriede, N.A.; Soranno, D.E.; Sander, V.; Perreau, T.; Starr, M.C.; Yuen, P.S.T.; Siskind, L.J.; Hutchens, M.P.; Davidson, A.J.; Burmeister, D.M.; et al. Experimental models of acute kidney injury for translational research. *Nat. Rev. Nephrol.* **2022**, *18*, 277–293. [CrossRef]
15. Drummond, B.E.; Chambers, B.E.; Wesselman, H.M.; Gibson, S.; Arceri, L.; Ulrich, M.N.; Gerlach, G.F.; Kroeger, P.T.; Leshchiner, I.; Goessling, W.; et al. *osr1* Maintains Renal Progenitors and Regulates Podocyte Development by Promoting *wnt2ba* via the Antagonism of *hand2*. *Biomedicines* **2022**, *10*, 2868. [CrossRef] [PubMed]

 *biomedicines*

Article

Cutaneous and Developmental Effects of *CARD14* Overexpression in Zebrafish

Avital Baniel [1,*], Limor Ziv [2], Zohar Ben-Moshe [3], Ofer Sarig [1], Janan Mohamad [1,4], Alon Peled [1], Gideon Rechavi [2], Yoav Gothilf [3] and Eli Sprecher [1,4]

[1] Division of Dermatology, Tel Aviv Sourasky Medical Center, Tel Aviv 64239, Israel
[2] Cancer Research Center, Sheba Medical Center, Ramat Gan 52620, Israel
[3] Department of Neurobiology, The George S. Wise Faculty of Life Sciences, Tel-Aviv University, Tel Aviv 6997801, Israel
[4] Department of Human Molecular Genetics and Biochemistry, Sackler Faculty of Medicine, Tel-Aviv University, Tel Aviv 6997801, Israel
* Correspondence: avital.baniel@gmail.com

Abstract: Background: Gain-of-function mutations in *CARD14* have recently been shown to be involved in the pathogenesis of psoriasis and pityriasis rubra pilaris (PRP). Those mutations were found to activate the NF-kB signaling pathway. Objective: Zebrafish is often used to model human diseases in general, and in skin disorders more particularly. In the present study, we aimed to examine the effect of *CARD14* overexpression in zebrafish with the aim to validate this model for future translational applications. Methods: We used light microscopy, scanning electron microscopy, histological analysis and whole mount in situ hybridization as well as real-time PCR to ascertain the effect of *CARD14* overexpression in the developing zebrafish. Results: Overexpression of human *CARD14* had a marked morphological and developmental effect on the embryos. Light microscopy demonstrated a characteristic cutaneous pattern including a granular surface and a spiky pigment pattern. In situ hybridization revealed keratinocytes of uneven size and shape. Scanning electron microscopy showed aberrant production of actin microridges and a rugged keratinocyte cell surface, reminiscent of the human hyperkeratotic phenotype. Developmentally, overexpression of *CARD14* had a variable effect on anterior-posterior axis symmetry. Similar to what has been observed in humans with psoriasis or PRP, NF-kB expression was higher in *CARD14*-overexpressing embryos compared to controls. Conclusions: Overexpression of *CARD14* results in a distinct cutaneous pattern accompanied by hyperactivation of the NF-kB pathway, suggesting that the zebrafish represents a useful system to model *CARD14*-associated papulosquamous diseases.

Keywords: zebrafish; *CARD14*; psoriasis

Citation: Baniel, A.; Ziv, L.; Ben-Moshe, Z.; Sarig, O.; Mohamad, J.; Peled, A.; Rechavi, G.; Gothilf, Y.; Sprecher, E. Cutaneous and Developmental Effects of *CARD14* Overexpression in Zebrafish. *Biomedicines* **2022**, *10*, 3192. https://doi.org/10.3390/biomedicines10123192

Academic Editor: James A. Marrs

Received: 29 September 2022
Accepted: 6 December 2022
Published: 8 December 2022

1. Introduction

Psoriasis and pityriasis rubra pilaris (PRP) are common papulosquamous diseases which bear some overlapping features but have been traditionally considered as distinct clinical and histopathological entities [1]. Many genetic factors have been implicated in the pathogenesis of psoriasis [2]. In PRP, familial disease is rarer, and is estimated to account for approximately 5% of the cases [3]. Of interest, the study of the genetic basis of both diseases has revealed that they may share a common etiology. Indeed, familial cases of either psoriasis or PRP have been shown to be caused by mutations in the same gene *CARD14*, encoding the caspase recruitment domain-containing protein 14 [4,5]. In fact *CARD14* maps to the PSORS2 psoriasis-associated locus [6,7]. *CARD14* mutations have been shown to cause a wide range of phenotypes, recently termed CARD14-associated papulosquamous eruption (CAPE) [8]. Most disease-causing mutations in *CARD14* affect the coiled-coil domain of the protein and are gain-of-function variants [9]. In contrast loss-of-function mutations in *CARD14* have been shown to cause an atopic dermatitis-like phenotype [9–12].

CARD14 is a member of the CAspase Recruitment Domain (CARD) family of proteins which mediate signal transduction in apoptotic and inflammatory processes. These proteins share a CARD domain comprising 6 or 7 anti-parallel alpha-helices which enable highly specific homophilic interactions between proteins involved in signal transduction [13,14]. CARD14 functions by activating the NF-kB signaling pathway as well as the p38 and JNK MAP kinase pathways, by forming a complex with BCL10 and the paracaspase MALT1 [15].

Despite its evolutionary distance, the zebrafish model shares many biological features with humans [16]. A homolog of *CARD14* has been found in several teleost fishes including the zebrafish [17], but its physiological function is not yet explored. Phylogenetic analysis of the two species results in a bootstrap value of 91% and the CARD domain, essential for binding BCL10 and activation of NFkB is conserved [17]. Since psoriasis- and PRP-causing *CARD14* mutations have been shown to result in a gain-of-function [4,18] and increased expression of CARD14 in the skin [4,15], we reasoned that overexpression of *CARD14* in zebrafish may result in a phenotype relevant to human diseases which may subsequently be of help in ascertaining the effect of potential treatments on CARD14-induced cell signaling.

2. Materials and Methods

2.1. DNA and RNA Injection to Zebrafish Embryos

A male and a female fish were brought together into a laying chamber towards the evening of the day prior to the injection. On the morning of the injection day, at first light, the embryos were collected as a single cell (0–30 min after fertilization). The embryos were arrayed on a 4% agarose injection plate.

For injection of human *CARD14* DNA (*hcard14*), the embryos were injected with *CARD14*-expression constructs of wildtype or PRP-causing mutation (p.Glu138del) which were generated as previously described [18]. Control embryos consisted of either mock-injected embryos or embryos injected with an empty DNA plasmid (pcDNA3.1).

For injection of zebrafish *card14* mRNA (*zfcard14*), the coding sequence was PCR-amplified from larval cDNA (F,TCGGATTTAGGGATTTCAGA ; R,TATCTGGGTGTCATGCCTCA) and cloned into pCS2+ vector based on NEBuilder® Assembly tool (http://nebuilder.neb.com, accessed on 3 June 2019) using the NEBuilder® HiFi DNA Assembly Cloning Kit (New England BioLabs, Ipswich, MA, USA). Control embryos consisted of either *gfp mRNA* or *card14 mRNA* containing a stop codon replacing the start codon. Introduction of the stop mutation was accomplished by the Q5® Site-Directed Mutagenesis Kit (New England BioLabs). Primers for mutagenesis were designed utilizing the NEBaseChanger® tool (http://nebasechanger.neb.com, accessed on 3 June 2019). Cloning and mutagenesis were validated by direct sequencing. DNA or RNA were loaded into an internal filament of a capillary that was stretched with a micro pipette puller. The capillary was then connected to the injection device (air pressure microinjector-PV830 Pneumatic PicoPump, Sarasota, FL, USA). Approximately 1 nL of DNA was injected into a fertilized egg at a single cell stage. All zebrafish procedures were approved by the Tel-Aviv University Animal Care Committee (04-18-051) and conducted in accordance with the National Council for Animal Experimentation, Ministry of Health, Israel.

2.2. Whole Mount In Situ Hybridization (WISH)

Larvae samples were fixed in 4% PFA overnight at 4 °C and subsequently dehydrated by serial dilutions of methanol (25%, 50%, 75%, 100%) in phosphate-buffered saline solution (PBST) supplemented with 0.1% Tween-20. Dehydrated larvae were maintained at $-20°C$. For WISH, larvae were rehydrated by a reversed gradient of methanol in PBST. Then, depigmentation was performed with 5% H_2O_2 in PBST. Permeabilization was achieved by 10 µg/mL of proteinase K in PBST. Then, larvae were fixed in 4% PFA for 30 min at room temperature. Fixed larvae were washed by PBST (3×) and incubated in a pre-hybridization buffer [1 mM EDTA, 5XSSC (0.75 M NaCl, 75 mM sodium citrate, pH 7.0.), 2% Roche blocking powder, 50% formamide, 0.1%, 1 mg/mL Torula yeast RNA, 0.1%

CHAPS, DEPC-treated ddH$_2$O, Triton-X, 50 mg/mL heparin) at 65 °C overnight. On the next day, riboprobes designed using Primer3 (https://primer3.ut.ee/, accessed on 1 July 2019) (Table S1) were added and incubation was pursued for additional 48 h, prior to washing out of unbound riboprobes by 2X SSC and 0.2X SSC (each solution 3 × 30 min).

2.3. Scanning Electron Microscopy (SEM)

The samples were fixed by Kranovsky fixative 4% PFA, 2% glutaraldehyde (GA) in 0.1 M Cacodilate (Caco) buffer. Subsequently, samples were stained with 1% OsO4 in Caco 0.1 M buffer, dehydrated in an ethanol gradient (50%, 70%, 96%, and 100%), and critical-point dried using a critical-point dryer BAL-TEC CPD 030 (Leica Bio-Systems, Wetzlar, Germany). The dried samples were then put on aluminum stabs covered with carbon tape and coated with a thin layer of gold/palladium alloy in Edwards sputter coater. Sample visualization was attained by secondary electron detector in a high-resolution Ultra 55 SEM (Zeiss, Jena, Germany).

2.4. Histology

Larvae samples were fixed in 4% PFA, dehydrated in 70% ethanol and embedded in paraffin. Cross sections (2 μm thick) were cut using a Shandon M1R rotary microtome (Marshall Scientific, Hampton, NH, USA) and stained by hematoxylin and eosin (H&E).

2.5. RNA Purification

Fifty-100 mg of tissue were homogenized with 1 mL of TRI Reagent® (Sigma-Aldrich, St. Louis, MO, USA) at room temperature for 5 min. Then, 0.2 mL of chloroform was added for every 1 mL of TRI Reagent®. Tubes were shaken for 15 s, incubated at room temperature for 15 min and then centrifuged at 12,000× g for 15 min at 4 °C. The supernatant was separated, 0.5 mL of isopropanol was added per 1 mL of TRI Reagent®. Samples were then incubated at room temperature for 10 min and centrifuged at 12,000× g for 10 min at 4 °C. The fluid was discarded and 1 mL of ethanol 75% was supplemented to the pellet for every 1 mL used in the preparation of the sample and then centrifuged at 7500× g for 5 min at 4 °C.

2.6. Real Time PCR

For quantitative real-time PCR (qPCR) analysis, cDNA was synthesized from 1000 ng of total RNA by the qScript kit (Quanta Biosciences, Gaithersburg, MD, USA). cDNA PCR amplification was performed using the PerfeCTa SYBR Green FastMix (Quanta Biosciences, Gaithersburg, MD, USA) on a StepOnePlus system (Applied Biosystems, Waltham, MA, USA) with gene-specific intron-crossing pairs of oligonucleotides (designed using primer3 (Table S2). Cycling conditions were: 95 °C, 20 s and then 95 °C, 3 s; 60 °C, 30 s for 40 cycles. Samples were analyzed in triplicates. For each primer set, standard curves were obtained with successive cDNA dilutions. Results were normalized to *ppiab* (NM_199957) mRNA levels. For list of primers, see Supplementary Material.

3. Results

3.1. Effect of CARD14 Overexpression on Zebrafish Larvae Development

Zebrafish embryos were injected with RNA expressing *zfcard14*, h*CARD14* and *gfp* as control. Overall, 197 larvae overexpressing either zfCARD14 (136) or hCARD14 (61) and 140 controls were examined. The larvae were examined from day 1 to 7 post fertilization. A markedly disturbed developmental pattern was observed, affecting mainly the caudal pole. Of note, the phenotype was abnormal albeit of variable severity with curved and irregularly shaped tails up to almost complete absence of tail development (Figure 1a–e). Cephalic development was normal at the exception of occasional asymmetrical eye development. Cardiac edema was also a common finding.

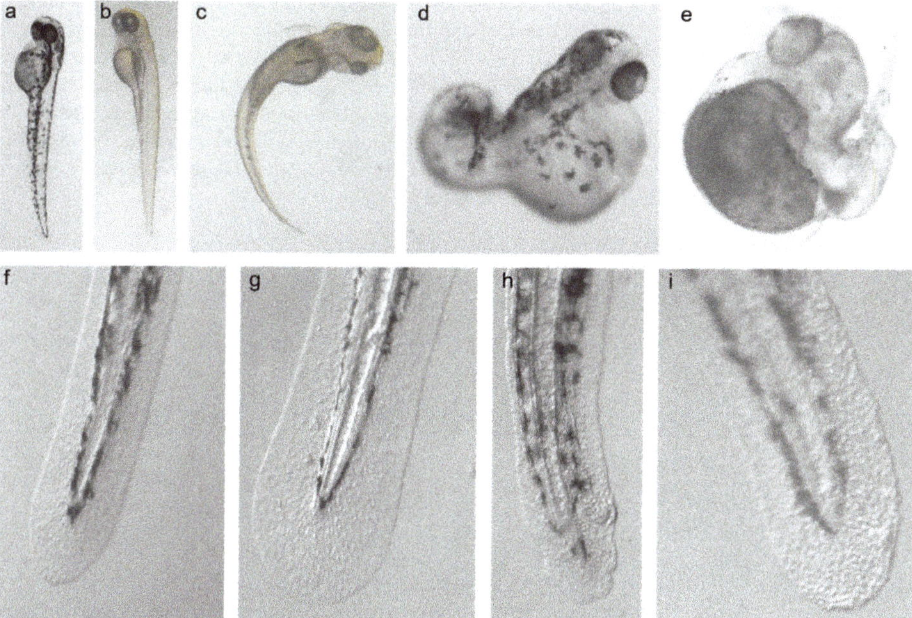

Figure 1. Developmental and cutaneous effects of *Card14* overexpression. Three days post-fertilization (dpf) uninjected (**a**) and *gfp*-RNA (**b**) injected larvae as compared to larvae injected with *zfCard14* RNA (**c–e**). The latter demonstrating a wide range of disturbed caudal development. Note asymmetric eye development in (**c,e**). Smooth surface, regular border and uniform pigmentation are seen in uninjected (**f**) and *gfp*-RNA injected (**g**) larvae; in contrast, irregular borders, spiky pigment and a "granular" surface are seen in larva injected with *zfcard14* (**h**) and *hcard14* (**i**) RNA.

3.2. Effect of CARD14 Overexpression on Zebrafish Larvae Cutaneous Morphology

Meticulous examination of skin morphology revealed distinctive cutaneous morphological changes. The skin border of larvae injected with *zfcard14* and *hCARD14* was irregular, showing a pattern of "hills and valleys". In addition, larvae skin contained clusters of granules as opposed to the smoother and more uniform appearance of the surface of control larvae (Figure 1f–i). In contrast with control larvae in which symmetrically organized pigmented globules with uniform borders were seen, in larvae overexpressing CARD14, an asymmetrical spiky pattern of pigment was often observed (Figure 1h). As is the rule in zebrafish embryo injections, developmental abnormalities were occasionally noted in controls to some extent, but none showed a conspicuous and redundant developmental or cutaneous pattern as in CARD14 overexpressing larvae. Embryos injected with a vector carrying a PRP-causing mutation (p.Glu138del) showed the same cutaneous and developmental pattern.

3.3. Effect of CARD14 Overexpression on Keratinocyte Morphology

To examine the effect of *CARD14* overexpression on epidermal cellular morphology, we used WISH to determine the expression of the cytokeratin 1-encoding gene. This gene is specifically expressed in the zebrafish developing epidermis and therefore allows for the delineation of keratinocyte morphology. A total of 43 larvae overexpressing *CARD14* and 32 controls were examined, all displaying relatively normal gross development. Keratinocytes of control larvae displayed equal distribution and uniform shape and size, whereas keratinocytes of *zfcard14*-overexpressing embryos show an unequal distribution and size. Clusters of cells could be seen in certain areas, and a range of cell size was seen, from small to "giant cells" (Figure 2a–f). To further characterize cutaneous cellular changes, we examined the larvae via scanning electron microscopy. Ten larvae overexpressing either *zfcard14* or *hCARD14* and 10 *gfp* injected controls were examined. Aberrant production of apical microridges was seen. In control embryos, an array of microridges overlie the cell surface that is smooth. In *zfcard14*-overexpressing embryos, microridge production is disturbed, and the cell surface displays a mountainous topography, protruding above cell surface, suggestive of hyperkeratosis (Figure 2g–j). Finally, larvae overexpressing *zfcard14* displayed thicker skin than control larvae, as revealed by measurements of the skin in larval histological sections (Figure 3).

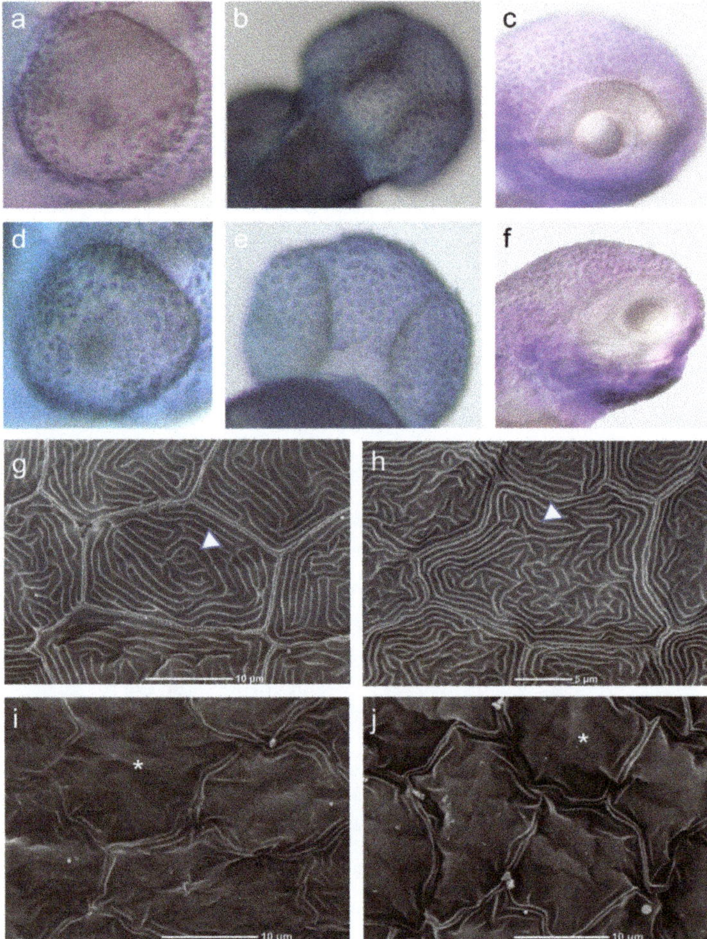

Figure 2. Keratinocyte morphology. Upper panel: Whole mount in situ hybridization with a probe recognizing cytokeratin1 mRNA (dark stain) at 2 dpf (**a,c,d,f**) and 3 dpf (**b,e**) demonstrating the distribution and morphology of keratinocytes over the eye (**a,d**) and over the head (**b,c,e–f**) areas. Control larvae were injected with an empty plasmid (**b**) or *gfp*RNA (**a,c**), and compared to larvae injected with *hCARD14* (**e**) or *zfcard14* (**d,f**). Note in both regions the uneven distribution and size as well as cell cluster formation as a consequence of *CARD14* overexpression as opposed to the even size and regular borders of control keratinocytes. Lower panel: Scanning electron microscopy of 2 dpf embryos. Uninjected (**g**) and *gfp*-RNA injected (**h**) embryos display a uniform and complete array of microridges (triangle), and smooth cell surface, while *zfcard14* (**i**) and *hCARD14* injected (**j**) embryos display uneven cell surface topography (asterisk).

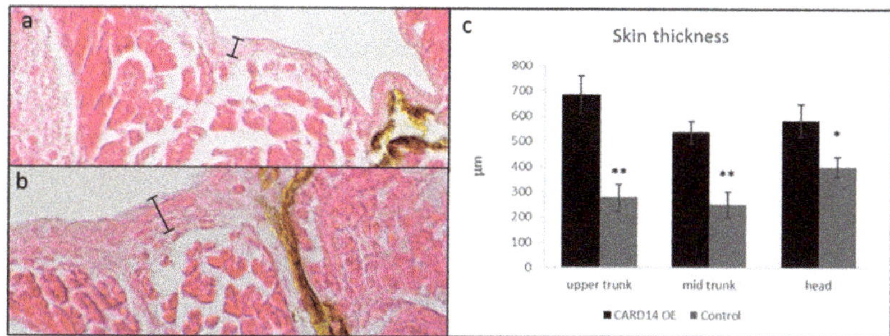

Figure 3. Histopathology at 3 dpf. H&E staining ×40 magnification. Skin of control larvae (**a**) injected with *zfcard14*-RNA containing a stop codon display a thinner epidermis than larvae injected with *zfcard14*-RNA (**b**). Graph (**c**) shows average of 6 measurements at 3 different anatomical sites: head at superior edge of eyes, superior and mid trunk (two-sided *t*-test; * $p < 0.05$, ** $p < 0.01$).

3.4. Over Expression of CARD14 Affects Nfκb and Planar Cell Polarity Signaling Pathways

Since gain of function mutations in *CARD14* up-regulate Nfκb activity in vitro and in vivo [4,18], we examined the effect of *CARD14* overexpression on this pathway activity in injected embryos. Using, qPCR, we observed a significantly higher expression of *nfkb2*, encoding a light polypeptide gene enhancer in B cells 2 (p49/p100), in injected embryos compared to controls. *ccl20b*, encoding a cytokine induced by Nfκb was up-regulated as well (Figure 4).

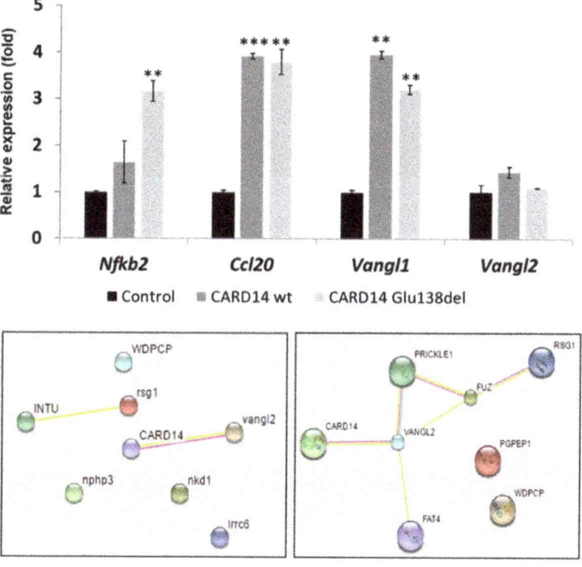

Figure 4. Effect of *CARD14* overexpression on gene expression. Embryos were injected with expression vectors carrying human wild type and PRP-causing *CARD14* mutation (p.Glu138del). Relative expression of *nfkb2*, *ccl20b*, *vangl1* and *vangl2* was ascertained 1 dpf using qPCR (upper panel). Results represent the mean of 3 replicates and are provided as gene expression relative to gene expression in control embryos ± standard error normalized to *ppiab* mRNA levels (two-sided *t*-test; ** $p < 0.01$, *** $p < 0.001$). Protein interaction (STRING, http://string-db.org, accessed on 17 May 2019) analysis of *CARD14* and planar cell polarity proteins in zebrafish (left lower panel) and human (right lower panel) is predictive of a functional interaction between *CARD14* and VANGL2.

The prominent asymmetry of the developing embryos overexpressing CARD14 prompted us to explore the effect of *CARD14* on genes encoding proteins regulating planar cell polarity. Querying the STRING protein interaction prediction program, we found a predicted direct interaction between *CARD14* and the Van Gogh-like protein 2 (VANGL2), a planar cell polarity protein involved in the WNT/PCP pathway (Figure 4). We therefore examined the expression of fish *vangl1*, *vangl2* and *daam1* encoding a protein (Dishevelled Associated Activator Of Morphogenesis 1) expressed downstream to *vangl2*. qPCR showed higher expression of *vangl1* and to a lesser extent *vangl2* in 1 dpf embryos overexpressing *CARD14*, compared to control embryos (Figure 4). Expression of *daam1* did not change. These results suggest CARD14 not only affects the skin, but may alter cephalo-caudal development via interaction with PCP pathway proteins.

4. Discussion

The discovery of the role played by CARD14 in familial psoriasis and PRP [4,5] has not only led to the elucidation of the role of this protein in numerous acquired forms of these disorders [5,18–28], but also positions CARD14 as a novel therapeutic and possibly pharmacogenetic target [15,29–31]. A practical, robust and rapid model for therapeutic screen is therefore urgently needed. In the present study, we ascertained the zebrafish for its ability to replicate the biological abnormalities seen in humans with gain-of-function mutations in *CARD14*.

Overexpression of *zfcard14* and *hCARD14* in zebrafish revealed marked and reproducible abnormalities in cutaneous morphology including a serrated skin line, spiky pigmentation and a granular surface reflecting the formation of cell aggregates, corresponding to hyperkeratosis. In situ hybridization revealed an uneven size and shape of keratinocytes. Not only were the morphological changes in the developing zebrafish skin reminiscent of the papulosquamous phenotype of psoriasis, gene expression was also consistent with the abnormalities seen in humans with gain-of-function mutations in *CARD14*, as *CARD14* overexpression was clearly associated with up-regulation of the NF-kB pathway in zebrafish as previously seen in humans [4,17].

Interestingly, large scale mutagenesis screens in zebrafish uncovered a mutation initially named *m14* which features keratinocyte hyperplasia resulting in cell aggregates. These findings, reminiscent of our own observations (see above), led investigators to rename this mutation *psoriasis*, because of the phenotypic similarity with the disease [32]. The gene harboring the *psoriasis* mutation has not yet been unequivocally identified. However it has been mapped at a distance of 8 Mb from the zebrafish *card14* gene [32], suggesting that *card14* may in fact be the gene harboring the *psoriasis* mutation.

Electronic microscopy revealed abnormal topography of keratinocytes' surface and disturbed microridge production. Microridges, are apical actin protrusions, widely found on vertebrate squamous epithelia [33]. The proposed function of microridges include mucous retention, membrane storage and abrasion resistance [34]. It is largely unknown how microridges are formed, but their formation has been linked to cell polarity pathways [35]. Accordingly, *CARD14* overexpression was associated with up-regulation of NFkB targets as seen in PRP patients [4], but also resulted in altered expression of 2 genes critical for planar cell polarity (PCP), vangl1 and vangl2.

VANGL1 and VANGL2 are evolutionary conserved PCP proteins shown to mediate global cell polarization in different species [36,37]. In zebrafish, *vangl2* has been shown to modulate morphogenetic movements during zebrafish gastrulation through inhibition of the Wnt-β-catenin pathway [38]. In mice, mutations in *Vangl1* and *Vangl2* result in abnormal polarity of cochlear hair cells [39,40]. In humans, mutations in both *VANGL1* and *VANGL2* cause lethal neural tube defects [41,42]. Last but not least, cutaneous effects have been linked to VANGL2 too, as *Vangl2* mutation causes polarization defects of hair growth and direction in mice [43]. VANGL2 was found to interact with proteins of the disheveled-wnt pathway through its PDZ binding domain [44], suggesting a possible interaction with CARD14 which possesses a PDZ domain. Injection of the human *VANGL* gene was able to partially rescue a pathological phenotype due to down-regulation of vangl2 in zebrafish, indicative of a high degree of functional redundancy of VANGL genes across evolution [45]. Of interest, cell polarity has been suggested to play a role in the pathogenesis of autoinflammatory disorders [46,47], although evidence on the role of polarization in autoimmune diseases is limited. Taken together, CARD14 may possibly play a role in cell polarization and subsequent cephalo-caudal embryonic movements via its interaction with vangl2. The mechanism by which CARD14 mediates polarity and possibly cell migration remains to be studied.

In summary, we have developed a practical model for *CARD14*-overexpression featuring morphological and molecular abnormalities of relevance to *CARD14*-associated human disorders. This model may in the future not only allow for novel therapeutics to be tested in a streamlined fashion, but may also shed new light on the pathogenesis of these conditions.

Supplementary Materials: The following supporting information can be downloaded at: https://www.mdpi.com/article/10.3390/biomedicines10123192/s1, Table S1: Riboprobes used for WISH; Table S2: Oligonucleotides used for qPCR.

Author Contributions: A.B. carried out the experiments and wrote the manuscript. L.Z. and Z.B.-M. contributed to zebrafish injections and experiments design. J.M. and A.P. helped with construct cloning. O.S. provided scientific supervision. Y.G. provided scientific supervision as head of zebrafish laboratory. G.R. provided scientific supervision as head of the research institution. E.S. provided scientific supervision as head of dermatology laboratory. All authors have read and agreed to the published version of the manuscript.

Funding: This research received no external funding.

Institutional Review Board Statement: All zebrafish procedures were approved by the Tel-Aviv University Animal Care Committee (04-18-051) and conducted in accordance with the National Council for Animal Experimentation, Ministry of Health, Israel.

Informed Consent Statement: Not applicable.

Data Availability Statement: The data presented in this study are available on request from the corresponding author.

Acknowledgments: In memory of Izhak Brickner, a man of sea with a vast soul, who skillfully performed our histological sections.

Conflicts of Interest: The authors declare no conflict of interest.

References

1. Nguyen, C.V.; Farah, R.S.; Maguiness, S.M.; Miller, D.D. Follicular Psoriasis: Differentiation from Pityriasis Rubra Pilaris-An Illustrative Case and Review of the Literature. *Pediatr. Dermatol.* **2017**, *34*, e65–e68. [CrossRef] [PubMed]
2. Tsoi, L.C.; Spain, S.L.; Ellinghaus, E.; Stuart, P.E.; Capon, F.; Knight, J.; Tejasvi, T.; Kang, H.M.; Allen, M.H.; Lambert, S.; et al. Enhanced meta-analysis and replication studies identify five new psoriasis susceptibility loci. *Nat. Commun.* **2015**, *6*, 7001. [CrossRef] [PubMed]
3. Vanderhooft, S.L.; Francis, J.S.; Holbrook, K.A.; Dale, B.A.; Fleckman, P. Familial pityriasis rubra pilaris. *Arch. Dermatol.* **1995**, *131*, 448–453. [CrossRef] [PubMed]
4. Fuchs-Telem, D.; Sarig, O.; van Steensel, M.A.; Isakov, O.; Israeli, S.; Nousbeck, J.; Richard, K.; Winnepenninckx, V.; Vernooij, M.; Shomron, N.; et al. Familial pityriasis rubra pilaris is caused by mutations in CARD14. *Am. J. Hum. Genet.* **2012**, *91*, 163–170. [CrossRef] [PubMed]
5. Jordan, C.T.; Cao, L.; Roberson, E.D.; Duan, S.; Helms, C.A.; Nair, R.P.; Duffin, K.C.; Stuart, P.E.; Goldgar, D.; Hayashi, G.; et al. Rare and common variants in CARD14, encoding an epidermal regulator of NF-kappaB, in psoriasis. *Am. J. Hum. Genet.* **2012**, *90*, 796–808. [CrossRef] [PubMed]
6. Helms, C.; Cao, L.; Krueger, J.G.; Wijsman, E.M.; Chamian, F.; Gordon, D.; Heffernan, M.; Daw, J.A.; Robarge, J.; Ott, J.; et al. A putative RUNX1 binding site variant between SLC9A3R1 and NAT9 is associated with susceptibility to psoriasis. *Nat. Genet.* **2003**, *35*, 349–356. [CrossRef]
7. Jordan, C.T.; Cao, L.; Roberson, E.D.; Pierson, K.C.; Yang, C.F.; Joyce, C.E.; Ryan, C.; Duan, S.; Helms, C.A.; Liu, Y.; et al. PSORS2 is due to mutations in CARD14. *Am. J. Hum. Genet.* **2012**, *90*, 784–795. [CrossRef]
8. Israel, L.; Mellett, M. Clinical and Genetic Heterogeneity of CARD14 Mutations in Psoriatic Skin Disease. *Front. Immunol.* **2018**, *9*, 2239. [CrossRef]

9. Mellett, M. Regulation and dysregulation of CARD14 signalling and its physiological consequences in inflammatory skin disease. *Cell. Immunol.* **2020**, *354*, 104147. [CrossRef]
10. DeVore, S.B.; Stevens, M.L.; He, H.; Biagini, J.M.; Kroner, J.W.; Martin, L.J.; Khurana Hershey, G.K. Novel role for caspase recruitment domain family member 14 and its genetic variant rs11652075 in skin filaggrin homeostasis. *J. Allergy Clin. Immunol.* **2022**, *149*, 708–717. [CrossRef]
11. Murase, Y.; Takeichi, T.; Akiyama, M. Aberrant CARD14 function might cause defective barrier formation. *J. Allergy Clin. Immunol.* **2019**, *143*, 1656–1657. [CrossRef] [PubMed]
12. Peled, A.; Sarig, O.; Sun, G.; Samuelov, L.; Ma, C.A.; Zhang, Y.; Dimaggio, T.; Nelson, C.G.; Stone, K.D.; Freeman, A.F.; et al. Loss-of-function mutations in caspase recruitment domain-containing protein 14 (CARD14) are associated with a severe variant of atopic dermatitis. *J. Allergy Clin. Immunol.* **2019**, *143*, 173–181. [CrossRef] [PubMed]
13. Scudiero, I.; Vito, P.; Stilo, R. The three CARMA sisters: So different, so similar: A portrait of the three CARMA proteins and their involvement in human disorders. *J. Cell. Physiol.* **2014**, *229*, 990–997. [CrossRef] [PubMed]
14. Jiang, C.; Lin, X. Regulation of NF-kappaB by the CARD proteins. *Immunol. Rev.* **2012**, *246*, 141–153. [CrossRef]
15. Afonina, I.S.; Van Nuffel, E.; Baudelet, G.; Driege, Y.; Kreike, M.; Staal, J.; Beyaert, R. The paracaspase MALT1 mediates CARD14-induced signaling in keratinocytes. *EMBO Rep.* **2016**, *17*, 914–927. [CrossRef] [PubMed]
16. Howe, K.; Clark, M.D.; Torroja, C.F.; Torrance, J.; Berthelot, C.; Muffato, M.; Collins, J.E.; Humphray, S.; McLaren, K.; Matthews, L.; et al. The zebrafish reference genome sequence and its relationship to the human genome. *Nature* **2013**, *496*, 498–503. [CrossRef] [PubMed]
17. Chang, M.X.; Chen, W.Q.; Nie, P. Structure and expression pattern of teleost caspase recruitment domain (CARD) containing proteins that are potentially involved in NF-kappaB signalling. *Dev. Comp. Immunol.* **2010**, *34*, 1–13. [CrossRef]
18. Li, Q.; Jin Chung, H.; Ross, N.; Keller, M.; Rews, J.; Kingman, J.; Sarig, O.; Fuchs-Telem, D.; Sprecher, E.; Uitto, J. Analysis of CARD14 Polymorphisms in Pityriasis Rubra Pilaris: Activation of NF-kappaB. *J. Investig. Dermatol.* **2015**, *135*, 1905–1908. [CrossRef]
19. Feng, C.; Wang, T.; Li, S.J.; Fan, Y.M.; Shi, G.; Zhu, K.J. CARD14 gene polymorphism c.C2458T (p.Arg820Trp) is associated with clinical features of psoriasis vulgaris in a Chinese cohort. *J. Dermatol.* **2016**, *43*, 294–297. [CrossRef]
20. Ammar, M.; Jordan, C.T.; Cao, L.; Lim, E.; Bouchlaka Souissi, C.; Jrad, A.; Omrane, I.; Kouidhi, S.; Zaraa, I.; Anbunathan, H.; et al. CARD14 alterations in Tunisian patients with psoriasis and further characterization in European cohorts. *Br. J. Dermatol.* **2016**, *174*, 330–337. [CrossRef]
21. Mossner, R.; Frambach, Y.; Wilsmann-Theis, D.; Lohr, S.; Jacobi, A.; Weyergraf, A.; Muller, M.; Philipp, S.; Renner, R.; Traupe, H.; et al. Palmoplantar Pustular Psoriasis Is Associated with Missense Variants in CARD14, but Not with Loss-of-Function Mutations in IL36RN in European Patients. *J. Investig. Dermatol.* **2015**, *135*, 2538–2541. [CrossRef] [PubMed]
22. Berki, D.M.; Liu, L.; Choon, S.E.; Burden, A.D.; Griffiths, C.E.; Navarini, A.A.; Tan, E.S.; Irvine, A.D.; Ranki, A.; Ogo, T.; et al. Activating CARD14 Mutations Are Associated with Generalized Pustular Psoriasis but Rarely Account for Familial Recurrence in Psoriasis Vulgaris. *J. Investig. Dermatol.* **2015**, *135*, 2964–2970. [CrossRef] [PubMed]
23. Sugiura, K.; Muto, M.; Akiyama, M. CARD14 c.526G>C (p.Asp176His) is a significant risk factor for generalized pustular psoriasis with psoriasis vulgaris in the Japanese cohort. *J. Investig. Dermatol.* **2014**, *134*, 1755–1757. [CrossRef] [PubMed]
24. Qin, P.; Zhang, Q.; Chen, M.; Fu, X.; Wang, C.; Wang, Z.; Yu, G.; Yu, Y.; Li, X.; Sun, Y.; et al. Variant analysis of CARD14 in a Chinese Han population with psoriasis vulgaris and generalized pustular psoriasis. *J. Investig. Dermatol.* **2014**, *134*, 2994–2996. [CrossRef] [PubMed]
25. Takeichi, T.; Sugiura, K.; Nomura, T.; Sakamoto, T.; Ogawa, Y.; Oiso, N.; Futei, Y.; Fujisaki, A.; Koizumi, A.; Aoyama, Y.; et al. Pityriasis Rubra Pilaris Type V as an Autoinflammatory Disease by CARD14 Mutations. *JAMA Dermatol.* **2017**, *153*, 66–70. [CrossRef] [PubMed]
26. Inoue, N.; Dainichi, T.; Fujisawa, A.; Nakano, H.; Sawamura, D.; Kabashima, K. CARD14 Glu138 mutation in familial pityriasis rubra pilaris does not warrant differentiation from familial psoriasis. *J. Dermatol.* **2016**, *43*, 187–189. [CrossRef] [PubMed]
27. Hong, J.B.; Chen, P.L.; Chen, Y.T.; Tsai, T.F. Genetic analysis of CARD14 in non-familial pityriasis rubra pilaris: A case series. *Acta Derm.-Venereol.* **2014**, *94*, 587–588. [CrossRef]
28. Eytan, O.; Qiaoli, L.; Nousbeck, J.; van Steensel, M.A.; Burger, B.; Hohl, D.; Taieb, A.; Prey, S.; Bachmann, D.; Avitan-Hersh, E.; et al. Increased epidermal expression and absence of mutations in CARD14 in a series of patients with sporadic pityriasis rubra pilaris. *Br. J. Dermatol.* **2014**, *170*, 1196–1198. [CrossRef]
29. Eytan, O.; Sarig, O.; Sprecher, E.; van Steensel, M.A. Clinical response to ustekinumab in familial pityriasis rubra pilaris caused by a novel mutation in CARD14. *Br. J. Dermatol.* **2014**, *171*, 420–422. [CrossRef]
30. Van Nuffel, E.; Schmitt, A.; Afonina, I.S.; Schulze-Osthoff, K.; Beyaert, R.; Hailfinger, S. CARD14-Mediated Activation of Paracaspase MALT1 in Keratinocytes: Implications for Psoriasis. *J. Investig. Dermatol.* **2016**, *137*, 569–575. [CrossRef]
31. Wu, K.C.; Reynolds, N.J. CARD14 mutations may predict response to antitumour necrosis factor-alpha therapy in psoriasis: A potential further step towards personalized medicine. *Br. J. Dermatol.* **2016**, *175*, 17–18. [CrossRef] [PubMed]
32. Webb, A.E.; Driever, W.; Kimelman, D. Psoriasis regulates epidermal development in zebrafish. *Dev. Dyn. Off. Publ. Am. Assoc. Anat.* **2008**, *237*, 1153–1164. [CrossRef]
33. Lam, P.Y.; Mangos, S.; Green, J.M.; Reiser, J.; Huttenlocher, A. In vivo imaging and characterization of actin microridges. *PLoS ONE* **2015**, *10*, e0115639. [CrossRef] [PubMed]

34. Sperry, D.G.; Wassersug, R.J. A proposed function for microridges on epithelial cells. *Anat. Rec.* **1976**, *185*, 253–257. [CrossRef]
35. Raman, R.; Damle, I.; Rote, R.; Banerjee, S.; Dingare, C.; Sonawane, M. aPKC regulates apical localization of Lgl to restrict elongation of microridges in developing zebrafish epidermis. *Nat. Commun.* **2016**, *7*, 11643. [CrossRef]
36. Antic, D.; Stubbs, J.L.; Suyama, K.; Kintner, C.; Scott, M.P.; Axelrod, J.D. Planar cell polarity enables posterior localization of nodal cilia and left-right axis determination during mouse and Xenopus embryogenesis. *PLoS ONE* **2010**, *5*, e8999. [CrossRef]
37. Torban, E.; Kor, C.; Gros, P. Van Gogh-like2 (Strabismus) and its role in planar cell polarity and convergent extension in vertebrates. *Trends Genet. TIG* **2004**, *20*, 570–577. [CrossRef]
38. Park, M.; Moon, R.T. The planar cell-polarity gene stbm regulates cell behaviour and cell fate in vertebrate embryos. *Nat. Cell Biol.* **2002**, *4*, 20–25. [CrossRef]
39. Torban, E.; Patenaude, A.M.; Leclerc, S.; Rakowiecki, S.; Gauthier, S.; Elfinger, G.; Epstein, D.J.; Gros, P. Genetic interaction between members of the Vangl family causes neural tube defects in mice. *Proc. Natl. Acad. Sci. USA* **2008**, *105*, 3449–3454. [CrossRef]
40. Montcouquiol, M.; Rachel, R.A.; Lanford, P.J.; Copeland, N.G.; Jenkins, N.A.; Kelley, M.W. Identification of Vangl2 and Scrb1 as planar polarity genes in mammals. *Nature* **2003**, *423*, 173–177. [CrossRef]
41. Lei, Y.P.; Zhang, T.; Li, H.; Wu, B.L.; Jin, L.; Wang, H.Y. VANGL2 mutations in human cranial neural-tube defects. *N. Engl. J. Med.* **2010**, *362*, 2232–2235. [CrossRef] [PubMed]
42. Kibar, Z.; Torban, E.; McDearmid, J.R.; Reynolds, A.; Berghout, J.; Mathieu, M.; Kirillova, I.; De Marco, P.; Merello, E.; Hayes, J.M.; et al. Mutations in VANGL1 associated with neural-tube defects. *N. Engl. J. Med.* **2007**, *356*, 1432–1437. [CrossRef] [PubMed]
43. Devenport, D.; Fuchs, E. Planar polarization in embryonic epidermis orchestrates global asymmetric morphogenesis of hair follicles. *Nat. Cell Biol.* **2008**, *10*, 1257–1268. [CrossRef] [PubMed]
44. Yao, R.; Natsume, Y.; Noda, T. MAGI-3 is involved in the regulation of the JNK signaling pathway as a scaffold protein for frizzled and Ltap. *Oncogene* **2004**, *23*, 6023–6030. [CrossRef]
45. Reynolds, A.; McDearmid, J.R.; Lachance, S.; De Marco, P.; Merello, E.; Capra, V.; Gros, P.; Drapeau, P.; Kibar, Z. VANGL1 rare variants associated with neural tube defects affect convergent extension in zebrafish. *Mech. Dev.* **2010**, *127*, 385–392. [CrossRef]
46. Ostensson, M.; Monten, C.; Bacelis, J.; Gudjonsdottir, A.H.; Adamovic, S.; Ek, J.; Ascher, H.; Pollak, E.; Arnell, H.; Browaldh, L.; et al. A possible mechanism behind autoimmune disorders discovered by genome-wide linkage and association analysis in celiac disease. *PLoS ONE* **2013**, *8*, e70174. [CrossRef]
47. De Luca, M.; Pellegrini, G.; Zambruno, G.; Marchisio, P.C. Role of integrins in cell adhesion and polarity in normal keratinocytes and human skin pathologies. *J. Dermatol.* **1994**, *21*, 821–828. [CrossRef]

Article

Acute Administration of Ethanol and of a D1-Receptor Antagonist Affects the Behavior and Neurochemistry of Adult Zebrafish

Tanya Scerbina [1,2] and Robert Gerlai [1,3,*]

1 Department of Psychology, University of Toronto, Mississauga, ON L5L 1C6, Canada
2 Council of Ministers of Education, Toronto, ON M4V 1N6, Canada
3 Department of Cell & Systems Biology, University of Toronto, Toronto, ON M5S 3G5, Canada
* Correspondence: robert.gerlai@utoronto.ca; Tel.: +1-905-569-4255

Abstract: Alcohol abuse represents major societal problems, an unmet medical need resulting from our incomplete understanding of the mechanisms underlying alcohol's actions in the brain. To uncover these mechanisms, animal models have been proposed. Here, we explore the effects of acute alcohol administration in zebrafish, a promising animal model in alcohol research. One mechanism via which alcohol may influence behavior is the dopaminergic neurotransmitter system. As a proof-of-concept analysis, we study how D1 dopamine-receptor antagonism may alter the effects of acute alcohol on the behavior of adult zebrafish and on whole brain levels of neurochemicals. We conduct these analyses using a quasi-inbred strain, AB, and a genetically heterogeneous population SFWT. Our results uncover significant alcohol x D1-R antagonist interaction and main effects of these factors in shoaling, but only additive effects of these factors in measures of exploratory behavior. We also find interacting and main effects of alcohol and the D1-R antagonist on dopamine and DOPAC levels but only alcohol effects on serotonin. We also uncover several strain dependent effects. These results demonstrate that acute alcohol may act through dopaminergic mechanisms for some but not all behavioral phenotypes, a novel discovery, and also suggest that strain differences may, in the future, help us identify molecular mechanisms underlying acute alcohol effects.

Keywords: alcohol abuse; alcoholism; ethanol; ethyl alcohol; dopamine; shoaling; zebrafish

Citation: Scerbina, T.; Gerlai, R. Acute Administration of Ethanol and of a D1-Receptor Antagonist Affects the Behavior and Neurochemistry of Adult Zebrafish. *Biomedicines* **2022**, *10*, 2878. https://doi.org/10.3390/biomedicines10112878

Academic Editors: James A. Marrs and Swapnalee Sarmah

Received: 12 October 2022
Accepted: 7 November 2022
Published: 10 November 2022

Publisher's Note: MDPI stays neutral with regard to jurisdictional claims in published maps and institutional affiliations.

1. Introduction

Abuse of alcohol (ethanol, ethyl alcohol or EtOH) represents a large societal problem worldwide [1–3]. Treatment options for alcoholism are limited [4–6] by our incomplete understanding of the mechanisms underlying alcohol's actions in the brain [7–10]. One factor that has been proposed to play roles in the development of chronic alcohol abuse is the initial response to acute alcohol exposure [11,12]. For example, people who better tolerate the acute effects of alcohol may be prone to drink larger amount of alcohol and may start to drink more often, and thus have a higher risk of developing chronic alcohol abuse and dependence [13,14]. Acute effects of alcohol depend upon a variety of factors including genetic differences among people [14]. However, the neurobiological mechanism underlying acute alcohol effects, or the genes involved in individual differences in response to acute alcohol consumption, are not fully understood.

Numerous animal models have been proposed to facilitate discovery of such mechanisms [15–17]. The zebrafish is a relative novice in this research, nevertheless, it has been proposed to be a promising model organism [18–20]. There are several reasons for this. The zebrafish has been found to possess numerous evolutionarily conserved features, from the nucleotide sequence of its genes [21–23], through its neurotransmitter systems [24] to its behavior [25,26]. Thus, it is considered to be a translationally relevant model organism for the analysis of human brain function and dysfunction [26–28]. The translational relevance is further increased when zebrafish results are compared and combined with those obtained with rodents and humans, and thus common overlapping features and mechanisms are identified [29]. The zebrafish is also considered to represent a reasonable compromise between system complexity (it is a vertebrate) and practical simplicity (it is small, easy to breed and cheap to keep in large numbers) [29]. Last, the method of alcohol delivery (and delivery of several other drugs or compounds) can be achieved in a non-invasive manner by immersing the fish in the solution [30,31]. For these reasons, we decided to use zebrafish and investigate the effects of acute alcohol administration using this model organism.

Alcohol is a complex drug from the perspective of pharmacological properties, as it directly interacts with a large number of molecular targets, and indirectly affects an even larger number of biochemical processes and neurobiological mechanisms [7–10]. Ideally, systematic large-scale mutation screens, drug/small molecule screens or comprehensive transcriptome analyses may be performed to discover the details of the complex mechanisms and effects of acute alcohol administration in the brain of vertebrates. Although such screens or systematic comprehensive studies are technically feasible, up to this date they have not been performed with any model organism for the analysis of acute alcohol effects. Nevertheless, the zebrafish has been employed in such large-scale comprehensive screening approaches aimed at other phenotypes and biological questions, particularly in the field of embryology [32,33]. In this paper, we describe results that represent a proof of principle, providing the first unequivocal piece of evidence that acute alcohol effects can be mediated by the dopaminergic neurotransmitter system in zebrafish, implying that comprehensive screening may also be feasible for mechanistic analysis of the actions of alcohol using this translationally relevant model organism.

Given the complex pharmacological profile of alcohol, the questions of what phenotype should one study to uncover acute alcohol effects, and what mechanisms, biochemical pathways or molecular targets, should one investigate are not simple to answer. We have decided to study behavior as the primary endpoint of acute alcohol administration-induced effects. Behavioral analysis may allow one to detect functional changes in the brain without having to have *a priory* information about where and what exactly these changes may be [26,34]. Furthermore, previously, we and others have found acute alcohol administration to alter a variety of behavioral phenotypes. For example, similarly to what has been shown in humans [35], acute alcohol administration has been found to alter responding to social cues, i.e., to the sight of conspecifics in zebrafish too. For example, shoaling (group forming) is dose dependently impaired by acute alcohol administration in zebrafish [31,36,37]. Similarly, locomotor activity (distance swum) is also significantly affected by acute alcohol, with lower doses increasing and higher doses decreasing activity in zebrafish [37,38], findings corresponding well to those obtained with humans [39].

As for mechanisms, we decided to focus on the dopaminergic system. There are multiple reasons for this. Alcohol has been shown to engage reward pathways in mammals, including humans [40], and the dopaminergic system has long been known to play crucial roles in reward [41]. In zebrafish, acute alcohol administration has been shown to increase dopamine and DOPAC (the metabolite of dopamine) levels in the brain [24,31,42–44], and has also been shown to lead to brighter coloration, which normally can be seen during spawning or fighting among zebrafish [37], resembling the euphoria induced by acute alcohol in humans [45]. Furthermore, while alcohol has been shown to disrupt shoaling [31,46,47], sight of conspecifics has been found to be rewarding in learning tasks [48–51] and to increase dopamine and DOPAC levels in the brain of zebrafish [52,53]. Additionally, acute administration of a dopamine D1-receptor (D1-R) antagonist has been shown to disrupt shoaling [50]. What is not known, however, is whether the dopaminergic neurotransmitter system is actually involved in acute alcohol administration induced behavioral changes. The current study is aimed at answering this question.

Thus, our working hypothesis is that, although alcohol is known to engage a variety of mechanisms in the brain, its disruptive effects on shoaling (social behavioral responses) may be mediated by the dopaminergic system. To test this possibility, we conducted a study with a 4 × 3 × 2 between subject experimental design. That is, we acutely exposed each experimental zebrafish to one of four alcohol concentrations (ranging between 0% and 1% vol/vol) and one of three dopamine D1-R antagonist concentrations (ranging between 0 and 1 mg/L), and we conducted these experiments on two genetically distinct zebrafish populations (strains). The reason for using two strains was that prior studies indicated strain differences in shoaling [54] as well as in alcohol induced behavioral and neurochemical changes [36,55–57]. We measured a number of behavioral responses, including distance to animated images of conspecifics (an artificial shoal) as well as measures of general activity. In addition, we also analyzed levels of neurochemicals from whole brain extracts, including levels of dopamine, DOPAC, serotonin and 5HIAA (the metabolite of serotonin). We investigated whether alcohol and/or the D1-R antagonist had main effects on these phenotypes and whether these effects were interacting or additive. We reasoned that if acute alcohol administration-induced effects on shoaling responses are mediated via the dopaminergic system, we should see significant interaction between the effects of alcohol and the D1-R antagonist. If, however, the effects of these two drugs are independent of each other, we should find additivity, i.e., lack of significant interaction. We emphasize that combined application of acute alcohol administration and dopamine-receptor antagonism has not been conducted before, and thus the question of whether alcohol can exert its behavioral and neurochemical effects via the dopaminergic system has not been answered, the goal of the current study.

2. Materials and Methods

2.1. Animals and Housing

All zebrafish tested in this study were bred, raised and maintained in the Gerlai Zebrafish Facility of the University of Toronto Mississauga as described before [49]. Briefly fertilized eggs were collected from multiple spawning zebrafish pairs and kept in small hatching tanks. At 5 days post-fertilization (dpf), the free-swimming fry were transferred to Aquaneering nursery tanks (Aquaneering Inc., San Diego, CA, USA) and fed artificial plankton (100 μm diameter fry food by Zeigler, Gardens, PA, USA). Starting at age 10 dpf fry were fed freshly hatched Artemia salina nauplii, and from 15 dpf onward, a mixture of crushed tetramin tropical flakes (Tetra Co., Melle, Germany) and spirulina flakes (Jehmco Inc., Lambertville, NJ, USA). At this age the small fish were transferred to 3 L holding tanks placed on the Aquaneering system rack. The Aquaneering rack uses a recirculating filtration system with mechanical (sponge), biological (fluidized bed), chemical (activated carbon) filter components as well as a UV sterilizing unit. In addition, to further improve water quality, 10% of the water on this rack was replaced automatically once a day. The system water had a pH of 7, and salinity was maintained at 300 μS. This level of salinity was achieved by reverse osmosis filtration and reconstitution of salt concentration by adding Instant Ocean Sea Salt. Light cycle was maintained at 12 h dark and 12 h light using ceiling mounted fluorescent light fixtures with lights turned on at 7:00 h.

We employed two genetically distinct populations of zebrafish, from here onward referred to as "strains": a genetically well-defined quasi-inbred strain, AB, and a genetically heterogeneous population we designate as SFWT (short-fin wild type). The reasons for choosing these two populations were as follows. AB is the most frequently employed strain in zebrafish research. It is a quasi-inbred strain with 80% of the loci in a homozygous form. Its widespread use in zebrafish research facilitates replicability and reproducibility across laboratories [58]. As we were interested in whether genotype-dependent differences exist in alcohol and/or D1-R antagonist induced responses, we employed another zebrafish strain SFWT. SFWT is a genetically heterogeneous population in which high genetic variance is expected among individuals, and high level of heterozygosity is expected across the genetic loci for each individual. This is because the SFWT zebrafish employed in this study are the first filial generation of parents (purchased from Big Als Aquarium Warehouse Mississauga, ON, Canada) that came from a commercial breeding facility in Singapore, a location close to the natural geographic distribution of zebrafish, and in which a large number of zebrafish breeders are employed. We chose this population as we expect it to represent species typical features of zebrafish better than potentially unique inbred strains. All experimental zebrafish (AB and SFWT) were bred, raised and maintained in the same room of the Gerlai Zebrafish Facility under identical conditions and at the same time. All experimental zebrafish were tested when fully grown, sexually mature, at their age of 4–6 months post-fertilization.

2.2. Experimental Design, Drug Treatment Procedure

As briefly mentioned above, we employed a $4 \times 3 \times 2$ between subject experimental design, with alcohol concentration having 4 levels (0.00 (control), 0.25, 0.50, 1.00 vol/vol% external bath), D1-receptor antagonist (SCH23390) having 3 levels (0.0 (control), 0.1, 1.0 mg/L external bath, and strain having 2 levels (AB, SFWT). For the behavioral analysis, sample size (n) of each of the 24 groups was 20, except for the following three groups: n(Strain AB, Alcohol 0.5%, D1-R antagonist 0.1 mg/L) = 21; n(Strain AB, Alcohol 1%, D1-R antagonist 0 mg/L) = 19; n(Strain AB, Alcohol 1%, D1-R antagonist 1 mg/L) = 21. For the analysis of neurochemicals sample size (n) of each of the group was 10. We chose these sample sizes because previously we have found significant behavioral and/or neurochemical changes induced by a variety of psychopharmacological studies we conducted, including those with alcohol and a D1-R antagonist [43,50].

Alcohol (anhydrous ethanol 99.95%) was obtained from the University of Toronto Med store. The dopamine D1-R specific antagonist SCH23390 (R-(+)-8-chloro-2,3,4,5-tetrahydro-3-methyl-5-phenyl-1H-3-benzazepine-7-ol) was obtained from Sigma-Aldrich, Oakville, ON, Canada (Cat # D054). Fish were randomly assigned to their respective treatment group, with each fish administered a given alcohol and D1-R antagonist only once (a between subject experimental design). The sex ratio for all groups was 50/50%, and sex was not found to have a significant main effect or interaction with any factors. Thus, data were pooled for sexes for all subsequent data analyses. Experimental zebrafish were first immersed in their respective D1-R antagonist solution for 30 min in a 300 mL beaker in which oxygenation was provided via an air-stone connected to an aquarium air-pump. Although ADME (absorption, distribution, metabolism and excretion) data are not available for most drugs in zebrafish, including for SCH23390, extrapolating from mammalian studies [59–62] as well as based upon previous zebrafish studies [43,50], we expected that 30 min long bath immersion in the above concentration of this D1-R antagonist should be sufficient to induce significant behavioral and neurochemical changes in adult zebrafish. Immediately following the immersion in the D1-R antagonist solution, zebrafish were placed in their respective alcohol bath concentration (also in 300 mL beakers equipped with oxygenating air-stone) for 60 min. The concentrations employed in this study correspond to prior concentrations of alcohol used acutely with adult zebrafish [31,36,38,43,44]. The length of immersion employed in the current study was based upon the previous finding showing that alcohol levels reached a steady maximal plateau in brain tissue after 60 min of bath immersion in adult zebrafish [63]. Furthermore, we expected alcohol to remain in the brain in sufficient amounts at least for the entire period of the 16 min long behavioral test session that immediately followed the immersion [63]. Immediately after the D1-R antagonist and alcohol bath immersion, zebrafish underwent behavioral testing. The temporal order of to which D1-R antagonist concentration and to which subsequent alcohol concentration zebrafish were exposed was randomized.

2.3. Behavioral Test and Procedure

The behavioral experimenter was blind to the strain origin (AB and SFWT appear identical) and to the bath concentration of D1-R antagonist and of alcohol the experimental zebrafish received. Furthermore, the temporal order of behavioral testing of the zebrafish followed the random temporal order of the drug concentrations. We employed a simple novel aquarium task described in a detailed manner previously [64]. Briefly, the experimental fish were placed singly in a 37.5 L glass aquarium (50 cm $\times$ 25 cm $\times$ 30 cm, length × width × height), flanked by an LCD computer monitor (17-inch Samsung Sync-Master 732 N) on each of its small sides. The aquarium was illuminated from above by a 15 W fluorescent light tube. The experimental fish were monitored and their behavior analyzed for a total of 16 min via a camcorder (JVC GZ-MG37u) that viewed the aquarium from the front. Video-recordings were transferred to a computer and later analyzed using Ethovision Color Pro Video-tracking software application (Version 3, Noldus Info Tech Wageningen, The Netherlands). For the first 8 min of the recording session, no stimuli were presented to the experimental zebrafish. For the second 8 min of the session, computer animated images (moving with a varying speed from 1 to 4 cm/s) of 5 zebrafish of the same size as the experimental fish were presented on one of the LCD monitors flanking the experimental aquarium using a custom software application (ZFT, developed in-house and first described by Saverino & Gerlai [65] (also see [51]). The side on which these animated conspecific images were shown randomly varied between experimental zebrafish. Using the Ethovision software application, we extracted numerous parameters of the swim path of the experimental zebrafish as follows. To evaluate shoaling, the distance of the experimental zebrafish from the stimulus screen that presented the conspecific images was measured in cm [64]. It is the distance between the center-point of the experimental fish and the glass side wall of the experimental aquarium flanked by the active computer monitor. We also measured intra-individual (temporal) variance of the distance of experimental zebrafish from the glass wall flanked by the computer monitor that showed the images. This variance reflects how consistently/inconsistently the experimental fish behaved, i.e., how much they changed their distance on the horizontal axis relative to the location of the animated conspecific images. In order to evaluate locomotory activity, we measured the total distance the fish swum. Bottom dwell time is often regarded as a measure of anxiety [66]. To evaluate this response, we quantified the distance between the experimental fish and the bottom of the tank. We also quantified the intra-individual variance of distance from bottom, a measure of vertical exploration [67], i.e., variability of the location of the fish along the vertical axis. Last, we quantified absolute turn angle, i.e., the change of direction of locomotion from one frame to the next irrespective of whether it was clock- or counter-clockwise.

2.4. Evaluation of Levels of Neurochemicals Using HPLC

Previously, we found acute administration of alcohol to significantly increase the level of dopamine and its metabolite, DOPAC, as well as of serotonin and its metabolite 5HIAA [24,31,44,50]. We also found shoaling images to induce a rapid increase of dopamine and DOPAC levels, but no changes in serotonin or 5HIAA levels [52]. Thus, we decided to measure whether alcohol and/or the D1-R antagonist alters the levels of these neurochemicals independently (additive effects) or in combination (interaction), a question that has not been investigated before. Another reason for their choice was practical. They could be measured from the same tissue homogenates using a single run using high-precision liquid chromatography (HPLC) with electrochemical detection [24]. Our HPLC methods have been described in detail elsewhere [24]. Briefly, we exposed a set of AB and SFWT zebrafish to alcohol and the D1-R antagonist exactly the same way as those fish we tested behaviorally. Additionally, the fish we used for HPLC analysis were of the same size and age as those we studied in the behavioral experiment. Experimental zebrafish were quickly decapitated using surgical scissors and their brains were removed and placed on dry ice. Subsequently brains were sonicated in artificial cerebrospinal fluid, and protein content of the sonicate was analyzed using a BioRad protein assay reagent (BioRad, Hercules, CA, USA). Next the sonicate was centrifuged and the supernatant was collected for HPLC analysis. We employed a BAS 460 Microbore HPLC with electrochemical detection (Bio-analytical Systems Inc., West Lafayette, IN, USA) using a Uniget C-18 reverse phase microbore column as the stationary phase (BASi, Cat no. 8949). The mobile phase consisted of buffer (0.1 M monochloro acetic acid, 0.5 mM Na-EDTA, 0.15 g/L Na-octylsulfonate and 10 nM sodium chloride, pH 3.1, obtained from Sigma), acetonitrile and tetrahydrofuran (obtained from

Fisher Scientific) at a ratio 94:3.5:0.7. The flow rate was set to 1.0 mL/min and the working electrode (Uniget 3 mm glass carbon, BAS P/N MF-1003) was set at 550 mV vs. Ag/Ag/Cl reference electrode. Detection gain was set to 1.0 nA, filter was at 0.2 Hz, and detection limit was set to 20 nA. Five μL of the sample supernatant was directly injected into the HPLC system for analysis. Standard dopamine, DOPAC, serotonin and 5HIAA (Sigma) were employed to quantify and identify the peaks on the chromatographs.

2.5. Data Calculation and Statistical Analysis

One may define social behavior as all behavioral responses that are induced by social stimuli. However, this may be a mistake. Social stimuli, such as the sight of animated conspecific images employed here, are expected to induce a variety of responses. For example, in addition to shoaling (swimming close to conspecifics) the sight of conspecifics may reduce anxiety/fear, i.e., antipredatory responses and thus, e.g., alter bottom dwelling [28,68–70], and/or may reduce exploratory behavior, e.g., locomotory activity measured by total distance swum and by amount of turning performed [64]. These three responses (shoaling, antipredatory responses and exploratory behavior) may represent independent behavioral strategies/responses/states elicited by social stimuli, and they thus may also have idiosyncratic underlying mechanisms. As a result, they may be influenced differently by alcohol and/or by the D1-R antagonist. To investigate this possibility, we decided to calculate the change in behaviors of zebrafish induced by the presentation of conspecifics, i.e., the temporal change from the pre-conspecific stimulus period to the conspecific-stimulus presentation period as follows. All behavioral measures were extracted from x-y coordinates measured 10 times a second using Ethovision Tracking software. The x-y coordinate data were first used to calculate the total distance swum, and the average of distance to stimulus, of variance of distance to stimulus, of distance to bottom, of variance of distance to bottom and of turn angle, respectively, recorded for the first 8 min and separately also for the second 8 min of the recording session. Then, we subtracted the thus calculated value obtained for each behavior for the first 8 min from the value obtained for that behavior of the second 8 min. That is, we calculated the change of behavior induced by the presence of animated conspecific images.

With our HPLC data, we conducted the following transformations before statistical analysis. We standardized the neurochemical amounts measured to the brain tissue sample weight. That is, we expressed our data as μg of neurochemical per mg of total brain protein.

Using SPSS (version 23 for the PC) we conducted three-factorial Variance Analyses (ANOVAs) to investigate whether the concentration of alcohol (factor 1), the concentration of D1-R antagonist (factor 2) and whether the strain origin of the fish (factor 3) had significant main effects on the above-described behavioral changes and on the relative neurochemical amounts, and whether there were significant interactions among/between these factors. Subsequently, we performed a Tukey Honestly Significant Difference (HSD) *post hoc* test for each strain separately, which allowed us to compare all 12 groups per strain with each other without inflating type-1 error. Although this type of *post hoc* test is excellent for determining whether particular groups or a group differ from others or another, given the large number of these groups in our study and given that not all group comparisons are informative, we focused on the pattern of results, and only present/discuss the most important significant differences among these groups identified by the Tukey HSD tests.

We rejected the null hypothesis (no significant effect/interaction or no difference between/among groups) when the probability of such null hypotheses was not larger than 5%, i.e., when $p \leq 0.05$. We note that the parametric statistical analyses we conducted here are insensitive to the violation of the criteria of homogeneity of variances and normality of distribution when the compared groups do not have grossly different sample sizes. Our sample sizes across the $4 \times 3 \times 2$ (i.e., 24) treatment/strain groups were nearly identical. Thus, we did not check for the above criteria and did not perform scale transformations.

3. Results

3.1. Behavioral Parameters

Reduction of distance to the stimulus screen in response to the presentation of animated images of conspecifics is considered a measure of the strength of shoaling behavior [51]. Here, we also found the distance to stimulus screen to significantly diminish as a result of the presentation of animated images of conspecifics, as shown by the negative values presented on Figure 1. Although all treatment groups for both strains show this reduction of distance, the figure also demonstrates that the amount of reduction of the distance differed across the treatment groups and strains. For example, higher alcohol concentrations appeared to diminish this reduction, i.e., impaired the shoaling response. Similarly, D1-R antagonist also appeared to exert a dose dependent effect, reducing shoaling. Last, the effects of alcohol and the D1-R antagonist do not appear to be additive. These observations were confirmed by the results of ANOVA. It found the effect of alcohol ($F(3, 457) = 13.91$, $p < 0.001$), and the effect of the D1-R antagonist ($F(2, 457) = 5.03$, $p = 0.007$) to be significant, but the effect of strain non-significant ($F(1, 457) = 2.79$, $p = 0.10$). The alcohol x D1-R antagonist interaction was also significant ($F(6, 457) = 4.32$, $p < 0.001$), but the D1-R antagonist x strain ($F(2, 457) = 1.98$, $p = 0.14$), the alcohol x strain ($F(3, 457) = 1.46$, $p = 0.23$) and the alcohol x D1-R antagonist x strain ($F(6, 457) = 0.96$, $p = 0.45$) interactions were all non-significant. Post hoc Tukey HSD test confirmed a significant dose dependent alcohol effect. For example, among the 0 mg/L D1-R antagonist treated fish, the highest alcohol concentration (1%) group fish showed significantly (0.05%) less shoaling response compared to the control (0%) alcohol treated fish and to those treated with the intermediate concentrations (0.5 or 0.25%) in both strains. The significant interaction between alcohol and D1-R antagonist treatments is also well illustrated if one compares the D1-R dose response of 0% alcohol treated fish (first three bars on the bar graphs of Figure 1) with the D1-R antagonist dose response of the 1% alcohol treated fish (last three bars of Figure 1). Tukey HSD revealed that for the 0% alcohol treated AB zebrafish increasing D1-R concentrations had significant ($p < 0.05$) shoaling response reducing effect. Whereas for the 1% alcohol treated AB zebrafish, this dose response was reversed, albeit without significant ($p > 0.05$) differences among the D1-R concentration groups. The findings were similar, but not identical, for SFWT zebrafish, where the D1-R antagonist dose response was found U-shaped in the 0% alcohol treated SFWT fish (with the 0.1 mg/L and 1 mg/L D1-R concentrations group differing significantly, $p < 0.05$) but quasi-linear for the 1% alcohol treated SFWT fish (with the 0 mg/L and 1 mg/L D1-R antagonist treated fish differing significantly, $p < 0.05$).

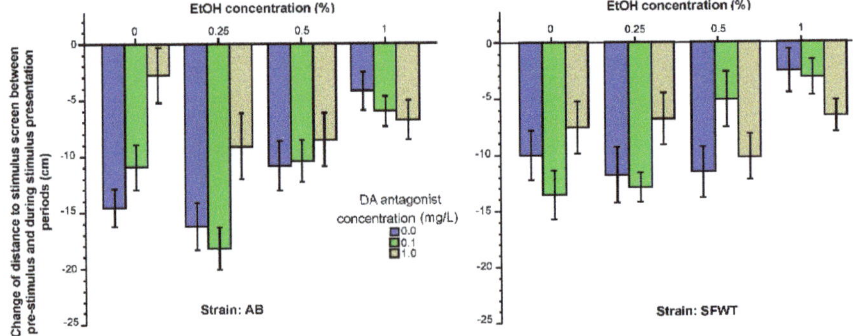

Figure 1. Change of distance to stimulus screen elicited by the presentation of animated conspecific images is influenced by the concentration of acute alcohol (X-axis) and the concentration of acute D1-R antagonist (legend) in an interactive manner. Mean ± S.E.M. are shown.

Intra-individual variance in the shoaling response (the variance of the change of distance from the no-stimulus period to the stimulus period) is shown in Figure 2. This variance quantifies how consistently/inconsistently the fish responded to the presentation of the animated conspecific images. Negative values shown on this figure mean that the variance was reduced by the shoaling images, i.e., the experimental fish remained at a more constant distance from the stimulus screen upon the presentation of the conspecific images compared to how they behaved before the images were shown. Figure 2 also suggests that alcohol reduced this variance (consistency) in a dose dependent manner. Similarly, the D1-R antagonist also reduced the variance with high doses having the strongest effects. Last, it appears that AB responded with more reduction of variance (increased their consistency of how far they swum from the images) in response to the animated conspecific images than SFWT did. These observations were confirmed by ANOVA. It showed that the effect of alcohol ($F_{(3, 457)} = 18.23$, $p < 0.001$), the effect of the D1-R antagonist ($F_{(2, 457)} = 5.20$, $p = 0.006$) and the effect of strain were all significant ($F_{(1, 457)} = 7.76$, $p = 0.006$). However, no interaction term was found significant (alcohol x D1-R antagonist ($F_{(6, 457)} = 0.23$, $p = 0.71$), alcohol x strain ($F_{(3, 457)} = 0.96$, $p = 0.41$), D1-R antagonist x strain ($F_{(2, 457)} = 0.64$, $p = 0.53$), alcohol x D1-R antagonist x strain ($F_{(6, 457)} = 0.90$, $p = 0.49$). Tukey HSD test confirmed that among the 0 mg/L D1-R antagonist treated zebrafish of both strains, exposure to the intermediate concentration of 0.25% alcohol reduced the variance of distance most (significantly ($p < 0.05$) more compared with other concentration groups) and the 1% alcohol treated zebrafish reduced their variance the least (significantly ($p < 0.05$) less than fish given 0.25% or 0% alcohol). Tukey HSD test also showed that the highest concentration of D1-R antagonist (1 mg/L) had the strongest effect leading to the least amount of variance reduction (significantly ($p < 0.05$) less compared to fish treated with 0 mg/L D1-R antagonist).

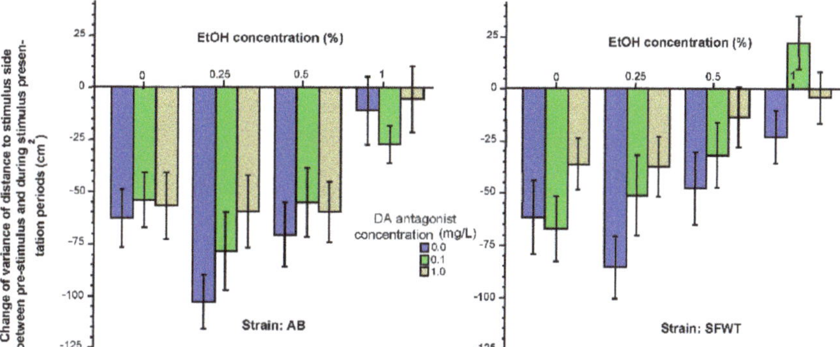

Figure 2. Change of temporal intra-individual variance of distance to stimulus screen (horizontal exploration) elicited by the presentation of animated conspecific images is influenced by the concentration of acute alcohol (X-axis) and the concentration of acute D1-R antagonist (legend) in an additive manner. Additionally, note the significant strain difference. Mean ± S.E.M. are shown.

Exploratory activity may also change when shoal mates are present, as their presence may diminish the need to find escape routes, search for predators or hiding places, and because food may also be more efficiently found by more eyes and lateral lines, i.e., by shoal-mates. Thus, although exploratory behavior is not a form of social behavior, it may alter in response to the sight of conspecifics. A measure of exploratory behavior is general locomotory activity. Here, we quantified total distance swum and expressed the values as the change of this distance induced by the images of conspecifics (Figure 3). Overall, the figure suggest that experimental zebrafish of both strains reduced their total distance in response to the shown images, but this reduction depended on alcohol and D1-R antagonist dose employed, with higher concentrations of these drugs diminishing the effect of the social stimuli (reduced change of distance). It also appears that AB zebrafish reduced their distance in response to the presentation of animated conspecific images more compared to SFWT zebrafish. These observations were confirmed by ANOVA, which showed that the effect of alcohol (F(3, 457) = 28.12, $p < 0.001$), the effect of the D1-R antagonist (F(2, 457) = 6.45, $p = 0.002$) and the effect of strain were all significant (F(1, 457) = 5.36, $p = 0.02$). However, no interaction term was found significant (alcohol x D1-R antagonist (F(6, 457) = 0.51, $p = 0.80$), alcohol x strain (F(3, 457) = 1.07, $p = 0.36$), D1-R antagonist x strain (F(3, 457) = 0.18, $p = 0.83$), alcohol x D1-R antagonist x strain (F(6, 457) = 1.35, $p = 0.23$). Post hoc Tukey HSD test showed that the highest alcohol dose group (1%) showed the least reduction of total distance (for example, among the 0 mg/L D1-R antagonist treated fish, the 1% alcohol treated fish significantly ($p < 0.05\%$) differed from fish of all other alcohol groups in both strains). It also showed the fish treated with the highest D1-R antagonist dose (1 mg/L) also reduced their distance the least.

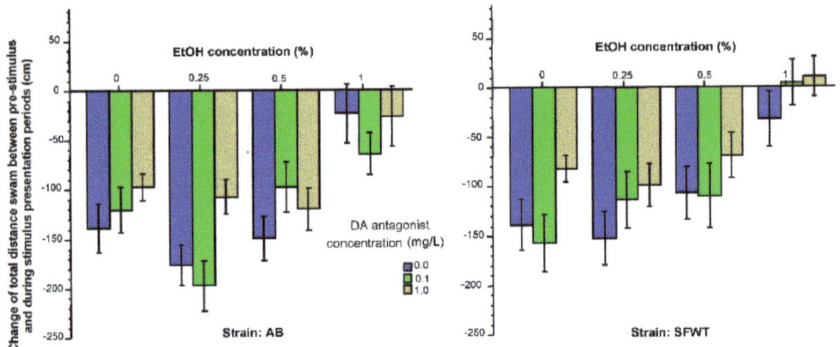

Figure 3. Change of total distance swum elicited by the presentation of animated conspecific images is influenced by the concentration of acute alcohol (X-axis) and the concentration of acute D1-R antagonist (legend) in an additive manner. Additionally, note the significant strain differences. Mean ± S.E.M. are shown.

The pattern of results was quite different for the behavioral measure distance to bottom. Here, again, we assumed that the appearance of conspecific images should alter the distance experimental fish swim from the bottom. Bottom dwell time has been suggested as a measure of anxiety [66], and presence of conspecifics has been shown to decrease anxiety [68]. Thus, we assumed that presentation of conspecifics should increase the distance to bottom. Figure 4 shows this is not what we found. The generally negative values shown on this figure suggest that zebrafish of most treatment groups decreased their distance, i.e., swam closer, to the bottom in response to the presentation of animated conspecific images. ANOVA found that alcohol had a significant effect on this response (F(3, 457) = 4.67, $p = 0.003$), but the effect of the D1-R antagonist was non-significant (F(2, 457) = 1.76, $p = 0.17$). The strain effect was found significant (F(1, 457) = 9.01, $p = 0.003$). However, no interaction term was found significant (alcohol x D1-R antagonist (F(6, 457) = 1.57, $p = 0.15$), alcohol x strain (F(3, 457) = 0.57, $p = 0.63$), D1-R antagonist x strain (F(3, 457) = 0.35, $p = 0.70$), alcohol x D1-R antagonist x strain (F(6, 457) = 1.05, $p = 0.39$). Tukey HSD test showed that the apparent differences seen among alcohol and D1-R treatment groups are non-significant ($p > 0.05$), except that the highest alcohol concentration (1%) treated fish changed their distance from bottom less compared to the lower-intermediate alcohol concentration (0.25%) treated fish.

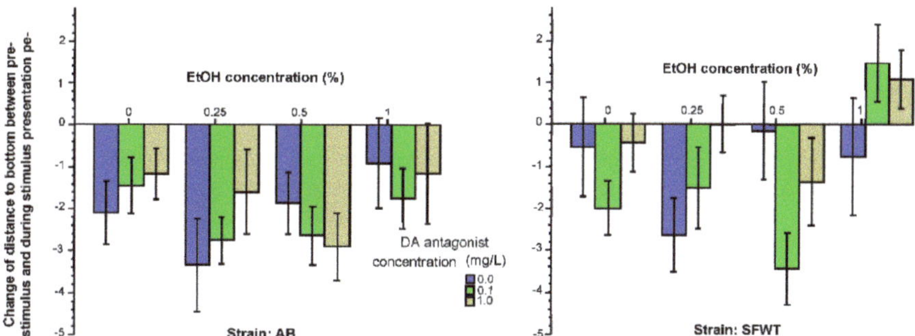

Figure 4. Change of distance to bottom elicited by the presentation of animated conspecific images is influenced by the concentration of acute alcohol (X-axis) and by the strain origin of the fish, but not by the concentration of acute D1-R antagonist (legend). Mean ± S.E.M. are shown.

Intra-individual temporal variance of distance to bottom reflects how much the fish move up and down in the water column, i.e., it is a measure of vertical exploration. Exploration is expected to diminish when fish become habituated to a novel environment, i.e., when the fish is experiencing less novelty induced anxiety [71]. As presence of conspecifics has been found to reduce anxiety [68], we expected the variance of distance to bottom to diminish in response to the presentation of animated conspecific images. This is exactly what we found, as demonstrated by the generally negative values (reduced variance) shown in Figure 5. From this figure it is also apparent that alcohol had a dose dependent effect, it diminished the reduction of variance to bottom. Although less obviously, but the D1-R antagonist also appeared to diminish this change. ANOVA found the alcohol ($F_{(3, 457)} = 21.64$, $p < 0.001$) and the D1-R antagonist effects significant ($F_{(2, 457)} = 5.76$, $p = 0.003$). The strain effect was non- significant ($F_{(1, 457)} = 0.42$, $p = 0.52$), and we found the interaction terms also non-significant (alcohol x D1-R antagonist ($F_{(6, 457)} = 1.27$, $p = 0.27$), alcohol x strain $F_{(6, 457)} = 1.57$, $p = 0.20$), D1-R antagonist x strain ($F_{(3, 457)} = 0.12$, $p = 0.89$), alcohol x D1-R antagonist x strain ($F_{(6, 457)} = 0.92$, $p = 0.48$). Tukey HSD test found that the highest alcohol concentration essentially abolished ($p < 0.05$) the change in vertical exploration induced by the animated conspecific images) irrespective of the D1-R antagonist dose employed.

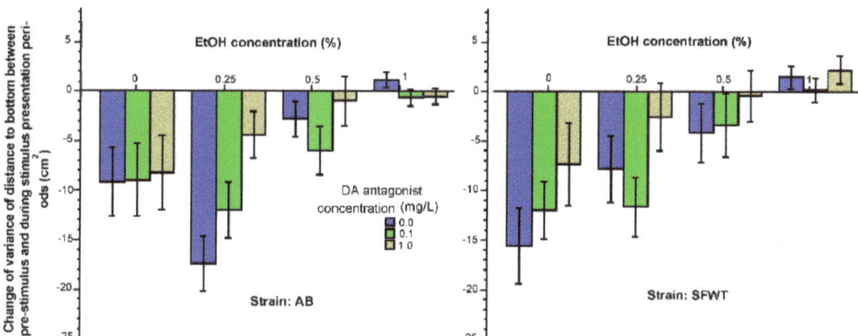

Figure 5. Change of intra-individual temporal variance of distance to bottom (vertical exploration) elicited by the presentation of animated conspecific images is influenced by the concentration of acute alcohol (X-axis) and by the concentration of acute D1-R antagonist (legend) in an additive manner. Mean ± S.E.M. are shown.

The last behavior we measured and analyzed was absolute turn angle, i.e., the amount of turning irrespective of its direction. Although not described before, we expected turning to increase in response to the animated conspecific images, as we expected the experimental fish to follow these images near the short side of the aquarium swimming back and forth. This is exactly what we found. The increased turn angle is well demonstrated by Figure 6, which shows values for most treatment groups to be positive (increase from the pre-stimulus period to the stimulus period). Furthermore, the highest alcohol dose (1%) appears to have abolished this increase and the D1-R antagonist also appears to have blunted the increase. ANOVA confirmed these observations and found the effect of alcohol ($F_{(3, 457)} = 6.80$, $p < 0.001$) and the effect of the D1-R antagonist ($F_{(2, 457)} = 4.47$, $p = 0.01$) significant. The strain effect was non-significant ($F_{(1, 457)} = 2.05$, $p = 0.15$). The double interaction terms were also non-significant (alcohol x D1-R antagonist $F_{(6, 457)} = 0.27$, $p = 0.95$; alcohol x strain $F_{(6, 457)} = 1.09$, $p = 0.35$; D1-R antagonist x strain $F_{(3, 457)} = 0.27$, $p = 0.77$). However, the triple interaction term, alcohol x D1-R antagonist x strain, was significant ($F_{(6, 457)} = 2.42$, $p = 0.026$). Tukey HSD test showed that in AB zebrafish, the highest alcohol concentration significantly ($p < 0.05$) blunted the turn angle increasing effect of the animated conspecific images, and a similar significant effect of the highest dose of the D1-R antagonist was also found ($p < 0.05$). However, this significant alcohol effect was absent in SFWT fish.

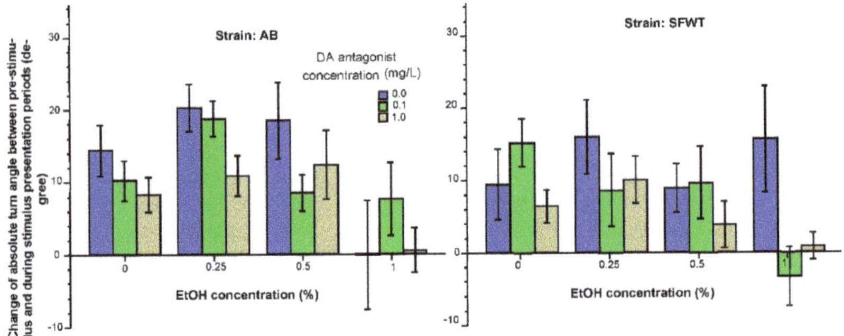

Figure 6. Change of absolute turn angle elicited by the presentation of animated conspecific images is influenced by the concentration of acute alcohol (X-axis) and by the concentration of acute D1-R antagonist (legend) in an additive manner. Mean ± S.E.M. are shown.

3.2. Neurochemical Parameters

We next analyzed the levels of neurochemicals, dopamine, DOPAC, serotonin and 5HIAA from whole brain extracts using HPLC. Dopamine levels (Figure 7) appeared to be elevated by acute exposure to alcohol. In AB strain zebrafish, alcohol had a linear dose dependent effect with the highest dose having the strongest effect. The D1-R antagonist on the other hand, reduced dopamine levels when applied at the highest concentration. It is also apparent that these two drugs did not have an additive effect (i.e., subtractive effect in this case) on dopamine levels. Another clearly observable finding is that both the absolute values and also the dose response curves for these drugs differ between AB and SFWT zebrafish. These observations were supported by the results of ANOVA. It found all main effects and interaction terms to be highly significant (alcohol $F_{(3, 216)} = 516.74$ $p < 0.001$; D1-R antagonist $F_{(2, 216)} = 282.88$, $p < 0.001$; strain $F_{(1, 216)} = 472.11$, $p < 0.001$; alcohol x D1-R antagonist $F_{(6, 216)} = 14.53$, $p < 0.001$; alcohol x strain $F_{(6, 216)} = 56.66$ $p < 0.001$; D1-R antagonist x strain $F_{(3, 216)} = 7.64$, $p = 0.001$; alcohol x D1-R antagonist x strain $F_{(6, 216)} = 10.59$, $p < 0.001$). Tukey HSD tests confirmed these findings and, for example, found the effect of alcohol on AB zebrafish exposed to 0 mg/L D1-R antagonist linearly dose dependent with every dose group significantly ($p < 0.05$) different from the other, and also every alcohol dose group exposed the 0.1 mg/L D1-R antagonist from each other. However, for the 1 mg/L D1-R antagonist exposed AB zebrafish the alcohol dose response was found non-linear, i.e., plateauing for the two highest alcohol concentrations. That is, these two highest alcohol concentration groups did not significantly differ from each other ($p > 0.05$) but did significantly ($p < 0.05$) differ from the group treated acutely with the lowest alcohol concentration (0.25%) and also from the control group (0% alcohol). A somewhat different dose group differences were found by Tukey HSD for the SFWT zebrafish. That is, the SFWT zebrafish alcohol dose response curves were found inverted U-shaped and not linear when the fish were exposed to no D1-R antagonist (0 mg/L) or the lower dose of this drug (0.1 mg/L), with all alcohol concentration groups significantly ($p < 0.05$) differing from each other. Similarly to the AB fish, however, the SFWT zebrafish exposed to the highest D1-R antagonist dose (1 mg/L) the alcohol dose response showed a plateau, with the two highest alcohol concentration groups having the most similar values, both significantly ($p < 0.05$) higher than the values obtained for fish unexposed to alcohol (0% alcohol group) or the lowest alcohol dose (0.25% alcohol).

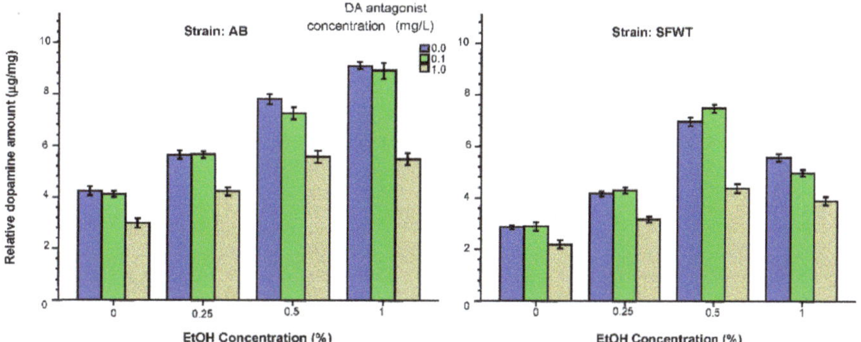

Figure 7. Dopamine levels relative to total brain protein weight are influenced by the concentration of acute alcohol (X-axis) and the concentration of acute D1-R antagonist (legend) in an interactive manner. Mean ± S.E.M. are shown. Additionally, note the significant strain difference.

DOPAC levels (Figure 8) were also elevated by alcohol, with higher concentrations having stronger effects in both strains. Similarly to dopamine, the effect of the lower dose of the D1-R antagonist (0.1 mg/L) appeared to be negligible and did not appreciably modify the effect of alcohol. However, the higher dose of this drug (1 mg/L D1-R antagonist) had a clearly observable DOPAC reducing effect blunting the alcohol induced increase, with slightly different alcohol dose response trajectories in AB and SFWT fish. These results were supported by ANOVA, which found the effect of alcohol ($F(3, 216) = 333.92$, $p < 0.001$), the effect of D1-R antagonist ($F(2, 216) = 106.61$, $p < 0.001$), the effect of strain ($F(1, 216) = 69.17$, $p < 0.001$) all significant. The interaction terms alcohol x D1-R antagonist ($F(6, 216) = 9.08$, $p < 0.001$), alcohol x strain ($F(6, 216) = 15.39$, $p < 0.001$) were also found significant. However, the D1-R antagonist x strain interaction ($F(3, 216) = 1.28$, $p = 0.28$) and the alcohol x D1-R antagonist x strain interaction ($F(6, 216) = 0.86$, $p = 0.53$) were non-significant. Tukey HSD tests showed that in both AB and SFWT zebrafish increasing alcohol concentrations resulted in significantly elevated DOPAC with each alcohol concentration group differing from all others ($p < 0.05$) for fish that were exposed either to no D1-R antagonist (0 mg/L) or the lower concentration (0.1 mg/L). However, for fish that were exposed to the high concentration of D1-R antagonist (1 mg/L) Tukey HSD found all alcohol exposed fish (for AB) or the two highest alcohol concentration exposed fish (SFWT) not to differ significantly ($p > 0.05$) from each other but to significantly ($p < 0.05$) differ from the 0% alcohol group (AB), or the 0% and 0.25% alcohol groups (SFWT).

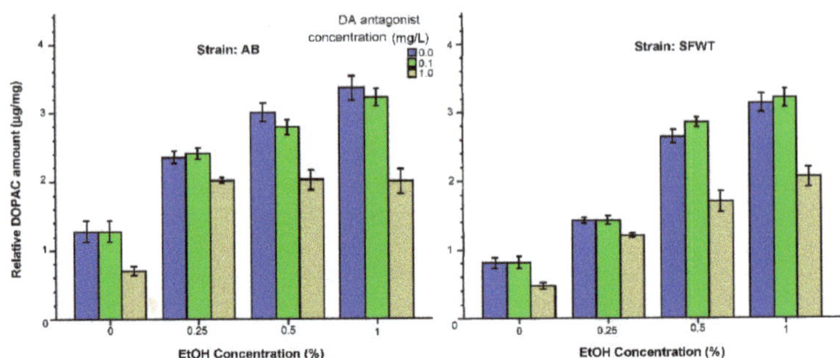

Figure 8. DOPAC levels relative to total brain protein weight are influenced by the concentration of acute alcohol (X-axis) and the concentration of acute D1-R antagonist (legend) in an interactive manner. Additionally, note the significant strain difference. Mean ± S.E.M. are shown.

The alcohol and D1-R dose response curves obtained for the levels of serotonin (Figure 9), extracted from whole brain tissue samples appear to be different compared to those obtained for dopamine and DOPAC. It appears that alcohol only exerted a robust effect at the highest (1%) concentration in AB zebrafish, but in SFWT such robust effect was seen at the 0.5% alcohol concentration. Additionally, differently from the dopamine and DOPAC results, the D1-R antagonist did not appear to have a robust effect in any strain of zebrafish exposed to any alcohol dose. ANOVA confirmed a significant alcohol effect $(F(3, 216) = 275.27, p < 0.001)$, but found the effects of D1-R antagonist $(F(2, 216) = 0.22, p = 0.80)$ and strain $(F(1, 216) = 0.82, p = 0.37)$ non-significant. The interaction term alcohol x D1-R antagonist reached the level of significance $(F(6, 216) = 2.20\ p = 0.04)$, and the alcohol x strain interaction was also significant $(F(6, 216) = 305.21, p < 0.001)$. However, the D1-R antagonist x strain interaction $(F(3, 216) = 0.05, p = 0.95)$ and the alcohol x D1-R antagonist x strain interaction $(F(6, 216) = 0.79, p = 0.58)$ were non-significant. Tukey HSD showed that the D1-R antagonist was essentially ineffective as no significant difference $(p > 0.05)$ among D1-R concentration groups were found for zebrafish within any given alcohol dose treatment set for either strain. As to the alcohol dose response curves, Tukey HSD confirmed that the 1% alcohol treated AB zebrafish significantly differed from all AB zebrafish alcohol groups, including the 0% alcohol control, all these other alcohol groups did not significantly differ from each other $(p > 0.05)$. Tukey HSD test also confirmed that at 0.5% alcohol treated SFWT zebrafish significantly differed $(p < 0.05)$ from SFWT zebrafish of all the other alcohol groups and these latter groups did not significantly differ from each other $(p > 0.05)$.

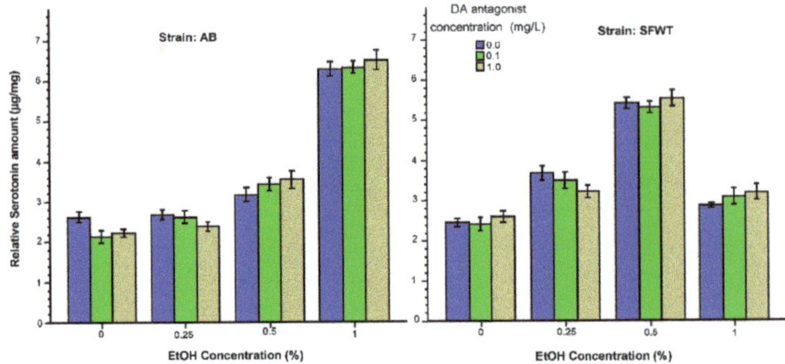

Figure 9. Serotonin levels relative to total brain protein weight are influenced by the concentration of acute alcohol (X-axis) but not by the concentration of acute D1-R antagonist (legend). Additionally note the strain dependent alcohol dose responses (alcohol x strain interaction). Mean ± S.E.M. are shown.

The last result we consider is the effects of alcohol and of the D1-R antagonist on 5HIAA (Figure 10). For this metabolite of serotonin, alcohol appeared to exert a linear dose dependent effect in AB zebrafish, with higher alcohol concentrations elevating 5HIAA levels more strongly. Whereas for the SFWT fish the dose response curve was inverted U-shaped, with the strongest alcohol effect obtained for the 0.5% intermediate alcohol dose. The D1-R antagonist appeared to affect only certain alcohol dose groups. ANOVA found a significant alcohol ($F(3, 216) = 166.07$, $p < 0.001$), D1-R antagonist ($F(2, 216) = 22.85$, $p < 0.001$) and strain effect ($F(1, 216) = 24.96$, $p < 0.001$). The interaction term, alcohol x D1-R antagonist, was non-significant ($F(6, 216) = 1.91$ $p = 0.08$), but the alcohol x strain interaction ($F(6, 216) = 40.30$, $p < 0.001$), D1-R antagonist x strain interaction ($F(3, 216) = 3.61$, $p = 0.03$) and the alcohol x D1-R antagonist x strain interaction ($F(6, 216) = 2.21$, $p = 0.04$) were significant. Tukey HSD test confirmed the linear alcohol dose dependent effect in AB fish, i.e., found significant ($p < 0.05$) differences between groups of AB zebrafish receiving the increasing alcohol dose under the same D1-R treatment. For example, in AB, the effect of D1-R treatment was found significant ($p < 0.05$) for the 0% alcohol treated fish, as the 1 mg/L (highest dose) D1-R antagonist dose group differed from the 0 mg/L D1-R antagonist dose group. For the 1% alcohol treated AB fish, again the highest D1-R dose group was found to significantly differ from the no D1-R treated group. In SFWT, the D1-R antagonist was found to have an effect only under the 0% alcohol and 0.5% alcohol treatment. For the former, Tukey HSD found the 0 mg/L and the 1 mg/L D1-R antagonist treated fish to significantly differ ($p < 0.05$), and for the latter the 1 mg/L D1-R antagonist treated fish to significantly differ ($p < 0.05$) from both the 0 and the 0.1 mg/L D1-R antagonist treated groups.

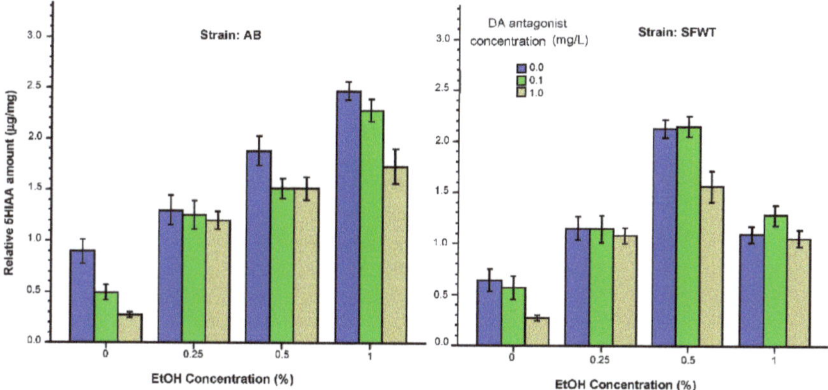

Figure 10. 5HIAA levels relative to total brain protein weight are influenced by the concentration of acute alcohol (X-axis) and by the concentration of acute D1-R antagonist (legend). Additionally, note the strain dependent alcohol dose responses (alcohol x strain interaction). Mean ± S.E.M. are shown.

4. Discussion

Zebrafish have previously been found to approach animated images of conspecifics upon their appearance and to stay in proximity of these images as long as they were being shown [64]. The strength of this shoaling response was found indistinguishable from that of responses elicited by live conspecifics or by video playback of previously recorded live conspecifics [72]. On the other hand, the response to animated conspecific images was clearly different from the response to animated images for which the pixels of the photos of conspecifics were scrambled and thus the images did not resemble a conspecific (or any other fish species) [52]. These scrambled images did induce a rapid approach (assumingly representing exploration of novel stimuli), but the experimental fish did not remain with these images, unlike when actual conspecific images were being shown [52]. Thus, we concluded that species-specific characteristics of the animated conspecific images must be present in order for experimental zebrafish to stay in close proximity to them [52,64,73]. In freely moving groups of zebrafish, members of the shoal were found to remain at a well-defined distance from each other, i.e., at a distance that is similar to how far a single experimental zebrafish swims from the animated images of conspecifics [73,74]. Thus, we concluded that the reduction of distance induced by animated images of conspecifics is an appropriate measure of the natural shoaling response of zebrafish [51]. In our current study, we found again that presentation of animated images of conspecifics reduced the distance between the experimental fish and the image presentation computer screen. In the past, we showed that acute exposure to alcohol disrupts shoaling in zebrafish [31,47]. This is what we observed in the current study too. At the highest dose (1%), acute alcohol almost completely abolished the shoaling response. We also knew from our past experiments that D1-R antagonism also disrupts shoaling [50], that conspecific images rapidly elevate dopamine and DOPAC levels in the brain of the exposed fish [52,53], and that acute alcohol also elevates dopamine and DOPAC levels [24,31,44]. We also discovered that acute alcohol increases tyrosine hydroxylase protein expression and dopamine synthesis in the zebrafish brain [42]. What we did not know is whether there is causal link between alcohol induced dopaminergic and behavioral responses. In the current study, we found proof for this causal link.

We argued that if acute alcohol and the D1-R antagonist acts independently, we should be able to observe their effects on behavioral and neurochemical phenotypes as being additive, i.e., without significant interaction. Our results obtained for the shoaling response (change of distance to stimulus screen induced by conspecific images, Figure 1) demonstrates that the effects of acute alcohol and of the D1-R antagonist were non-additive, as we found a highly significant interaction between these two drugs. Importantly, the alcohol x D1-R antagonist interaction did not depend upon the strain origin of the fish, i.e., it was evident in both strains. For example, while in fish treated with 0% alcohol the D1-R antagonist dose dependently reduced the shoaling response, the trend was reversed in 1% alcohol treated fish, in which the highest D1-R antagonist dose treated fish showed the strongest shoaling response. The alcohol x D1-R antagonist interaction thus means that these two drugs do not exert their effects on the shoaling response via two independent (parallel) biochemical pathways or neurobiological mechanisms, but rather, alcohol and the D1-R antagonist converge on the same biological mechanism. Given that the D1-R antagonist employed here are thought to be a dopamine receptor specific drug in mammals and in zebrafish that thus influences the dopaminergic system [43,61], we conclude that acute alcohol affects shoaling via a dopaminergic mechanism in zebrafish.

However, this finding does not mean that acute alcohol exerts its effects solely via dopaminergic mechanisms, or the effects of alcohol on all conspecific image induced behavioral responses are mediated by the dopaminergic system. For example, for the variance of the shoaling response (Figure 2, variance of the change of distance to the stimulus screen induced by the conspecific images), we found significant alcohol and D1-R antagonist effects, but these effects did not interact. First, we emphasize that "variance" here does not mean differences among individuals. Intra-individual variance here represents the time dependent changes in the distance of the single fish to the stimulus screen. It is, in essence, a measure of exploration of the conspecific images on the horizontal axis. High variance means the experimental fish often went closer and other times went further away from the conspecific images. The lack of alcohol x D1-R antagonist interaction, coupled with the significant main effects of these factors means that the effects of these two drugs acted independently on this measure of exploratory behavior. Alcohol increased the intra-individual temporal variance of shoaling response in a U-shaped dose dependent manner with the highest variance induced by the low dose (0.25%) alcohol and the smallest variance by the highest dose (1%) alcohol. Whereas the D1-R antagonist reduced the intra-individual variance only if administered at the highest dose (1 mg/L). Again, importantly, these independent drug effects and dose response curves replicated well in both strains, AB and the SFWT. Thus, not only the lack of alcohol x D1-R interaction but also the trajectory of the dose response curves differed between the shoaling response and its intra-individual temporal variance (horizontal exploration). We speculate that such psychopharmacological differences between these two behavioral measures (the shoaling response and horizontal exploration behavior) may represent different underlying neurobiological mechanisms. Furthermore, because the effect of alcohol was independent of the effect of the D1-R antagonist, and because alcohol is known to act through a variety of mechanisms, we argue that the horizontal exploration behavior was influenced by alcohol likely via non-dopaminergic mechanisms.

To further strengthen this argument, one may examine the results we obtained for the behavioral measures, total distance swum or the intra-individual variance of distance from bottom. We first emphasize that again, just like in the above behaviors, we are discussing the change in these behaviors induced by the shoaling images. The statistical results as well as the dose response curves obtained for these two behavioral measures (Figures 3 and 5) are practically identical to those discussed above for the horizontal exploration behavior (Figure 2). For example, the change in the total distance swum was affected significantly by alcohol and the D1-R antagonist, but these effects were independent of each other, i.e., were additive (subtractive in this case, to be precise). This is not surprising because distance traveled is often used as a measure of exploratory behavior in studies of fish [75–77] and mammals [78]. The intra-individual variance of distance from bottom, represents vertical exploration. Thus, we conclude that horizontal exploration (intra-individual variance of distance to the stimulus screen) vertical exploration (intra-individual distance to the bottom) and general exploratory behavior (total distance swum) are likely to be affected by alcohol and the D1-R antagonist via mechanisms that are, at least partially, different from those underlying shoaling behavior itself, and that alcohol affected these measures of exploratory behavior via non-dopaminergic mechanisms.

Examination of the results we obtained for the behavioral measure, distance to bottom reveals yet another different pattern of drug effects. Bottom dwell time is often regarded (although not without controversies) as a measure of anxiety in zebrafish [66]. Sight of conspecifics are expected to reduce anxiety and thus increase distance to bottom. We found the opposite, i.e., increased distance from bottom induced by the conspecific images, yet another indication that perhaps bottom dwell time, or distance from bottom, is not a reliable measure of anxiety in zebrafish [79,80]. Irrespective of whether distance to bottom is or is not an appropriate anxiety indicator, we found alcohol but not the D1-R antagonist to exert a significant effect on this response, yet again suggesting underlying mechanisms at least partially distinct both from those of shoaling and of exploratory behavior.

The last behavior we consider in this discussion is the change in turn angle. This behavioral response was affected significantly by both alcohol and the D1-R antagonist, but again not in interaction with each other. Additionally, importantly, the dose response curves of the drugs are unique for this behavior too: only the highest concentration of each of these two drugs had a significant effect in AB zebrafish, and only the highest concentration of the D1-R antagonist had an effect in SFWT, but alcohol was ineffective in this latter strain at all the concentrations examined.

We emphasize that all the above discussed behavioral effects reflect the change between the first and the second 8 min of the recording session, i.e., between the periods during which shoaling images were absent (first period) and the period during which they were present (second period). Although all these responses represent a change induced by the presentation of conspecifics, as we argued above, they may not all represent social behavior, i.e., shoaling. For example, while the reduction of distance to stimulus screen in response to conspecific images is indeed a shoaling response, alteration in vertical and horizontal exploration or distance traveled in general, represents changes in exploratory behavior. Our current findings of idiosyncratic drug effects showing, for example, alcohol x D1-R antagonist interaction for shoaling, but no such interaction and different dose response curves for exploratory behaviors, thus further strengthens the argument that these are distinct behavioral responses not just based upon their phenotypical appearance (and perhaps functional meaning), but also in terms of the mechanisms underlying them. Although these mechanisms are unknown at this point, our current results suggest that at least for shoaling, alcohol induced changes are mediated by the dopaminergic neurotransmitter system. Our neurochemical results are in line with this conclusion as they also revealed D1-R antagonist dependent alcohol effects.

Analysis of the effects of alcohol and the D1-R antagonist showed that both drugs significantly affected dopamine levels (Figure 7): alcohol significantly increased and the D1-R antagonist significantly decreased it. Importantly, the effects of these two drugs were not additive. A similar alcohol x D1-R antagonism interaction was also found for DOPAC (Figure 8), dopamine's metabolite. Among the 0 mg/L and 0.1 mg/L D1-R antagonist treated AB zebrafish alcohol had a linear dose dependent effect but for the AB fish treated with the highest D1-R antagonist dose (1 mg/L) the alcohol effect plateaued at the 0.5% alcohol dose. A similar plateauing effect is also evident in AB fish for DOPAC. Such interactive effects of alcohol and the D1-R antagonist thus may explain the similar drug interaction found for the shoaling response. However, we also note the dose response trajectories obtained for dopamine and DOPAC (Figures 7 and 8) do not precisely mirror those obtained for the shoaling response (Figure 1), suggesting that very likely the latter is also influenced by mechanisms other than the dopaminergic system.

In the past, we found shoaling images not to alter serotonin and 5HIAA levels in the brain of zebrafish [52]. We also found acute alcohol to increase the levels of these neurochemicals [31]. Importantly, here we show that although alcohol at the highest concentration in AB (1%), or at the intermediate concentration (0.5%) in SFWT, does increase serotonin levels, the D1-R antagonist had no effect. Given that here we found the D1-R antagonist to significantly affect shoaling but not serotonin levels, and given that shoaling images were previously not found to alter serotonin levels, we conclude that shoaling does not depend upon serotonin levels. Nevertheless, we cannot fully exclude that the serotoninergic neurotransmitter system influences alcohol induced changes in shoaling. This is because although the dose response trajectories for shoaling and for 5HIAA do differ, the levels of the latter neurochemical (the metabolite of serotonin) were influenced by alcohol and the D1-R antagonist in an interactive manner.

The last point we wish to discuss is the strain differences and the interactions of strain effect with alcohol and/or with the D1-R antagonist. In the current study, strain effects or strain x D1-R antagonist and/or strain x alcohol interactions were not found for the shoaling response but were found for other behaviors. While strain differences and strain interactions in these latter behaviors do complicate the interpretation of our results as they mean that alcohol and/or D1-R effects are not faithfully replicated between the two strains, they also represent a potentially interesting starting point for future studies. Given that fish of the AB and SFWT strains were bred, raised, maintained and tested under identical conditions, the observed phenotypical differences between them must be the result of the genetic differences between the strains. Such genetic differences may be exploited in a variety of ways for the investigation of mechanisms underlying acute alcohol administration induced changes in brain function and behavior. For example, one may conduct a linkage analysis and identify the locus/loci of genes that may underlie differential alcohol effects, or, for example, one could also conduct gene expression analyses to identify differentially expressed genes involved in alcohol effects.

5. Conclusions

In this proof-of-principle study, we obtained the first piece of evidence that the shoaling response of zebrafish induced by the presentation of conspecific images is impaired by alcohol via the dopaminergic system. We also found that other behavioral responses to the shoaling stimulus are influenced by alcohol in a manner that is independent of dopaminergic mechanisms. Our study was not designed to pinpoint such mechanisms, or to essay the potentially large number of alcohol targets that could be involved. Nevertheless, given the powerful recombinant DNA methods and neuroscience techniques already developed for the zebrafish ([81]; for near comprehensive list of methods and applications also see [79]), our results highlight the possibility that this species may be an excellent model organism with which the mechanistic underpinnings of the effects of alcohol and of shoaling and other behaviors may be explored.

Author Contributions: R.G. formulated the concept and designed the study. T.S. conducted the experiments and collected the data. R.G. analyzed the data. T.S. and R.G. wrote the paper. All authors have read and agreed to the published version of the manuscript.

Funding: RG is funded by NSERC (Canada) Discovery Grant #311637, the University of Toronto Mississauga Distinguished Professorship Award, and the UTM Provostial Grant.

Institutional Review Board Statement: The study was conducted in accordance with local (University of Toronto), Provincial (Ontario) and Federal (Canada) guidelines and laws for the ethical and humane use of animals in research. The studies described in this paper have been approved (20011370, 17-Feb-22) by the University of Toronto Mississauga Local Animal Care Committee (LACC).

Data Availability Statement: The data present in this study are available from the corresponding author upon request.

Acknowledgments: We would like to thank Diptendu Chatterjee for his technical help with HPLC.

Conflicts of Interest: The authors declare no conflict of interest.

References

1. Park, S.H.; Kim, D.J. Global and regional impacts of alcohol use on public health: Emphasis on alcohol policies. *Clin. Mol. Hepatol.* **2020**, *26*, 652–661. [CrossRef]
2. Laramée, P.; Kusel, J.; Leonard, S.; Aubin, H.J.; François, C.; Daeppen, J.B. The economic burden of alcohol dependence in Europe. *Alcohol Alcohol.* **2013**, *48*, 259–269. [CrossRef] [PubMed]
3. Navarro, H.J.; Doran, C.M.; Shakeshaft, A.P. Measuring costs of alcohol harm to others: A review of the literature. *Drug Alcohol Depend.* **2011**, *114*, 87–99. [CrossRef] [PubMed]
4. Anton, R.F.; Kranzler, H.R.; Meyer, R.E. Neurobehavioral aspects of the pharmacotherapy of alcohol dependence. *Clin. Neurosci.* **1995**, *3*, 145–154.
5. Ray, L.A.; Bujarski, S.; Grodin, E.; Hartwell, E.; Green, R.; Venegas, A.; Lim, A.C.; Gillis, A.; Miotto, K. State-of-the-art behavioral and pharmacological treatments for alcohol use disorder. *Am. J. Drug Alcohol Abuse* **2019**, *45*, 124–140. [CrossRef]
6. Morgane, G.L.; Delphine, L.G.; Guillaume, K.; Yves, L.J. The Place of Pharmacotherapy in Alcohol Use Disorder Management in Family Practice-A Systematic Review. *Curr. Pharm. Des.* **2021**, *27*, 2737–2745. [CrossRef] [PubMed]
7. Samson, H.H.; Harris, R.A. Neurobiology of alcohol abuse. *Trends Pharmacol. Sci.* **1992**, *13*, 206–211. [CrossRef]
8. Nevo, I.; Hamon, M. Neurotransmitter and neuromodulatory mechanisms involved in alcohol abuse and alcoholism. *Neurochem. Int.* **1995**, *26*, 305–336. [CrossRef]
9. Ward, R.J.; Lallemand, F.; de Witte, P. Biochemical and neurotransmitter changes implicated in alcohol-induced brain damage in chronic or 'binge drinking' alcohol abuse. *Alcohol Alcohol.* **2009**, *44*, 128–135. [CrossRef]
10. Abrahao, K.P.; Salinas, A.G.; Lovinger, D.M. Alcohol and the Brain: Neuronal Molecular Targets, Synapses, and Circuits. *Neuron* **2017**, *96*, 1223–1238. [CrossRef]
11. de Wit, H.; Phillips, T.J. Do initial responses to drugs predict future use or abuse? *Neurosci. Biobehav. Rev.* **2012**, *36*, 1565–1576. [CrossRef] [PubMed]
12. Hinckers, A.S.; Laucht, M.; Schmidt, M.H.; Mann, K.F.; Schumann, G.; Schuckit, M.A.; Heinz, A. Low level of response to alcohol as associated with serotonin transporter genotype and high alcohol intake in adolescents. *Biol. Psychiatry* **2006**, *60*, 282–287. [CrossRef]
13. Elvig, S.K.; McGinn, M.A.; Smith, C.; Arends, M.A.; Koob, G.F.; Vendruscolo, L.F. Tolerance to alcohol: A critical yet understudied factor in alcohol addiction. *Pharmacol. Biochem. Behav.* **2021**, *204*, 173155. [CrossRef] [PubMed]
14. Parker, C.C.; Lusk, R.; Saba, L.M. Alcohol Sensitivity as an Endophenotype of Alcohol Use Disorder: Exploring Its Translational Utility between Rodents and Humans. *Brain Sci.* **2020**, *10*, 725. [CrossRef]
15. Crabbe, J.C. Genetic animal models in the study of alcoholism. *Alcohol. Clin. Exp. Res.* **1989**, *13*, 120–127. [CrossRef]
16. Crabbe, J.C. Use of animal models of alcohol-related behavior. *Handb. Clin. Neurol.* **2014**, *125*, 71–86. [CrossRef] [PubMed]
17. Lovinger, D.M.; Crabbe, J.C. Laboratory models of alcoholism: Treatment target identification and insight into mechanisms. *Nat. Neurosci.* **2005**, *8*, 1471–1480. [CrossRef]
18. Gerlai, R. Embryonic alcohol exposure: Towards the development of a zebrafish model of fetal alcohol spectrum disorders. *Dev. Psychobiol.* **2015**, *57*, 787–798. [CrossRef]
19. Tran, S.; Gerlai, R. Recent advances with a novel model organism: Alcohol tolerance and sensitization in zebrafish (*Danio rerio*). *Prog. Neuro-Psychopharmacol. Biol. Psychiatry* **2014**, *55*, 87–93. [CrossRef]

20. Tran, S.; Facciol, A.; Gerlai, R. The zebrafish, a novel model organism for screening compounds affecting acute and chronic ethanol induced effects. *Int. Rev. Neurobiol.* **2016**, *126*, 467–484. [CrossRef]
21. Howe, K.; Clark, M.D.; Torroja, C.F.; Torrance, J.; Berthelot, C.; Muffato, M.; Collins, J.E.; Humphray, S.; McLaren, K.; Matthews, L.; et al. The zebrafish reference genome sequence and its relationship to the human genome. *Nature* **2013**, *496*, 498–503. [CrossRef] [PubMed]
22. Reimers, M.J.; Hahn, M.E.; Tanguay, R.L. Two zebrafish alcohol dehydrogenases share common ancestry with mammalian class I, II, IV, and V alcohol dehydrogenase genes but have distinct functional characteristics. *J. Biol. Chem.* **2004**, *279*, 38303–38312. [CrossRef]
23. Postlethwait, J.H. The zebrafish genome: A review and msx gene case study. *Genome Dyn.* **2006**, *2*, 183–197. [CrossRef]
24. Chatterjee, D.; Gerlai, R. High Precision Liquid Chromatography Analysis of Dopaminergic and Serotoninergic Responses to Acute Alcohol Exposure in Zebrafish. *Behav. Brain Res.* **2009**, *200*, 208–213. [CrossRef] [PubMed]
25. Kalueff, A.V.; Stewart, A.M.; Gerlai, R. Zebrafish as an emerging model for studying complex brain disorders. *Trends Pharmacol. Sci.* **2014**, *35*, 63–75. [CrossRef] [PubMed]
26. Gerlai, R. Zebrafish Phenomics: Behavioral Screens and Phenotyping of Mutagenized Fish. *Curr. Opin. Behav. Sci.* **2014**, *2*, 21–27. [CrossRef]
27. Gerlai, R. Using Zebrafish to Unravel the Genetics of Complex Brain Disorders. *Curr. Top. Behav. Neurosci.* **2012**, *12*, 3–24. [CrossRef] [PubMed]
28. Gerlai, R. High-throughput Behavioral Screens: The First Step towards Finding Genes Involved in Vertebrate Brain Function Using Zebrafish. *Molecules* **2010**, *15*, 2609–2622. [CrossRef]
29. Gerlai, R. Evolutionary conservation, translational relevance and cognitive function: The future of zebrafish in behavioral neuroscience. *Neurosci. Biobehav. Rev.* **2020**, *116*, 426–435. [CrossRef]
30. Bailey, C.D.C.; Gerlai, R.; Cameron, N.M.; Marcolin, M.L.; McCormick, C.M. Preclinical methodological approaches investigating the effects of alcohol on perinatal and adolescent neurodevelopment. *Neurosci. Biobehav. Rev.* **2020**, *116*, 436–451. [CrossRef]
31. Gerlai, R.; Chatterjee, D.; Pereira, T.; Sawashima, T.; Krishnannair, R. Acute and Chronic alcohol dose: Population differences in behavior and neurochemistry of zebrafish. *Genes Brain Behav.* **2009**, *8*, 586–599. [CrossRef] [PubMed]
32. Love, D.R.; Pichler, F.B.; Dodd, A.; Copp, B.R.; Greenwood, D.R. Technology for high-throughput screens: The present and future using zebrafish. *Curr. Opin. Biotechnol.* **2004**, *15*, 564–571. [CrossRef] [PubMed]
33. Bugel, S.M.; Tanguay, R.L.; Planchart, A. Zebrafish: A marvel of high-throughput biology for 21st century toxicology. *Curr. Environ. Health Rep.* **2014**, *1*, 341–352. [CrossRef] [PubMed]
34. Gerlai, R. Phenomics: Fiction or the Future? *Trends Neurosci.* **2002**, *25*, 506–509. [CrossRef]
35. Attwood, A.S.; Munafò, M.R. Effects of acute alcohol consumption and processing of emotion in faces: Implications for understanding alcohol-related aggression. *J. Psychopharmacol.* **2014**, *28*, 719–732. [CrossRef]
36. Gerlai, R.; Prajapati, S.; Ahmad, F. Differences in acute alcohol induced behavioral responses among zebrafish populations. *Alcohol. Clin. Exp. Res.* **2008**, *32*, 1763–1773. [CrossRef]
37. Gerlai, R.; Lahav, M.; Guo, S.; Rosenthal, A. Drinks like a fish: Zebra fish (*Danio rerio*) as a behavior genetic model to study alcohol effects. *Pharmacol. Biochem. Behav.* **2000**, *67*, 773–782. [CrossRef]
38. Tran, S.; Gerlai, R. Time-course of behavioural changes induced by ethanol in zebrafish (*Danio rerio*). *Behav. Brain Res.* **2013**, *252*, 204–213. [CrossRef]
39. Hendler, R.A.; Ramchandani, V.A.; Gilman, J.; Hommer, D.W. Stimulant and sedative effects of alcohol. *Curr. Top. Behav. Neurosci.* **2013**, *13*, 489–509. [CrossRef]
40. Charlet, K.; Beck, A.; Heinz, A. The dopamine system in mediating alcohol effects in humans. *Curr. Top. Behav. Neurosci.* **2013**, *13*, 461–488. [CrossRef]
41. Wise, R.A.; Rompre, P.P. Brain dopamine and reward. *Annu. Rev. Psychol.* **1989**, *40*, 191–225. [CrossRef]
42. Tran, S.; Facciol, A.; Nowicki, M.; Chatterjee, D.; Gerlai, R. Acute alcohol exposure increases tyrosine hydroxylase protein expression and dopamine synthesis in zebrafish. *Behav. Brain Res.* **2017**, *317*, 237–241. [CrossRef] [PubMed]
43. Tran, S.; Nowicki, M.; Muraleetharan, A.; Chatterjee, D.; Gerlai, R. Differential effects of acute administration of SCH-23390, a D1-receptor antagonist, and of ethanol on swimming activity, anxiety-related responses, and neurochemistry of zebrafish. *Psychopharmacology* **2015**, *232*, 3709–3718. [CrossRef] [PubMed]
44. Chatterjee, D.; Shams, S.; Gerlai, R. Chronic and acute alcohol administration induced neurochemical changes in the brain: Comparison of distinct zebrafish populations. *Amino Acids* **2014**, *46*, 921–930. [CrossRef]
45. Avery, D.H.; Overall, J.E.; Calil, H.M.; Hollister, L.E. Alcohol-induced euphoria: Alcoholics compared to nonalcoholics. *Int. J. Addict.* **1982**, *17*, 823–845. [CrossRef] [PubMed]
46. Miller, N.; Greene, K.; Dydynski, A.; Gerlai, R. Effects of nicotine and alcohol on zebrafish (*Danio rerio*) shoaling. *Behav. Brain Res.* **2012**, *240*, 192–196. [CrossRef]
47. Ariyasiri, K.; Gerlai, R.; Kim, C.H. Acute ethanol induces behavioral changes and alters c-fos expression in specific brain regions, including the mammillary body, in zebrafish. *Prog. Neuropsychopharm. Biol. Psychiatry* **2021**, *109*, 110264. [CrossRef]
48. Al-Imari, L.; Gerlai, R. Sight of conspecifics as reward in associative learning tasks for zebrafish (*Danio rerio*). *Behav. Brain Res.* **2008**, *189*, 216–219. [CrossRef]

29. Sison, M.; Gerlai, R. Associative learning performance is impaired in zebrafish (*Danio rerio*) by the NMDA-R antagonist MK-801. *Neurobiol. Learn. Mem.* **2011**, *96*, 230–237. [CrossRef]
30. Scerbina, T.; Chatterjee, D.; Gerlai, R. Dopamine receptor antagonism disrupts social preference in zebrafish, a strain comparison study. *Amino Acids* **2012**, *43*, 2059–2072. [CrossRef]
31. Gerlai, R. Animated Images in the analysis of zebrafish behaviour. *Curr. Zool.* **2017**, *63*, 35–44. [CrossRef]
32. Saif, M.; Chatterjee, D.; Buske, C.; Gerlai, R. Sight of conspecific images induces changes in neurochemistry in zebrafish. *Behav. Brain Res.* **2013**, *243*, 294–299. [CrossRef]
33. Fernandes, Y.; Rampersad, M.; Gerlai, R. Embryonic alcohol exposure impairs the dopaminergic system and social behavioural responses in adult zebrafish. *Int. J. Neuropsychopharmacol.* **2015**, *18*, pyu089. [CrossRef]
34. Mahabir, S.; Chatterjee, D.; Buske, C.; Gerlai, R. Maturation of shoaling in two zebrafish strains: A behavioral and neurochemical analysis. *Behav. Brain Res.* **2013**, *247*, 1–8. [CrossRef] [PubMed]
35. Mahabir, S.; Chatterjee, D.; Gerlai, R. Strain dependent neurochemical changes induced by embryonic alcohol exposure in zebrafish. *Neurotoxicol. Teratol.* **2014**, *41*, 1–7. [CrossRef]
36. Pan, Y.; Chaterjee, D.; Gerlai, R. Strain dependent gene expression and neurochemical levels in the brain of zebrafish: Focus on a few alcohol related targets. *Physiol. Behav.* **2012**, *107*, 773–780. [CrossRef]
37. Pannia, E.; Tran, S.; Rampersad, M.; Gerlai, R. Acute ethanol exposure induces behavioural differences in two zebrafish (*Danio rerio*) strains: A time course analysis. *Behav. Brain Res.* **2013**, *259*, 174–185. [CrossRef] [PubMed]
38. Gerlai, R. Reproducibility and replicability in zebrafish behavioral neuroscience research. *Pharmacol. Biochem. Behav.* **2019**, *178*, 30–38. [CrossRef] [PubMed]
39. Cameron, D.L.; Williams, J.T. Dopamine D1 receptors facilitate transmitter release. *Nature* **1993**, *366*, 344–347. [CrossRef]
40. Darbin, O.; Risso, J.J.; Rostain, J.C. Dopaminergic control of striatal 5-HT level at normobaric condition and at pressure. *Undersea Hyperb. Med.* **2010**, *37*, 159–166.
41. Steketee, J.D. Injection of SCH 23390 into the ventral tegmental area blocks the development of neurochemical but not behavioral sensitization to cocaine. *Behav. Pharmacol.* **1998**, *9*, 69–76.
42. Kurata, K.; Shibata, R. Effects of D1 and D2 antagonists on the transient increase of dopamine release by dopamine agonists by means of brain dialysis. *Neurosci. Lett.* **1991**, *133*, 77–80. [CrossRef]
43. Tran, S.; Chatterjee, D.; Gerlai, R. An integrative analysis of ethanol tolerance and withdrawal in zebrafish (*Danio rerio*). *Behav. Brain Res.* **2015**, *276*, 161–170. [CrossRef] [PubMed]
44. Fernandes, Y.; Gerlai, R. Long-term behavioral changes in response to early developmental exposure to ethanol in zebrafish. *Alcohol. Clin. Exp. Res.* **2009**, *33*, 601–609. [CrossRef] [PubMed]
45. Saverino, C.; Gerlai, R. The social zebrafish: Behavioral responses to conspecific, heterospecific, and computer animated fish. *Behav. Brain Res.* **2008**, *191*, 77–87. [CrossRef]
46. Egan, R.J.; Bergner, C.L.; Hart, P.C.; Cachat, J.M.; Canavello, P.R.; Elegante, M.F.; Elkhayat, S.I.; Bartels, B.K.; Tien, A.K.; Tien, D.H.; et al. Understanding behavioral and physiological phenotypes of stress and anxiety in zebrafish. *Behav. Brain Res.* **2009**, *205*, 38–44. [CrossRef]
47. Ahmed, O.; Seguin, D.; Gerlai, R. An automated predator avoidance task in zebrafish. *Behav. Brain Res.* **2011**, *216*, 166–171. [CrossRef]
48. Shams, S.; Seguin, D.; Facciol, A.; Chatterjee, D.; Gerlai, R. Effect of Social Isolation on Anxiety-Related Behaviors, Cortisol, and Monoamines in Adult Zebrafish. *Behav. Neurosci.* **2017**, *131*, 492–504. [CrossRef]
49. Soares, M.C.; Gerlai, R.; Maximino, C. The integration of sociality, monoamines, and stress neuroendocrinology in fish models: Applications in the neurosciences. *J. Fish Biol.* **2018**, *93*, 170–191. [CrossRef]
50. Gerlai, R. Fear responses and antipredatory behavior of zebrafish: A translational perspective. In *Behavioral and Neural Genetics of Zebrafish*, 1st ed.; Gerlai, R., Ed.; Elsevier: Amsterdam, The Netherlands; Academic Press: Cambridge, MA, USA, 2020; pp. 155–173.
51. Wong, K.; Elegante, M.; Bartels, B.; Elkhayat, S.; Tien, D.; Roy, S.; Goodspeed, J.; Suciu, C.; Tan, J.; Grimes, C.; et al. Analyzing habituation responses to novelty in zebrafish (*Danio rerio*). *Behav. Brain Res.* **2010**, *208*, 450–457. [CrossRef]
52. Qin, M.; Wong, A.; Seguin, D.; Gerlai, R. Induction of social behaviour in zebrafish: Live versus computer animated fish as stimuli. *Zebrafish* **2014**, *11*, 185–197. [CrossRef]
53. Miller, N.; Gerlai, R. From Schooling to Shoaling: Patterns of Collective Motion in Zebrafish (*Danio rerio*). *PLoS ONE* **2012**, *7*, e48865. [CrossRef]
54. Buske, C.; Gerlai, R. Shoaling develops with age in Zebrafish (*Danio rerio*). *Prog. Neuro-Psychopharmacol. Biol. Psychiatry* **2011**, *35*, 1409–1415. [CrossRef] [PubMed]
55. Nabinger, D.D.; Altenhofen, S.; Peixoto, J.V.; da Silva, J.M.K.; Gerlai, R.; Bonan, C.D. Feeding status alters exploratory and anxiety-like behaviors in zebrafish larvae exposed to quinpirole. *Prog. Neuro-Psychopharm. Biol. Psychiatry* **2020**, *108*, 110179. [CrossRef] [PubMed]
56. Gerlai, R.; Crusio, W.E. Organization of motor and posture patterns in paradise fish (*Macropodus opercularis*): Environmental and genetic components of phenotypical correlation structures. *Behav. Genet.* **1995**, *25*, 385–396. [CrossRef]
57. Blaser, R.; Gerlai, R. Behavioral phenotyping in Zebrafish: Comparison of three behavioral quantification methods. *Behav. Res. Methods* **2006**, *38*, 456–469. [CrossRef]

78. Gerlai, R.; Roder, J. Abnormal exploratory behaviour in transgenic mice carrying multiple copies of the human gene for S100ß. *J. Psychiatry Neurosci.* **1995**, *2*, 105–112.
79. Gerlai, R. *Behavioral and Neural Genetics of Zebrafish*, 1st ed.; Elsevier: Amsterdam, The Netherlands; Academic Press: Cambridge, MA, USA, 2020; p. 614.
80. Gerlai, R. Zebrafish antipredatory responses: A future for translational research? *Behav. Brain Res.* **2010**, *207*, 223–231. [CrossRef]
81. Postlethwait, J.H.; Talbot, W.S. Zebrafish genomics: From mutants to genes. *Trends Genet.* **1997**, *13*, 183–190. [CrossRef]

Article

osr1 Maintains Renal Progenitors and Regulates Podocyte Development by Promoting *wnt2ba* via the Antagonism of *hand2*

Bridgette E. Drummond [1], Brooke E. Chambers [1], Hannah M. Wesselman [1], Shannon Gibson [1], Liana Arceri [1], Marisa N. Ulrich [1], Gary F. Gerlach [1], Paul T. Kroeger [1], Ignaty Leshchiner [2], Wolfram Goessling [2] and Rebecca A. Wingert [1,2,*]

[1] Department of Biological Sciences, Center for Stem Cells and Regenerative Medicine, Center for Zebrafish Research, University of Notre Dame, Notre Dame, IN 46556, USA
[2] Brigham and Women's Hospital, Genetics and Gastroenterology Division, Harvard Medical School, Harvard Stem Cell Institute, Boston, MA 02215, USA
* Correspondence: rwingert@nd.edu; Tel.: +1-574-631-0907

Citation: Drummond, B.E.; Chambers, B.E.; Wesselman, H.M.; Gibson, S.; Arceri, L.; Ulrich, M.N.; Gerlach, G.F.; Kroeger, P.T.; Leshchiner, I.; Goessling, W.; et al. *osr1* Maintains Renal Progenitors and Regulates Podocyte Development by Promoting *wnt2ba* via the Antagonism of *hand2*. *Biomedicines* **2022**, *10*, 2868. https://doi.org/10.3390/biomedicines10112868

Academic Editors: James A. Marrs and Swapnalee Sarmah

Received: 2 October 2022
Accepted: 5 November 2022
Published: 9 November 2022

Publisher's Note: MDPI stays neutral with regard to jurisdictional claims in published maps and institutional affiliations.

Abstract: Knowledge about the genetic pathways that control nephron development is essential for better understanding the basis of congenital malformations of the kidney. The transcription factors Osr1 and Hand2 are known to exert antagonistic influences to balance kidney specification. Here, we performed a forward genetic screen to identify nephrogenesis regulators, where whole genome sequencing identified an *osr1* lesion in the novel *oceanside* (*ocn*) mutant. The characterization of the mutant revealed that *osr1* is needed to specify not renal progenitors but rather their maintenance. Additionally, *osr1* promotes the expression of *wnt2ba* in the intermediate mesoderm (IM) and later the podocyte lineage. *wnt2ba* deficiency reduced podocytes, where overexpression of *wnt2ba* was sufficient to rescue podocytes and *osr1* deficiency. Antagonism between *osr1* and *hand2* mediates podocyte development specifically by controlling *wnt2ba* expression. These studies reveal new insights about the roles of Osr1 in promoting renal progenitor survival and lineage choice.

Keywords: kidney; podocyte; nephron; development; zebrafish; *osr1*; *wnt2ba*; *hand2*

1. Introduction

The kidney is the organ that cleanses our blood and initiates the process of waste excretion. The portions of the kidney that make this possible are the nephrons, which are composed of a blood filter, tubule, and collecting duct. The blood filter itself is composed of several cellular features including a capillary bed with a fenestrated endothelium. Octopus-like epithelial cells known as podocytes are situated in opposition to a specialized glomerular basement membrane (GBM) surrounding the capillaries [1]. Filtration is accomplished due to the layered ultrastructure of fenestrated epithelium, GBM, and podocytes, which keeps large bulky particles from entering the tubule [2,3]. Podocytes form elaborate cellular extensions and are connected to adjacent podocytes through cell membrane-based protein interactions that create a specialized barrier known as the slit diaphragm [2]. The slit diaphragm allows small or appropriately charged molecules to pass, which initiates a filtrate product that flows into the tubule and is subsequently augmented by specialized solute transporters arranged in a segmental pattern to create a concentrated waste product [1,4]. Within the nephron tubule, the proximal segments perform the bulk of reabsorption, particularly of organic molecules, while the distal segments fine tune the amount of water within the filtrate [5]. Human kidneys, each composed of around a million nephrons, filter all of the blood in our body almost 30 times daily to produce 1–2 quarts of urine (NIDDK). Damage to the specialized cells of the kidney is nearly permanent, as the human kidney has limited regenerative capacity. Therapeutic interventions that can reverse renal damage for patients with acquired kidney diseases and birth defects are urgently needed [6–8]. Expanding our understanding about renal development pathways proffers a valuable avenue for addressing these needs.

Early in embryogenesis, the intermediate mesoderm (IM) gives rise to the earliest form of the kidney, or pronephros. The paraxial mesoderm (PM) and lateral plate mesoderm (LPM) flank the developing IM [9,10]. While functional in lower vertebrates, the pronephros is vestigial in mammals [11]. This structure degenerates to give rise to the mesonephros, which is the terminal kidney in fish and amphibians, but in mammals, it is followed by the metanephros [12]. While humans are born with a static number of nephrons in their metanephros [13,14], teleost fishes such as the zebrafish (*Danio rerio*) continue to form nephrons throughout their lifetime and upon injury [15,16].

Despite these differences, zebrafish do exhibit fundamental genetic and morphological similarities in kidney organogenesis to mammals. For example, the mammalian renal progenitor markers *Lim homeobox 1 (LHX1)* and *paired box gene 2 (PAX2)* are orthologous to *LIM homeobox 1a (lhx1a)* and *paired box 2a (pax2a)*, which are also renal progenitor markers in zebrafish [17,18]. Zebrafish podocytes resemble their mammalian counterparts and express markers including *Wilms tumor 1a, Wilms tumor 1b, nephrosis 1, congenital Finnish type (nephrin), nephrosis 2, idiopathic,* and *steroid-resistant (podocin) (wt1a/b, nphs1,* and *nphs2)* that closely correspond to the human homologs *WT1, NPHS1,* and *NPHS2,* respectfully [19–22]. Further, the zebrafish nephron exhibits a conserved collection of solute transporter genes that are arranged into two proximal and two distal segments similar to other vertebrates including mammals (Figure 1A) [23–25]. The genetic conservation combined with the simplicity of the two-nephron and single blood filter pronephros makes the embryonic zebrafish kidney an accessible and powerful genetic model for gaining insight into the many puzzles and complexities of kidney development.

A critical regulator of kidney development in both zebrafish and mammals is the zinc-finger transcription factor *odd skipped-related 1 (osr1)*; in mice, *Osr1* is one of the earliest markers of the IM. Fate-mapping studies have shown that *Osr1*+ cells differentiate into renal progenitors and renal-associated vasculature [26]. *Osr1*−/− mice fail to express renal progenitors or develop metanephric kidneys, which contributes to embryonic lethality [27,28]. Similar to mouse studies, zebrafish *osr1* is an initial marker of the IM [26]. Further, knock down of *osr1* causes edema, disrupts glomerular morphogenesis, and reduces proximal tubules in both zebrafish and *Xenopus* [29]. Subsequent studies have confirmed these findings [10,30–33], though the intriguing observation that *osr1* knockdown causes the kidney structure to be lost in the region abutting somites 3–5 is not fully understood. In humans, mutations in *OSR1* have been clinically linked to hypomorphic kidneys, making the continued study of this factor and its genetic regulatory network a necessity [34].

Here, we report the zebrafish *oceanside (ocn)* mutation, which was identified from a forward genetic screen for defects in kidney development based on its striking reduction in podocytes and anterior pronephros tubules [35,36], a phenotype resembling other podocyte mutants [37]. Whole-genome sequencing revealed a novel causative lesion: a premature stop codon in exon 2 of *osr1*. In addition to *ocn*−/− recapitulating previously observed alterations in mesoderm-derived tissues, reductions in nephron tubules and podocytes were rescued with ectopic *osr1* cRNA. Interestingly, *osr1* was not needed to establish the renal progenitor field but was needed to maintain the renal progenitors, as they underwent apoptosis in the absence of Osr1. We also found that developing podocytes expressed *wnt2ba* and that this expression was significantly decreased in *ocn*−/−. A loss of *wnt2ba* led to a reduction in podocytes, and ectopic *wnt2ba* partially restored these cells in *ocn*−/−. Further, we placed *wnt2ba* downstream of the antagonistic influences exerted by *osr1* and *hand2* during renal progenitor ontogeny. Together, these data illuminate novel functions of Osr1 that are essential for forwarding our understanding of kidney development and may have important implications for congenital renal defects and diseases.

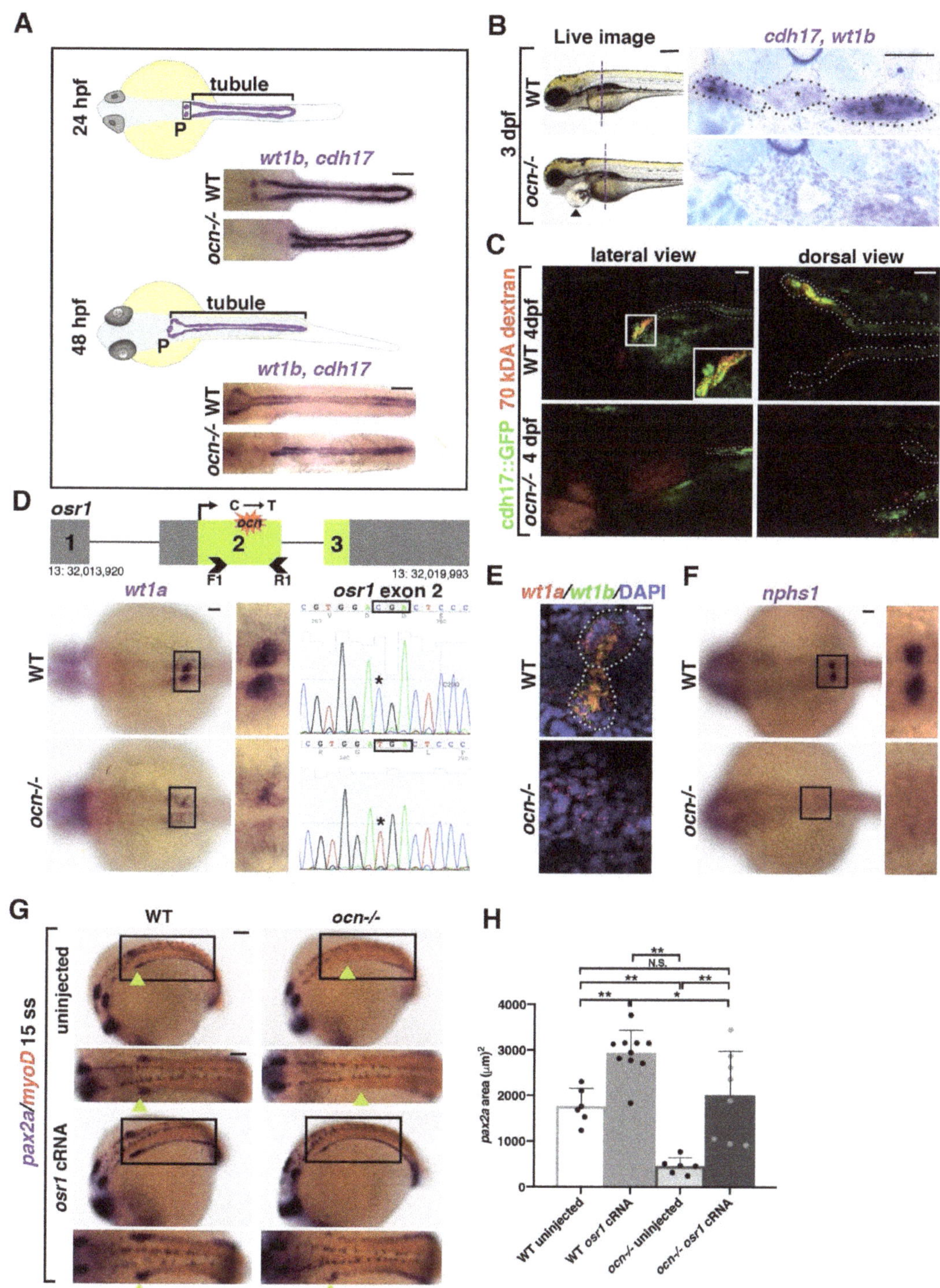

Figure 1. The ENU mutant *ocn* has a proximally abrogated pronephros due to a lesion in the gene *osr1*. (**A**) At 24 hpf, the zebrafish pronephros contains two clusters of podocytes (P) and two nephron

tubules. By 48 hpf, the pronephros is functional as the podocyte progenitors have migrated to the midline and fused. In *ocn*−/− mutants, podocyte progenitors (*wt1b*) are reduced at both stages. The pronephric tubules (*cdh17*) were truncated at 24 hpf, which became more dramatic at 48 hpf. Scale bar is 50 μm. (**B**) A live time course of *ocn* revealed pericardial edema beginning at 72 hpf, as indicated by black arrow heads. This fluid imbalance was symptomatic of organ dysfunction. Scale bar is 70 μm. WISH experiments to view podocytes and tubules (*wt1b*, *cdh17*) were also conducted at 72 hpf. JB-4 serial sectioning was conducted on three WT and three *ocn*−/− embryos to examine the anterior pronephros; the location is marked by the dashed vertical line. WT siblings had an intact pronephros (dotted outline), including a glomerulus (asterisk) with two tubules. Mutant sections of this same region had no discernable blood filter or tubule structure. Scale bar is 50 μm. (**C**) At 48 hpf, *ocn*::cdh17::GFP embryos were injected with 70 kDA rhodamine dextran (red). These embryos were assessed at 96 hpf. Nephron tubules are shown by the dotted outline. WT siblings exhibited no edema and appeared to uptake the dextran in the proximal region, as indicated by yellow coloration (inset). However, in mutants with pericardial edema and truncated tubules, there was no evidence of dextran within the tubule, suggesting that active uptake was not occurring in these mutants. Scale bar is 15 μm for lateral images and 50 μm for dorsal views. (**D**) After assessment of the genetic candidates obtained via whole genome sequencing, *osr1* appeared to be an attractive possibility due to a C to T SNP that was predicted to cause a premature stop codon. The predicted lesion (red shape) is located in exon 2 of *osr1*. We designed primers that flanked exon 2 (arrow heads) for Sanger sequencing. Embryos with reduced *wt1a* WISH staining exhibited a "TGA" codon within exon 2 of *osr1* that is normally a "CGA" codon in WT embryos. Scale bar is 30 μm. (**E**) To confirm that *wt1a*-podocytes were reduced in *ocn*−/−, FISH with *wt1a* and *wt1b* was performed at 24 hpf. There were little to no double-positive cells seen in genotype-confirmed mutants, whereas both clusters of *wt1a/b* podocytes were evident in WT embryos. Scale bar is 10 μm. (**F**) The slit diaphragm marker *nphs2* was similarly reduced in 24 hpf *ocn* mutants. Scale bar is 50 μm. (**G,H**) At the 15 ss, *pax2a* marks the developing IM, the beginning of which is shown with green arrowheads. In *ocn*−/−, the anterior region of *pax2a* is decreased. When *ocn*−/− was injected with *osr1* cRNA, *pax2a* expression was restored. Interestingly, *pax2a* was significantly expanded in WT embryos injected with *osr1*. Absolute area measurements of *pax2a* were taken from somites 1–5. *p*-values: ** $p < 0.001$, * $p < 0.05$, N.S. = not significant. Scale bar is 50 μm.

2. Materials and Methods

2.1. Creation and Maintenance of Zebrafish Lines

Zebrafish were housed in the Center for Zebrafish Research in the Freimann Life Science Center at the University of Notre Dame. All experiments and protocols used in this study were approved by the Institutional Animal Care and Use Committee (IACUC) with protocol numbers 16-07-325 and 19-06-5412. We performed an ENU haploid genetic screen as described [35,36].

2.2. Live Imaging and Dextran Injections

Embryos were grown in E3 medium at approximately 28 °C. For live imaging, embryos were placed in a solution of 2% methylcellulose/E3 and 0.02% tricaine and placed in a glass depression slide. For dextran injection experiments, embryos were also incubated with 0.0003% phenylthiourea (PTU) in E3 to inhibit pigment development. At 3 dpf, *ocn*::cdh17::GFP animals were anesthetized and injected with 40 kDA rhodamine-dextran. Embryos were then examined and imaged 24 h after injection.

2.3. WISH, FISH, IF, Sectioning and Image Acquisition

WISH was performed as described in previous studies [23,24,38–40]. For each marker, embryos from at least 3 sets of adult parents (e.g., 3 biological replicates) were assessed and a minimum 5 mutants and 5 siblings were imaged for each experiment. FISH was performed as described [41,42]. Immunofluorescence (IF) was performed as previously described [37,42]. Embryos from WISH experiments were embedded in JB-4 plastic blocks and cut to obtain 4 μm sections that were counterstained with methylene blue (0.5%). Alcian blue staining was performed as described [36].

2.4. Genotyping

Direct genotyping on *ocn* fin clips and embryos was carried out by PCR amplification of exon 2 of *osr1*: forward primer: CCCCATTCACTTTGCCACGCTGCACCTTTTC, reverse primer: CTGTGGTCTCTCAGGTGGTCCTGCCTCCTAAA). Dilutions of purified PCR products were then subjected to Sanger sequencing by the Genomics Core at University of Notre Dame using the forward primer.

2.5. Morpholinos and RT-PCR

osr1 morpholino (ATCTCATCCTTACCTGTGGTCTCTC) was first described in [30] and was designed to block the splice donor site of exon 2. We used the primers GTGACTGTATCTGAATCCTCTTATTTTGGATCGTCTCGCTTCACAAAGAACTG and CTGTAGGCTATGGAAGTTTGCCTTTTCAGGAAGCTCTTTGGTCAG to perform RT-PCR as described in [37] to confirm the interruption of exon 2 splicing activity. *wnt2ba* splice-blocking morpholino (CTGCAGAAACAAACAGACAATTAAG) was previously utilized in [32], and the following primers were used to amplify the entire transcript by RT-PCR: forward primer: ATGCCAGAGTGTGATGGAGTTGGGTGCGCGTCGCCGGCGC, reverse primer: GCTGGAGCGAGACCACACTGTGTTCGGCCGC, and additionally looked for the presence of intronic sequence with intronic forward primer: ATCACAGGGGTATCATTATCACAAAAAATTGTAAATAAATG. While this splice-blocking morpholino was the primary method of *wnt2ba* knockdown, a *wnt2ba* ATG morpholino, ACCCAACTCCATCACACTCTGGCAT [43], was used to confirm phenotypes seen with the splicing morpholino. The *hand2* ATG morpholino (CCTCCAACTAAACTCATGGCGACAG) was used as described in [44].

2.6. Statistics and Measurements

Absolute domain lengths and area measurements were taken from five representative embryos per control or experimental treatment group, performed in triplicate using Fiji ImageJ. Averages, standard deviations, and unpaired Student's *t*-tests were then calculated in Microsoft Excel and GraphPad Prism. In experiments where *osr1* cRNA was used, body axis measurements were taken for injected and uninjected embryos. The tubule measurements for each group were divided by the body length to discern what percentage of the body length was occupied by the kidney. To normalize the data, these percentages were subjected to arcsine degree transformation and then run through a Student's *t*-test to determine significance.

3. Results

3.1. ocn Encodes a Premature Stop Codon in osr1 and Mutants Exhibits Defective Podocyte and Pronephric Tubule Development

A forward genetic haploid screen was performed to identify regulators of nephrogenesis using the zebrafish pronephros model [35,36]. The *ocn* mutant was isolated due to its loss of podocytes and abrogation of the anterior pronephros (Figure 1A). Whole-mount in situ hybridization (WISH) was performed to delineate the two pronephros tubules based on the expression of transcripts encoding *cadherin 17* (*cdh17*) and the podocytes based on *wt1b* expression at 24 and 48 h post fertilization (hpf) (Figure 1A). Both tubule length and podocyte area were significantly reduced in *ocn* mutants at these time points compared with wild-type (WT) embryos (Supplemental Figure S1A). By 72 hpf, *ocn* mutants exhibited dramatic pericardial edema that progressed in severity through 120 hpf and was ultimately lethal (Figures 1B and S1B). Since the kidneys play a major role in fluid homeostasis, this phenotype was a probable indicator of renal dysfunction.

To explore this further, WISH staining to assess tubule and podocyte morphology was conducted on 72 hpf *ocn* and WT embryos. The animals were embedded in JB-4 plastic resin and serially sectioned. In WT embryos, the blood filter could be detected as a mass of dense capillaries containing glomerular podocytes (*wt1b+*) that were flanked by *cdh17+* tubules (Figure 1B). In *ocn* mutants, however, both the midline glomerulus structure and the flanking proximal tubules were abrogated (Figure 1B). Instead, a dilated dorsal aorta was identified in this region (Figure 1B). While it was clear that the proximal pronephros was absent in *ocn* mutants, it was uncertain if this truncated kidney retained any functionality. Therefore, kidney functionality was assessed using an endocytosis assay whereby 70 kDA rhodamine-dextran was injected into the vasculature of *ocn::cdh17::GFP* embryos, which exhibited a pronephric truncation that phenocopied WISH experiments at 3 days post fertilization (dpf) onwards (Supplemental Figure S1C). Transgenic animals were injected with rhodamine-dextran at 48 hpf and then assessed at 48 h post injection (hpi). While dextran was endocytosed in the proximal tubules of non-edemic WT siblings, there was no dextran uptake observed within the truncated tubules of the edemic *ocn* mutant embryos (Figure 1C). Additionally, we assessed epithelial polarity through the immunofluorescence (IF) staining of Na-K-ATPase, which marks these transporters localized along the basolateral sides of kidney epithelial cells, and aPKC, which marks the apical epithelial side [45]. This experiment revealed a similar reduction in tubule and podocytes in *ocn* as seen with our WISH experiments using the markers *cdh17/wt1b* (Supplemental Figure S1D). Together, this provided strong evidence that the stunted pronephros in *ocn−/−* was indeed functionally defective.

Next, to identify the causative lesion in *ocn*, whole-genome sequencing was conducted on pools of genomic DNA collected from 24 hpf WISH-identified putative mutants and WT siblings [46,47]. The analysis of the sequencing was performed using SNPtrack software, whereby we detected a strong candidate SNP that was centrally located on chromosome 13 (Supplemental Figure S2A). Specifically, the putative SNP encoded a missense C to T mutation and was predicted to result in an amino acid substitution from an arginine to a premature stop codon in exon 2 of *osr1* (Figure 1D and Figure S2A).

To further assess if the predicted stop codon in exon 2 of *osr1* was linked with the *ocn* phenotype, we performed additional genotyping analysis. For this, genomic DNA was isolated from individual embryos that had been identified as *ocn* mutants or WTs based on WISH with the podocyte marker *wt1a* at 24 hpf, and PCR was performed to amplify exon 2 of *osr1* followed by direct Sanger sequencing (Figure 1D). Out of 20 *ocn* embryos with reduced *wt1a* staining, all 20 were homozygous for the C to T mutation in exon 2 of *osr1* (Figure 1D). Protein alignment showed that zebrafish and human OSR1 protein are 264 amino acids (aa) and 266 aa in length, respectively (Supplemental Figure S2B). While they exhibit 77% conservation in overall aa sequence, the three zinc-finger domains responsible for DNA binding activity are 100% conserved across humans, mice and zebrafish (Supplemental Figure S2B). The *osr1* genetic lesion in *ocn−/−* placed a premature stop codon at residue 165 before all three zinc-finger domains (Supplemental Figure S2B). This suggested that the truncated Osr1 protein produced in *ocn−/−* would not contain any functional domains and would thus be unable to act as a targeted transcription factor. Next, we verified the effectiveness of a splice-blocking *osr1* morpholino with RT-PCR (Supplemental Figure S3). The *osr1* morphants had a decrease in podocytes and proximal tubule length that phenocopied *ocn−/−* and was consistent with previously reported phenotypes [10,30–32] (Supplemental Figure S3).

Previous literature has indicated that *osr1* acts to restrict venous development in order to promote other mesodermal fates such as the kidney and the pectoral fins [10,30,32]. At 4 dpf, Alcian blue staining indicated that *ocn−/−* possessed shorter, malformed pectoral fins (Supplemental Figure S1E). The fin bud area, which gives rise to pectoral fins, was significantly reduced in *ocn−/−* mutants compared with siblings as seen by the marker *MDS1 and EVI1 complex locus* (*mecom*) at 24 hpf (Supplemental Figure S1F). Additionally, Alcian blue staining revealed altered craniofacial cartilage formation in mutants, which fits with previous literature placing *osr1* as a regulator of palatogenesis in zebrafish and mice [48,49] (Supplemental Figure S1E). In sum, these mesodermal phenotypes were consistent with *osr1* deficiency.

Next, we evaluated other aspects of pronephros development. As *wt1a* expression appeared to be severely diminished and also disorganized in *ocn*−/−, we evaluated additional markers to better understand the features of podocyte lineage development in mutants. Podocytes were examined at 24 hpf using a *wt1a/wt1b* double fluorescent in situ (FISH). While clusters of *wt1a+/wt1b+* podocytes were visible in siblings, mutants had a scarcity of double-positive cells (Figure 1E). There was also a dearth of *nphs1+* cells, which is a marker of the podocyte slit diaphragm and suggested that podocyte differentiation was also disrupted (Figure 1F). Additionally, *ocn*−/− embryos displayed diminished *pax2a* expression at the 15 ss compared to siblings (Figure 1G), a characteristic of *osr1* morphants in previous studies as well [10,30,32].

To test whether the mutation in *osr1* was the specific cause of this phenotype, we performed rescue studies. Injection with *osr1* capped RNA (cRNA) rescued this domain in *ocn* mutants and expanded it in WT siblings (Figure 1G,H). In sum, *ocn* mutants reciprocated a multitude of mesodermal phenotypes seen in *osr1* literature in zebrafish and across taxa. Given the ability of *osr1* cRNA to rescue key mesodermal phenotypes in *ocn*−/− and the catastrophic nature of the *osr1* mutation, we concluded that Osr1 deficiency is responsible for the *ocn* phenotype.

3.2. Kidney Progenitors Are Specified in osr1 Deficient Animals, but Subsequently Undergo Apoptosis

Previous studies suggest that the anterior pronephros abrogation in *osr1* zebrafish morphants is due to a fate change where blood/vasculature and endoderm form instead of renal progenitors [30,33,50]. Interestingly, in the *Osr1* mouse knockout model, there was an increase in apoptosis that occurred within the kidney tissue [27]. However, in both models, renal progenitors are initially established [27,30]. Thus, we next sought to delineate the cellular dynamics of renal progenitor specification in our *ocn* mutant model and to address if alterations in proliferation or apoptosis occur during pronephros development in the absence of *osr1*.

To investigate this, we first performed WISH studies. The LPM is marked by *T-cell acute lymphocytic leukemia 1 (tal1)*, and gives rise to hemangioblasts [51,52]. The IM and hemangioblast domains at the 7 ss were not noticeably different between WT and *ocn*−/− embryos, as indicated by the markers *pax2a*, and *tal1* (Figure 2A). However, as previously noted, by the 15 ss, there was a decrease in the anterior-most domain of *pax2a* expression in *ocn*−/− embryos (Figure 1G). To further assess the anterior *pax2a+* cells between the 7 ss and 15 ss, we performed double FISH studies in WT and *ocn*−/− embryos to assess *pax2a* and *tal1* expression. DAPI staining was also utilized to discern features such as the trunk somites, which allowed for accurate staging. Embryos were flat-mounted and imaged as previously described (Supplemental Figure S2A) [39]. Further, IF was also performed on these samples with anti-caspase-3 antibody to assess the number apoptotic bodies or anti-pH3 to identify proliferating cells. In our analysis, we focused on the changes to these markers within somites 1–5, as the IM adjacent to somite 3 gives rise to podocytes [21].

Beginning at the 7 ss, *ocn*−/− embryos exhibited a significant increase in the number of caspase-3+ fragments within the combined *tal1* and *pax2a* fields near somites 1–5 (Figure 2B). However, by the 15 ss, few apoptotic fragments were visible in the area of interest, with no significant differences between WT and *ocn*−/− embryos (Figure 2B). Similar to the 7 ss, we found a significant increase in the number of total caspase-3+ fragments at the 10 ss in *ocn*−/− mutants and *osr1* morphants, while the WT siblings had little to no apoptosis occurring in this area (Figure 2C–E). Interestingly, most of the apoptosis that occurred in mutants and morphants happened within the *pax2a* kidney field specifically (Figure 2E). Another finding of note was that the caspase-3+ fragment number was not significantly different between *osr1* morphants and *ocn*−/− for either assessment (Figure 2D,E).

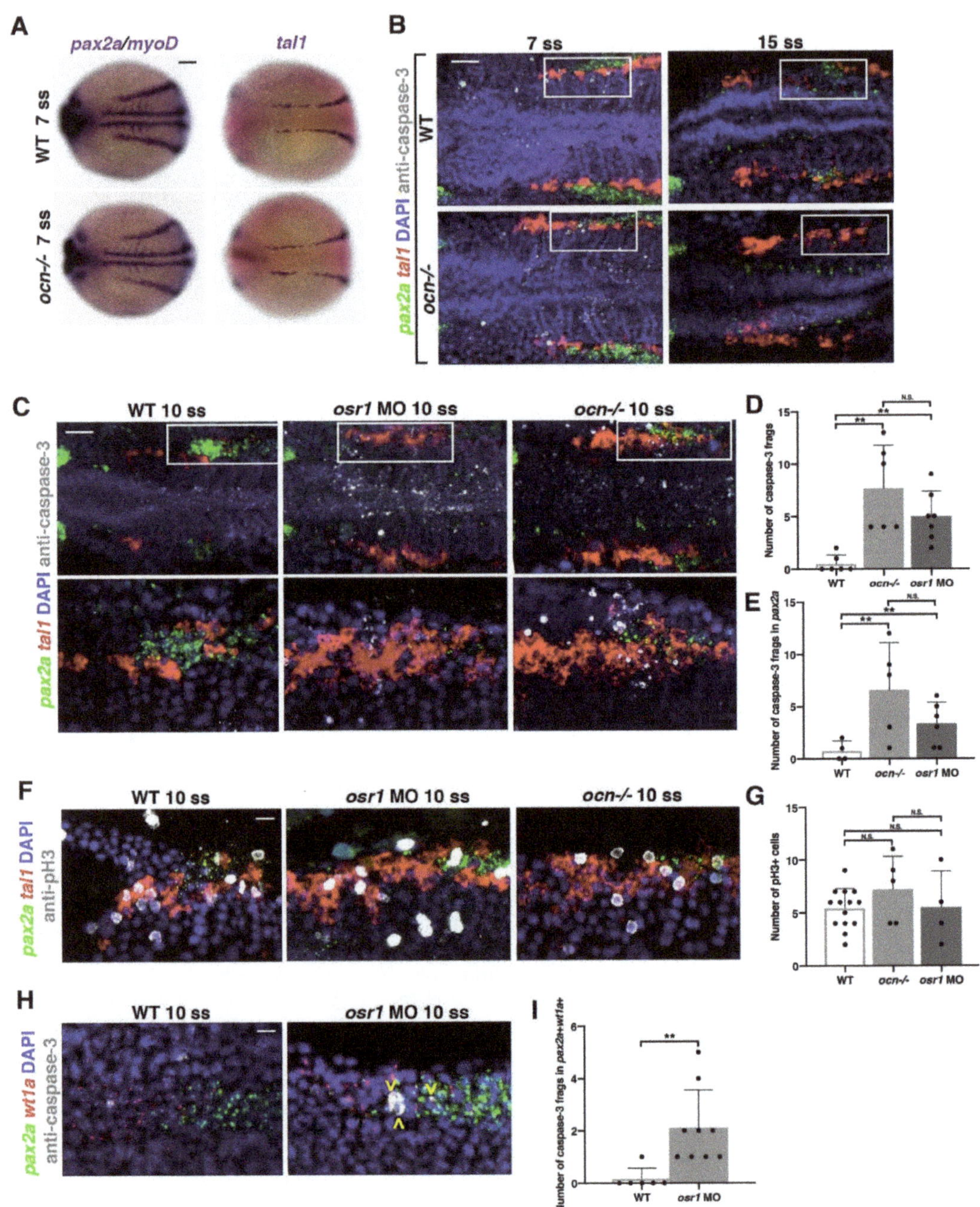

Figure 2. *osr1* is required to maintain and promote kidney development at the expense of hemangioblasts. (**A**) Although *pax2a* is restricted in 15 ss mutants, at 7 ss, *ocn−/−* embryos had a *pax2a* domain that appeared to occupy the same domain as WT siblings. Similarly, the hemangioblast marker *tal1* appeared mostly WT in *ocn−/−* at 7 ss. Scale bar is 50 μm (**B**) FISH with probes for *pax2a* (green) and *tal1* (red) and ICC using anti-caspase-3 (white) to mark apoptotic cells was conducted at 7 ss and 15 ss. The number of fragments in the combined *pax2a* and *tal1* fields from somites 1–5 were

increased in mutants at 7 ss but not at 15 ss. Scale bar is 35 μm (**C–E**) At 10 ss, little to no caspase-3 fragments were seen in *pax2a* or *tal1* domains from somites 1–5, but a significant number were seen in *osr1* morphants and *ocn−/−*. The *tal1* domain was also expanded in both loss-of-function models. Scale bar is 35 μm. (**F,G**) ICC with the proliferative cell marker anti-pH3 was also conducted. Despite the expansion in *tal1* in *osr1*-deficient models, there was no significant change in the number of proliferating cells. Scale bar is 10 μm. (**H,I**) FISH experiments were conducted to assess changes in apoptosis in *wt1a+/pax2a+* podocyte progenitors. There was a significant increase in the number of apoptotic fragments within this field in mutants compared with WT siblings. Scale bar is 10 μm. A minimum of three individuals were assessed for each group across experiments. Photos are max intensity projections from z-stacks, and each side of mesoderm was quantified individually. *p*-values: ** $p < 0.001$, N.S. = not significant.

To further understand the cell dynamics across this time course, absolute area measurements of *pax2a* and *tal1* were taken at 7, 10, and 15 ss from somites 1–5 in WT and *ocn−/−*. Surprisingly, the area of the *tal1* domain was already expanded at 7 ss and continued to expand through the 15 ss (Supplemental Figure S4). However, across the 3 time points examined, a reduction in *pax2a* area was only significantly different between WT and *ocn−/−* at 15 ss (Supplemental Figure S4). Further, although the *tal1* field was expanded in *ocn−/−* embryos at the 10 ss, there was no significant difference in proliferating pH3+ cells between WT and *osr1* loss of function models (Figure 2F,G). Additionally, no significant changes in proliferation were seen between mutants and WTs at the 8 ss (Supplemental Figure S4).

To determine if apoptosis was occurring within podocyte progenitors in the *pax2a* kidney field, we performed an additional FISH with *wt1a* and *pax2a* at 10 ss. During this time point, while *pax2a* expression begins adjacent to somite 3, *wt1a* was expressed from somites 1–3 (Figure 2H). Similar to the *pax2a* domain, the *wt1a* domain did not appear to be reduced at this time point, though it did become restricted and disorganized by 24 hpf (Figure 1D). We found a significant increase in caspase-3+ fragments that were double-positive for *wt1a* and *pax2a* in *osr1* morphants compared with WTs (Figure 2H,I). These results demonstrated that abnormal apoptosis occurred in podocyte progenitors due to the loss of *osr1*. In sum, *osr1* is not needed to initiate the *pax2a* progenitor pool, but it is needed to maintain this population, including the podocyte precursors, during pronephros development.

3.3. Ectopic osr1 Is Transiently Sufficient to Rescue Renal Progenitors

Our observation that *pax2a+* renal progenitors arise in *ocn* mutants, but are not maintained, is consistent with previous data that *osr1* knockdown leads to a reduced *pax2a+* renal progenitor field by the 14 ss [30]. As *pax2a* expression marks both podocyte and tubule precursors [31], we hypothesized that *osr1* is likely needed for podocyte and tubule progenitor maintenance. To determine this, we performed a series of rescue studies in our *ocn* mutants to test if one or both of these compartments requires *osr1* for its maintenance.

First, we tested whether the overexpression of *osr1* mRNA was sufficient to rescue podocytes in *ocn* mutants by assessing *wt1b* expression, which specifically marks the podocyte lineage [20,21]. The provision of *osr1* mRNA robustly rescued the development of *wt1b+* podocytes in *ocn* mutants at the 15 ss (Figure 3A,B). However, by the 22 ss, we were only able to achieve a partial podocyte rescue, though tubules within the same individuals appeared to be WT in length (Figure 3C,D). Consistent with this, we were unable to obtain a podocyte rescue at 24 hpf (data not shown), though again we could achieve a rescue of the truncated tubules (Figure 3E,F). Interestingly, the overexpression of *osr1* was sufficient to induce ectopic *cdh17+* cells in about 55% of injected embryos (Figure 3G). It should be noted that *osr1* cRNA did lead to a decrease in body axis length when compared with uninjected WTs and mutants, which in turn affected pronephros length (Supplemental Figure S5). Despite this, the percentage of kidney length to body length was not significantly different between embryos injected with *osr1* cRNA and uninjected animals.

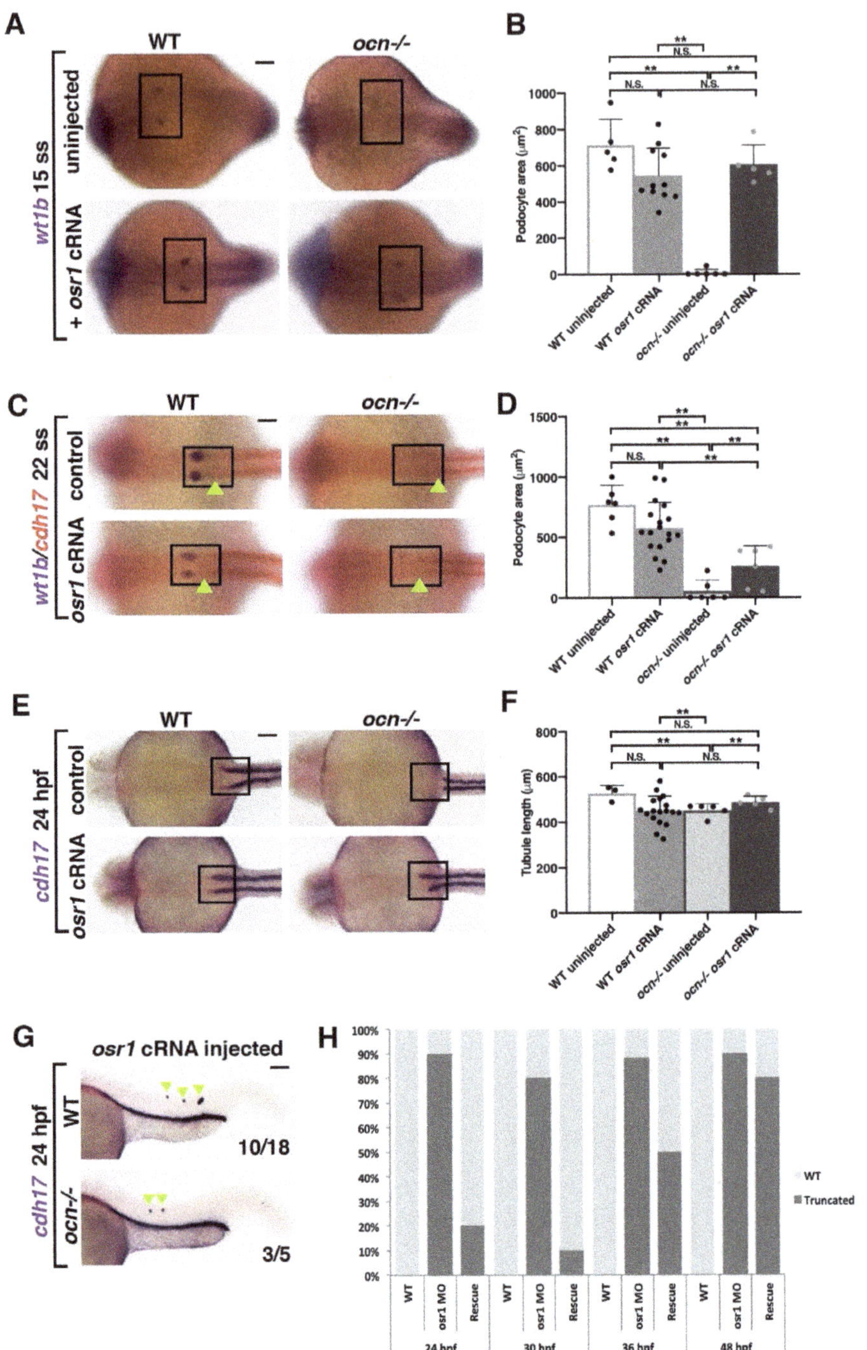

Figure 3. *osr1* is required for the continued development of kidney lineages. (**A,B**) Embryos from *ocn*
incrosses were injected at the 1-cell stage with 50 pgs of *osr1* cRNA and examined. At 15 ss, podocytes
(*wt1b*) were robustly rescued in *ocn−/−*. (**C,D**) However, by 22 ss, podocytes were only partially
rescued in *ocn−/−* injected embryos, while tubule lengths in the same animals were not significantly
different from WTs, as shown by the green arrowheads. (**E–G**) Truncated tubule was still able to be

rescued in mutants at 24 hpf. Further, overexpression of *osr1* induced ectopic tubule formation (green arrowheads). (**H**) A rescue time course was conducted with *osr1* MO and *osr1* cRNA to determine when the *osr1* cRNA dosage became insufficient to rescue. While there was a 90% rescue rate at 24 hpf, by 48 hpf, this rate had dropped to 20%. This indicated that continued *osr1* is needed for normal tubule development. *p*-values: ** $p < 0.001$, N.S. = not significant. For tubule rescue at 24 hpf, p-values were obtained from arcsin transformed kidney to body percentage calculations for each group. Scale bar is 50 μm for all images.

We also conducted a rescue time course by co-injecting *osr1* MO and *osr1* cRNA and performed WISH using *cdh17* to assess the tubules during a number of stages. While 90% of animals injected with both constructs exhibited a WT tubule length at 24–27 ss, by 36 hpf, only 50% showed a rescue (Figures 3H and S5B). At 48 hpf, only 20% of injected embryos had a WT length pronephric tubules while 80% had a unilateral or bilateral reduction (Figures 3H and S5B). Together, this indicated that the pronephros requires a continued presence of *osr1* to maintain the tubule population as development progressed.

3.4. wnt2ba Is a Novel Podocyte Marker and Regulator

Given the importance of *osr1* to podocyte development and maintenance, we wanted to identify downstream factors that promote podocytes. It was previously shown that the canonical Wnt ligand *wingless-type MMTV integration site family, member 2Ba* (*wnt2ba*) is expressed in a similar proximal swath of IM as *osr1* [32]. We observed a similar expression pattern of *wnt2ba* in the anterior IM as early as 13 ss (Figure 4A). To specifically determine which cells *wnt2ba* was expressed in, we conducted FISH studies. At the 20–22 ss, *wnt2ba* transcripts were colocalized in cells within the anterior most region of *pax2a+* and *wt1b+* IM (Figures 4B and S6). At 15 ss, *wnt2ba* transcripts were also colocalized with *wt1a/b+* podocyte progenitor cells (Supplemental Figure S6). At 24 hpf, *wnt2ba* was expressed in both *wt1b+* podocyte precursor cells and in the neighboring cells of the IM (Figure 4C). By 48 hpf, *wnt2ba* was restricted to the podocytes and overlapped precisely with *wt1b* expression (Figure 4C). Taken together, we conclude that *wnt2ba* is a novel podocyte marker, thereby extending prior observations [32]. We also examined the expression of the zebrafish *wnt2ba* paralogue *wnt2bb* at 24 hpf using FISH and determined that these transcripts by comparison were located anterior to both the podocyte and kidney fields (Supplemental Figure S7).

Given its expression in podocyte progenitors, we hypothesized that Wnt2ba might have roles in podocyte specification or differentiation. To explore whether *wnt2ba* is needed for proper podocyte formation, we performed *wnt2ba* loss of function studies. We first verified a morpholino that blocked splicing at exon 1, as well as a morpholino that targeted the start site (Supplemental Figure S8). When *wnt2ba* morphants were examined at 24 hpf, there was a significant reduction in the expression of *wt1b* and *nphs1* that corresponded to a smaller podocyte area and net cell number (Figure 4D–H). We also found that the area of *wt1a+/wt1b+* co-expressing podocytes was decreased in *wnt2ba* morphants at 24 hpf (Figure 4I,J). Furthermore, the decrease in podocyte number in *wnt2ba* morphants occurred between the 15 and 22 ss, suggesting *wnt2ba* is required for maintaining the podocyte lineage (Supplemental Figure S9). In contrast, *wnt2ba* morphants showed no discernable changes in the development or maintenance of the *cdh17+* nephron tubule (Supplemental Figure S8E). Collectively, these data lead us to conclude that *wnt2ba* is a significant regulator of podocyte ontogeny.

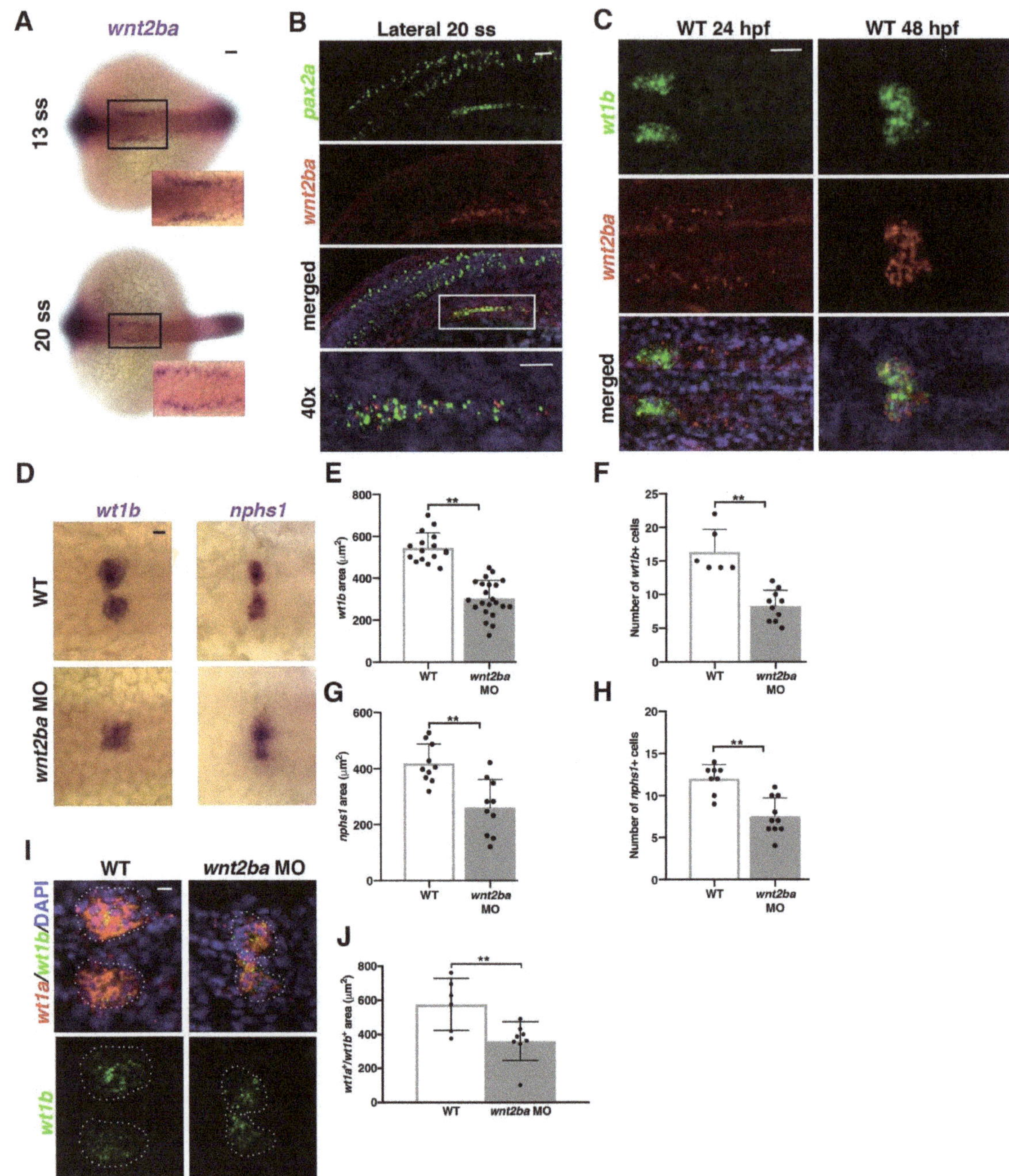

Figure 4. *wnt2ba* is a podocyte marker and regulator. (**A**) *wnt2ba* is expressed in bilateral stripes as early as 13 ss. Scale bar is 30 μm. (**B**) *wnt2ba* (red) is expressed within the anteriormost region of the IM, as shown by colocalization with *pax2a* (green). White box denotes area of colocalization, which is magnified in bottom panel. DAPI (blue) marks nuclei. Scale bar is 15 μm. (**C**) *wnt2ba* (red) also colocalized with the podocyte marker *wt1b* at 24 hpf, though at this time point it was also expressed

in the putative neck segment domain. By 48 hpf, the *wnt2ba* domain was specified to the podocytes. Scale bar is 30 μm. (**D–H**) Podocyte area and cell number was assessed in *wnt2ba* morpholino-injected animals and determined to be reduced compared with WT controls. Both *wt1b* and *nphs1* showed a significant decrease in domain area in *wnt2ba* morphants compared with WT embryos. (**I,J**) FISH with *wt1a* and *wt1b* was performed at 24 hpf. There was a significant area reduction in the podocyte domain seen in *wnt2ba* morphants compared to WTs. Scale bar is 10 μm. *p*-values: ** *p* < 0.001. Scale bar is 30 μm.

3.5. osr1 Promotes wnt2ba in the Podocyte Developmental Pathway

Previous research has demonstrated that *osr1* morphants exhibited a dramatic decrease in the *wnt2ba* pronephric domain, though *wnt2ba* morphants had no notable change in *osr1* expression [32]. They postulated that *osr1* acts to promote *wnt2ba* in the IM, so which allows for proper pectoral fin development to occur [32]. This led us to hypothesize that this same genetic cascade in the IM promotes the formation of proximal pronephric tissues, such as the podocytes, and is dysfunctional in *ocn*−/−.

In congruence with this prior study, we found that *wnt2ba* is significantly reduced in *ocn*−/− at both 15 ss and is almost completely absent by 24 hpf (Figure 5A). We also observed that *wnt2ba* and *osr1* transcripts were colocalized in a population of presumptive IM cells at 15 ss and 22 ss, putting them in the right place and the right time to interact (Figure 5B and Figure S10). The overexpression of *wnt2ba* led to an increase in podocyte number and domain area in injected WT embryos, as seen with an increase in the markers *wt1b* and *nphs1* (Figure 5C–H). While injection with *osr1* MO alone leads to diminished podocytes, coinjection of *wnt2ba* cRNA with *osr1* MO led to a rescue in podocyte area and cell count (Figure 5C–H). Together, this indicates that *wnt2ba* is sufficient to drive podocyte development and does so downstream of *osr1*.

3.6. hand2 Suppresses Podocyte Development by Restricting wnt2ba Expression and Podocyte Development

The bHLH transcription factor *heart and neural crest derivatives expressed 2 (hand2)* has been shown to be antagonistic to *osr1* in early mesoderm development [10]. The loss of *osr1* leads to decreases in podocytes and tubules and an increase in hemangioblasts; in contrast, the loss of *hand2* results in expansions in renal cells at the expense of vasculature [10]. The concomitant knockdown of *osr1* and *hand2* rescues tubule development [10] and podocyte development [33].

When we knocked down *hand2* using an ATG morpholino, we observed a separation in the *myosin light chain 7 (myl7)* heart field at 22 ss that matched previously observed phenotypes [44] (Supplemental Figure S11). The knockdown of *hand2* also caused a significant increase in *wt1b*+ podocyte domain area and cell number (Figure 6A–C). While uninjected *ocn*−/− embryos had few to no podocytes, injecting *ocn-/-* with *hand2* MO resulted in an expansion in podocyte number and area that was significantly different from mutants (Figure 6A–C). Similarly, *hand2* morphants had a significantly larger *wnt2ba* domain, and *hand2/osr1* MO coinjection rescued the usually abrogated *wnt2ba* domain (Figure 6D,E). This indicated that an imbalance of *hand2* and *osr1* leads to changes in *wnt2ba* expression, which alters podocyte development.

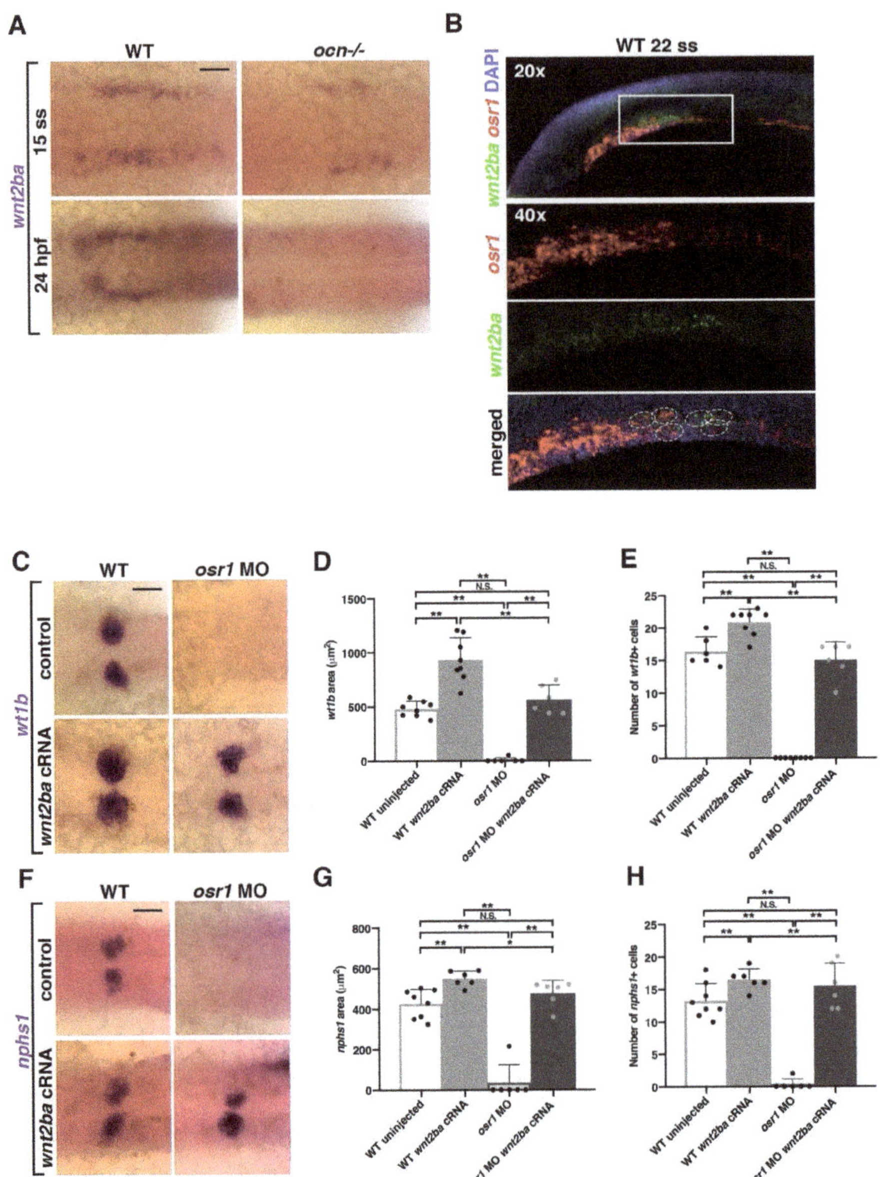

Figure 5. *wnt2ba* is sufficient for podocyte development downstream of *osr1*. (**A**) When *wnt2ba* was assessed at 15 ss, in *ocn−/−* and WT siblings, it was evident that staining was reduced in mutants. By 24 hpf, *wnt2ba* staining was almost completely absent, tracking with the loss of other podocyte markers in *ocn−/−*. Scale bar is 30 μm. (**B**) FISH experiments showed that *osr1* and *wnt2ba* colocalized in cells at 22 ss. (**C–H**) Embryos were injected with *osr1* MO and/or *wnt2ba* cRNA at the one-cell stage, and podocytes and the developing slit diaphragm were visualized at 24 hpf using *wt1b* and *nphs1*, respectively. Embryos injected with *osr1* MO alone showed few podocyte or slit diaphragm cells, while embryos injected with *wnt2ba* cRNA alone had an increased podocyte area. Coinjected embryos had a partial rescue of podocytes, indicating that *wnt2ba* is a downstream factor in the podocyte pathway. A minimum of 5 individuals were imaged for quantification. *p*-values ** $p < 0.001$, * $p < 0.05$, N.S. = not significant. Scale bar is 30 μm.

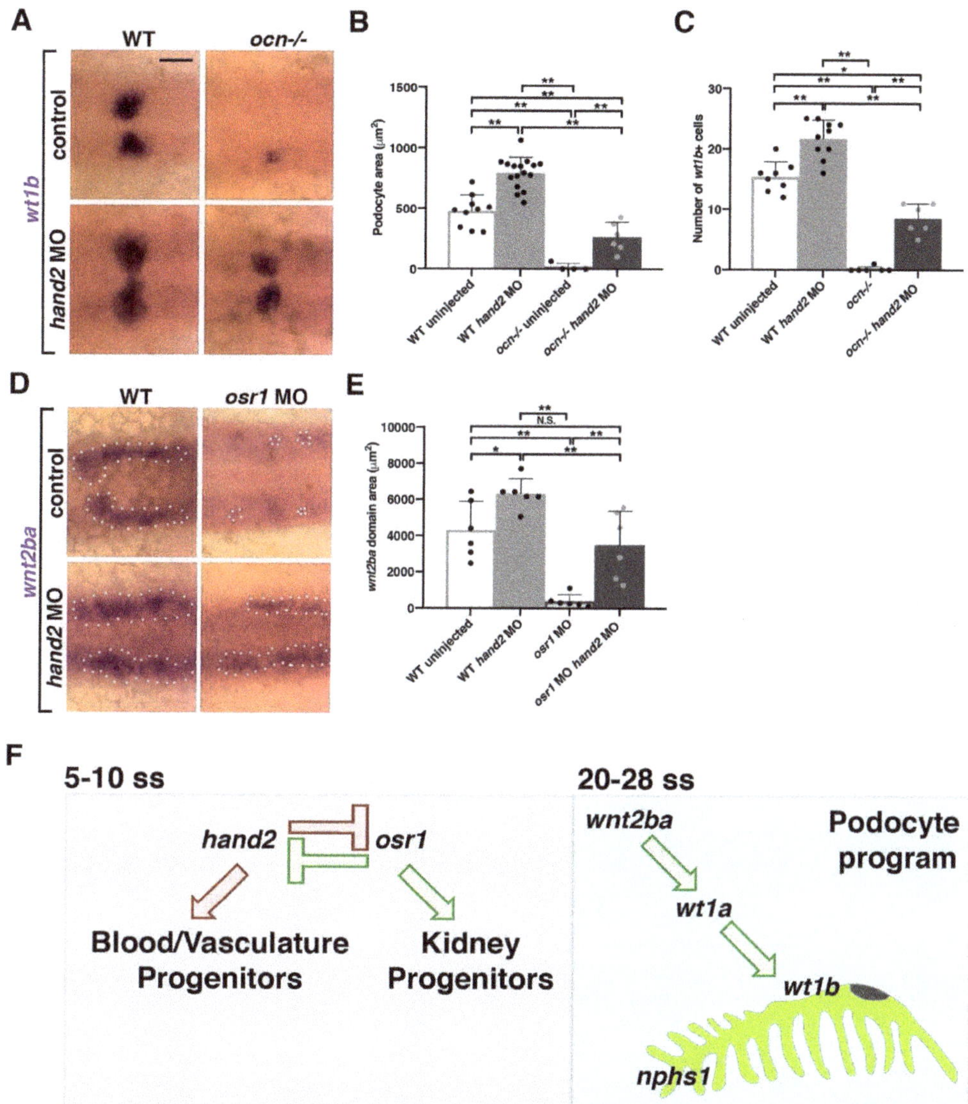

Figure 6. Acting in opposition to *osr1*, *hand2* inhibits *wnt2ba*-driven podocyte development (**A–C**) Embryos from *ocn* incrosses were injected with *hand2* MO. Podocyte area and cell number were partially rescued in *ocn−/−* injected embryos and were expanded in WT injected siblings. This signified that *hand2* suppresses podocyte formation. (**D,E**) Similarly, *wnt2ba* expression was rescued in *osr1/hand2* morphants compared with uninjected WT controls. Injection of *hand2* MO alone led to a significant increase in the *wnt2ba* domain. All images were at 24 hpf, scale bar is 30 µm. A minimum of 5 individuals were imaged for quantification. *p*-values: ** $p < 0.001$, * $p < 0.05$, N.S. = not significant. (**F**) *osr1* promotes podocyte and kidney lineages and suppress blood and vasculature, while *hand2* acts in opposition. Imbalance of either of these factors leads to changes in mesodermal fates. Podocyte development is one example of a mesodermal fate that is altered by imbalance of *osr1/hand2*. This is because the downstream factor *wnt2ba* is decreased without *osr1* yet increased in the absence of *hand2*. *wnt2ba* endorses the podocyte factor *wt1a/b*, which has been shown to be required for formation of podocytes and the slit diaphragm (*nphs1*).

4. Discussion

While dozens to hundreds of podocyte diseases and maladies have been characterized, the genetic explanations for their origins and progression are lacking. One reason for this is that there are relatively few factors that are known to promote the development of these specialized epithelia. Continuing to identify these factors is critical for future diagnostics and treatments for podocytopathy. In this study, we have both reexamined a previous factor shown to promote podocyte fates, *osr1*, and also identified a new downstream regulator, *wnt2ba*.

In this study, *ocn* was identified as a mutant of interest in a forward genetic screen due to displaying pericardial edema and decreases in podocytes and proximal tubules. Through whole-genome sequencing, we determined that *ocn*−/− harbors a SNP in exon 2 that leads to a premature stop in *osr1*. This SNP was confirmed as the causative lesion in *ocn*−/− when *osr1* cRNA could rescue each of these phenotypes. Upon confirmation that *ocn* was an *osr1* mutant line, we next sought to fully assess how *osr1* loss of function impacts kidney development in the context of a zebrafish mutant. We found that *osr1* is needed to maintain renal progenitors and inhibit the development of hemangioblasts.

Further, we established a genetic pathway controlled by *osr1* that regulates podocyte survival by promoting *wnt2ba* expression. We found that *wnt2ba* is an IM/podocyte marker that is likewise diminished in *ocn*−/−. Loss and gain of *wnt2ba* lead to a decrease and increase in podocyte area, demonstrating that *wnt2ba* is both necessary and sufficient to drive podocyte development. Notably, *wnt2ba* can rescue podocytes in an *osr1*-deficient background, which places this factor downstream of *osr1* (Figure 6F). Finally, the *osr1*/*wnt2ba* podocyte pathway is negatively regulated by *hand2* (Figure 6F).

4.1. Osr1 Acts to Promote Podocytes

The earliest known podocyte marker in zebrafish is *wt1a*, though the paralogue, *wt1b* that appears at 12 ss is expressed in a more specific territory [20,53]. It has also been suggested that *wt1a* is more dominant than *wt1b*, as knockdown of *wt1a* leads to loss of *nphs1/2*, while knockdown of *wt1b* causes less dramatic podocyte phenotypes [54]. Zebrafish literature has shown that *osr1* morphants exhibit reductions in *wt1b*, *lhx1a*, and *nphs1/2* at 24 hpf that have also been observed in *ocn*−/− [29–31]. However, the relationship between *wt1a* and *osr1* has yet to be fully understood. Tomar et al. [31] placed *wt1a* upstream of *osr1* due to *osr1* being reduced in *wt1* morpholino-injected embryos and *wt1a* expression being interpreted as unchanged in *osr1* morpholino-injected animals. However, in our studies, *osr1* morphants did exhibit alterations in *wt1a*+ cell organization and a restriction in domain that phenocopies *ocn*−/−. Mouse studies have shown that *WT1*+/−; *OSR1*+/- mice exhibit smaller kidneys, suggesting that these factors act cooperatively in kidney and podocyte development [55]. If *osr1* and *wt1a* did have a similar synergistic relationship in zebrafish kidney development, this would also explain reports that *wt1* morphants exhibit a loss of podocytes and proximal tubules reminiscent of *osr1* loss of function models [31]. While there are currently limitations in using anti-Osr1 antibodies in any in vivo model, progress in this area is needed in order to ascertain if *wt1a* and *osr1* are directly interacting during kidney development.

4.2. Osr1 Is Needed for Kidney Cell Maintenance

While *ocn*−/− exhibit normal patterning of IM early in development, by the time specification to pronephros is beginning to occur around 15 ss, the anterior domain is absent. Our experiments demonstrated that this is due to two events: (1) an expansion of hemangioblasts and (2) the apoptosis of podocyte progenitors in this region. Work in chick and mouse has shown that while mesonephric tissues and markers are present, apoptosis occurs within the nephrogenic mesenchyme that keeps metanephric tissues from forming in *Osr1* knockout animals [27,28]. Further, previous studies have shown that *Osr1* acts synergistically with factors such as *Wt1* and *Six2* to renew renal stem cell pools to inhibit premature differentiation and thus cell death [55]. A similar apoptosis event has not been recorded in *osr1* loss of function zebrafish models prior to this study, and we hypothesize that *osr1* plays a similar role in progenitor self-renewal in zebrafish.

The expansion in the hemangioblast domain in *osr1* morphants has been documented by other groups, where it was suggested that *pax2a*+ cells were converting to *tal*+ cells [30]. Additionally, the expansion in vessel progenitors has been reported in an *osr1* TALEN mutant [33]. However, our results add one further element to these early events, as we have captured cell apoptosis in *pax2a*+ cells of *osr1* mutant embryos. Further, our studies have revealed that the timing of the *pax2a* domain decrease and hemangioblast domain increase is not equivalent. The hemagioblasts expand hours prior to the loss of the anterior IM domain. We postulate that *osr1* may inhibit hemangioblast formation either indirectly or in an independent mechanism than it uses to promote IM and podocytes.

4.3. Wnt2ba Is a Novel Regulator of Podocyte Development

Finally, *wnt2ba* is a ligand that functions in the canonical Wnt/beta-catenin pathway. As a member of this pathway, *wnt2ba* acts to promote cell growth, differentiation, and migration during development. In regard to kidney development, *Wnt2b* can be detected in the kidney stroma in mice as early as E11.5 [56,57] and in humans WNT2B is expressed in fetal kidney stroma [58]. In addition, cells expressing Wnt2b promote ureteric branching in culture [56]. *Wnt2/2b* is also paramount to normal lung and pectoral fin development in both aquatic and mammalian species [32,59]. Interestingly, *osr1* has been shown to act downstream of retinoic acid signaling yet upstream of *wnt2b* in both pectoral fin development in zebrafish [32] and in lung progenitor specification in foregut endoderm in *Xenopus* [60]. However, our study has both evaluated the role of *wnt2ba* as a regulator of kidney development and placed its function downstream of *osr1* to specifically promote the podocyte lineage. Further, we show that *osr1* promotes *wnt2ba* expression during podocyte development through a mechanism involving the inhibition of *hand2*. In synchrony with our data, a recent report similarly concluded that the reciprocal antagonism between *osr1* and *hand2* is essential for the normal emergence of *wt1b*+ podocyte precursors [33]. Future work to assess whether Osr1 directly binds the *wnt2ba* promotor, and the identity of other regulatory factor(s), will be absolutely critical in order to decipher the underlying molecular mechanisms of the genetic relationships reported in the present work. Furthermore, the identification of other candidate Osr1 targets will be crucial in expanding our knowledge about the roles of this critical gene.

We show in the present study that *wnt2ba* is a regulator of podocyte development but that loss of *wnt2ba* does not cause compelling changes in either the PCT segment or the nephron tubule length. Another study by the team of Lyons et al. [61] showed that the broad inhibition of Wnt signaling through the heat-shock activation of *dkk1* led to an abrogation in the zebrafish pronephros that resembles *osr1* loss of function models. Wnt ligands are highly regionalized to allow for precise regulation during tissue development [57,62]. Our findings that *wnt2ba* is restricted to the podocytes by 48 hpf could reflect regional specificity. This suggests that there are other Wnt ligands and receptors that act to regulate certain kidney lineages in zebrafish development. The loss of one or more of these factors in combination with *wnt2ba* could lead to an anterior truncation of the pronephros that resembles the experiments from Lyons et al. [61]. Future studies are needed to discern these factors and additional downstream targets of both *wnt2ba* and *osr1*.

Taken together, these results have allowed us to garner new insights into podocyte development in zebrafish. By selecting *ocn* as a mutant of interest from our ENU screen, we have discovered an *osr1* mutant and confirmed its significance in zebrafish pronephros development in an unbiased manner. We have expanded on previous findings by demonstrating that *osr1* is required to inhibit apoptosis in specified kidney precursors, and later for nephron cell maintenance. We have also ascertained new roles for *osr1* in promoting *wnt2ba* expression, which it does in part through the antagonism of *hand2*. Finally, our results show that *wnt2ba* mitigates podocyte development downstream of the *osr1/hand2* interaction. Given how little is known about CAKUT and kidney agenesis, findings from genetics studies such as the present work are crucial to furthering our understanding about the causes and solutions to these disease states.

5. Conclusions

The *ocn* zebrafish mutant provides a new vertebrate genetic model to expand our understanding about the roles of *osr1* during kidney development. During the genesis of the zebrafish pronephros, a deficiency of *osr1* causes an abrogation of podocyte and proximal tubule lineages, which are specified but subsequently undergo apoptosis in the absence of *osr1*. Furthermore, we conclude that *osr1* regulates podocyte survival by promoting the expression of *wnt2ba*, a factor that is both necessary and sufficient for podocyte ontogeny. Finally, the function of *wnt2ba* in the podocyte developmental program is impacted by the antagonistic interactions between *osr1* and the transcription factor *hand2*.

Supplementary Materials: The following supporting information can be downloaded at: https://www.mdpi.com/article/10.3390/biomedicines10112868/s1, Figure S1: Additional *ocn* phenotypes, Figure S2: A genetic lesion in the *osr1* transcription factor results in truncated osr1 protein in the *ocn* mutant line; Figure S3: Verification of *osr1* splice-blocking morpholino phenotypes; Figure S4: Additional analysis of cell dynamics in *osr1*-deficient models; Figure S5: *osr1* cRNA injection additional analysis; Figure S6: *wnt2ba* colocalizes with early, developing podocytes; Figure S7: *wnt2bb* does not colocalize with podocytes; Figure S8: Assessment of *wnt2ba* MO through RT-PCR analysis; Figure S9: *wnt2ba* knockdown causes decreased podocytes at 22 ss; Figure S10: Additional stages of *wnt2ba* and *osr1* co-localization; Figure S11: *hand2* MO replicates previous studies.

Author Contributions: Conceptualization, R.A.W. and B.E.D.; methodology, R.A.W. and B.E.D.; validation, R.A.W. and B.E.D.; formal analysis, R.A.W. and B.E.D.; investigation, B.E.D., B.E.C., H.M.W., S.G., L.A., M.N.U., G.F.G., P.T.K., I.L. and W.G.; writing—original draft preparation, R.A.W and B.E.D.; writing—review and editing, R.A.W., B.E.D., B.E.C., H.M.W. and S.G.; supervision, R.A.W.; project administration, R.A.W.; funding acquisition, R.A.W. and B.E.D. All authors have read and agreed to the published version of the manuscript.

Funding: This research was funded by NIH Grant R01DK100237 to R.A.W., start-up funds from the University of Notre Dame College of Science to R.A.W., a National Science Foundation Graduate Research Fellowship DGE-1313583 awarded to B.E.D. We also thank Elizabeth and Michael Gallagher for their generous gift to the University of Notre Dame to support stem cell research. The funders had no role in the study design, data collection and analysis, decision to publish, or manuscript preparation.

Institutional Review Board Statement: The study was approved and conducted according to the guidelines of the University of Notre Dame Institutional Animal Care and Use Committee under protocol numbers 16-07-325 and 19-06-5412.

Informed Consent Statement: Not applicable.

Data Availability Statement: All data are contained in this article and the Supplementary Materials.

Acknowledgments: We thank the staffs of the Department of Biological Sciences and Center for Stem Cells and Regenerative Medicine and particularly express our gratitude to staff of the Center for Zebrafish Research at the University of Notre Dame for their dedication to and care for our zebrafish aquarium. R.A.W. thanks G.R.W. for unwavering support and encouragement and B.C., K.P., and M.M. for their support and advice. Finally, we thank all the past and current members of our lab for their support, discussions, and insights about the fascinating topic of kidney development.

Conflicts of Interest: The authors declare no conflict of interest.

Abbreviations

cadherin 17 (*cdh17*); *centrin 4* (*cetn4*); chronic kidney disease (CKD); *chloride channel K* (*clcnk*); congenital anomalies of the kidney and urinary tract (CAKUT); corpuscle of Stannius (CS); days post fertilization (dpf); distal early (DE) segment; distal late (DL) segment; fluorescent in situ hybridization (FISH); hours post fertilization (hpf); intermediate mesoderm (IM); interrenal gland (IR); *LIM homeobox 1a* (*lhx1a*); multiciliated cells

(MCCs); N-ethyl-N-nitrosourea (ENU); *nephrosis 1, congenital Finnish type (nephrin); nephrosis 2 (nphs1), idiopathic, steroid-resistant (podocin)(nphs2), odd-skipped related transcription factor 1 (osr1); odd-skipped related transcription factor 2 (osr2); outer dense fiber of sperm tails 3B (odf3b); paired box 2a (pax2a);* proximal convoluted tubule (PCT); proximal straight segment (PST); *renal outer medullary potassium, channel 2 (romk2); solute carrier family 4 (sodium bicarbonate cotransporter), member 4a (slc4a4a); solute carrier family 9, subfamily A (slc9a3); solute carrier family 12 (slc12a1); solute carrier family 12 sodium/chloride transporter, member 3 (slc12a3); solute carrier family 20, member 1a (slc20a1a); transient receptor potential cation channel, subfamily M, member 7 (trpm7);* somite stage (ss); whole mount in situ hybridization (WISH); *Wilms tumor 1a (wt1a); Wilms tumor 1b (wt1b); wingless-type MMTV integration site family, member 2Ba (wnt2ba); wingless-type MMTV integration site family, member 2Bb (wnt2bb);* Zebrafish International Research Center (ZIRC).

References

1. Ichimura, K.; Kakuta, S.; Kawasaki, Y.; Miyaki, T.; Nonami, T.; Miyazaki, N.; Nakao, T.; Enomoto, S.; Arai, S.; Koike, M.; et al. Morphological process of podocyte development revealed by block-face scanning electron microscopy. *J. Cell Sci.* **2017**, *130*, 132–142. [CrossRef] [PubMed]
2. Grahammer, F. New structural insights into podocyte biology. *Cell Tissue Res.* **2017**, *369*, 5–10. [CrossRef] [PubMed]
3. Pavenstädt, H.; Kriz, W.; Kretzler, M. Cell biology of the glomerular podocyte. *Physiol Rev.* **2003**, *83*, 253–307. [CrossRef]
4. Garg, P. A review of podocyte biology. *Am. J. Nephrol.* **2018**, *47*, 3–13. [CrossRef] [PubMed]
5. Zhuo, J.L.; Li, X.C. 2013. Proximal nephron. *Compr. Physiol.* **2013**, *3*, 1079. [PubMed]
6. Wiggins, R.C. The spectrum of podocytopathies: A unifying view of glomerular diseases. *Kidney Int.* **2007**, *71*, 1205–1214. [CrossRef]
7. Romagnani, P.; Remuzzi, G.; Glassock, R.; Levin, A.; Jager, K.J.; Tonelli, M.; Massy, Z.; Wanner, C.; Anders, H. Chronic kidney disease. *Nat. Rev. Dis. Primers* **2017**, *3*, 17088. [CrossRef]
8. Reiser, J.; Sever, S. Podocyte biology and pathogenesis of kidney disease. *Annu. Rev. Med.* **2013**, *64*, 357–366. [CrossRef]
9. Gerlach, G.F.; Wingert, R.A. Kidney organogenesis in the zebrafish: Insights into vertebrate nephrogenesis and regeneration. *Wiley Interdiscip. Rev. Dev. Biol.* **2013**, *2*, 559–585. [CrossRef]
10. Perens, E.A.; Garavito-Aguilar, Z.V.; Guio-Vega, G.P.; Peña, K.T.; Schindler, Y.L.; Yelon, D. Hand2 inhibits kidney specification while promoting vein formation within the posterior mesoderm. *eLife* **2016**, *5*, e19941. [CrossRef]
11. Little, M.H.; McMahon, A.P. Mammalian kidney development: Principles, progress, and projections. *Cold Spring Harb. Perspect. Biol.* **2012**, *4*, a008300. [CrossRef] [PubMed]
12. McMahon, A.P. Development of the Mammalian Kidney. *Curr. Top. Dev. Biol.* **2016**, *117*, 31. [PubMed]
13. Luyckx, V.A.; Shukha, K.; Brenner, B.M. Low nephron number and its clinical consequences. *Rambam Maimonides Med. J.* **2011**, *2*, e0061. [CrossRef] [PubMed]
14. Little, M.H. Growing kidney tissue from stem cells: How far from "party trick" to medical application? *Cell Stem Cell* **2016**, *18*, 695–698. [CrossRef] [PubMed]
15. Drummond, B.E.; Wingert, R.A. Insights into kidney stem cell development and regeneration using zebrafish. *World J. Stem Cells* **2016**, *8*, 22–31. [CrossRef]
16. Drummond, I.A.; Davidson, A.J. Zebrafish kidney development. *Methods Cell Biol.* **2016**, *134*, 391–429.
17. Diep, C.Q.; Ma, D.; Deo, R.C.; Holm, T.M.; Naylor, R.W.; Arora, N.; Wingert, R.A.; Bollig, F.; Djordjevic, G.; Lichman, B.; et al. Identification of adult nephron progenitors capable of kidney regeneration in zebrafish. *Nature* **2011**, *470*, 95–100. [CrossRef]
18. Naylor, R.W.; Przepiorski, A.; Ren, Q. HNF1B is essential for nephron segmentation during nephrogenesis. *J. Am. Soc. Nephrol.* **2013**, *24*, 77–87. [CrossRef]
19. Hsu, H.; Lin, G.; Chung, B. Parallel early development of zebrafish interrenal glands and pronephros: Differential control by Wt1 and Ff1b. *Development* **2003**, *130*, 2107–2116. [CrossRef]
20. Bollig, F.; Mehringer, R.; Perner, B.; Hartung, C.; Schäfer, M.; Schartl, M.; Volff, J.; Winkler, C.; Englert, C. Identification and comparative expression analysis of a second Wt1 gene in zebrafish. *Dev. Dyn.* **2006**, *235*, 554. [CrossRef]
21. O'Brien, L.L.; Grimaldi, M.; Kostun, Z.; Wingert, R.A.; Selleck, R.; Davidson, A.J. Wt1a, Foxc1a, and the Notch Mediator Rbpj physically interact and regulate the formation of podocytes in zebrafish. *Dev. Biol.* **2011**, *358*, 318–330. [CrossRef] [PubMed]
22. Zhu, X.; Chen, Z.; Zeng, C.; Wang, L.; Xu, F.; Hou, Q.; Liu, Z. Ultrastructural characterization of the pronephric glomerulus development in zebrafish. *J. Morphol.* **2016**, *277*, 1104–1112. [CrossRef] [PubMed]
23. Wingert, R.A.; Selleck, R.; Yu, J.; Song, H.; Chen, Z.; Song, A.; Zhou, Y.; Thisse, B.; Thisse, C.; McMahon, A.P.; et al. The *cdx* genes and retinoic acid control the positioning and segmentation of the zebrafish pronephros. *PLoS Genet.* **2007**, *3*, 1922–1938. [CrossRef] [PubMed]
24. Wingert, R.A.; Davidson, A.J. Zebrafish nephrogenesis involves dynamic spatiotemporal expression changes in renal progenitors and essential signals from retinoic acid and *irx3b*. *Dev. Dyn.* **2011**, *240*, 2011–2027. [CrossRef] [PubMed]

25. Desgrange, A.; Cereghini, S. Nephron patterning: Lessons from Xenopus, zebrafish, and mouse studies. *Cells* **2015**, *4*, 483–499. [CrossRef]
26. Mugford, J.W.; Sipilä, P.; McMahon, J.A.; McMahon, A.P. Osr1 expression demarcates a multi-potent population of intermediate mesoderm that undergoes progressive restriction to an Osr1-dependent nephron progenitor compartment within the mammalian kidney. *Dev. Biol.* **2008**, *324*, 88–98. [CrossRef]
27. James, R.G.; Kamei, C.N.; Wang, Q.; Jiang, R.; Schultheiss, T.M. Odd-Skipped related 1 is required for development of the metanephric kidney and regulates formation and differentiation of kidney precursor cells. *Development* **2006**, *133*, 2995–3004. [CrossRef]
28. Wang, Q.; Lan, Y.; Cho, E.; Maltby, K.M.; Jiang, R. Odd-Skipped Related 1 (Odd 1) is an essential regulator of heart and urogenital development. *Dev. Biol.* **2005**, *288*, 582–594. [CrossRef]
29. Tena, J.J.; Neto, A.; de la Calle-Mustienes, E.; Bras-Pereira, C.; Casares, F.; Gómez-Skarmeta, J.L. Odd-Skipped genes encode repressors that control kidney development. *Dev. Biol.* **2007**, *301*, 518–531. [CrossRef]
30. Mudumana, S.P.; Hentschel, D.; Liu, Y.; Vasilyev, A.; Drummond, I.A. Odd skipped related1 reveals a novel role for endoderm in regulating kidney versus vascular cell fate. *Development* **2008**, *135*, 3355–3367. [CrossRef]
31. Tomar, R.; Mudumana, S.P.; Pathak, N.; Hukriede, N.A.; Drummond, I.A. Osr1 is Required for Podocyte Development Downstream of Wt1a. *J. Am. Soc. Neph.* **2014**, *25*, 2539–2545. [CrossRef] [PubMed]
32. Neto, A.; Mercader, N.; Gómez-Skarmeta, J.L. The Osr1 and Osr2 genes act in the pronephric anlage downstream of retinoic acid signaling and upstream of Wnt2b to maintain pectoral fin development. *Development* **2012**, *139*, 301–311. [CrossRef] [PubMed]
33. Perens, E.A.; Diaz, J.T.; Quesnel, A.; Crump, J.G.; Yelon, D. osr1 couples intermediate mesoderm cell fate with temporal dynamics of vessel progenitor cell differentiation. *Development* **2021**, *148*, dev198408. [CrossRef] [PubMed]
34. Zhang, Z.; Iglesias, D.; Eliopoulos, N.; El Kares, R.; Chu, L.; Romagnani, P.; Goodyer, P. A variant OSR1 allele which disturbs OSR1 mRNA expression in renal progenitor cells is associated with reduction of newborn kidney size and function. *Hum. Mol. Genet.* **2011**, *20*, 4167–4174. [CrossRef] [PubMed]
35. Kroeger, P.T.; Poureetezadi, S.J.; McKee, R.; Jou, J.; Miceli, R.; Wingert, R.A. Production of haploid zebrafish embryos by in vitro fertilization. *J. Vis. Exp.* **2014**, *89*, 51708. [CrossRef]
36. Chambers, B.E.; Gerlach, G.F.; Clark, E.G.; Chen, K.H.; Levesque, A.E.; Leshchiner, I.; Goessling, W.; Wingert, R.A. Tfap2a is a novel gatekeeper of nephron differentiation during kidney development. *Development* **2019**, *146*, dev172387. [CrossRef]
37. Kroeger, P.T.; Drummond, B.E.; Miceli, R.; McKernan, M.; Gerlach, G.F.; Marra, A.N.; Fox, A.; McCampbell, K.K.; Leshchiner, I.; Rodriguez-Mari, A.; et al. The zebrafish kidney mutant *zeppelin* reveals that *brca2/fancd1* is essential for pronephros development. *Dev. Biol.* **2017**, *428*, 148–163. [CrossRef]
38. Li, Y.; Cheng, C.N.; Verdun, V.A.; Wingert, R.A. Zebrafish nephrogenesis is regulated by interactions between retinoic acid, *mecom* and Notch signaling. *Dev. Biol.* **2014**, *386*, 111–122. [CrossRef]
39. Cheng, C.N.; Li, Y.; Marra, A.N.; Verdun, V.; Wingert, R.A. Flat mount preparation for observation and analysis of zebrafish embryo specimens stained by whole mount in situ hybridization. *J. Vis. Exp.* **2014**, *89*, 51604. [CrossRef]
40. Cheng, C.N.; Wingert, R.A. Nephron proximal tubule patterning and corpuscles of Stannius formation are regulated by the *sim1* transcription factor and retinoic acid in zebrafish. *Dev. Biol.* **2015**, *399*, 100–116. [CrossRef]
41. Marra, A.N.; Ulrich, M.; White, A.; Springer, M.; Wingert, R.A. Visualizing multiciliated cells in the zebrafish through a combined protocol of whole mount fluorescent in situ hybridization and immunofluorescence. *J. Vis. Exp.* **2017**, *129*, 56261. [CrossRef] [PubMed]
42. Marra, A.N.; Chambers, B.E.; Chambers, J.M.; Drummond, B.E.; Adeeb, B.D.; Wesselman, H.M.; Morales, E.E.; Handa, N.; Pettini, T.; Ronshaugen, M.; et al. Visualizing gene expression during zebrafish pronephros development and regeneration. *Methods Cell Biol.* **2019**, *154*, 183–215. [PubMed]
43. Wakahara, T.; Kusu, N.; Yamauchi, H.; Kimura, I.; Konishi, M.; Miyake, A.; Itoh, N. Fibin, a novel secreted lateral plate mesoderm signal, is essential for pectoral fin bud initiation in zebrafish. *Dev. Biol.* **2007**, *303*, 527–535. [CrossRef] [PubMed]
44. Maves, L.; Tyler, A.; Moens, C.B.; Tapscott, S.J. Pbx acts with Hand2 in early myocardial differentiation. *Dev. Biol.* **2009**, *333*, 409–418. [CrossRef]
45. Gerlach, G.F.; Wingert, R.A. Zebrafish pronephros tubulogenesis and epithelial identity maintenance are reliant on the polarity proteins prkc iota and zeta. *Dev. Biol.* **2014**, *396*, 183–200. [CrossRef]
46. Leshchiner, I.; Alexa, K.; Kelsey, P.; Adzhubei, I.; Austin-Tse, C.A.; Cooney, J.D.; Anderson, H.; King, M.J.; Stottmann, R.W.; Garnaas, M.K.; et al. Mutation mapping and identification by whole-genome sequencing. *Genome Res.* **2012**, *22*, 1541–1548. [CrossRef]
47. Ryan, S.; Willer, J.; Marjoram, L.; Bagwell, J.; Mankiewicz, J.; Leshchiner, I.; Goessling, W.; Bagnat, M.; Katsanis, N. Rapid identification of kidney cyst mutations by whole exome sequencing in zebrafish. *Development* **2013**, *140*, 4445–4451. [CrossRef]
48. Swartz, M.E.; Sheehan-Rooney, K.; Dixon, M.J.; Eberhart, J.K. Examination of a palatogenic gene program in zebrafish. *Dev. Dyn.* **2011**, *240*, 2204–2220. [CrossRef]
49. Liu, H.; Lan, Y.; Xu, J.; Chang, C.; Brugmann, S.A.; Jiang, R. Odd-Skipped related-1 controls neural crest chondrogenesis during tongue development. *Proc. Natl. Acad. Sci. USA* **2013**, *110*, 18555–18560. [CrossRef]
50. Terashima, A.V.; Mudumana, S.P.; Drummond, I.A. Odd skipped related 1 is a negative feedback regulator of nodal-induced endoderm development. *Dev. Dyn.* **2014**, *243*, 1571–1580. [CrossRef]

51. Gering, M.; Rodaway, A.R.; Göttgens, B.; Patient, R.K.; Green, A.R. The SCL gene specifies haemangioblast development from early mesoderm. *EMBO J.* **1998**, *17*, 4029–4045. [CrossRef] [PubMed]
52. Liao, E.C.; Paw, B.H.; Oates, A.C.; Pratt, S.J.; Postlethwait, J.H.; Zon, L.I. SCL/Tal-1 transcription factor acts downstream of cloche to specify hematopoietic and vascular progenitors in zebrafish. *Genes Dev.* **1998**, *12*, 621–626. [CrossRef] [PubMed]
53. Kroeger, P.T.; Wingert, R.A. Using zebrafish to study podocyte genesis during kidney development and regeneration. *Genesis* **2014**, *52*, 771–792. [CrossRef] [PubMed]
54. Perner, B.; Englert, C.; Bollig, F. The Wilms tumor genes Wt1a and Wt1b control different steps during formation of the zebrafish pronephros. *Dev. Biol.* **2007**, *309*, 87–96. [CrossRef] [PubMed]
55. Xu, J.; Liu, H.; Chai, H.; Lan, Y.; Jiang, R. Osr1 interacts synergistically with Wt1 to regulate kidney organogenesis. *PLoS ONE* **2016**, *11*, e0159597. [CrossRef]
56. Lin, Y.; Liu, A.; Zhang, S.; Ruusunen, T.; Kreidberg, J.A.; Peltoketo, H.; Drummond, I.; Vainio, S. Induction of ureter branching as a response to Wnt-2b signaling during early kidney organogenesis. *Dev. Dyn.* **2001**, *222*, 26–39. [CrossRef]
57. Iglesias, D.M.; Hueber, P.; Chu, L.; Campbell, R.; Patenaude, A.; Dziarmaga, A.J.; Quinlan, J.; Mohamed, O.; Dufort, D.; Goodyer, P.R. canonical WNT signaling during kidney development. *Am. J. Physiol. Ren. Physiol.* **2007**, *293*, 494. [CrossRef]
58. Combes, A.N.; Combes, A.N.; Zappia, L.; Er, P.X.; Oshlack, A.; Little, M.H. Single-cell analysis reveals congruence between kidney organoids and human fetal kidney. *Genome Med.* **2019**, *11*, 3. [CrossRef]
59. Goss, A.M.; Tian, Y.; Tsukiyama, T.; Cohen, E.D.; Zhou, D.; Lu, M.M.; Yamaguchi, T.P.; Morrisey, E.E. Wnt2/2b and B-Catenin signaling are necessary and sufficient to specify lung progenitors in the foregut. *Dev. Cell* **2009**, *17*, 290–298. [CrossRef]
60. Rankin, S.A.; Gallas, A.L.; Neto, A.; Gómez-Skarmeta, J.L.; Zorn, A.M. Suppression of Bmp4 signaling by the zinc-finger repressors Osr1 and Osr2 is required for Wnt/B-Catenin-mediated lung specification in Xenopus. *Development* **2012**, *139*, 3010–3020. [CrossRef]
61. Lyons, J.P.; Miller, R.K.; Zhou, X.; Weidinger, G.; Deroo, T.; Denayer, T.; Park, J.; Ji, H.; Hong, J.Y.; Li, A.; et al. Requirement of Wnt/B-Catenin signaling in pronephric kidney development. *Mech. Dev.* **2009**, *126*, 142–159. [CrossRef] [PubMed]
62. Verkade, H.; Heath, J.K. Wnt signaling mediates diverse developmental processes in zebrafish. *Methods Mol. Biol.* **2008**, *469*, 225. [PubMed]

 *biomedicines*

Article

Comparative Transcriptome Analysis Provides Novel Molecular Events for the Differentiation and Maturation of Hepatocytes during the Liver Development of Zebrafish

Yasong Zhao [1,3,†], Xiaohui Li [4,†], Guili Song [1], Qing Li [1], Huawei Yan [2,*] and Zongbin Cui [1,2,*]

[1] State Key Laboratory of Freshwater Ecology and Biotechnology, Institute of Hydrobiology, Chinese Academy of Sciences, Wuhan 430072, China

[2] Guangdong Provincial Key Laboratory of Microbial Culture Collection and Application, State Key Laboratory of Applied Microbiology Southern China, Institute of Microbiology, Guangdong Academy of Sciences, Guangzhou 510070, China

[3] College of Advanced Agricultural Sciences, University of Chinese Academy of Sciences, Beijing 100049, China

[4] Yangtze River Fisheries Research Institute, Chinese Academy of Fishery Sciences, Wuhan 430223, China

[*] Correspondence: yanhw@mail3.sysu.edu.cn (H.Y.); cuizb@gdim.cn (Z.C.); Tel.: +86-27-68780090 (Z.C.)

[†] These authors contributed equally to this work.

Citation: Zhao, Y.; Li, X.; Song, G.; Li, Q.; Yan, H.; Cui, Z. Comparative Transcriptome Analysis Provides Novel Molecular Events for the Differentiation and Maturation of Hepatocytes during the Liver Development of Zebrafish. *Biomedicines* 2022, 10, 2264. https://doi.org/10.3390/biomedicines10092264

Academic Editor: James A. Marrs

Received: 23 August 2022
Accepted: 9 September 2022
Published: 13 September 2022

Publisher's Note: MDPI stays neutral with regard to jurisdictional claims in published maps and institutional affiliations.

Abstract: The liver plays an essential role in multiple biological functions including metabolism, detoxification, digestion, coagulation, and homeostasis in vertebrates. The specification and differentiation of embryonic hepatoblasts, the proliferation of hepatocytes, and the hepatic tissue architecture are well documented, but molecular events governing the maturation of hepatocytes during liver development remain largely unclear. In this study, we performed a comparative transcriptome analysis of hepatocytes that were sorted by flow cytometry from developing zebrafish embryos at 60, 72, and 96 hpf. We identified 667 up-regulated and 3640 down-regulated genes in hepatocytes between 60 and 72 hpf, 606 up-regulated and 3924 down-regulated genes between 60 and 96 hpf, and 1693 up-regulated genes and 1508 down-regulated genes between 72 and 96 hpf. GO enrichment analysis revealed that key biological processes, cellular components, and molecular functions in hepatocytes between 60 to 72 hpf, such as cell cycle, DNA replication, DNA repair, RNA processing, and transcription regulation, are mainly associated with the proliferation of hepatocytes. In addition to biological processes, cellular components, and molecular functions for cell proliferation, molecular functions for carbohydrate metabolism were enriched in hepatocytes during 72 to 96 hpf. KEGG enrichment analysis identified key signaling pathways, such as cell cycle, RNA degradation, ubiquitin-mediated proteolysis, ErbB and Hedgehog signaling, basal transcription factors, Wnt signaling, and glycan degradation, which are closely associated with cell proliferation or carbohydrate metabolism in hepatocytes between 60 to 72 hpf. Newly enriched signaling pathways in hepatocytes during 72 to 96 hpf include metabolisms of pyrimidine, purine, nicotinate and nicotinamide, *caffeine*, glycine, serine and threonine, ABC transporters, and p53 signaling that function in metabolisms of lipid, protein and energy, cellular secretion, or detoxification, indicating the functional maturation of hepatocytes between 72 to 96 hpf. These findings provide novel clues for further understanding the functional differentiation and maturation of hepatocytes during liver development.

Keywords: zebrafish; liver; hepatocyte; transcriptome; signaling pathways

1. Introduction

The liver is a vital organ with multiple biological functions, including metabolism, detoxification, digestion, and homeostasis in vertebrates [1]. As the metabolic center of the body, liver consists of multiple cell types in which hepatocytes (HCs) account for approximately 70% of the total liver cell population and carry out most functions of the liver, including the metabolism of lipids and drugs; storage of amino acids, iron, and glycogen; and production of clotting factors [2,3]. Therefore, the development, functions, and diseases of liver in vertebrates have been popular research topics [4–6].

Previous studies have revealed that molecular mechanisms of early liver development are conserved in vertebrates [7–9]. The factors encoding by homologous genes can regulate hepatic patterning in both zebrafish and mice [10,11]. Many highly-conserved and tightly-controlled signaling pathways, such as Wnt/β-catenin, Fgf, and Bmp signaling pathways play major roles during liver development [12–14]. A series of liver-specific factors, pan-endodermal factors, and factors from the surrounding mesodermal tissues together formed a genetic network and precisely control the process of liver organogenesis [15,16].

In mice, the differentiation of hepatoblasts initiates at approximately E11.5–E13.5 and terminates at E18.5 [15,17]. During this stage, Notch and TGFβ signals promote the maturation of cholangiocytes and inhibit the specification of hepatoblasts toward hepatocytes [18]. The maturation of hepatocytes is a gradual process that begins immediately after the differentiation of bipotential hepatoblasts and uninterruptedly until postnatally to develop the organizational architectural features of the liver [19]. Wnt/β-catenin signaling was responsible for the differentiation and expansion of hepatocytes, as well as managing the accomplishment of hepatocyte zonation. *Wnt1* and *Lhx2* deletion embryos showed an ectopic activation state with ECM deposition, fibrosis, and the smaller liver phenotype due to hepatocyte proliferation abnormalities [20]. Other factors that promote hepatocyte maturation include the cytokine oncostatin M (OSM), hepatocyte growth factor (HGF), and the continuing inhibition of Notch and TGF signaling [21,22].

Zebrafish (*Danio rerio*) is an excellent model organism to study both liver development and regeneration [23,24]. The zebrafish liver, as other vertebrates, develops from the endoderm, but it lacks hematopoietic function, and mutations in genes associated with early liver development do not result in hematopoietic dysfunction [25].

The process of zebrafish liver organogenesis can be divided into four main phases including hepatoblasts specification, budding, differentiation, and outgrowth [26,27]. Zebrafish liver morphogenesis started at 22 hpf and ended at 96 hpf, which can be distinguished by morphological changes and the expression patterns of molecular markers, such as *hhex*, *proxl*, *foxa3*, and *gata6* [28]. The mesodermal signals, such as Fgf, Bmp, and Wnt signaling are closely involved in the control of hepatoblast specification, which can be marked by *prox 1* and *hhex* at 22 hpf in zebrafish [29,30]. The differentiation and maturation of hepatocytes and outgrowth of zebrafish liver occur between 50 and 96 hpf, which are marked by *vitamin D binding protein (gc)*, *fatty acid binding protein 10a (fabp10a)*, *ceruloplasmin (cp)*, and the dramatic increase in liver size to establish the liver functions [31–33]. Moreover, Hnf members are a group of transcription factors that are enriched in the liver when compared to other organs, of which Hnf1 and Hnf6 are essential factors in hepatocyte maturation and outgrowth [34,35]. Although a large body of studies have focused on liver development, including the specification of hepatoblasts, the proliferation of hepatocytes, and the hepatic tissue architecture, the molecular events underlying the differentiation of hepatocyte functions during outgrowth and maturation of liver remain largely unclear.

Generation of multiple transgenic zebrafish lines carrying a fluorescent liver, such as *Tg(fabp10a:dsRed;ela3l:EGFP)* zebrafish [36], which can express the red fluorescent protein under hepatocyte-specific promoter *fabp10a* (fatty acid binding protein 10a, also called *l-fabp*), allows the *dynamic* study of early developmental of liver in vivo [37]. In this study, hepatocytes with red fluorescence were sorted with a flow cytometry from developing embryos at 60, 72, and 96 hpf of the *Tg(fabp10a:dsRed;ela3l:EGFP)* line, followed by a high-throughput RNA-seq analysis. The comparison of transcriptome profiling for hepatocyte populations from three time points revealed multiple signaling pathways and biological processes that are associated with the functional differentiation and maturation of hepatocytes.

2. Materials and Methods

2.1. Maintenance of Zebrafish

Wild type AB strain and *Tg(fabp10a:dsRed;ela3l:EGFP)* line were maintained and bred in a circulating water system at 28 °C. Embryos at different developmental stages were determined according to hours post-fertilization (hpf). The collection and culture of embryos were performed following the previous methods [38].

2.2. Preparation of Zebrafish Liver Cell Suspension

Approximately 4000 embryos were collected from the hybrid offspring of wild-type (WT) females and transgenic zebrafish males with a red fluorescent liver. The zebrafish larvae at 20 hpf were cultured in 0.3 × Danieau's solution (17 mM NaCl, 0.2 mM KCl, 0.12 mM MgSO$_4$, 0.18 mM Ca(NO$_3$)$_2$, 1.5 mM HEPES, pH 7.6) containing 0.003% (w:v) 1-phenyl 2-thiourea (PTU). At 60, 72 or 96 hpf, after anesthetized in 0.3 × Danieau's solution containing 0.016% Tricaine mesylate (Tricaine methanesulfonate, MS-222) (Sigma, St. Louis, MO, USA), approximately 2000 fluorescent zebrafish embryos were selected under a fluorescence microscope and washed three times with 0.3 × Danieau's solution. Then, 50 larvae fish were put into a 2 mL centrifuge tube, followed by the removal of excess liquid, addition of 1 mL trypsin-EDTA solution, digestion in a water bath at 28 °C for 10 min, and pipetting up and down with a P200 pipette every 3 min. After digestion, Hi-FBS at a final concentration of 5% was added to stop the reaction and the tubes were placed in a water bath at 28 °C for 5–7 min. Next, the above digestion solution was filtered into a new 1.5 mL centrifuge tube using a 200-mesh sieve and then centrifuged at 310× *g* for 5 min. Finally, the zebrafish hepatocytes were resuspended in PBS (Hyclone, #SH30256.01, Logan, UT, USA). All steps of the experiments were performed on ice unless otherwise noted [39].

2.3. Isolation of Fluorescent Cells in Zebrafish Liver

Hoechest 33342 (Beyotime, #C1025, Shanghai, China) is a membrane-permeable fluorescent dye, which shows a bright-blue signal in the nucleus of apoptotic cells. The wavelength of the fluorescent dye is different from the wavelength of the red fluorescence of hepatocytes, which is convenient for removing dead cells before sorting as previously described [40]. After primary sorting, recovery, and secondary sorting with a flow cytometry (BD Biosciences, Franklin Lakes, NJ, USA), the cells were sorted into FACSmax solution (AMSBIO, Abingdon, UK) and 10–20 μL was used for cell counting with a Neubauer cell counter slide to check the purity of desired cells. Finally, we isolated red-fluorescent hepatocytes with a purity of more than 95%. Samples for RNA-seq analysis at each time point have three replicates.

2.4. RNA Extraction

RNA extraction was conducted immediately after the fluorescence-activated cell sorting (FACS). Sorted hepatocytes from three time points were transferred directly into the lysis buffer supplied by the RNeasy Micro Kit (Qiagen, #74004, Düsseldorf, Germany) and then performed according to the manufacturer's instructions [41]. RNA was isolated from cell lysates right away. RNA degradation and contamination were detected using 1% agarose gel electrophoresis. A NanoPhotometer NP60 Spectrometer (Implen GmbH, München, Germany) was used to measure the purity and concentration of RNA. Following the manufacturer's instructions, the Direct-zol RNA MiniPrep Kit (ZYMO RESEARCH, #R2050, Irvine, CA, USA) was used to purify RNA [42].

2.5. Library Construction and High-Throughput Sequencing

The construction of libraries and high-throughput sequencing were performed using an Illumina Genome Analyzer IIx platform (GA IIx, Illumina, San Diego, CA, USA) in the Analysis and Testing Center of the Institute of Hydrobiology, Chinese Academy of Sciences (http://www.ihb.ac.cn/fxcszx/, accessed on 8 November 2021). Purified mRNA samples were fragmented into small pieces and double-stranded cDNA was synthesized using random hexamer primers. Synthetic cDNAs were end-repaired, phosphorylated, 3'-adenylated, adaptor-ligated, and PCR-amplified to construct sequencing libraries. Three independent biological replicates for samples from three time points were used for library construction.

2.6. Bioinformatic Analysis

The raw reads contain adapters or low-quality bases. Therefore, the reads were filtered with PRINSEQ (version 0.20.4, Schmieder R, San Diego, CA, USA) to obtain high quality clean reads [43]. Based on the reference-based approach, we used StringTie (version 1.3.1, StringTie, Baltimore, MD, USA) software to assemble the mapped reads of each sample [44,45]. The differentially expression analysis of data for samples between two time points was performed following the previous methods [46]. Genes with a fold change ≥ 2 and a q-value ≤ 0.05 were considered significantly and differentially expressed. After comparison with the zebrafish reference genomic data (*Danio rerio*. GRCz11, version-103), an enrichment analysis of differentially expressed genes (DEGs) identified the signal transduction pathways and metabolic pathways. GO (Gene ontology) enrichment analyses were conducted using Cytoscape-v3.8.2 plugins ClueGO-v2.5.8. KEGG (Kyoto Encyclopedia of Genes and Genomes) enrichment analyses were performed using the clusterProfiler package. Calculation and creation of Jaccard Coefficient (JC) and network of hub genes and pathways were performed as previously described [47,48].

2.7. Statistical Analysis

Statistical analysis of data was performed using GraphPad Prism 8.3.0 software (GraphPad Software, San Diego, CA, USA).

3. Results

3.1. The Isolation of Zebrafish Hepatocytes

The transgenic zebrafish line *Tg(fabp10a:dsRed;ela3l:EGFP)* was used to isolate hepatocytes labeled by DsRed, a red-fluorescent protein. During embryonic development, the size of developing zebrafish liver increased rapidly from 60 to 96 hpf as shown by the red fluorescence area (Figure 1A). Developing embryos at 60, 72, or 96 hpf were digested into individual cells. The embryonic cells were sorted with flow cytometry to obtain dsRed-expressing hepatocytes at 60, 72, or 96 hpf, followed by RNA-seq (Figure 1B). Population 1 (P1) represents a population of cells that have been removed from cellular debris. After primary sorting and recovery, we obtained population 2 (P2), which excluded the dead cells stained with Hoechest. The number of sorted hepatocytes continued to increase with the outgrowth of liver from 60 to 96 hpf (Supplementary Figure S1). We collected cell population 3 (P3) at 60, 72, or 96 hpf, in which the red-fluorescent hepatocytes take up to more than 95%. The number of sorted hepatocytes for further analysis were 193, 200, and 3491, which continued to increase with the outgrowth of liver from 60 to 96 hpf (Figure 1C)

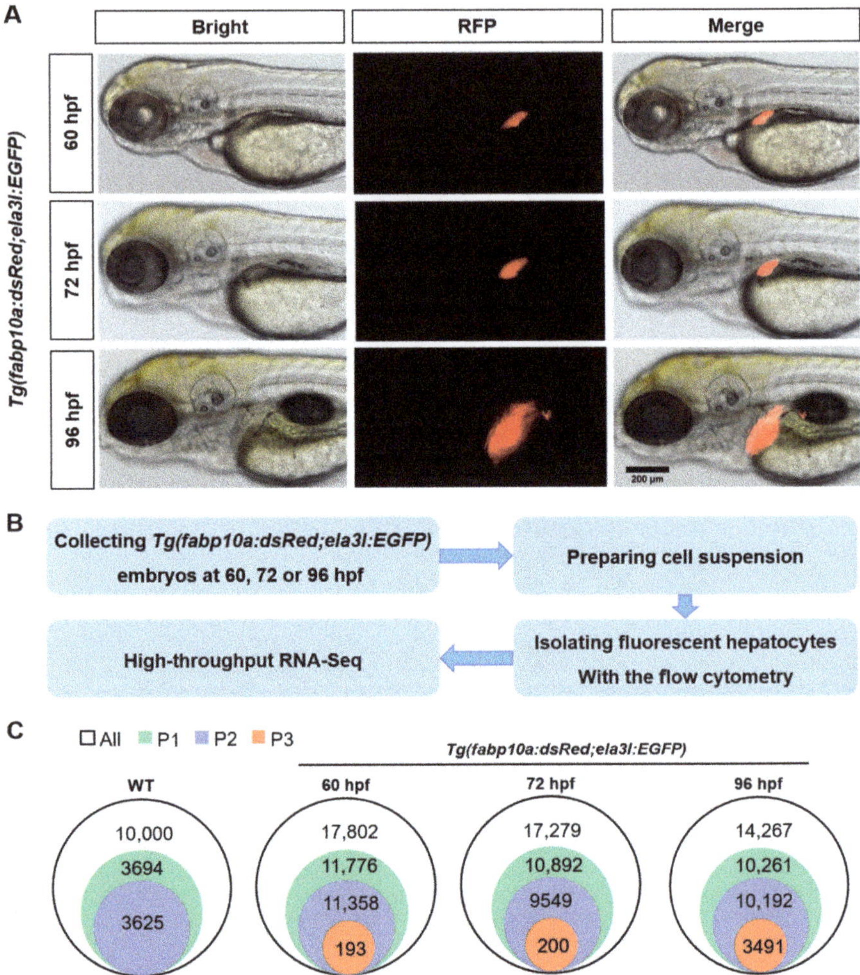

Figure 1. The isolation of zebrafish hepatocytes. (**A**) The size changes of red-fluorescent liver at 60, 72 and 96 hpf in transgenic zebrafish *Tg(fabp10a:dsRed;ela3l:EGFP)*. Scale bar, 200 μm. (**B**) The technical roadmap for isolation of zebrafish hepatocytes. (**C**)The cell populations of hepatocytes sorted from *Tg(fabp10a:dsRed;ela3l:EGFP)* embryos at 60, 72 and 96 hpf.

3.2. Quality Analysis of the Transcriptome Data for Zebrafish Hepatocytes

To understand biological processes, cellular functions and signaling pathways that are associated with the functional differentiation and maturation of hepatocytes in developing embryo at 60, 72, and 96 hpf, nine samples of P3 hepatocytes were sorted for construction of RNA libraries and subsequcent high-throughput RNA-seq sequencing and samples at each time point contain three independent biological replicates.

RNA-seq analysis generated 27.2249–40.4443 million pairs (M) of total reads for each of the samples and approximately 72.32–77.77% of the processed reads were mapped to the reference genome of zebrafish and unique mapped genes accounted for more than 90% of total mapped genes (Figure 2A). The Q20 and Q30 of the three groups at 60, 72, and 96 hpf were all above 85% and the GC content was 46.89% at 60 hpf, 49.31% at 72 hpf and 46.77% at 96 hpf, respectively (Table 1). These data demonstrated the relatively high quality of the RNA sequencing. The first (PC1) and second (PC2) principal component analysis (PCA showed a variation of 59.08% and 18.82%, indicating a clear separation of genes at different time periods during the early development of the zebrafish liver (Figure 2B).

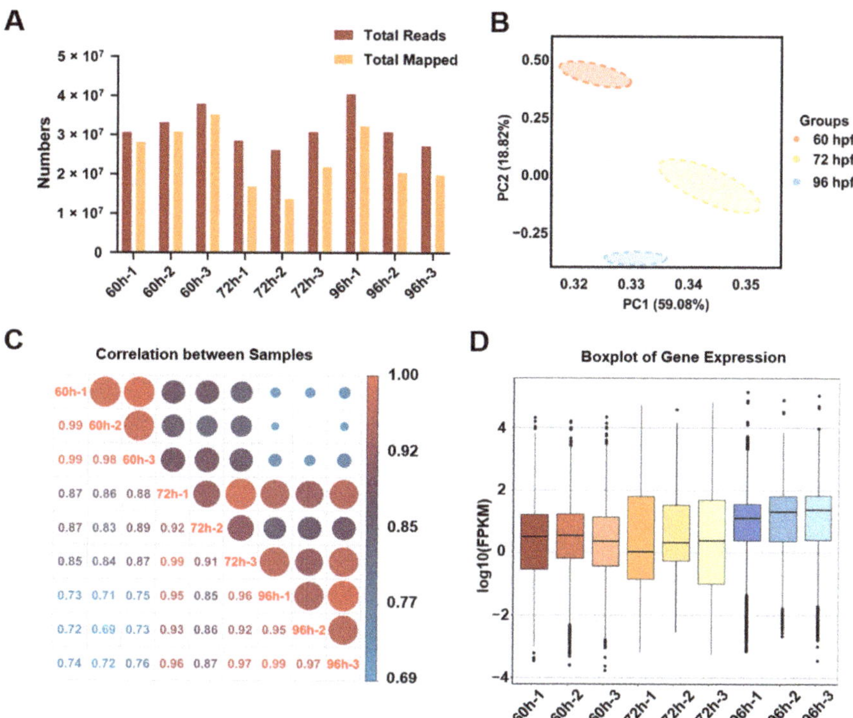

Figure 2. The quality analysis of transcriptome data for zebrafish hepatocytes. (**A**) The RNA-seq data of sorted liver cells at three time points. (**B**) The principal component analysis (PCA) of samples at three time points. (**C**) The Pearson's correlation coefficient matrix of RNA-seq between samples at three time points. (**D**) The boxplots of gene expression in samples at three time points.

Table 1. Statistics of RNA-seq output data.

Sample Names	Total Reads (M)	Q20 (%)	Q30 (%)	GC (%)	Read Length (bp) (bp (bp))
60h-1	30.5984 M	92.18	86.74	47.10	114
60h-2	33.1771 M	92.12	86.62	46.76	116
60h-3	37.8892 M	91.96	86.47	46.81	115
72h-1	28.4021 M	92.08	86.67	49.80	110
72h-2	26.1058 M	91.28	86.10	48.58	106
72h-3	30.7598 M	91.99	86.55	49.56	111
96h-1	40.4443 M	92.63	87.45	49.58	107
96h-2	30.7934 M	91.55	86.18	44.64	111
96h-3	27.2249 M	89.38	83.54	46.10	108

Notes: M, million pairs; bp, base pair.

To evaluate the similarity between samples collected at the same time points, we calculated the correlation between different samples. The closer correlation coefficient between samples gets to 1, the higher similarity between samples is, and the fewer differentially expressed genes between samples. We found that Pearson's correlation between samples at the same time point was 0.91–0.99 and the correlation between 60 and 72 hpf was higher than between 60 and 96 hpf (Figure 2C), which was consistent with the data of

PCA. Boxplot comparison of the distributions of gene expression data after normalization showed that the means and ranges of gene expression in each sample exhibit a uniformity of the expression distribution (Figure 2D).

3.3. Differentially Expressed Genes in Hepatocytes of Developing Liver

We then performed a Venn diagram analysis to identify differentially expressed genes (DEGs) for three time periods 60–72 hpf, 72–96 hpf, and 60–96 hpf in heptocytes of developing liver. A total of 7255 DEGs were found, including up-regulated (log2foldchange ≥ 1, adjusted *p*-value ≤ 0.05) and down-regulated (log2foldchange ≤ −1, adjusted *p*-value ≤ 0.05) between two time points were listed in Supplementary Table S1. We found 667 up-regulated and 3640 down-regulated genes in heptocytes betweent 60 and 72 hpf, 1693 up-regulated and 1508 down-regulated genes between 72 and 96 hpf, and 606 up-regulated and 3924 down-regulated genes between 60 and 96 hpf (Figure 3A). Among these DEGs, 673 DEGs were specifically detected between 60 and 72 hpf, 633 DEGs between 72 and 96 hpf, and 1179 DEGs between 60 and 96 hpf (Figure 3B).

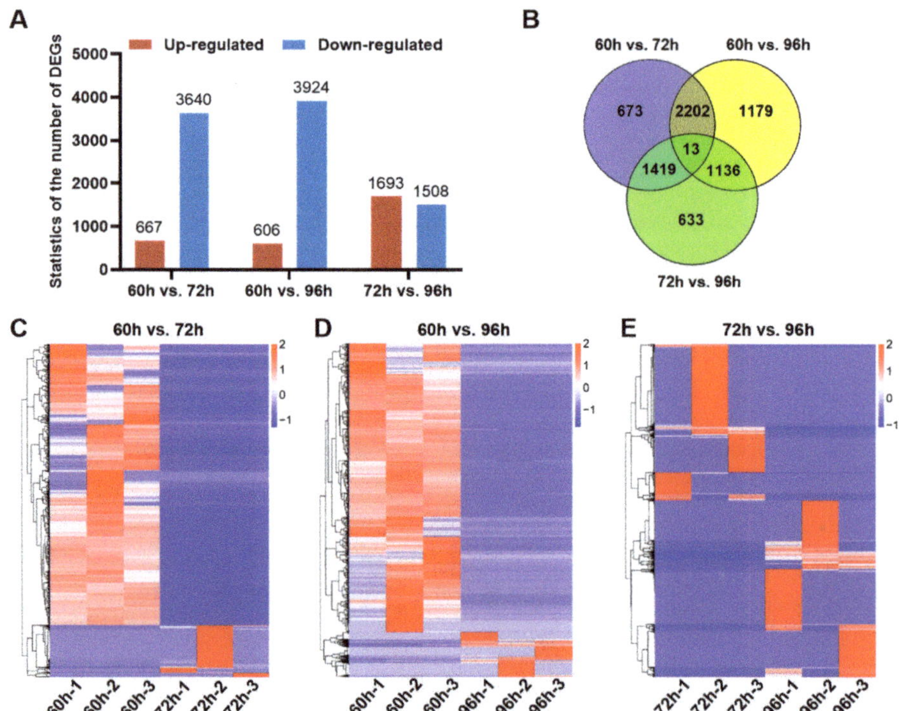

Figure 3. The differentially expressed genes analysis in early liver developmental stages. (**A**) Statistic of differentially expressed genes (DEGs) in hepatocytes between two time points. (**B**) The Venn diagram analysis of DEGs in hepatocytes among three time points. Heatmaps of DEGs between two time points, including 60 and 72 hpf (**C**), 60 and 96 hpf (**D**), as well as 72 and 96 hpf (**E**).

As shown in cluster heatmaps, a striking difference in the expression of genes can be found in hepatocytes between two time points and most DEGs at 72 hpf or 96 hpf were down-regulated in comparison with the expression of corresponding genes at 60 hpf (Figure 3C–E), suggesting that the differentiation of hepatocyte functions occurred from 60 to 96 hpf.

3.4. GO Enrichment Analysis of Specifically Expressed Genes in Hepatocytes during Liver Development

To further explore the differences in biological processes (BP), cellular composition (CC) and molecular functions (MF) in heptocytes from 60 to 96 hpf, all DEGs were divided into seven groups (a–g) (Figure 4A; Table S2). DEGs in groups a and b represent genes specifically expressed in hepatocytes from 60 to 72 hpf, which account for 39.63% of all DEGs. DEGs in groups f and g were specifically expressed in heptocytes from 72 to 96 hpf, which account for 24.39% of total DEGs. DEGs in groups c, d, and e showed no significance in hepatocytes from 60 to 72 hpf, but DEGs in group c stand for a significant difference in hepatocytes from 60 to 96 hpf, suggesting that DEGs in group c are also associated with functional differention of hepatocytes. Thus, GO enrichment analysis of DEGs in groups (a + b, f + g, c + f + g, d + e, b + e, c + f) were performed.

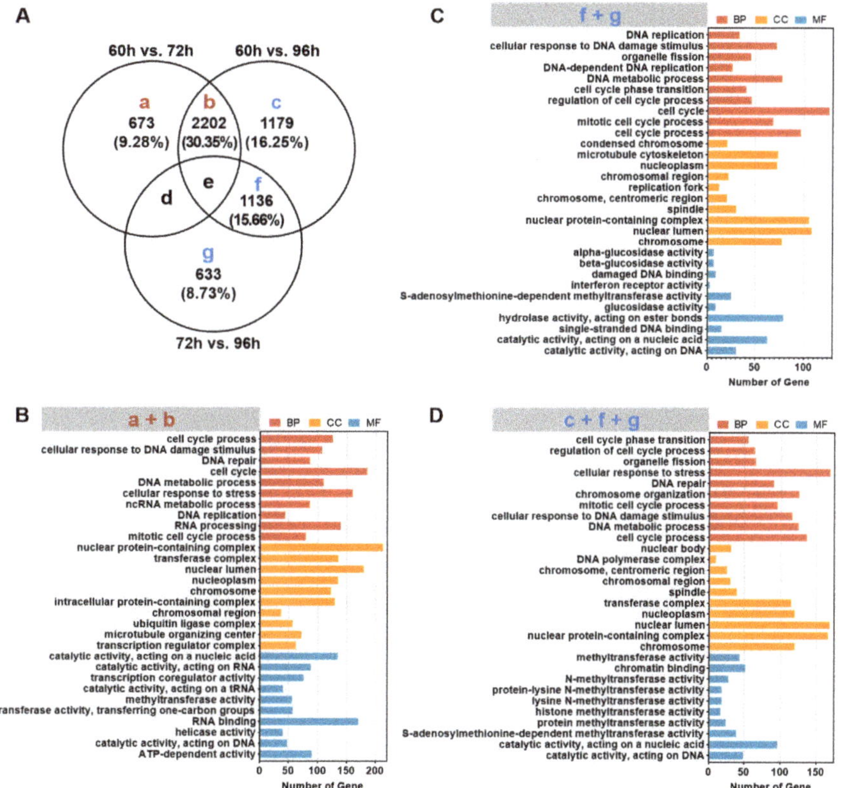

Figure 4. GO enrichment analysis of genes specifically expressed during different time periods of liver development. (**A**) The Venn diagram analysis of differentially expressed genes. a–g: different letters represent genes specifically expressed in different Venn groups. (**B**) GO enrichment analysis of genes specifically expressed in groups a and b that represent a class of genes specifically expressed in hepatocytes from 60 to 72 hpf. (**C**) GO enrichment analysis of genes specifically expressed in groups f and g that represent a class of genes specifically expressed in hepatocytes from 72 to 96 hpf. (**D**) GO enrichment analysis of genes specifically expressed in groups c, f, and g that represent all differentially expressed genes in hepatocytes between 60–72 hpf and 72–96 hpf.

Top 10 GO terms and genes related to biological processes, cellular composition and molecular function in different groups were listed in Supplementary Table S3. The top 10 GO terms enriched from genes specifically expressed in groups a and b are primarily cell cycle, cellular response to stress, DNA replication, DNA repair and RNA processing in BP, nuclear protein-containing complex and nuclear lumen in CC, and catalytic activity on DNA and RNA and RNA binding in MF (Figure 4B), which are closely associated with cell proliferation. Most BPs, CCs, and MFs related to cell proliferation were also shared in hepatocytes among groups (d + e, b + e, c + f) at 60–72 hpf, 72–96 hpf and 60–96 hpf (Supplementary Figure S1). Moreover, lipid biosynthetic process, ATP dimethylallyltransferase activity and ADP dimethylallyltransferase activity were shared in hepatocytes between 60–72 hpf and 72–96 hpf (groups d + e, Supplementary Figure S2). These data suggest that functions of cell proliferation, lipid synthesis, and energy metabolism are developed in hepatocytes from 60 to 96 hpf.

In addition to DEGs related to cell proliferation, the genes specifically expressed in groups f and g were associated with molecular functions of hydrolase activity on ester bonds, glucosidase activity for carbohydrate metabolism and S-adenosylmethionine-dependent methyltransferase activity in hepatocytes between 72 to 96 hpf (Figure 4C). Moreover, multiple methyltransferase activities were enriched in hepatocytes from 72 to 96 hpf (groups f + g + c, Figure 4D) and from 60 to 96 hpf (groups c + f, Supplementary Figure S2).

3.5. Enrichment of KEGG Pathways and Hub Genes Associated with the Proliferation and Maturation of Hepatocytes

KEGG enrichment analysis was performed to reveal the functional characteristics of DEGs in hepatocytes of developing liver (Supplementary Figure S3; Supplementary Table S4). The distances between different signaling pathways were calculated by Jaccard coefficient according to the proportion of shared genes to obtain the signaling networks of DEGs in different groups. Most of genes specifically expressed in groups a and b from 60 hpf to 72 hpf were enriched in signaling pathways, such as cell cycle, RNA degradation, ubiquitin-mediated proteolysis, ErbB, Hedgehog, basal transcription factors, Wnt, and glycan degradation (groups a + b; Figure 5A), which are closely associated with cell proliferation or carbohydrate metabolism in hepatocytes between 60 to 72 hpf. The ErbB signaling pathway, ubiquitin mediated proteolysis, and cell cycle were the top hub pathways in the network of KEGG enrichment signaling pathways for groups a and b (Figure 5B; Table 2; Supplementary Table S5).

Table 2. Top 10 hub pathways of genes included in a and b ranked by MCC method.

Rank	Signaling Pathway	Score
1	Ubiquitin mediated proteolysis	4
1	Cell cycle	4
3	Homologous recombination	2
3	ErbB signaling pathway	2
3	mRNA surveillance pathway	2
3	Hedgehog signaling pathway	2
7	RNA degradation	1
7	Base excision repair	1
7	Fanconi anemia pathway	1
7	Nucleocytoplasmic transport	1

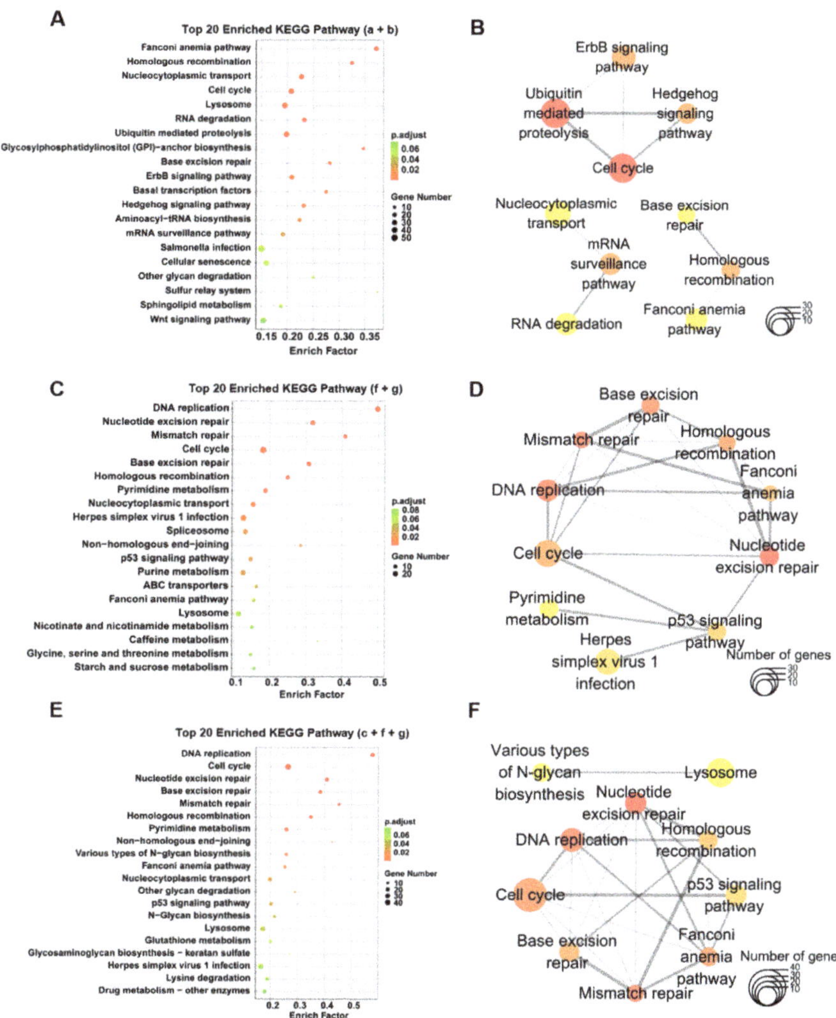

Figure 5. Hub signaling pathways from KEGG enrichment analysis of genes specifically expressed in hepatocytes at different stages. Dot plots of KEGG signaling pathways (**A**) and networks of top 10 hub pathways (**B**) for genes specifically expressed from 60 to 72 hpf (a and b). Dot plots of KEGG signaling pathways (**C**) and networks of top 10 hub pathways (**D**) for genes specifically expressed from 72 to 96 hpf (f and g). Dot plots of KEGG signaling pathways (**E**) and networks of top 10 hub pathways (**F**) for genes specifically expressed from 60 to 96 hpf (c, f and g). Node color stands for the enrichment p-value in the pathway.

In addition to signaling pathways for cell proliferation, newly enriched signaling pathways in hepatocytes between 72 to 96 hpf include metabolisms of pyrimidine, purine, nicotinate and nicotinamide, caffeine, glycine, serine and threonine, ABC transporters, and p53 signaling (groups f + g; Figure 5C), which function in metabolisms of lipid, protein and energy, cellular secretion, and detoxification, indicating the functional maturation of hepatocytes between 72 to 96 hpf. The top hub pathways include the cell cycle, DNA replication, pyrimidine metabolism, and p53 signaling (Figure 5D; Table 3; Supplementary Table S5). Similar signaling pathways were overrepresented in genes specifically expressed in groups c, f, and g (Figure 5E) and the top hub signaling pathways include various types of N-glycan biosynthesis, DNA replication and repair, and p53 signaling (Figure 5F; Table 4; Supplementary Table S5).

Table 3. Top 10 hub pathways of genes included in f and g ranked by MCC method.

Rank	Signaling Pathway	Score
1	Nucleotide excision repair	74
2	Mismatch repair	72
2	DNA replication	72
4	Base excision repair	49
5	Homologous recombination	48
6	Cell cycle	26
7	Fanconi anemia pathway	24
8	p53 signaling pathway	5
9	Herpes simplex virus 1 infection	3
10	Pyrimidine metabolism	2

Table 4. Top 10 hub pathways of genes included in c, f, and g ranked by MCC method.

Rank	Signaling Pathway	Score
1	Nucleotide excision repair	102
2	Mismatch repair	96
2	DNA replication	96
4	Cell cycle	54
4	Fanconi anemia pathway	54
6	Base excision repair	49
7	Homologous recombination	48
8	p53 signaling pathway	7
9	Lysosome	3
9	Various types of N-glycan biosynthesis	3

The hub genes of these KEGG pathways were examined with CytoHubba. In hepato-cytes from 60 to 72 hpf, the hub genes (cul3b, cbl, mgrn1a, mdm2, cdc23, cul1b, smurf2, anapc7fb, xw11b, and cdc16) in groups a and b were clustered into ErbB signaling pathway, cell cycle, hedgehog signaling pathway, and ubiquitin-mediated proteolysis (Figure 6A,B). In hepatocytes from 72 to 96 hpf, the hub genes (pole4, rfc5, rpa1, rfc3, lig1, pcna, pole3, rpa2, pole2, and pold1) in groups f and g were clustered into DNA replication, homologous recombination, mismatch repair, nucleotide excision repair, base excision repair and cell cycle (Figure 6C,D). Similar to those in groups f and g, the hub genes (pole2, rfc3, pole, pcna, pold1, pole3, rpa2, rfc5, lig1, and rpa1) in groups c, f, and g were clustered into nucleotide excision repair, cell cycle, DNA replication, base excision repair, mismatch repair, and fanconi anemia pathway (Figure 6E,F).

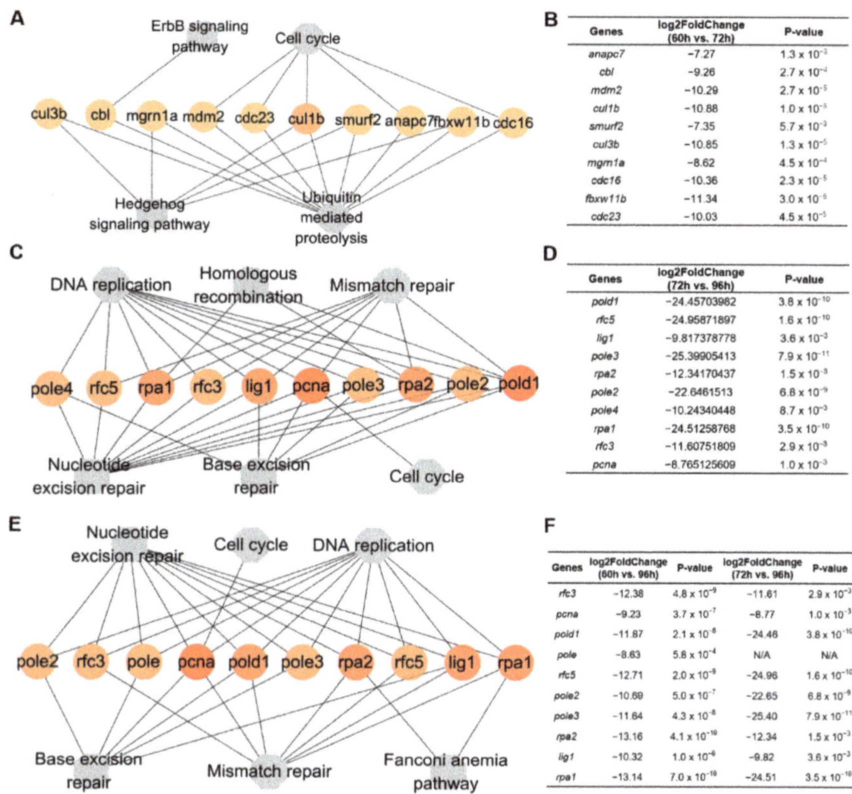

B

Genes	log2FoldChange (60h vs. 72h)	P-value
anapc7	−7.27	1.3×10^{-3}
cbl	−9.26	2.7×10^{-4}
mdm2	−10.29	2.7×10^{-5}
cul1b	−10.88	1.0×10^{-5}
smurf2	−7.35	5.7×10^{-3}
cul3b	−10.85	1.3×10^{-5}
mgm1a	−8.62	4.5×10^{-4}
cdc16	−10.36	2.3×10^{-5}
fbxw11b	−11.34	3.0×10^{-6}
cdc23	−10.03	4.5×10^{-5}

D

Genes	log2FoldChange (72h vs. 96h)	P-value
pold1	−24.45703982	3.8×10^{-10}
rfc5	−24.95871897	1.6×10^{-10}
lig1	−9.817378778	3.6×10^{-3}
pole3	−25.39905413	7.9×10^{-11}
rpa2	−12.34170437	1.5×10^{-3}
pole2	−22.6461513	6.8×10^{-9}
pole4	−10.24340448	8.7×10^{-3}
rpa1	−24.51258768	3.5×10^{-10}
rfc3	−11.60751809	2.9×10^{-3}
pcna	−8.765125609	1.0×10^{-3}

F

Genes	log2FoldChange (60h vs. 96h)	P-value	log2FoldChange (72h vs. 96h)	P-value
rfc3	−12.38	4.8×10^{-9}	−11.61	2.9×10^{-3}
pcna	−9.23	3.7×10^{-7}	−8.77	1.0×10^{-3}
pold1	−11.87	2.1×10^{-8}	−24.46	3.8×10^{-10}
pole	−8.63	5.8×10^{-4}	N/A	N/A
rfc5	−12.71	2.0×10^{-9}	−24.96	1.6×10^{-10}
pole2	−10.69	5.0×10^{-7}	−22.65	6.8×10^{-9}
pole3	−11.64	4.3×10^{-8}	−25.40	7.9×10^{-11}
rpa2	−13.16	4.1×10^{-10}	−12.34	1.5×10^{-3}
lig1	−10.32	1.0×10^{-6}	−9.82	3.6×10^{-3}
rpa1	−13.14	7.0×10^{-10}	−24.51	3.5×10^{-10}

Figure 6. Hub genes within KEGG pathways. (**A**) Networks of 10 hub genes mapped to top four pathways in groups a and b from 60 to 72 hpf. (**B**) The fold changes of hub genes in groups a and b. (**C**) Networks of 10 hub genes mapped to top 6 pathways from 72 to 96 hpf in groups f and g. (**D**) The fold changes of hub genes in groups f and g. (**E**) Networks of 10 hub genes mapped to top 6 pathways from 60 to 96 hpf in groups c, f, and g. (**F**) The fold changes of hub genes in groups c, f, and g.

3.6. Dynamic Changes of DEGs in Hepatocytes during Liver Development

The transcriptome data were normalized by z-score and analyzed with fuzzy c-means clustering to classify the dynamic trends of DEGs in hepatocytes at 60, 72, and 96 hpf during liver development of zebrafish. The 7255 DEGs from 3 time periods 60–72 hpf, 72–96 hpf, and 60–96 hpf in heptocytes were categorized into 9 distinct clusters of which each cluster exhibited distinct expression patterns (Figure 7A).

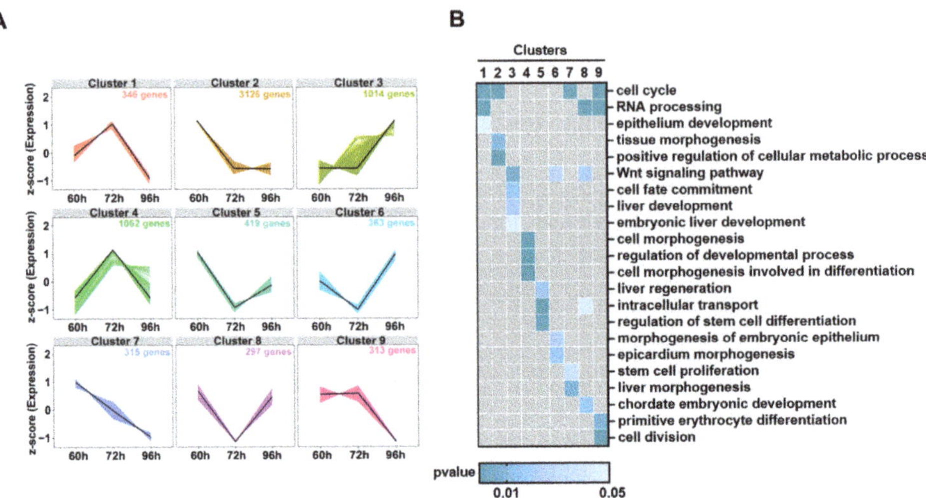

Figure 7. Dynamic changes of DEGs in hepatocytes from 60 to 96 hpf (**A**) Hierarchical clustering of changes in DEGs. (**B**) Functional annotations of different clusters by GO analysis. Heatmaps of biological processes were displayed according to their statistical significance ($p < 0.05$) and locations in the GO tree.

GO enrichment analysis revealed that genes of nine clusters were associated with distinct biological processes (Figure 7B; Supplementary Table S6). DEGs in clusters 1 ($n = 346$, 4.8%) and 4 ($n = 1062$, 19.4%) were upregulated at 72 hpf and then gradually downregulated at 96 hpf in comparison with those at 60 hpf. DEGs in cluster 1 were enriched in biological processes of cell cycle, RNA processing, and epithelium development while DEGs in cluster 4 were overrepresented in biological processes of cell morphogenesis, regulation of developmental process, and cell morphogenesis involved in differentiation.

DEGs in clusters 2 ($n = 3126$, 43.1%) were significantly downregulated at 72 hpf and maintained at a low expression level at 96 hpf when compared to those at 60 hpf. These DEGs are involed in cell cycle, tissue morphogenesis, and positive regulation of cellular metabolic process.

DEGs in cluster 3 ($n = 1014$, 14%) were upregulated at 96 hpf when compared with those at 60 and 72 hpf, which were associated with Wnt signaling pathway, cell fate commitment and liver development. DEGs in Clusters 5 ($n = 419$, 5.8%), 6 ($n = 363$, 5.0%) and 8 ($n = 297$, 4.1%) showed a "V" pattern of expression from 60 to 90 hpf, which were specifically associated with liver regeneration, intracellular transport, regulation of stem cell differentiation, Wnt signaling pathway, epicardium morphogenesis, and RNA processing. DEGs in Cluster 7 ($n = 315$, 4.3%) were continually downregulated from 60 to 96 hpf, which were highly enriched in cell cycle, stem cell proliferation and liver morphogenesis. DEGs in cluster 9 ($n = 313$, 4.3%) were almost unaltered from 60 to 72 hpf but downregulated from 72 to 96 hpf and these DEGs were associated with cell cycle, RNA processing, cell division, and primitive erythrocyte differentiation.

The process of cell cycle process appeared in four clusters (1, 2, 7 and 9), RNA processing in three clusters (1, 8, and 9), Wnt signaling pathway in three clusters (3, 6, and 8), and intracellular transport in two clusters (5 and 8), indicating that these biological processes are important in the proliferation and functional maturation of hepatocytes from 60 to 96 hpf.

4. Discussion

The liver is an essential organ in the body and performs a number of crucial activities, such as detoxification, metabolism, and homeostasis in vertebrates [1]. Liver diseases are becoming a worldwide problem that is threatening the health of humans [49]. Zebrafish (*Danio rerio*) are now commonly used in research on embryonic development, liver regeneration, and diseases [50,51]. In zebrafish, liver is an accessory organ of the foregut and liver morphogenesis can be divided into four phases, including the specification of hepatoblasts, the budding, differentiation, and outgrowth of hepatocytes [52,53]. The budding phase occurs at 24 hpf and ends at 50 hpf to form the hepatic duct. During the subsequent growth phase, the size, shape, and placement of liver began to extend across the midline ventral to esophagus and forms the architecture [54]. Studies in mammals indicated that liver development began with the appearance of liver buds, to the formation of liver progenitor cells, followed by the proliferation, differentiation, and migration of hepatic progenitor cells, and finally to the formation of liver, undergoing a complex process of cell signal regulation [12,21,53]. Furthermore, extrinsic signaling pathways and cell-autonomous transcription factors tightly regulate liver organogenesis [55]. Although the developmental patterns of liver in vertebrates are well established, biological processes and signaling pathways controlling the proliferation and maturation of hepatocytes remain largely unknown. In this study, we isolated hepatocytes from the development of embryos of *Tg(fabp10a:dsRed;ela3l:EGFP)* zebrafish at 60, 72, and 96 hpf and performed a comparative transcriptome analysis of these three hepatocyte populations. We identified a large number of DEGs, which are overrepresented in processes and signaling pathways associated with hepatocyte proliferation and function maturation.

A previous study with inflammation models CCl4 and partial hepatectomy has shown that HNF4, CAR, and Krüppel-like factors MafF and ELK1 were conserved as key regulators of hepatoblasts [56]. From 60 to 90 hpf, many GO terms were associated with the proliferation of hepatocytes, such as cell cycle, cellular response to stress, DNA replication, DNA repair and RNA processing in BP, nuclear protein-containing complex and nuclear lumen in CC, and catalytic activity on DNA and RNA binding in MF. In addition to cell proliferation, lipid biosynthetic process, ATP dimethylallyltransferase activity, and ADP dimethylallyltransferase activity were shared in hepatocytes between 60–72 hpf and 72–96 hpf, indicating that hepatocytes from 60 to 90 hpf are still proliferating and functions of lipid synthesis and energy metabolism are established in hepatocytes from 60 to 96 hpf.

The liver development process involved in many pathways, such as bone morphogenetic protein (BMP), transforming growth factor β (TGFβ), Wnt, and Hippo and Notch signaling pathways in mammals [21]. The Wnt signaling pathway tightly controls embryogenesis, including hepatobiliary development, maturation, and zonation, and it can increase glucose metabolism in hepatocellular carcinoma cells [12,57]. The Wnt signal inhibitor IWR-1 can also significantly influence the development of zebrafish liver, which leads to liver dysplasia [58]. In this study, KEGG enrichment analysis indicated that most genes specifically expressed in hepatocytes from 60 to 72 hpf were enriched in signaling pathways, such as cell cycle, RNA degradation, ubiquitin-mediated proteolysis, ErbB, Hedgehog, basal transcription factors, Wnt, and glycan degradation. The ErbB family of proteins consist of four protein kinases involved in multiple signaling pathways, such as cell proliferation, differentiation, and apoptosis. Overexpression of ErbB2 promotes breast cancer cells to grow rapidly [59]. Moreover, a previous study has revealed that smn1, gemin3, and gemin5 were linked to a common set of genetic pathways, such as ErbB and tp53 pathways, which can affect the regeneration of liver [60]. Therefore, the liver between 60 and 72 hpf continues to grow and hepatocytes have developed the function of carbohydrate metabolism.

Metabolisms of lipid, protein, and energy were found to be closely related to the establishment of hepatocyte functions, which can prevent the accumulation of lipid droplets and provide the nutrients required in this process [16,61,62]. Albumin and urea secretion, glycogen storage, and metabolic activity of cytochrome P450 enzymes represent functional features of mature hepatocytes [18,63]. Genome-wide characterization of ESC-derived hepatocyte-like cells indicated that some genes are associated with metabolic processes such as small molecule metabolic processes or secondary metabolic processes [64]. Some transcription factors, such as FOXA1/2/3, HNF4α, and CEBPA can maintain hepatocyte maturation through a combined action [65]. In this study, we found that, in addition to signaling pathways for cell proliferation and DNA replication, newly enriched signaling pathways in hepatocytes between 72 to 96 hpf include metabolisms of pyrimidine, purine, nicotinate and nicotinamide, caffeine, glycine, serine and threonine, ABC transporters, and p53 signaling, which are known to function in metabolisms of lipid, protein and energy, cellular secretion, and detoxification. Moreover, the genes specifically expressed in hepatocytes from 72 to 96 hpf were enriched in molecular functions of hydrolase activity on ester bonds, glucosidase activity for carbohydrate metabolism, and *S-adenosylmethionine dependent methyltransferase activity*. Thus, hepatocytes between 72 to 96 hpf are functionally matured.

To further understand the regulatory mechanisms of hepatocyte maturation, we classified DEGs into nine dynamic clusters by z-score standardization and fuzzy c-means clustering analysis. We found that several important pathways for embryonic development function during hepatocyte maturation from 60 to 96 hpf. For example, the Wnt signaling pathway is known to function in liver development [12] and DEGs, such as *wnt7bb*, *rspo3*, *wnt6b*, *tmem88a*, and *wnt2ba* are enriched in Cluster 3 ($n = 1014$, 14%), 6 ($n = 363$, 5.0%) and 8 ($n = 297$, 4.1%) in hepatocytes from 60 to 90 hpf. Meanwhile, tissue morphogenesis, liver morphogenesis, and liver regeneration were found in Clusters 2 ($n = 3126$, 43.1%) and Clusters 4 ($n = 1062$, 19.4%), in which *bmpr2b*, *gata6*, *bmper*, *smc2*, and *smc5* were overrepresented. It is known that Bmpr2b, a bone morphogenetic protein receptor that can mediate the BMP signaling pathway, plays an indispensable role in the developmental process of the liver [66]. However, functional mechanisms underlying most DEGs, biological processes and signaling pathways found in the study remain to be further investigated.

5. Conclusions

Comparative transcriptome analysis has uncovered a significant difference in hepatocytes between 60–72 hpf and 72–96 hpf in the numbers, types, and expression levels of transcripts. Hepatocytes from 60 to 90 hpf proliferate and establish the functions of lipid synthesis and energy metabolism. Hepatocytes between 60 to 72 hpf developed the function of carbohydrate metabolism. Hepatocytes between 72 to 96 hpf are functionally matured due to the establishment of functions in metabolisms of lipid, protein and energy, cellular secretion, and detoxification. These findings provide novel information to further understand the mechanisms controlling the proliferation and maturation of hepatocytes during liver development.

Supplementary Materials: The following supporting information can be downloaded at: http //www.mdpi.com/article/10.3390/biomedicines10092264/s1, Figure S1: Flow cytometric analysis and the sorting of hepatocytes from embryos at different time points; Figure S2: GO enrichment analysis of genes expressed during different time periods of liver development; Figure S3: Dot plot of KEGG enrichment analysis of genes specifically expressed in hepatocytes at different stages; Table S1: Results of differentially expressed genes in different groups by DEseq2; Table S2: VENN analysis of differentially expressed genes; Table S3: GO enrichment analysis of differentially expressed genes; Table S4: KEGG enrichment analysis of differentially expressed genes; Table S5: Jaccard similarity index of signaling pathways; Table S6: The results of cluster for changes in differentially expressed genes at all three time points.

Author Contributions: Conceptualization, Z.C. and Y.Z.; methodology, X.L. and Y.Z.; software, Y.Z. and X.L.; formal analysis, X.L. and Q.L.; investigation, Y.Z. and X.L.; resources, Q.L. and G.S.; data curation, Y.Z.; writing—original draft preparation, Y.Z.; writing—review and editing, X.L. and Z.C.; visualization, X.L. and G.S.; supervision, Z.C.; project administration, Z.C.; funding acquisition, Z.C. and H.Y. All authors have read and agreed to the published version of the manuscript.

Biomedicines **2022**, 10, 2264

Funding: This research was funded by the National Key R & D Program of China (2018YFA0800503), the Special Fund Project for Guangdong Academy of Sciences to Build Domestic First-class Research Institutions (2021GDASYL-20210102003), and the National Natural Science Foundation of China (31871463 and 31571504). The funders played no role in the design of the study; the collection, analysis, and interpretation of data; and preparation of the manuscript.

Institutional Review Board Statement: The animal study (Y813125501) was reviewed and approved by the Institutional Animal Care and Use Committee of Institute of Hydrobiology.

Informed Consent Statement: Not applicable.

Data Availability Statement: Data in this study are contained within the article and supplementary material. The sequencing data of this study was submitted to NCBI Sequence Read Archive under BioProject accession number PRJNA849172.

Acknowledgments: We thank Yan Wang, Zhixian Qiao, and Fang Zhou at the Analysis and Testing Center of Institute of Hydrobiology, Chinese Academy of Sciences for technical support.

Conflicts of Interest: The authors declare that they have no conflict of interest.

References

1. Wu, J.; Lu, C.; Ge, S.; Mei, J.; Li, X.; Guo, W. Igf2bp1 is required for hepatic outgrowth during early liver development in zebrafish. *Gene* **2020**, *744*, 144632. [CrossRef] [PubMed]
2. Ding, C.; Li, Y.; Guo, F.; Jiang, Y.; Ying, W.; Li, D.; Yang, D.; Xia, X.; Liu, W.; Zhao, Y.; et al. A Cell-type-resolved Liver Proteome. *Mol. Cell. Proteom.* **2016**, *15*, 3190–3202. [CrossRef] [PubMed]
3. Field, H.A.; Ober, E.A.; Roeser, T.; Stainier, D.Y. Formation of the digestive system in zebrafish. I. Liver morphogenesis. *Dev. Biol.* **2003**, *253*, 279–290. [CrossRef]
4. Chu, J.; Sadler, K.C. New school in liver development: Lessons from zebrafish. *Hepatology* **2009**, *50*, 1656–1663. [CrossRef]
5. Sheaffer, K.L.; Kaestner, K.H. Transcriptional networks in liver and intestinal development. *Cold Spring Harb. Perspect. Biol.* **2012**, *4*, a008284. [CrossRef]
6. Madakashira, B.P.; Zhang, C.; Macchi, F.; Magnani, E.; Sadler, K.C. Nuclear Organization during Hepatogenesis in Zebrafish Requires Uhrf1. *Genes* **2021**, *12*, 1081. [CrossRef]
7. Chang, C.; Hu, M.; Zhu, Z.; Lo, L.J.; Chen, J.; Peng, J. liver-enriched gene 1a and 1b encode novel secretory proteins essential for normal liver development in zebrafish. *PLoS ONE* **2011**, *6*, e22910. [CrossRef]
8. Gruppuso, P.A.; Sanders, J.A. Regulation of liver development: Implications for liver biology across the lifespan. *J. Mol. Endocrinol.* **2016**, *56*, R115–R125. [CrossRef]
9. Wu, Z.; Guan, K.L. Hippo Signaling in Embryogenesis and Development. *Trends Biochem. Sci.* **2021**, *46*, 51–63. [CrossRef]
10. Zaret, K.S. Regulatory phases of early liver development: Paradigms of organogenesis. *Nat. Rev. Genet.* **2002**, *3*, 499–512. [CrossRef]
11. Si-Tayeb, K.; Lemaigre, F.P.; Duncan, S.A. Organogenesis and development of the liver. *Dev. Cell* **2010**, *18*, 175–189. [CrossRef] [PubMed]
12. Perugorria, M.J.; Olaizola, P.; Labiano, I.; Esparza-Baquer, A.; Marzioni, M.; Marin, J.J.G.; Bujanda, L.; Banales, J.M. Wnt-beta-catenin signalling in liver development, health and disease. *Nat. Rev. Gastroenterol. Hepatol.* **2019**, *16*, 121–136. [CrossRef] [PubMed]
13. Tsai, S.M.; Liu, D.W.; Wang, W.P. Fibroblast growth factor (Fgf) signaling pathway regulates liver homeostasis in zebrafish. *Transgenic Res.* **2013**, *22*, 301–314. [CrossRef] [PubMed]
14. Kirchgeorg, L.; Felker, A.; van Oostrom, M.; Chiavacci, E.; Mosimann, C. Cre/lox-controlled spatiotemporal perturbation of FGF signaling in zebrafish. *Dev. Dyn.* **2018**, *247*, 1146–1159. [CrossRef]
15. Gordillo, M.; Evans, T.; Gouon-Evans, V. Orchestrating liver development. *Development* **2015**, *142*, 2094–2108. [CrossRef]
16. Trefts, E.; Gannon, M.; Wasserman, D.H. The liver. *Curr. Biol.* **2017**, *27*, R1147–R1151. [CrossRef]
17. Yang, L.; Wang, W.H.; Qiu, W.L.; Guo, Z.; Bi, E.; Xu, C.R. A single-cell transcriptomic analysis reveals precise pathways and regulatory mechanisms underlying hepatoblast differentiation. *Hepatology* **2017**, *66*, 1387–1401. [CrossRef]
18. Ge, C.; Ye, J.; Weber, C.; Sun, W.; Zhang, H.; Zhou, Y.; Cai, C.; Qian, G.; Capel, B. The histone demethylase KDM6B regulates temperature-dependent sex determination in a turtle species. *Science* **2018**, *360*, 645–648. [CrossRef]
19. Zorn, A.M. *Liver Development*; StemBook: Cambridge, MA, USA, 2008.
20. Yin, C.; Evason, K.J.; Asahina, K.; Stainier, D.Y. Hepatic stellate cells in liver development, regeneration, and cancer. *J. Clin. Investig.* **2013**, *123*, 1902–1910. [CrossRef]
21. Campbell, S.A.; Stephan, T.L.; Lotto, J.; Cullum, R.; Drissler, S.; Hoodless, P.A. Signalling pathways and transcriptional regulators orchestrating liver development and cancer. *Development* **2021**, *148*, dev199814. [CrossRef]
22. Shin, D.; Monga, S.P. Cellular and molecular basis of liver development. *Compr. Physiol.* **2013**, *3*, 799–815. [CrossRef] [PubMed]

23. Sadler, K.C.; Krahn, K.N.; Gaur, N.A.; Ukomadu, C. Liver growth in the embryo and during liver regeneration in zebrafish requires the cell cycle regulator, uhrf1. *Proc. Natl. Acad. Sci. USA* **2007**, *104*, 1570–1575. [CrossRef] [PubMed]
24. Macchi, F.; Sadler, K.C. Unraveling the Epigenetic Basis of Liver Development, Regeneration and Disease. *Trends Genet.* **2020**, *36*, 587–597. [CrossRef]
25. Sadler, K.C.; Amsterdam, A.; Soroka, C.; Boyer, J.; Hopkins, N. A genetic screen in zebrafish identifies the mutants vps18, nf2 and foie gras as models of liver disease. *Development* **2005**, *132*, 3561–3572. [CrossRef] [PubMed]
26. Niu, X.; Shi, H.; Peng, J. The role of mesodermal signals during liver organogenesis in zebrafish. *Sci. China Life Sci.* **2010**, *53*, 455–461. [CrossRef]
27. Huang, H.; Ruan, H.; Aw, M.Y.; Hussain, A.; Guo, L.; Gao, C.; Qian, F.; Leung, T.; Song, H.; Kimelman, D.; et al. Mypt1-mediated spatial positioning of Bmp2-producing cells is essential for liver organogenesis. *Development* **2008**, *135*, 3209–3218. [CrossRef]
28. Wang, S.; Miller, S.R.; Ober, E.A.; Sadler, K.C. Making It New Again: Insight into Liver Development, Regeneration, and Disease from Zebrafish Research. *Curr. Top. Dev. Biol.* **2017**, *124*, 161–195. [CrossRef]
29. Ober, E.A.; Verkade, H.; Field, H.A.; Stainier, D.Y. Mesodermal Wnt2b signalling positively regulates liver specification. *Nature* **2006**, *442*, 688–691. [CrossRef]
30. Cheng, Y.C.; Wu, T.S.; Huang, Y.T.; Chang, Y.; Yang, J.J.; Yu, F.Y.; Liu, B.H. Aflatoxin B1 interferes with embryonic liver development: Involvement of p53 signaling and apoptosis in zebrafish. *Toxicology* **2021**, *458*, 152844. [CrossRef]
31. Noel, E.S.; Reis, M.D.; Arain, Z.; Ober, E.A. Analysis of the Albumin/alpha-Fetoprotein/Afamin/Group specific component gene family in the context of zebrafish liver differentiation. *Gene Expr. Patterns* **2010**, *10*, 237–243. [CrossRef]
32. Morrison, J.K.; DeRossi, C.; Alter, I.L.; Nayar, S.; Giri, M.; Zhang, C.; Cho, J.H.; Chu, J. Single-cell transcriptomics reveals conserved cell identities and fibrogenic phenotypes in zebrafish and human liver. *Hepatol. Commun.* **2022**, *6*, 1711–1724. [CrossRef] [PubMed]
33. He, J.; Chen, J.; Wei, X.; Leng, H.; Mu, H.; Cai, P.; Luo, L. Mammalian Target of Rapamycin Complex 1 Signaling Is Required for the Dedifferentiation from Biliary Cell to Bipotential Progenitor Cell in Zebrafish Liver Regeneration. *Hepatology* **2019**, *70*, 2092–2106. [CrossRef] [PubMed]
34. Sun, Z.; Hopkins, N. vhnf1, the MODY5 and familial GCKD-associated gene, regulates regional specification of the zebrafish gut, pronephros, and hindbrain. *Genes Dev.* **2001**, *15*, 3217–3229. [CrossRef] [PubMed]
35. Matthews, R.P.; Lorent, K.; Russo, P.; Pack, M. The zebrafish onecut gene hnf-6 functions in an evolutionarily conserved genetic pathway that regulates vertebrate biliary development. *Dev. Biol.* **2004**, *274*, 245–259. [CrossRef] [PubMed]
36. Korzh, S.; Pan, X.; Garcia-Lecea, M.; Winata, C.L.; Pan, X.; Wohland, T.; Korzh, V.; Gong, Z. Requirement of vasculogenesis and blood circulation in late stages of liver growth in zebrafish. *BMC Dev. Biol.* **2008**, *8*, 84. [CrossRef]
37. Her, G.M.; Chiang, C.-C.; Chen, W.-Y.; Wu, J.-L. In vivo studies of liver-type fatty acid binding protein (L-FABP) gene expression in liver of transgenic zebrafish (Danio rerio). *FEBS Lett.* **2003**, *538*, 125–133. [CrossRef]
38. Li, X.; Song, G.; Zhao, Y.; Zhao, F.; Liu, C.; Liu, D.; Li, Q.; Cui, Z. Claudin7b is required for the formation and function of inner ear in zebrafish. *J. Cell Physiol.* **2018**, *233*, 3195–3206. [CrossRef]
39. Manoli, M.; Driever, W. Fluorescence-activated cell sorting (FACS) of fluorescently tagged cells from zebrafish larvae for RNA isolation. *Cold Spring Harb. Protoc.* **2012**, *2012*, pdb-prot069633. [CrossRef]
40. Teng, F.; Xu, Z.; Chen, J.; Zheng, G.; Zheng, G.; Lv, H.; Wang, Y.; Wang, L.; Cheng, X. DUSP1 induces apatinib resistance by activating the MAPK pathway in gastric cancer. *Oncol. Rep.* **2018**, *40*, 1203–1222. [CrossRef]
41. Asai, S.; Ianora, A.; Lauritano, C.; Lindeque, P.K.; Carotenuto, Y. High-quality RNA extraction from copepods for Next Generation Sequencing: A comparative study. *Mar. Genom.* **2015**, *24 Pt 1*, 115–118. [CrossRef]
42. Rio, D.C.; Ares, M., Jr.; Hannon, G.J.; Nilsen, T.W. Purification of RNA using TRIzol (TRI reagent). *Cold Spring Harb. Protoc.* **2010**, *2010*, pdb-prot5439. [CrossRef] [PubMed]
43. Schmieder, R.; Edwards, R. Quality control and preprocessing of metagenomic datasets. *Bioinformatics* **2011**, *27*, 863–864. [CrossRef] [PubMed]
44. Pertea, M.; Pertea, G.M.; Antonescu, C.M.; Chang, T.C.; Mendell, J.T.; Salzberg, S.L. StringTie enables improved reconstruction of a transcriptome from RNA-seq reads. *Nat. Biotechnol.* **2015**, *33*, 290–295. [CrossRef]
45. Pertea, M.; Kim, D.; Pertea, G.M.; Leek, J.T.; Salzberg, S.L. Transcript-level expression analysis of RNA-seq experiments with HISAT, StringTie and Ballgown. *Nat. Protoc.* **2016**, *11*, 1650–1667. [CrossRef]
46. Long, Y.; Yan, J.; Song, G.; Li, X.; Li, X.; Li, Q.; Cui, Z. Transcriptional events co-regulated by hypoxia and cold stresses in Zebrafish larvae. *BMC Genom.* **2015**, *16*, 385. [CrossRef] [PubMed]
47. Hu, Y.; Pan, Z.; Hu, Y.; Zhang, L.; Wang, J. Network and Pathway-Based Analyses of Genes Associated with Parkinson's Disease. *Mol. Neurobiol.* **2017**, *54*, 4452–4465. [CrossRef]
48. Ge, G.; Long, Y.; Song, G.; Li, Q.; Cui, Z.; Yan, H. Transcriptomic Profiling Revealed Signaling Pathways Associated with the Spawning of Female Zebrafish under Cold Stress. *Int. J. Mol. Sci.* **2022**, *23*, 7494. [CrossRef]
49. Asrani, S.K.; Devarbhavi, H.; Eaton, J.; Kamath, P.S. Burden of liver diseases in the world. *J. Hepatol.* **2019**, *70*, 151–171. [CrossRef]
50. Goessling, W.; Sadler, K.C. Zebrafish: An Important Tool for Liver Disease Research. *Gastroenterology* **2015**, *149*, 1361–1377. [CrossRef]
51. Wilkins, B.J.; Pack, M. Zebrafish models of human liver development and disease. *Compr. Physiol.* **2013**, *3*, 1213–1230. [CrossRef]
52. Li, X.; Song, G.; Zhao, Y.; Ren, J.; Li, Q.; Cui, Z. Functions of SMC2 in the Development of Zebrafish Liver. *Biomedicines* **2021**, *9*, 1240. [CrossRef] [PubMed]

3. Ober, E.A.; Lemaigre, F.P. Development of the liver: Insights into organ and tissue morphogenesis. *J. Hepatol.* **2018**, *68*, 1049–1062. [CrossRef] [PubMed]
4. Kotiyal, S.; Fulbright, A.; O'Brien, L.K.; Evason, K.J. Quantifying Liver Size in Larval Zebrafish Using Brightfield Microscopy. *J. Vis. Exp.* **2020**, *2*, e60744. [CrossRef] [PubMed]
5. Aiello, N.M.; Stanger, B.Z. Echoes of the embryo: Using the developmental biology toolkit to study cancer. *Dis. Model. Mech.* **2016**, *9*, 105–114. [CrossRef]
6. Godoy, P.; Widera, A.; Schmidt-Heck, W.; Campos, G.; Meyer, C.; Cadenas, C.; Reif, R.; Stober, R.; Hammad, S.; Putter, L.; et al. Gene network activity in cultivated primary hepatocytes is highly similar to diseased mammalian liver tissue. *Arch. Toxicol.* **2016**, *90*, 2513–2529. [CrossRef]
7. Fan, Q.; Yang, L.; Zhang, X.; Ma, Y.; Li, Y.; Dong, L.; Zong, Z.; Hua, X.; Su, D.; Li, H.; et al. Autophagy promotes metastasis and glycolysis by upregulating MCT1 expression and Wnt/beta-catenin signaling pathway activation in hepatocellular carcinoma cells. *J. Exp. Clin. Cancer Res.* **2018**, *37*, 9. [CrossRef]
8. Liu, Y.; Guo, J.; Zhang, J.; Deng, Y.; Xiong, G.; Fu, J.; Wei, L.; Lu, H. Chlorogenic acid alleviates thioacetamide-induced toxicity and promotes liver development in zebrafish (Danio rerio) through the Wnt signaling pathway. *Aquat. Toxicol.* **2022**, *242*, 106039. [CrossRef]
9. Pramanik, A.; Laha, D.; Dash, S.K.; Chattopadhyay, S.; Roy, S.; Das, D.K.; Pramanik, P.; Karmakar, P. An in-vivo study for targeted delivery of copper-organic complex to breast cancer using chitosan polymer nanoparticles. *Mater Sci. Eng. C* **2016**, *68*, 327–337. [CrossRef]
10. Pei, W.; Xu, L.; Chen, Z.; Slevin, C.C.; Pettie, K.P.; Wincovitch, S.; Program, N.C.S.; Burgess, S.M. A subset of SMN complex members have a specific role in tissue regeneration via ERBB pathway-mediated proliferation. *NPJ Regen. Med.* **2020**, *5*, 6. [CrossRef]
11. Nguyen, P.; Leray, V.; Diez, M.; Serisier, S.; Le Bloc'h, J.; Siliart, B.; Dumon, H. Liver lipid metabolism. *J. Anim. Physiol. Anim. Nutr.* **2008**, *92*, 272–283. [CrossRef]
12. Reinke, H.; Asher, G. Circadian Clock Control of Liver Metabolic Functions. *Gastroenterology* **2016**, *150*, 574–580. [CrossRef] [PubMed]
13. Pan, T.; Tao, J.; Chen, Y.; Zhang, J.; Getachew, A.; Zhuang, Y.; Wang, N.; Xu, Y.; Tan, S.; Fang, J.; et al. Robust expansion and functional maturation of human hepatoblasts by chemical strategy. *Stem Cell Res. Ther.* **2021**, *12*, 151. [CrossRef] [PubMed]
14. Godoy, P.; Schmidt-Heck, W.; Natarajan, K.; Lucendo-Villarin, B.; Szkolnicka, D.; Asplund, A.; Bjorquist, P.; Widera, A.; Stober, R.; Campos, G.; et al. Gene networks and transcription factor motifs defining the differentiation of stem cells into hepatocyte-like cells. *J. Hepatol.* **2015**, *63*, 934–942. [CrossRef] [PubMed]
15. Tachmatzidi, E.C.; Galanopoulou, O.; Talianidis, I. Transcription Control of Liver Development. *Cells* **2021**, *10*, 2026. [CrossRef] [PubMed]
16. Herrera, B.; Addante, A.; Sanchez, A. BMP Signalling at the Crossroad of Liver Fibrosis and Regeneration. *Int. J. Mol. Sci.* **2017**, *19*, 39. [CrossRef]

 *biomedicines* 

Article

Zebrafish: A Useful Animal Model for the Characterization of Drug-Loaded Polymeric NPs

Sara Bozzer [1,*], Luca De Maso [1], Maria Cristina Grimaldi [1], Sara Capolla [2], Michele Dal Bo [2], Giuseppe Toffoli [2] and Paolo Macor [1,*]

[1]   Department of Life Sciences, University of Trieste, 34127 Trieste, Italy
[2]   Experimental and Clinical Pharmacology Unit, Centro di Riferimento Oncologico di Aviano (CRO), IRCCS, 33081 Aviano, Italy
*   Correspondence: sara.bozzer@phd.units.it (S.B.); pmacor@units.it (P.M.)

Abstract: The use of zebrafish (ZF) embryos as an in vivo model is increasingly attractive thanks to different features that include easy handling, transparency, and the absence of adaptive immunity until 4–6 weeks. These factors allow the development of xenografts that can be easily analyzed through fluorescence techniques. In this work, ZF were exploited to characterize the efficiency of drug-loaded polymeric NPs as a therapeutical approach for B-cell malignancies. Fluorescent probes, fluorescent transgenic lines of ZF, or their combination allowed to deeply examine biodistribution, elimination, and therapeutic efficacy. In particular, the fluorescent signal of nanoparticles (NPs) was exploited to investigate the in vivo distribution, while the colocalization between the fluorescence in macrophages and NPs allows following the elimination pathway of these polymeric NPs. Xeno-transplanted human B-cells (Nalm-6) developed a reproducible model useful for demonstrating drug delivery by polymeric NPs loaded with doxorubicin and, as a consequence, the arrest of tumor growth and the reduction in tumor burden. ZF proved to be a versatile model, able to rapidly provide answers in the development of animal models and in the characterization of the activity and the efficacy of drug delivery systems.

Keywords: zebrafish; polymeric NPs; doxorubicin

Citation: Bozzer, S.; De Maso, L.; Grimaldi, M.C.; Capolla, S.; Dal Bo, M.; Toffoli, G.; Macor, P. Zebrafish: A Useful Animal Model for the Characterization of Drug-Loaded Polymeric NPs. *Biomedicines* **2022**, *10*, 2252. https://doi.org/10.3390/biomedicines10092252

Academic Editors: James A. Marrs and Swapnalee Sarmah

Received: 19 August 2022
Accepted: 9 September 2022
Published: 11 September 2022

Publisher's Note: MDPI stays neutral with regard to jurisdictional claims in published maps and institutional affiliations.

1. Introduction

Although most human pathologies have been modeled using mammalian systems such as mice, in recent years, attention has focused on the tropical freshwater fish Danio rerio (zebrafish, ZF) as an outstanding tool for studying human diseases [1]. ZF models are now a well-known option for implementing personalized medicine strategies, along with other models of patient-derived xenografts or patient-derived organoids [2–6]. ZF models are small and robust, cheap to maintain, and a single matching produces hundreds of eggs that develop extremely rapidly. An incomparable and unique feature is their optical transparency, which is important for easy visualization of the xenotransplanted cells or the biodistribution of the subject matter of research, or both simultaneously, with the aid of fluorescence microscope and with high throughput results [7].

Various injection sites were tested [8–10], but the yolk sac [8,11,12] has been shown to be an ideal approach for localized xenotransplantation in 2-day-old embryos, and in parallel, the Cuvier's duct is the best option for delivering substances into the embryo's bloodstream [13]. On this basis, ZF represents an innovative tool in the research landscape to study cancer diseases [14], including pediatric cancers such as pediatric B-cell malignancies. Although the treatment of pediatric B-cell malignancies can be considered a success story, with current overall survival (OS) of ~80% in the United States, the therapy-related side effects are still alarming [15]. For this reason, in the last years, researchers' interest has more frequently been focused on the development of a strategy that combines the knowledge about drugs with newborn nano-carriers for effective and selective drug delivery

In this context, polymeric NPs with well-defined size and shape, such as those synthesized using the polymers polylactide-*co*-glycolide acid (PLGA) and poly (vinyl alcohol) (PVA) [16] can improve the drug delivery process, thanks to the encapsulation of the drug that protects it until the nano-vector reaches the target through the enhancer permeability and retention (EPR) effect and releases its contents, leading to a reduction in the severe side effects associated with the use of chemotherapeutic agents [17,18]. In this context, ZF is an ideal candidate to rapidly evaluate xenograft tumor development, including the development of a B-cell malignancy model implanted in a large number of animals, and to investigate and potentially compare novel therapeutic approaches during their initial characterization [19].

In the present study, we propose a fluorescent-based quantification method for the setup of a B-cell malignancy model in ZF embryos using Nalm-6 cells, a B-acute lymphoblastic leukemia (ALL)-like cell line. The developed Nalm-6 cell line model in ZF was employed to investigate the capability of PLGA-PVA polymeric NPs to reach the tumor site, as well as the killing capability of doxorubicin-loaded PLGA-PVA polymeric NPs. We found that PLGA-PVA polymeric NPs distribute in the ZF bloodstream and reach the tumor, and doxorubicin-loaded PLGA-PVA polymeric NPs are capable of killing Nalm-6 cells, thus reducing tumor cell burden.

2. Materials and Methods

2.1. PLGA-PVA Polymeric NPs Synthesis

PLGA-PVA polymeric NPs were produced in our laboratory with small modifications to the Vasir and Labhasetwar protocol [16]. Firstly, the PLGA (Sigma, Saint Louis, MO, USA) solution was prepared by dissolving 30 mg of PLGA in 1 mL of chloroform (Sigma) in a 5 mL glass vial with magnetic stirring. During this process, the solution for the aqueous core of NPs was prepared, and fluorescein-5-isothiocyanate (FITC)-conjugated Bovine Serum Albumin (BSA, Sigma) or doxorubicin (Pfizer Inc. New York, NY, USA) was mixed in Tris-EDTA buffer. Finally, the PVA (Sigma) solution was set; 0.2 g of PVA was sprinkled slowly over 10 mL of cold Tris-EDTA buffer, centrifuged at $200 \times g$ for 10 min at 4 °C, and 10 μL of chloroform was then added to saturate the PVA solution.

The solution for the core was added to the PLGA solution in two aliquots of 100 μL each, vortexed for 1 min after each addition, and sonicated. This primary emulsion was added in two portions to 6 mL of PVA solution and vortexed for 1 min after each addition. The resulting emulsion was stirred overnight (at RT) to allow chloroform to evaporate. NPs were recovered by ultracentrifugation at $11,000 \times g$ for 20 min at 4 °C, and the pellet was resuspended in 5 mL of Tris-EDTA buffer. The sample was washed again and resuspended in H_2O MilliQ filtered 0.2 μm.

2.2. PLGA-PVA Polymeric NPs Characterization

NPs (5 μL) were diluted in 995 μL of H_2O MilliQ filtered 0.2 μm and then analyzed through Dynamic Light Scattering (DLS). Instead, for morphological analysis, NPs (10 μL) were diluted 1:1 *v/v* with H_2O MilliQ filtered 0.2 μm, and a drop of the sample was then deposited on a carbon screen coated with copper; after evaporation of the excess water, transmission electron microscopy (TEM) analysis was performed.

The NPs encapsulation efficiency was indirectly quantified by exploiting the intrinsic fluorescence of the FITC-BSA (maximum excitation/emission 495/521 nm) or doxorubicin (maximum excitation/emission 470/560 nm). The fluorescence signal corresponding to the unencapsulated compound was subtracted from that relating to the starting amount added. Then, an interpolation analysis with the FITC-BSA or doxorubicin standard curve was performed. The fluorescence signal was acquired with the ChemiDoc Imaging System (Bio-Rad, Hercules, CA, USA). The encapsulation efficiency was extrapolated by setting as 100% reference the fluorescent signal given by the starting amount of the compounds.

For NPs binding/internalization studies, 250,000 Nalm-6 cells were centrifuged at 400 rcf for 5 min and resuspended in 500 μL of culture medium (RPMI-1640, Sigma;

supplemented with 10% of Fetal Bovine Serum, Sigma; 100 U/mL Penicillin/0.1 mg/mL Streptomycin, Sigma; 1% L-Glutamine, Sigma). Nalm-6 cells were incubated at 37 °C with increasing amounts (1, 2, and 4 µL) of FITC -BSA (Sigma)-NPs under shaking (800 rpm). At the end of incubation, cells were washed twice in Phosphate-Buffered Saline (PBS) and resuspended with 1% Paraformaldehyde (Sigma) diluted in PBS supplemented with 2% of BSA (Sigma), 0.7 mM $CaCl_2$, and 0.7 mM $MgCl_2$. The binding/internalization on the surface of cells was evaluated by flow cytometric analysis performed by an Attune® NxT Acoustic Focusing flow cytometer (Thermo Fisher Scientific, Waltham, MA, USA), acquiring 10,000 events; data were analyzed with Attune NxT Software (version 2.7, Thermo Fisher Scientific, Waltham, MA, USA).

2.3. MTT Viability Assay

Nalm-6 cells (200,000/200 µL of culture medium, RPMI-1640, Sigma; supplemented with 10% of Fetal Bovine Serum, Sigma; 100 U/mL Penicillin/0.1 mg/mL Streptomycin, Sigma; 1% L-Glutamine, Sigma) were incubated for 24 h at 37 °C under shaking (800 rpm) with free drugs (1µM of doxorubicin, Pfizer) or NPs (1µM of encapsulated drug for doxorubicin-loaded NPs and the same amount for FITC-BSA-NPs). Then, cells were resuspended in 200 µL of clear culture medium. Later, 20 µL of MTT (3-(4,5-Dimethylthiazol-2-yl)-2,5-Diphenyltetrazolium Bromide, MTT, Sigma) was added, and samples were incubated for 4 h at 37 °C under shaking (800 rpm). Samples were then centrifuged for 3 min at 20,000 rcf. The supernatant was discarded, and the deposited crystals were solubilized in 200 µL of Dimethylsulfoxide (DMSO, Sigma). The optical density (OD) was measured at 570 nm with ELISA Reader TECAN Infinite M200. The percentage of viable cells was calculated using untreated cells as a reference for 100% viable cells.

2.4. Cell Labeling

Nalm-6 cells were labeled with CellTrace™ Calcein Red-Orange-AM (Thermo-Fisher Scientific, maximum excitation/emission 577/590 nm) according to the manufacturer's instructions. Afterward, for the CD19 antigen expression analysis, cells were incubated with the primary mouse anti-human CD19 antibody (Immunotools, Gladiolenweg, Friesoythe Germany, final concentration 5 ng/µL). The secondary Alexa 488-conjugated anti-mouse antibody (2 ng/µL, Invitrogen, Carlsbad, CA, USA) was used to reveal bound antibodies. The cell viability and the antigen expression were evaluated by an Attune® NxT Acoustic Focusing flow cytometer (Thermo Fisher Scientific), acquiring 10,000 events; data were analyzed with Attune NxT Software. The same analysis was performed by immunofluorescence; cells were cytocentrifuged on a slide, and nuclei were stained with 4′,6-diamidino-2-phenylindole (DAPI, Sigma). Slides and images were analyzed, respectively, using fluorescence microscope Nikon Eclipse Ti-E live system and Image-J software (version 2.3.0/1.53f, GNU General Public License, Bethesda, MD, USA).

2.5. In Vivo Studies

All experimental procedures involving animals were done after Ministerial Approval 04086.N.SGL.

Zebrafish eggs were placed in E3 Medium supplemented with methylene blue 0.5% and incubated at 28 °C, and 24 h after fertilization (hpf), the eggs were manually dechorionated. Embryos were then placed in E3 Medium supplemented with 1-phenyl 2-thiourea (PTU, Sigma, final concentration 0.2 mM) to inhibit the production of melanin.

2.5.1. Biodistribution Studies

NPs biodistribution studies were performed by injecting NPs (4.6 nL/embryo) in the duct of Cuvier of anesthetized embryos (tricaine, Sigma, final concentration 0.02% using capillary glasses and a Nanoject II Auto-Nanoliter Injector (Drummond Scientific Co., Broomall, PA, USA). The entire process was conducted using a SteREO Microscope Discovery.V8 (Zeiss, Oberkochen, Germany, UE). At 24 h post-injection (hpi), the biodistri bution of the NPs was evaluated using the fluorescence microscope Nikon Eclipse T*i*-E live system; then, images were analyzed with Image-J software.

2.5.2. Xenograft Model

Forty-eight hpf embryos were anesthetized using tricaine (Sigma) and placed on agarose plates, and the excess water was removed to facilitate injection. Then, ~500 cells/embryo were injected in a final volume of 4.6 nL (final concentration 0.1 cell/μL) using Nanoject II Auto-Nanoliter Injector (Drummond Scientific). The localized model was set up by injecting cells in the perivitelline space and the diffused ones by injecting cells in the duct of Cuvier. The entire process was conducted using a SteREO Microscope Discovery.V8 (Zeiss). After cell injection, the embryos were kept at 30 °C and evaluated using the fluorescence microscope Nikon Eclipse T*i*-E live system. Images were analyzed with Image-J software.

3. Results

3.1. NPs Synthesis and Characterization

In order to evaluate the capability of PLGA-PVA polymeric NPs to be internalized by cells of the leukemia cell line model Nalm-6, the first type of NPs was produced. FITC-BSA-NPs (BSA-NPs) consisted of two parts: the external shell and the aqueous core. The shell represents the outer layer material, and it was composed of PLGA (30 mg/mL) and PVA (2% *w/v*). Instead, the core represents the inner material, consisting of FITC-BSA (20 mg/mL in a water buffer). Firstly, BSA-NPs were morphologically characterized, showing round shapes, a diameter lower than 350 nm, and a negative surface charge. The average diameter, polydispersity index (PDI), and zeta potential values of these NPs are reported in Figure 1A.

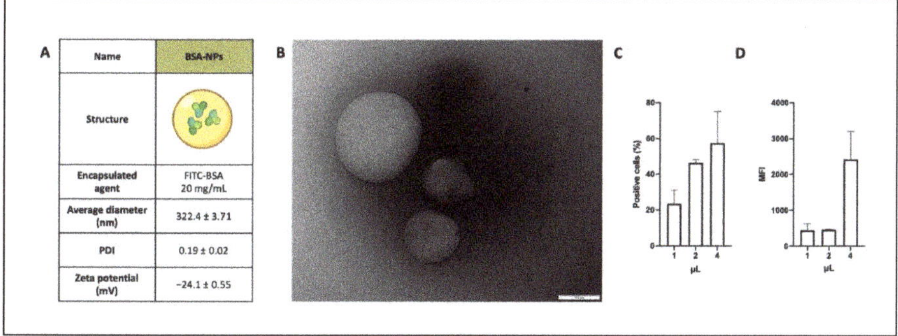

Figure 1. NPs' characterization: (**A**) Physicochemical characteristics of BSA-NPs, which are composed of PLGA-PVA polymers and filled with FITC-conjugated BSA. (**B**) TEM image of NPs (scale bar 100 nm). Nalm-6 cells were incubated with different amounts of NPs and analyzed by flow cytometry to evaluate their binding/internalization. (**C**) The percentage of positive cells and (**D**) Mean Fluorescence Intensity (MFI) are reported in the histograms. Data are reported as the mean ± SD.

In vitro tests were performed to assess NPs binding and internalization on malignant B-cells. Different amounts of BSA-NPs were incubated for 1 h with Nalm-6 cells and analyzed by flow cytometry. The binding/internalization of fluorescent BSA-NPs was evaluated by comparing the percentage of positive cells in Figure 1C, demonstrating a dose-related binding/internalization, which increased from 23.44% for 1 μL to 46.29% (2 μL) and to 57.46% for 4 μL. The Mean Fluorescence Intensity (MFI, Figure 1D) values of cells incubated with NPs were compared with the autofluorescence of cells. At 1 h of incubation, low amounts of NPs (1 and 2 μL) interact with cells five times less than the higher amount tested (438 and 459 vs. 2423).

3.2. NPs Biodistribution in Healthy ZF

To evaluate the biodistribution of BSA-NPs in ZF, embryos were manually dechorionated 24 h post-fertilization (hpf), and their pigmentation was inhibited using PTU, an
enzyme that can affect the conversion of tyrosine to melanin. BSA-NPs were micro-injected
into the duct of Cuvier in at least 20 animals per group 48 hpf and were analyzed by
fluorescence microscopy over the next three days. BSA-NPs were visualized by following
the FITC-BSA encapsulated inside their core (Figure 2).

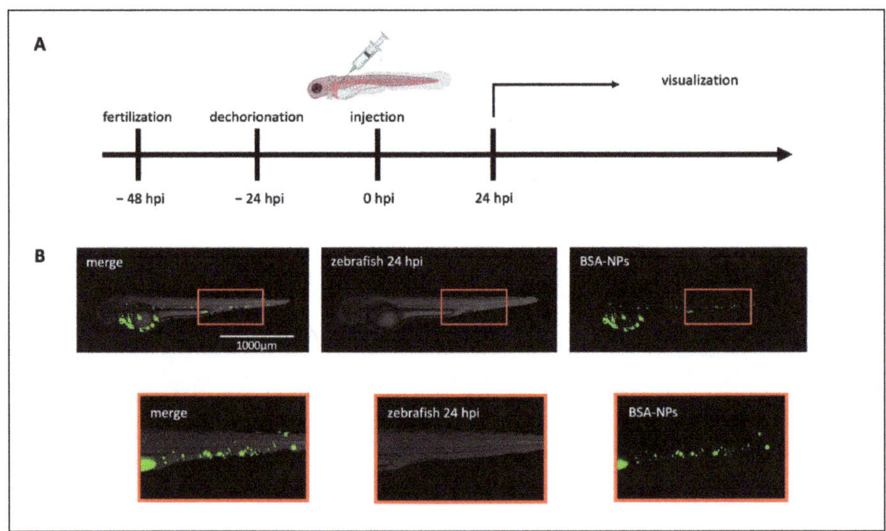

Figure 2. Zebrafish as an animal model for biodistribution studies: (**A**) Timeline of the experimental
procedure. ZF were followed for a total of 5 days. The reference point was set at the moment of
the injection (0 h post-injection, hpi), which was performed 48 h post-fertilization. Embryos were
manually dechorionated and incubated with PTU to prevent the formation of pigmented areas in
the body 24 h prior to the injection (−24 hpi). (**B**) BSA-NPs were injected in the duct of Cuvier and
analyzed 24 hpi demonstrating an accumulation of fluorescent BSA-NPs in the tail of ZF, known as
posterior blood island (PBI), which is a macrophage-rich area. Magnification 40×. Scale bar 1000 μm.

Twenty-four hours post-injection (hpi), NPs were already visible throughout the ZF
blood circulation in the entire bloodstream, and an accumulation of NPs was appreciable in
the tail of the fish, where the lower velocity of blood circulation and its flatness facilitate
visualization of circulating NPs. Moreover, this area is known as the posterior blood island
(PBI), and it is known as a macrophage-rich area [20].

3.3. Elimination of NPs

To evaluate the interaction between NPs and ZF's macrophages, the transgenic line
of ZF Tg(mpeg1:mCherry) was employed. This transgenic line possesses red-fluorescent
macrophages already visible in the PBI 24 hpf. Firstly, ZF were randomly assigned to
three different groups and then analyzed through fluorescence microscopy. A Region Of
Interest (ROI, Figure 3A) was drawn in the tail of the fish in correspondence with the PBI,
the red fluorescence in this area was quantified in each larvae, and the corrected total cell
fluorescence (CTCF) was calculated as follows:

$$CTCF = Integrated\ Density - [(Area\ of\ Selected\ Cells) \times (Mean\ Fluorescence\ of\ Background\ Readings)]$$

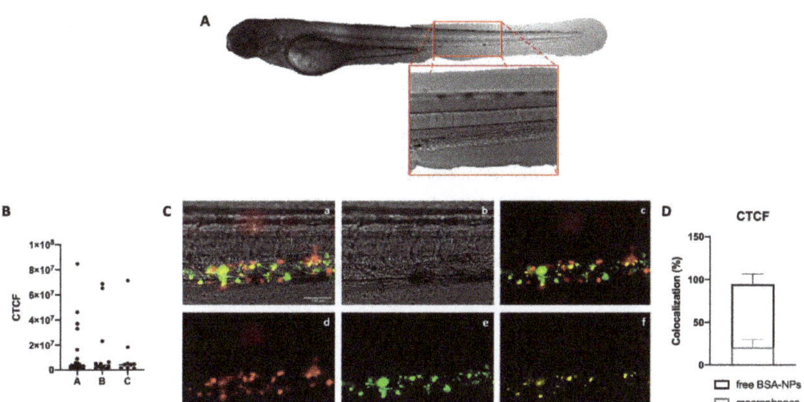

Figure 3. Zebrafish as a model for elimination studies: (**A**) Representation of the ROI chosen for the analysis of fluorescent signals in the Tg(mpeg1:mCherry) given by the presence of BSA-NPs. (**B**) Data analysis of the CTCF of the red-fluorescent areas (macrophages) of zebrafish randomly assigned to three different groups, suggesting that the fluorescent signal of macrophages is comparable between different animals. Data are reported as the median. (**C**) BSA-NPs were injected in the duct of Cuvier of transgenic line Tg(mpeg1:mCherry), and 24 hpi, ZF were analyzed through fluorescence microscopy. The upper left panel (**a**) represents the merge between macrophages (red, **d**), BSA-NPs (green, **e**), and the zebrafish's tail (gray, **b**). In the merge (**c**) between macrophages (red, **d**) and BSA-NPs (green, **e**), it is possible to observe yellow spots given by their colocalization, better appreciable isolating them (**f**), representing BSA-NPs engulfed by macrophages. (**D**) Data analysis of the CTCF of the colocalized areas. All data are reported as the mean ± SD. Magnification 200×. Scale bar 100 μm.

The data highlight that the three groups of ZF possess comparable amounts of red fluorescence and, as a consequence, comparable amounts of macrophages (Figure 3B). Therefore, 48 hpf, BSA-NPs were injected in the duct of Cuvier of the Tg(mpeg1:mCherry) transgenic line of ZF. As previously described, a ROI was drawn in the tail of the ZF, and fluorescent signals were analyzed. As shown in Figure 3C, in the selected ROI set in the PBI (panel b), the concomitant presence of macrophages (red fluorescence, panel d) and BSA-NPs (green fluorescence, panel e) was detected. The colocalization value was then compared to the fluorescent signal given by free BSA-NPs, highlighting that at 24 hpi, 20% of NPs were eliminated through macrophages, whereas 80% of them were free and still able to exploit their function (Figure 3D).

3.4. Setup of a Tumor-Bearing Model of ZF

A localized xenograft model using Nalm-6 cells was initially set up to test whether the tumor burden represented by the number of Nalm-6 cells could be easily determined from the fluorescent signal given by the fluorescent-labeled cells. In particular, B-cells were labeled with the Calcein-AM (Figure 4A), a cell-permeant dye used to determine cell viability in most eukaryotic cells, and their fluorescence was verified through flow cytometry (Figure 4B) and fluorescent microscopy (Figure 4C). Following the experimental timeline (Figure 4D), 48 hpf, ~500 Calcein-AM-labeled Nalm-6 cells were injected into the perivitelline space, resulting in the localization of tumor cells in the ventral thoracic area (Figure 4E).

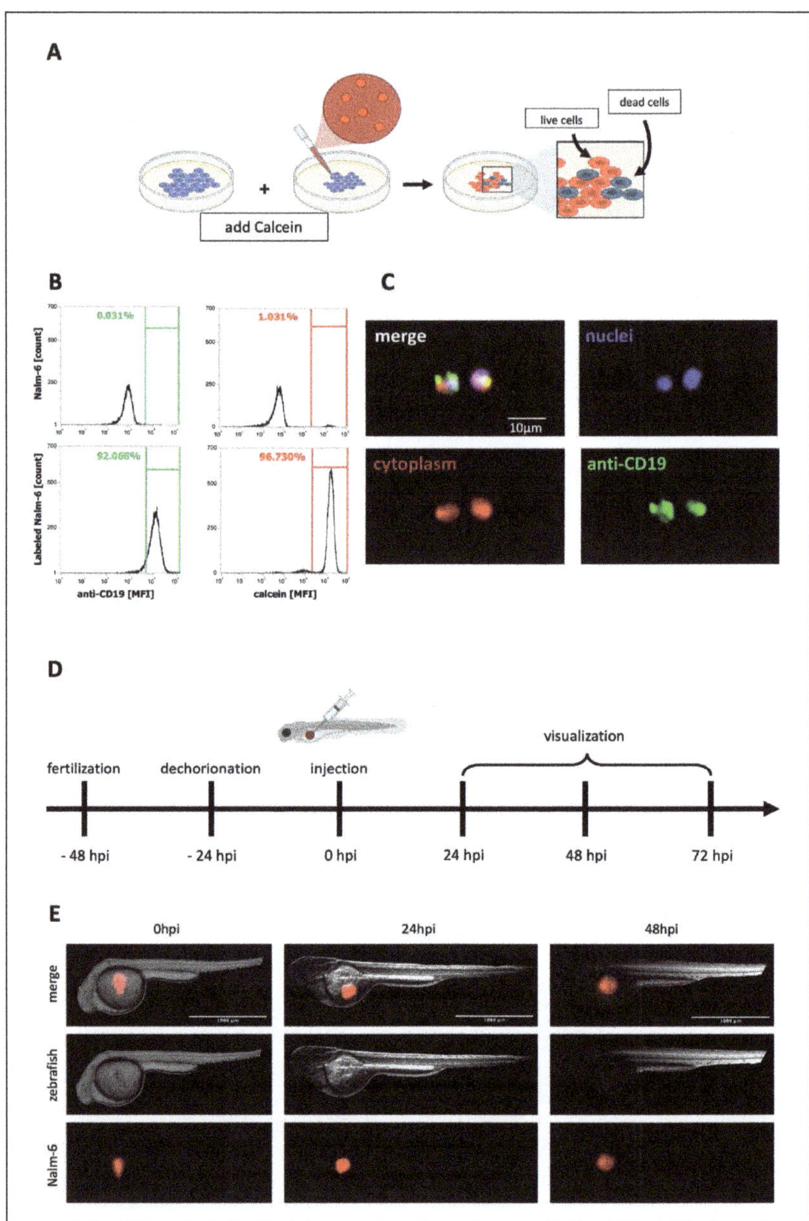

Figure 4. Localized xenograft zebrafish model: (**A**) Nalm-6 cells were labeled with Calcein-AM. After the internalization of Calcein-AM, the non-fluorescent Calcein-AM is converted in living cells to red-fluorescent Calcein-AM. Nalm-6 cells were also labeled with DAPI and anti-CD19 antibody (as described in materials and methods) and analyzed by (**B**) flow cytometric analysis and (**C**) immunofluorescence microscopy. Magnification 200×. Scale bar 10 µm. (**D**) Timeline of the experimental procedure. (**E**) Calcein-AM-labeled Nalm-6 cells were injected at 48 hpf in the perivitelline space of larvae. ZF were analyzed immediately after the injection (left panels), 24 hpi (central panels), and 48 hpi (right panels). Magnification 40×. Scale bar 1000 µm.

ZF were followed over time, and the red-fluorescent Calcein-AM emission in live cells was analyzed through fluorescence microscopy (Figure 4E) immediately after cells' injection (left panels), 24 hpi (central panels), and 48 hpi (right panels). The Nalm-6 tumor mass did not lose fluorescence over time, indicating that the cells were still alive. In addition, the Nalm-6 tumor mass appeared selectively localized in the flank of the ZF, over the yolk sac, and Nalm-6 cells did not appear to move through the ZF body.

3.5. Doxorubicin-Loaded NPs Synthesis and Characterization

PLGA-PVA NPs filled with doxorubicin (doxorubicin-NPs) were produced and consisted of two parts, as previously described for BSA-NPs: the external shell and the inner core, which was composed of doxorubicin (25 mg/mL in a water buffer). Doxorubicin-NPs were tested in vitro, showing round shapes, a diameter lower than 300 nm, and a small negative surface charge. The average diameter, polydispersity index (PDI), and zeta potential values of these types of NPs are reported in Figure 5A.

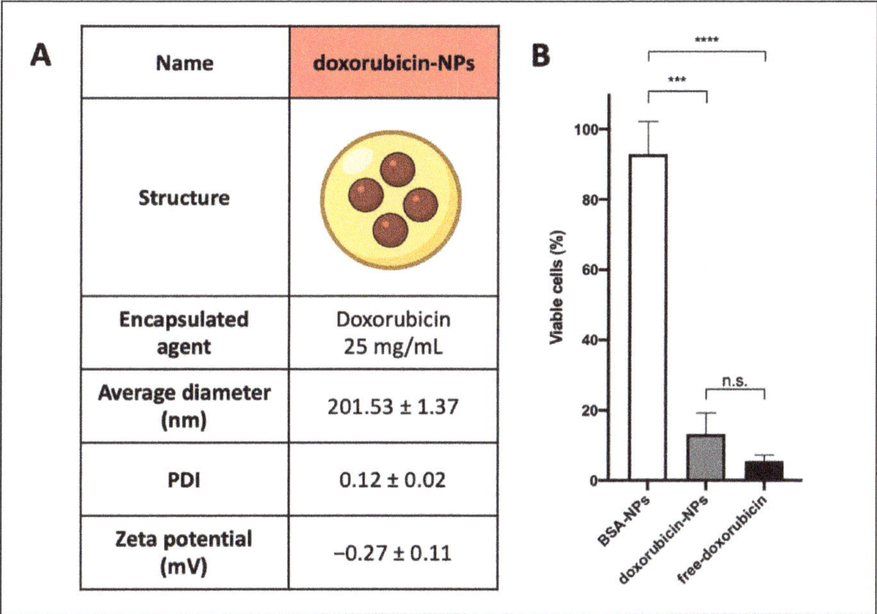

Figure 5. Drug-loaded NPs: (**A**) Physicochemical characteristics of doxorubicin-NPs, which are composed of PLGA-PVA polymers and filled with doxorubicin. (**B**) Nalm-6 cells were incubated with free doxorubicin, doxorubicin-NPs, or BSA-NPs for 24 h. Samples were then analyzed by evaluating MTT assay. Data are reported as the mean ± SD. $p \leq 0.001 = $ ***, $p \leq 0.0001 = $ ****, n.s. not significant.

To investigate the capability of drug-loaded NPs to kill Nalm-6 cells, they were tested in a viability assay using MTT evaluation (Figure 5B). Nalm-6 cells were incubated with doxorubicin-NPs (1 µM doxorubicin final concentration) and compared to cells incubated with the free drug (doxorubicin, 1 µM final concentration) or BSA-NPs. As evidenced in Figure 5B, the efficacy of doxorubicin-NPs (1 µM final concentration of the loaded drug) was comparable to that obtained with the free drug, and in parallel, it was demonstrated that this effect was obtained by the encapsulated drug and not due to polymers of the NPs' structure, because BSA-NPs did not affect cell viability.

3.6. Doxorubicin-Loaded NPs Arrest Tumor Growth

To evaluate the therapeutic efficacy of drug-loaded NPs and polymer toxicity, a ZF model of B-cell pathology was developed, with the specific aim to obtain a distributed model and not a localized one. As described in Figure 6A, ~500 Calcein-AM-labeled Nalm-6 cells were injected into the perivitelline space of ZF embryos at 48 hpf. BSA-NPs or doxorubicin-NPs (4,6 nL) were injected into the duct of Cuvier 4 h after the cells took root. ZF were analyzed immediately after the injection of NPs (Figure 6B, upper panels) and 24 hpi (Figure 6B, lower panels) through fluorescence microscopy.

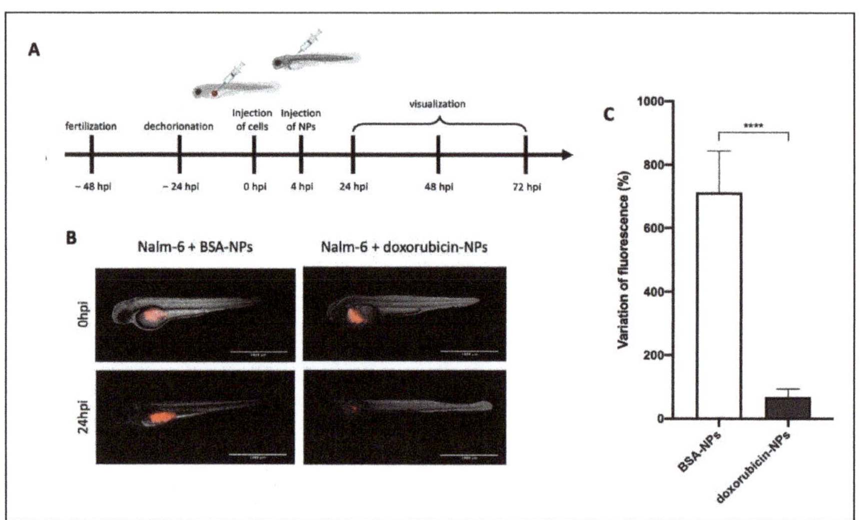

Figure 6. Evaluation of the therapeutic effect of doxorubicin-NPs: (**A**) Timeline of the experimental procedure. (**B**) Nalm-6 cells (labeled with red-fluorescent Calcein-AM) were injected into the perivitelline space of zebrafish embryos 48 hpf, and 4 h later, BSA-NPs (left panels) or doxorubicin-NPs (right panels) were injected in the duct of Cuvier. Zebrafish were analyzed immediately after the injection of NPs (upper panels) and 24 hpi (lower panels). Magnification 40×. Scale bar 1000 µm. (**C**) Data analysis of the variation of cells' fluorescence after treatment with NPs expressed as a percentage of untreated animals. All data are reported as the mean ± SD. $p \leq 0.0001$ = ****.

As described in Figure 6C, the CTCF of red-fluorescent Nalm-6 cells was calculated at 0 hpi and compared to the CTCF at 24 hpi. Treatment with BSA-NPs was not able to reduce the signal corresponding to the Nalm-6 tumor burden. In this case, the variation of the fluorescent signal given by NPs highlighted the safety of the polymers in vivo. On the other hand, treatment with doxorubicin-NPs was able to decrease the fluorescent signal corresponding to the Nalm-6 tumor burden ($p \leq 0.0001$, comparison between BSA-NPs and doxorubicin-NPs). This result demonstrated the ability of doxorubicin-NPs to arrest tumor growth in a localized model of B-cell malignancy.

4. Discussion

In this study, a fluorescent-based quantification method was developed to evaluate the number of tumor cell line cells in ZF models. In particular, a localized xenograft model of B-ALL in ZF was developed using Nalm-6 cells. In this context, a variety of features made ZF an excellent model organism. These include primarily economic advantages due to its small size requiring low space and maintenance costs [21]. Moreover, the genome is completely sequenced, shows a high level of similarity with humans (approximately 70%) and is easily manipulated [22]. This made it possible to create transgenic or mutant ZF lines to facilitate the observation of internal structures or biological processes. An example is the Casper mutants, which maintain the body transparency of the embryonic stage until adulthood. There are also reporter lines, such as tg (fli1a:eGFP) and mpeg1.1:mCherry which specifically respectively label endothelial cells and macrophages [23]. From the reproductive perspective, ZF represents an excellent experimental system since it can lay about 200 eggs per mating. Thus, having a large number of embryos guarantees the possibility of conducting large studies. There are also significant development considerations that exhibit optimal experimental properties [21].

Xenotransplantation represents a way to model tumor development and then study possible therapeutic approaches. ZF embryos represent a powerful model for cancer research, with a growing appreciation for their efficiency. Particularly, the transparent body wall and the absence of the immune response make ZF embryos optimal for xenograft as a tool to evaluate cancer progression and drug screening. Initial evidence of this was reported in 2005, when Lee et al. engrafted a human melanoma cell line in ZF, demonstrating the survival and migration of exogenous cells [23]. Over the years, ZF then acquired importance, joining the common murine models. Although the latter remain the "gold standard", limitations such as high costs, greater complexity, and the need for immunosuppression for xenograft make this system less flexible [24]. In contrast, there are advantages to the experimental practicality of ZF. Firstly, being transparent, the clear and simple observation of transplanted cells is guaranteed by employing microscopy techniques. Indeed, immediately after the injection, it is possible to observe the embryos under a microscope and follow them over time to evaluate any changes (e.g., tumor mass formation). Moreover, the use of transgenic lines and labeled transplanted cells help the study of the developmental process of the tumor [21]. All these observations demonstrate the interesting role of ZF xenotransplantation in studying tumor development and validating therapy efficacy. Starting from these considerations, ZF embryos can be used to test "next-generation" approaches such as NPs, which owe their success in drug delivery to the possibility of overcoming problems in cancer therapy-related, off-target side effects.

The treatment of pediatric cancers has been a success story, with current OS of ~80% in the United States. Nonetheless, this success has occurred at a significant price; with increased long-term cancer survivorship, there are also side effects. The most relevant is cardiovascular toxicity, which became apparent soon after the widespread use of anthracyclines in the 1970s. Several years after their discovery, these drugs continue to evoke considerable interest in basic science and clinical trial research [25–27]. In fact, anthracycline chemotherapy regimens play a prominent role in many cancer treatments, e.g., 50 to 60% of childhood cancer survivors are treated with an anthracycline regimen to the point that anthracyclines are listed among the World Health Organization (WHO) model list of essential medicines [28].

Despite their widely acknowledged efficacy, significant restrictions are associated with anthracycline treatment; the chemotherapy intensity has been raised to the limit of tolerance; therefore, there is a need for novel therapeutic approaches that are able to further improvement in outcomes and reduction in adverse effects [29].

NPs represent an alternative approach that is supposed to be more specific thanks to the possibility of treating the pathology through encapsulated chemotherapy that reaches the desired site, where they release the content.

The nanostructures developed in this study were made of PLGA-PVA polymers, which have been widely investigated to formulate biodegradable devices for the sustained delivery of drugs, proteins, and nucleic acids. Their biodegradability, biocompatibility, and safety profile are some of the main features that make these polymers optimal, even in vivo. PLGA is a copolymer consisting of two different monomer units, poly (glycolic acid) (PGA) and poly (lactic acid) (PLA), linked together, and the result is a linear, amorphous aliphatic polyester product. In vivo, the polymer undergoes degradation by hydrolysis with the following formation of the original monomers (i.e., lactic acid and glycolic acid), which are endogenous monomers also produced in normal physiological conditions. Thus, they are easily processed via metabolic pathways such as the Krebs cycle and removed as carbon dioxide and water, causing minimal systemic toxicity. Its success is particularly related to its continued drug release compared to conventional devices and to the negative charge, which is also crucial because it strongly influences the interaction between NPs and cells [30]. Another synthetic and biocompatible polymer extensively studied is PVA. PVA is frequently used as an emulsifier in the formulation of PLGA-PVA NPs due to its ability to form an interconnected structure with the PLGA, helping to achieve NPs that are relatively uniform and small [31,32]. In addition to the above-mentioned reasons, the simple and reproducible synthesis process and the possibility of surface functionalization (i.e., with targeting mechanisms) led us to focus on these nanodevices. Moreover, PLGA-PVA NPs keep water-soluble drugs/compounds trapped in the aqueous inner core, making these NPs an optimal delivery system.

PLGA-PVA NPs are spherical nano-sized core/shell structures that bind B-cells in a dose-dependent manner. When injected into the bloodstream of ZF embryos through the duct of Cuvier, which is an embryonic vein structure collecting all venous blood and leads directly to the heart's sinus venosus, NPs are broadly distributed in the embryo's body with a predilection for a region in the tail of the fish, known as the posterior blood island (PBI) This flat area is characterized by a reduced speed of the bloodstream, which facilitates the visualization of circulating NPs. Since the PBI is known to be a macrophage-rich area [20], the different accumulation of NPs is probably due to the macrophages' engulfment. To confirm this point, the transgenic ZF line Tg (mpeg1:mCherry) was previously exploited to visualize the interactions between macrophages and other immune cells or pathogens in vivo and to reexamine macrophage roles in inflammation, wound healing, and development, as well as their interactions with other cell types (e.g., vasculature, muscle) in vivo [33]; more interestingly, they represent a useful model in the study of the interaction between macrophages and NPs injected into the bloodstream. This approach allowed us to clarify that ~20% of PLGA-PVA NPs colocalize with the macrophages, indicating that this population of cells is implicated in their elimination, and this is probably due to the high surface-area-to-volume ratio. Moreover, when administered in vivo, NPs interact with circulating proteins (e.g. albumin, complement proteins, fibrinogen, and immunoglobulins) that are attracted to their surface to form a coating layer called protein corona; this affects the biological identity of NPs and, consequently, their functionality [34]. As a consequence, NPs became more visible to phagocytic cells, which recognize the materials via ligand–receptor interactions leading to the rapid elimination of NPs. On the other hand, the remaining ~80% of NPs remain free to execute their duty at the moment of the analysis.

In vivo experiments are strongly recommended to verify NPs' behavior in a complex environment; the optimal way to assess the efficacy of drug-loaded NPs is represented by observation of tumor growth arrest and a reduction in its burden. Therefore, the ZF embryo represents an interesting animal model considering that it embodies a versatile platform for xenotransplantation without risk of rejection. On the other hand, this feature also represents a limitation. Indeed, due to the lack of an adaptative immune system in this developmental stage, which is a difference from other, more complex animal models, such as mice and rats, ZF embryos do not allow us to evaluate the key role of adaptive immunity in the tumor microenvironment and cancer progression [7].

A diffused model of ALL is obviously a more realistic scenario; however, to better quantify the effect of drug-loaded NPs, a localized one is recommended. Therefore, a localized xenograft model was set up by injecting cells in the perivitelline space (at the margin between the yolk sac and the embryonic cell mass). The model clearly allowed the visualization of tumor cells localized in the ventral thoracic area and the verification of the tumor's dimensions through the fluorescent signal given by viable cells.

For what concerns the treatment of ALL, anthracyclines are a well-known class of chemotherapeutics that act mainly by intercalating DNA and interfering with DNA metabolism and RNA production. Two major dose-limiting toxicities of anthracyclines include myelosuppression and cardiotoxicity [26]. The PEGylated liposomal doxorubicin formulation "Doxil" was the first US Food and Drug Administration (FDA)-approved liposome chemotherapeutic agent in 1995. It has shown highly selective tumor localization and excellent pharmacokinetic properties in clinical applications. Starting from this point anthracyclines were chosen as candidates to be encapsulated inside nano-devices. Among anthracyclines, doxorubicin is widely used; however, its clinical use is restricted due to the severe risk to develop cardiotoxicity [26]. To specifically address this point, PLGA-PVA NPs loaded with doxorubicin were produced; their in vitro characterization demonstrated that the features of the NPs were not influenced by the presence of the drug in the core and that the cytotoxic activity of the drug was also maintained after its loading into the nano-devices.

The therapeutic efficacy of drug-loaded NPs was easily verified in a localized model of human leukemia in ZF embryos. It was possible to immediately visualize the local distribution of the cells and to measure the tumor burden. The potential therapeutic effect of drug-loaded NPs was obtained in only 24 h. Tumor-bearing ZF embryos were analyzed immediately after the injection of NPs and 24 hpi through fluorescence microscopy. The variation of the fluorescent signal highlights that the tumor burden in ZF that received drug-loaded NPs was reduced by more than 85% in comparison with ZF treated with NPs. Therefore, these data confirm the safety of the polymers in vivo and the therapeutic efficacy of drug-loaded NPs in a localized model of B-cell malignancy.

5. Conclusions

In this paper, we investigated ZF as a 3.0 animal model, resulting in a useful tool to study and characterize polymeric NPs. Our results highlighted that ZF can have a pivotal role in the study of the biodistribution, elimination, and therapeutic efficacy of doxorubicin-loaded polymeric NPs, which represent a promising tool for the treatment of B-cell malignancies.

More generally, ZF embryos can be suitable for the study of other nano-carriers to evaluate their toxicity, analyze their specificity, easily compare different therapeutic approaches, and choose the most effective. ZF represents a relevant model of B-cell malignancies, but the same approach can be transferred to the development of human–ZF models of other tumors, which could be useful in the development of specific drug-loaded polymeric NPs.

Author Contributions: Conceptualization, S.B., M.D.B. and P.M.; NP synthesis and characterization, L.D.M. and M.C.G.; data collection and assay development, S.B. and M.C.G.; data analysis and interpretation, S.B., M.C.G. and S.C.; writing—original draft preparation, S.B., M.D.B. and P.M.; writing—review and editing, S.B., L.D.M., M.C.G., M.D.B., G.T. and P.M.; supervision, P.M.; funding acquisition, G.T. and P.M. All authors have read and agreed to the published version of the manuscript.

Funding: This work was supported by the Italian Association for Cancer Research (AIRC), and by Fondo di Ricerca di Ateneo 2018-University of Trieste to P.M. and the Italian Ministry of Health (Ricerca Corrente) to G.T.

Institutional Review Board Statement: All experimental procedures involving animals were done according to the Italian Ministry of Health approval 04086.N.SGL.

Data Availability Statement: Data are contained within the article.

Conflicts of Interest: The authors declare no conflict of interest.

References

Goldsmith, J.R.; Jobin, C. Think small: Zebrafish as a model system of human pathology. *J. Biomed. Biotechnol.* **2012**, *2012*, 817341. [CrossRef] [PubMed]

Hu, Y.L.; Qi, W.; Han, F.; Shao, J.Z.; Gao, J.Q. Toxicity evaluation of biodegradable chitosan nanoparticles using a zebrafish embryo model. *Int. J. Nanomed.* **2011**, *6*, 3351–3359.

Jing, L.; Zon, L.I. Zebrafish as a model for normal and malignant hematopoiesis. *Dis. Model. Mech.* **2011**, *4*, 433–438. [CrossRef] [PubMed]

Ghotra, V.P.; He, S.; De Bont, H.; van der Ent, W.; Spaink, H.P.; van de Water, B.; Snaar-Jagalska, B.E.; Danen, E.H.J. Automated whole animal bio-imaging assay for human cancer dissemination. *PLoS ONE* **2012**, *7*, e31281. [CrossRef] [PubMed]

Rizeq, B.R.; Younes, N.N.; Rasool, K.; Nasrallah, G.K. Synthesis, Bioapplications, and Toxicity Evaluation of Chitosan-Based Nanoparticles. *Int. J. Mol. Sci.* **2019**, *20*, 5776. [CrossRef] [PubMed]

6. Di Franco, G.; Usai, A.; Funel, N.; Palmeri, M.; Montesanti, I.E.R.; Bianchini, M.; Gianardi, D.; Furbetta, N.; Guadagni, S.; Vasile, E.; et al. Use of zebrafish embryos as avatar of patients with pancreatic cancer: A new xenotransplantation model toward personalized medicine. *World J. Gastroenterol.* **2020**, *26*, 2792–2809. [CrossRef]

7. Bozzer, S.; Bo, M.D.; Toffoli, G.; Macor, P.; Capolla, S. Nanoparticles-Based Oligonucleotides Delivery in Cancer: Role of Zebrafish as Animal Model. *Pharmaceutics* **2021**, *13*, 1106. [CrossRef]

8. Haldi, M.; Ton, C.; Seng, W.L.; McGrath, P. Human melanoma cells transplanted into zebrafish proliferate, migrate, produce melanin, form masses and stimulate angiogenesis in zebrafish. *Angiogenesis* **2006**, *9*, 139–151. [CrossRef]

9. Pruvot, B.; Jacquel, A.; Droin, N.; Auberger, P.; Bouscary, D.; Tamburini, J.; Muller, M.; Fontenay, M.; Chluba, J.; Solary, E. Leukemic cell xenograft in zebrafish embryo for investigating drug efficacy. *Haematologica* **2011**, *96*, 612–616. [CrossRef]

10. He, S.; Lamers, G.E.; Beenakker, J.W.M.; Cui, C.; Ghotra, V.P.; Danen, E.H.; Meijer, A.H.; Spaink, H.P.; Snaar-Jagalska, B.E. Neutrophil-mediated experimental metastasis is enhanced by VEGFR inhibition in a zebrafish xenograft model. *J. Pathol.* **2012**, *227*, 431–445. [CrossRef]

11. Harfouche, R.; Basu, S.; Soni, S.; Hentschel, D.M.; Mashelkar, R.A.; Sengupta, S. Nanoparticle-mediated targeting of phosphatidylinositol-3-kinase signaling inhibits angiogenesis. *Angiogenesis* **2009**, *12*, 325–338. [CrossRef] [PubMed]

12. Zhao, H.; Tang, C.; Cui, K.; Ang, B.T.; Wong, S.T. A screening platform for glioma growth and invasion using bioluminescence imaging. *J. Neurosurg.* **2009**, *111*, 238–246. [CrossRef]

13. Letrado, P.; de Miguel, I.; Lamberto, I.; Díez-Martínez, R.; Oyarzabal, J. Zebrafish: Speeding up the Cancer Drug Discovery Process. *Cancer Res.* **2018**, *78*, 6048–6058. [CrossRef]

14. Siegel, R.L.; Miller, K.D.; Fuchs, H.E.; Jemal, A. Cancer Statistics, 2021. *CA Cancer J. Clin.* **2021**, *71*, 7–33. [CrossRef] [PubMed]

15. McEachron, T.A.; Helman, L.J. Recent Advances in Pediatric Cancer Research. *Cancer Res.* **2021**, *81*, 5783–5799. [CrossRef]

16. Vasir, J.K.; Labhasetwar, V. Preparation of biodegradable nanoparticles and their use in transfection. *Cold Spring Harb. Protoc.* **2008**, *2008*, pdb-prot4888. [CrossRef] [PubMed]

17. Maeda, H.; Nakamura, H.; Fang, J. The EPR effect for macromolecular drug delivery to solid tumors: Improvement of tumor uptake, lowering of systemic toxicity, and distinct tumor imaging in vivo. *Adv. Drug Deliv. Rev.* **2013**, *65*, 71–79. [CrossRef] [PubMed]

18. Danhier, F. To exploit the tumor microenvironment: Since the EPR effect fails in the clinic, what is the future of nanomedicine? *J. Control. Release* **2016**, *244 Pt A*, 108–121. [CrossRef]

19. Cabezas-Sainz, P.; Pensado-López, A.; Sáinz, J.B.; Sánchez, L. Modeling Cancer Using Zebrafish Xenografts: Drawbacks for Mimicking the Human Microenvironment. *Cells* **2020**, *9*, 1978. [CrossRef]

20. Warga, R.M.; Kane, D.A.; Ho, R.K. Fate mapping embryonic blood in zebrafish: Multi- and unipotential lineages are segregated at gastrulation. *Dev. Cell* **2009**, *16*, 744–755. [CrossRef]

21. Khan, F.R.; Alhewairini, S.S. Zebrafish (*Danio rerio*) as a Model Organism. In *Current Trends in Cancer Management*; IntechOpen: London, UK, 2019.

22. Rothenbucher, T.S.P.; Ledin, J.; Gibbs, D.; Engqvist, H.; Persson, C.; Hulsart-Billström, G. Zebrafish embryo as a replacement model for initial biocompatibility studies of biomaterials and drug delivery systems. *Acta Biomater.* **2019**, *100*, 235–243. [CrossRef] [PubMed]

23. Zhang, B.; Xuan, C.; Ji, Y.; Zhang, W.; Wang, D. Zebrafish xenotransplantation as a tool for in vivo cancer study. *Fam. Cancer* **2015**, *14*, 487–493. [CrossRef]

24. Veinotte, C.J.; Dellaire, G.; Berman, J.N. Hooking the big one: The potential of zebrafish xenotransplantation to reform cancer drug screening in the genomic era. *Dis. Model. Mech.* **2014**, *7*, 745–754. [CrossRef]

25. Bhagat, A.; Kleinerman, E.S. Anthracycline-Induced Cardiotoxicity: Causes, Mechanisms, and Prevention. *Adv. Exp. Med. Biol.* **2020**, *1257*, 181–192. [PubMed]

26. Cardinale, D.; Iacopo, F.; Cipolla, C.M. Cardiotoxicity of Anthracyclines. *Front. Cardiovasc. Med.* **2020**, *7*, 26. [CrossRef] [PubMed]

27. Narezkina, A.; Narayan, H.K.; Zemljic-Harpf, A.E. Molecular mechanisms of anthracycline cardiovascular toxicity. *Clin. Sci.* **2021**, *135*, 1311–1332. [CrossRef]

28. McGowan, J.V.; Chung, R.; Maulik, A.; Piotrowska, I.; Walker, J.M.; Yellon, D.M. Anthracycline Chemotherapy and Cardiotoxicity. *Cardiovasc. Drugs Ther.* **2017**, *31*, 63–75. [CrossRef]

29. Inaba, H.; Mullighan, C.G. Pediatric acute lymphoblastic leukemia. *Haematologica* **2020**, *105*, 2524–2539. [CrossRef]

30. Mirakabad, F.S.T.; Nejati-Koshki, K.; Akbarzadeh, A.; Yamchi, M.R.; Milani, M.; Zarghami, N.; Zeighamian, V.; Rahimzadeh, A.; Alimohammadi, S.; Hanifehpour, Y.; et al. PLGA-based nanoparticles as cancer drug delivery systems. *Asian Pac. J. Cancer Prev.* **2014**, *15*, 517–535. [CrossRef]

31. Sahoo, S.K.; Panyam, J.; Prabha, S.; Labhasetwar, V. Residual polyvinyl alcohol associated with poly (d,l-lactide-co-glycolide) nanoparticles affects their physical properties and cellular uptake. *J. Control. Release* **2002**, *82*, 105–114. [CrossRef]

32. Danhier, F.; Ansorena, E.; Silva, J.M.; Coco, R.; Le Breton, A.; Préat, V. PLGA-based nanoparticles: An overview of biomedical applications. *J. Control. Release* **2012**, *161*, 505–522. [CrossRef] [PubMed]

33. Ellett, F.; Pase, L.; Hayman, J.W.; Andrianopoulos, A.; Lieschke, G.J. mpeg1 promoter transgenes direct macrophage-lineage expression in zebrafish. *Blood* **2011**, *117*, e49–e56. [CrossRef] [PubMed]

34. Zeng, L.; Gao, J.; Liu, Y.; Gao, J.; Yao, L.; Yang, X.; Liu, X.; He, B.; Hu, L.; Shi, J.; et al. Role of protein corona in the biological effect of nanomaterials: Investigating methods. *TrAC Trends Anal. Chem.* **2019**, *118*, 303–314. [CrossRef]

Article

Effect of Lipopolysaccharides on Liver Tumor Metastasis of *twist1a*/*kras*^{V12} Double Transgenic Zebrafish

Jeng-Wei Lu [1,2,*], Liang-In Lin [2,3], Yuxi Sun [1,4], Dong Liu [4] and Zhiyuan Gong [1,*]

1 Department of Biological Sciences, National University of Singapore, Singapore 117543, Singapore; e0437708@u.nus.edu
2 Department of Clinical Laboratory Sciences and Medical Biotechnology, National Taiwan University, Taipei 10048, Taiwan; lilin@ntu.edu.tw
3 Department of Laboratory Medicine, National Taiwan University Hospital, Taipei 10048, Taiwan
4 Brain Research Center, School of Life Sciences, Southern University of Science and Technology, Shenzhen 518055, China; liud@sustech.edu.cn
* Correspondence: jengweilu@gmail.com (J.-W.L.); dbsgzy@nus.edu.sg (Z.G.); Tel.: +65-6516-2860 (Z.G.)

Abstract: The poor prognosis of patients diagnosed with hepatocellular carcinoma (HCC) is directly associated with the multi-step process of tumor metastasis. *TWIST1*, a basic helix-loop-helix (bHLH) transcription factor, is the most important epithelial-mesenchymal transition (EMT) gene involved in embryonic development, tumor progression, and metastasis. However, the role that *TWIST1* gene plays in the process of liver tumor metastasis in vivo is still not well understood. Zebrafish can serve as a powerful model for cancer research. Thus, in this study, we crossed *twist1a+* and *kras+* transgenic zebrafish, which, respectively, express hepatocyte-specific mCherry and enhanced green fluorescent protein (EGFP); they also drive overexpression of their respective transcription factors. This was found to exacerbate the development of metastatic HCC. Fluorescence of mCherry and EGFP-labeled hepatocytes revealed that approximately 37.5% to 45.5% of the *twist1a+*/*kras+* double transgenic zebrafish exhibited spontaneous tumor metastasis from the liver to the abdomen and tail areas, respectively. We also investigated the inflammatory effects of lipopolysaccharides (LPS) on the hepatocyte-specific co-expression of *twist1a+* and *kras+* in double transgenic zebrafish. Following LPS exposure, co-expression of *twist1a+* and *kras+* was found to increase tumor metastasis by 57.8%, likely due to crosstalk with the EMT pathway. Our results confirm that *twist1a* and *kras* are important mediators in the development of metastatic HCC. Taken together, our in-vivo model demonstrated that co-expression of *twist1a+*/*kras+* in conjunction with exposure to LPS enhanced metastatic HCC offers a useful platform for the study of tumor initiation and metastasis in liver cancer.

Keywords: liver tumor metastasis; lipopolysaccharides; *twist1a*; *kras*^{V12}; transgenic zebrafish

Citation: Lu, J.-W.; Lin, L.-I.; Sun, Y.; Liu, D.; Gong, Z. Effect of Lipopolysaccharides on Liver Tumor Metastasis of *twist1a/kras*^{V12} Double Transgenic Zebrafish. *Biomedicines* **2022**, *10*, 95. https://doi.org/10.3390/biomedicines10010095

Academic Editors: James A. Marrs and Swapnalee Sarmah

Received: 10 November 2021
Accepted: 30 December 2021
Published: 2 January 2022

Publisher's Note: MDPI stays neutral with regard to jurisdictional claims in published maps and institutional affiliations.

1. Introduction

Hepatocellular carcinoma (HCC) is the fifth most common cancer in the world and a major threat to human health [1–3]. Despite substantial progress in the treatment of HCC in recent years, 600,000 people still die from this disease annually, making it the third leading cause of cancer-related deaths worldwide [4,5]. Liver resection and transplantation are the most common HCC treatment methods; however, high recurrence and metastasis rates still lead to poor prognosis in HCC patients [6]. Therefore, early diagnosis and treatment of HCC are critical. A small number of genes have been linked to the occurrence and metastasis of HCC; however, elucidating the mechanisms which underlie liver tumor metastasis remains a pressing issue [7].

The significance of epithelial-mesenchymal transition (EMT) in the process of cancer metastasis has been explored previously. A large number of studies have found that EMT plays a key role in tumor invasion and metastasis, and *TWIST1* has been identified as an important regulating factor in the EMT process. *TWIST1* is a basic helix-loop-helix (bHLH) transcription factor and is also one of the most important EMT genes involved in embryonic development, tumor progression, and metastasis [8,9]. The fact that EMT has been shown to regulate various biological processes in tumors, including drug resistance, demonstrates its complexity [10]. *TWIST1* negatively regulates *E-cadherin* and positively regulates *Vimentin*. Both of these proteins are important to EMT induction. Dysregulation of *TWIST1* expression is associated with the *E-cadherin*-mediated loss of intercellular adhesion, activation of mesenchymal markers, and induction of cell motility [11]. Nonetheless, many aspects of EMT remain unclear and require further study. In particular, research is required to identify and understand EMT-related genes.

Abnormal expression of *TWIST1* has been frequently observed in many types of cancers. The upregulation of *TWIST1* in HCC cell lines promotes the proliferation, cell migration, invasion, and metastasis of cancer cells [7,12–14]. Overexpression of *TWIST1* is also associated with shorter overall survival in HCC patients [15]. Research into *TWIST1* has broad applications and potential therapeutic value for HCC. However, the relationship between *TWIST1* and the proto-oncogene *K-RAS*, which is a member of the RAS protein family and is mutated in a high percentage of human liver cancers, in HCC is unclear.

RAS proteins are a family of small molecular switches regulated by guanosine triphosphate, which can transmit signals from the cell membrane to the nucleus and activate a variety of signaling pathways involved in cell proliferation, transformation, and tumor progression. RAS family proteins include H-RAS, N-RAS, and K-RAS [16–18]. Many single-point mutations in *RAS* genes result in the constitutive activation of *RAS* with impaired GTPase activity, which leads to continuous stimulation of cell proliferation. The frequency of these gene mutations varies in different tumor types. In total, approximately 30% of human tumors have *RAS* gene mutations, and these mutations most commonly occur in the *K-RAS* gene [17]. For example, *K-RAS* mutations have been identified in 77% of human liver cancers, which is higher than the incidence of mutations in *H-RAS* and *N-RAS* in these cancers. The activation of RAS protein signals, which leads to the proliferation and transformation of hepatocytes, has also been observed in human HCC specimens [16,18,19].

Many microenvironmental inflammatory factors have been identified as potential therapeutic targets for HCC [20]. Aspirin, a non-steroidal anti-inflammatory drug, has been shown to reduce the risk of HCC and improve survival [21]. In addition, upregulation of Toll-like receptor (TLR) signaling, which is associated with inflammation-related cancers, has been found to play a key role in the prognosis of chronic and inflammatory diseases that lead to HCC [22]. Lipopolysaccharides (LPS), which are large molecules composed of lipids and polysaccharides that exist in the outer membrane of gram-negative bacteria function by binding to toll-like receptor 4 (TLR4). In HCC, the cooperation of TLR4 and toll-like receptor 9 (TLR9) may activate the signal transducer and activator of transcription 3 (STAT3) [23–26]. Exposure to LPS leads to tumor growth and angiogenesis in HCC via the TLR4 receptor in vivo. The signaling which occurs following induction by LPS also promotes EMT in HCC [25,27,28].

The occurrence of tumors can be clearly divided into three independent stages: tumor initiation, progression, and metastasis [29,30]. In previous research, we characterized novel roles of *twist1a* and *xmrk* (an activated epidermal growth factor receptor (EGFR) homolog) in tumorigenesis and metastasis and proposed a new animal model for screening anti-tumor metastasis drugs [31–33]. However, no reports pertaining to the use of animal models in the study of how *TWIST1* and *K-RAS* affect the initiation and maintenance of liver tumorigenesis have been published.

The use of zebrafish in the study of liver disease and HCC has recently become more widespread [34]. Thus, in the current study, we first investigated the potential relevance of *twist1a* and *kras* in liver tumors using a zebrafish model. We also explored the in-vivo mechanism which underlies the effects of LPS on liver tumors in *kras* or *twist1a*/*kras* transgenic zebrafish. Specifically, we were interested in how LPS promotes tumor progression and metastasis in these zebrafish. Results of this study helped to elucidate a new molecular mechanism of HCC and provided new insights pertaining to potential therapeutic targets against HCC.

2. Materials and Methods

2.1. Zebrafish Husbandry and Maintenance

All experimental protocols and procedures involving zebrafish were approved by the Institutional Animal Care and Use Committee (IACUC) of the National University of Singapore and National Taiwan University. These experiments were also conducted in accordance with the "Guide for the Care and Use of Laboratory Animals" of the National Institutes of Health. All zebrafish embryos and larvae were maintained in E3 medium. Adult zebrafish were maintained in a recirculating aquatic system at 28 °C with a 14-h light/10-h dark cycle in accordance with standard practice. Dry pellets (GEMMA Micro 150 and 300, Skretting Nutreco, Tooele, UT, USA) were fed to adult zebrafish twice a day at a designated amount of approximately 3% body mass and directly proportional to the density of zebrafish within the tank. The zebrafish were fed lab-grown brine shrimp or commercial fish feed three times per day [35,36].

2.2. Generation of fabp10a:twist1a/kras Double Transgenic Zebrafish

In brief, Both of fabp10a:mCherry-T2A-twist1a-ER^{T2} (abbreviated as *twist1a+*) and fabp10a:rtTA2s-M2;TRE2:EGFP-kras^{V12} (abbreviated as *kras+*) transgenic zebrafish lines were generated in our previous work [31,37]. (These transgenic lines respectively expressed hepatocyte-specific *twist1*a and *kras^{V12}*). The wild-type AB zebrafish strain was used as a control. To establish a fabp10a:twist1a/kras^{V12} double transgenic zebrafish (abbreviated as *twist1a+/kras+*), we crossed *twist1a+* and *kras+* transgenic zebrafish and then selected positive F1 larvae, which were maintained under the zebrafish husbandry conditions described above until reaching the adult stage and undergoing further research.

2.3. RNA Isolation and Reverse Transcription PCR (RT-PCR)

Total RNA was extracted from primary liver tumors, metastatic liver tumors, and adjacent normal tissue of adult zebrafish using the RNeasy Mini Kit (Qiagen, Hilden, Germany); 1 µg RNA was then reverse transcribed into complementary DNA (cDNA) using the QuantiTect Whole Transcriptome Kit (Qiagen, Hilden, Germany). We amplified cDNA templates via polymerase chain reaction (PCR) using exTEN 2× PCR Master Mix (Axil Scientific, Singapore, Singapore). To assess liver markers, expression of *fabp10a* (Primers: forward, CCAGTGACAGAAATCCAGCA; reverse, GTTCTGCAGACCAGCTTTCC), *tfa* (Primers: forward, TGCAGAAAAAGCTGGTGATG; reverse, ACAGCATGAACTGGCACTTG), and *actin* (Primers: forward, CTCCATCATGAAGTGCGACGT; reverse, CAGACGGAGTATTTGCGCTCA) internal control in adult primary liver tumors, metastatic liver tumors, and adjacent normal tissue, we employed RT-PCR according to the following protocol: 1 µL of cDNA was amplified for 1 cycle at 95 °C for 5 min; followed by 35 cycles at 95 °C for 10 s, 58 °C for 30 s, and 68 °C for 1 min. The cDNA was then incubated at 68 °C for an additional 7 min to allow for synthesis completion. The resulting PCR products were subjected to 1.0% agarose gel electrophoresis, in which actin was used as the internal control for the cDNA assay, in accordance with published primers and protocols [33,38,39].

2.4. Induction of Transgene Expression Using Doxycycline and 4-Hydroxytamoxifen Treatment

At 5 days post-fertilization (dpf), larvae were screened for positive mCherry and/or EGFP fluorescence (to identify *twist1a+* and/or *kras+* transgenic zebrafish, respectively) using a fluorescence stereo microscope (SMZ18, Nikon, Tokyo, Japan). Doxycycline (Dox, Sigma-Aldrich, St. Louis, MO, USA) and 4-Hydroxytamoxifen (4-OHT, Sigma-Aldrich, St. Louis, MO, USA) were used to respectively induce *kras* and *twis1ta* expression. Induction studies involving 3- to 5-month post-fertilization (mpf) adult fish were performed in a 5-L tank that contained fresh water (changed every other day). To maintain tumor growth and induce metastasis over the long-term, *twist1a+*, *kras+*, and *twist1a+/kras+* transgenic zebrafish, as well as their wild-type siblings, were treated with 20 μg/mL Dox and 1 μg/mL 4-OHT for 2 and 4 weeks.

2.5. Induction of Transgene Expression and LPS Exposure in Transgenic Zebrafish

Each treatment group included 20 larvae that were incubated in $1\times$ E3 medium and treated with 20 μg/mL Dox alone or with 20 μg/mL Dox + 40 ng/mL LPS (catalog number L4391; Sigma-Aldrich, St. Louis, MO, USA) and then maintained in 6-well plates for 3 days. Adult zebrafish were treated with 10 μg/mL Dox alone or with 10 μg/mL Dox + 40 ng/mL LPS for 2 weeks. Following that, the double expression of *twist1a+/kras+* in transgenic zebrafish was induced via treatment with 10 μg/mL Dox and 1 μg/mL 4-OHT in 5-L tank. The *twist1a+/kras+* treatment group was also exposed to 40 ng/mL LPS for 4 weeks. For these experiments, fresh water, Dox, 4-OHT, and LPS were changed every other day. The mortality of adult zebrafish was determined daily, and samples were collected to study long-term treatment effects.

2.6. Collection of Tissue, Paraffin Sectioning, and Histochemical Analysis

In accordance with published protocols [33,36,40], all zebrafish samples were collected following euthanization at 2- or 4-weeks post-induction (wpi). Liver tissues were then fixed and embedded in paraffin for histological analysis. Specifically, 5-μm sections were deparaffined, rehydrated, and examined using the EnVision™+ Dual Link System (Dako, Carpinteria, CA, USA) according to previous methodologies for immunohistochemistry (IHC) analysis. Primary antibodies included rabbit anti-PCNA (Dilution: 1:500; Catalog Number: FL-261, Santa Cruz, CA, USA), rabbit anti-Caspase-3 (Dilution: 1:200; Catalog Number: C92-065, BD Biosciences, San Diego, CA, USA), mouse anti-E-cadherin (Dilution: 1:200; Catalog Number: 610188, BD Biosciences, San Diego, CA, USA), and mouse anti-Vimentin (Dilution: 1:200; Catalog Number: 610188, Abcam, Cambridge, MA, USA), which were used to stain hepatic tissues of zebrafish at 4 °C overnight. After washing with $1\times$ phosphate-buffered saline (PBS), peroxidase activity was detected by incubating tissue sections at room temperature with a universal secondary biotinylated antibody for 30 min and then adding Dako diaminobenzidine (DAB) substrate for development. Tissue sections were counterstained with Mayer's hematoxylin before being dehydrated, cleared, and mounted with slide covers. An Axio Imager Z2 microscope (Zeiss LSM 880, Goettingen, Germany) was used to visualize the sections. Images were analyzed with constant acquisition setting (microscope, magnification, light intensity, exposure time) using a 200× or 400× microscope objective. The results of histochemical analysis and larvae measurement were evaluated by two independent senior scientists or pathologists in single-blind manner to evaluate.

2.7. Statistical Analysis

All statistical analyses were performed using GraphPad Prism 9 software (GraphPad Software, La Jolla, CA, USA), as previously described [36]. For all in-vivo experiments, a two-tailed unpaired Student's t-test or one-way analysis of variance (ANOVA) was applied to compare experimental and control groups. To determine the overall survival of zebrafish, survival rates were derived using the Kaplan-Meier method and log-rank test. Quantification of the IHC data using Image J software (NIH, Bethesda, MD, USA). Significance was set at a *p*-value of 0.05 or less.

3. Results

3.1. Phenotype of twist1a+/kras+ Double Transgenic Zebrafish and Liver Tumor Metastasis Induced by Dox Treatment

For induction studies, 3 to 5 months of mpf adult zebrafish were used for treatment experiments. To comprehensively elucidate tumor progression and metastatic development in *twist1a+/kras+* double transgenic zebrafish, experiments in this study employed *twist1a+*, *kras+*, and *twist1a+/kras+* transgenic zebrafish, as well as their non-transgenic wild-type siblings. All zebrafish were treated with Dox and 4-OHT. Long-term treatment samples were collected and investigated at 2 and 4 weeks (Supplementary Figure S1A).

To study how *twist1a+* and *kras+* affected tumor growth and liver tumor metastasis over the long term, all zebrafish groups (*twist1a+*, *kras+*, *twist1a+/kras+*, and wild-type) were treated with 20 µg/mL Dox and 1 µg/mL 4-OHT for 4 weeks. Compared with wild-type and *twist1a+* control groups, an enlarged abdomen and obvious liver overgrowth were observed in both *kras+* and *twist1a+/kras+* zebrafish at 2 and 4 wpi. Hematoxylin and eosin (H&E) staining revealed that all liver tumors in the *kras+* or *twist1a+/kras+* zebrafish ranged from normal liver morphology to HCC and included the following classes: normal, hyperplasia (HP), hepatocellular adenoma (HCA), and HCC (Figure 1A,G). Zebrafish hepatocellular neoplasms have similar histological characteristics to human hepatocellular neoplasms during the growth process. Therefore, the classification of zebrafish liver neoplasm types was based on the criteria for rodent hepatocellular neoplasms and established criteria as previously studied [41–44]. The classification criteria for liver neoplasm types are as follows: (1) The normal liver has a typical two-cell hepatic plate structure, uniform shape, size and clear boundaries of the cell. (2) The hyperplasia maintains hepatic plate arrangement with an increased prominent nucleus. (3) Unclear hepatic plates with clear cell boundary and relatively uniformed cell shape were found at HCA. (4) HCC was characterized by loss of cell boundaries and hepatic plate structure, increased mitotic cells and multiple nucleus. Furthermore, the body lengths of *kras+* and *twist1a+/kras+* transgenic zebrafish were significantly larger than those of the wild-type control; however, there was no significant difference in body weights among all groups (Figure 1B,C). A significantly higher mortality was observed in *kras+* or *twist1a+/kras+* zebrafish (Figure 1D). The representative images and percentages of *kras+* and *twist1a+/kras+* zebrafish exhibiting HCA or HCC development were as follows: HCA (2 wpi: 2/8; 1/11, respectively; 4 wpi: 3/8; 1/8, respectively) or HCC (2 wpi: 2/8; 3/11, respectively; 4 wpi: 3/8; 2/8, respectively). Some *twist1a+/kras+* zebrafish also showed evidence of metastatic HCC at 2 and 4 wpi (2 wpi: 5/11; 4 wpi: 3/8) (Figure 1E–G).

3.2. Detection of fabp10a and tfa Expression in Primary and Metastatic Liver Tumors Tissues from twist1a+/kras+ Double Transgenic Zebrafish

To determine the expression of *fabp10a* and *tfa* at metastatic tumor cells, we collected primary liver tumors, metastatic HCC tissues, and adjacent normal muscle tissues on zebrafish body from *twist1a+/kras+* transgenic zebrafish following treatment with 20 µg/mL Dox and 1 µg/mL 4-OHT. At 2 wpi, EGFP and mCherry fluorescence signal of *twist1a+/kras+* zebrafish revealed evidence of metastatic HCC (Figure 2A). To identify the expression of *fabp10a* and *tfa* at tumors, we semi-quantified the mRNA expression of two zebrafish liver markers (*fabp10a* and *tfa*) in primary liver tumors tissues, metastatic HCC tissues, and adjacent normal muscle tissues using semi-quantitative RT-PCR. We found that *fabp10a* and *tfa* genes were expressed in both primary tumors and metastatic HCC tissues, confirming that metastatic HCC may come from the liver. Note that mRNA expression of *fabp10a* and *tfa* was not observed in adjacent normal muscle tissues. *Actin* and non-template samples respectively served as positive and negative controls (Figure 2B).

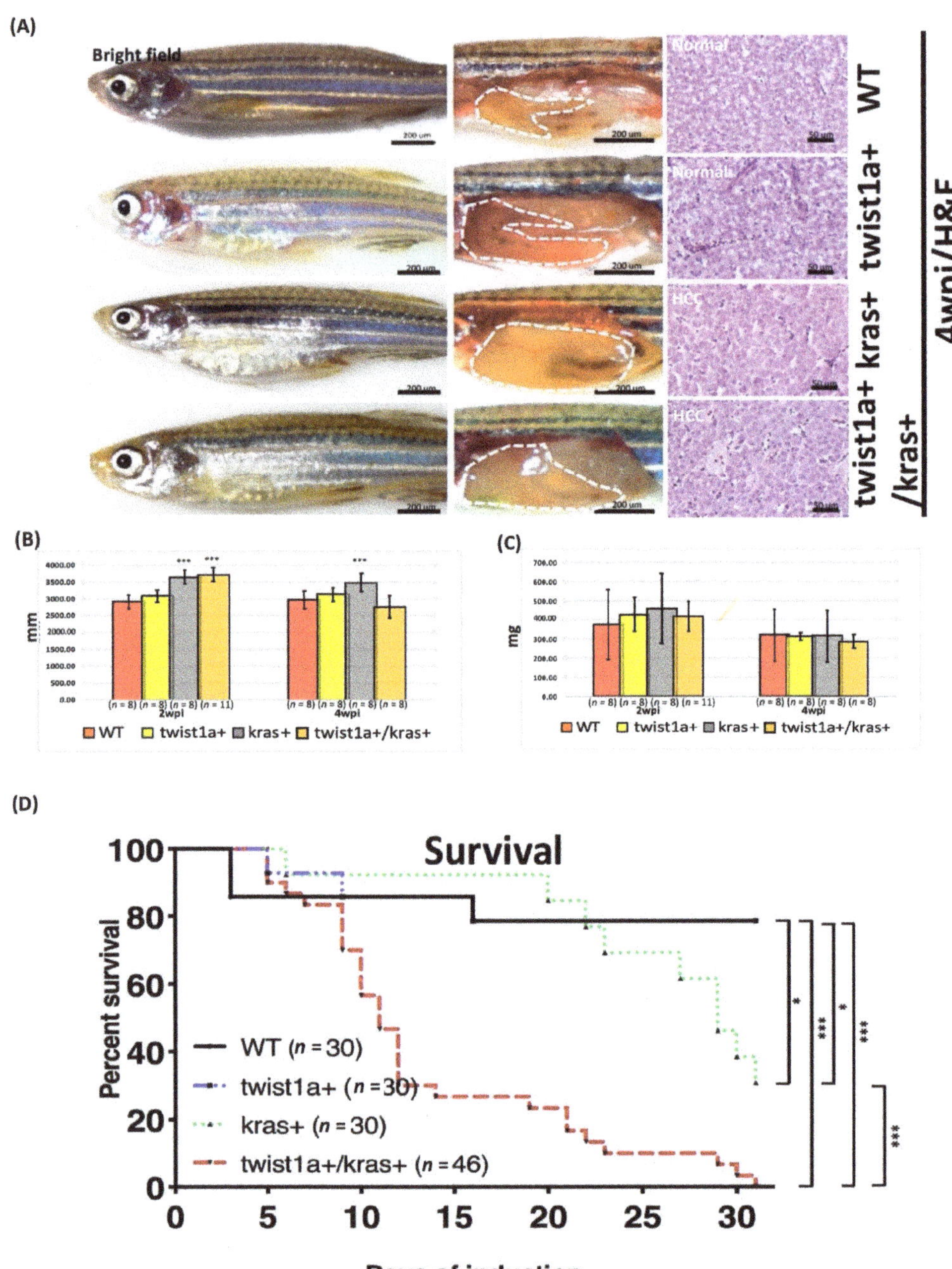

Figure 1. *Cont.*

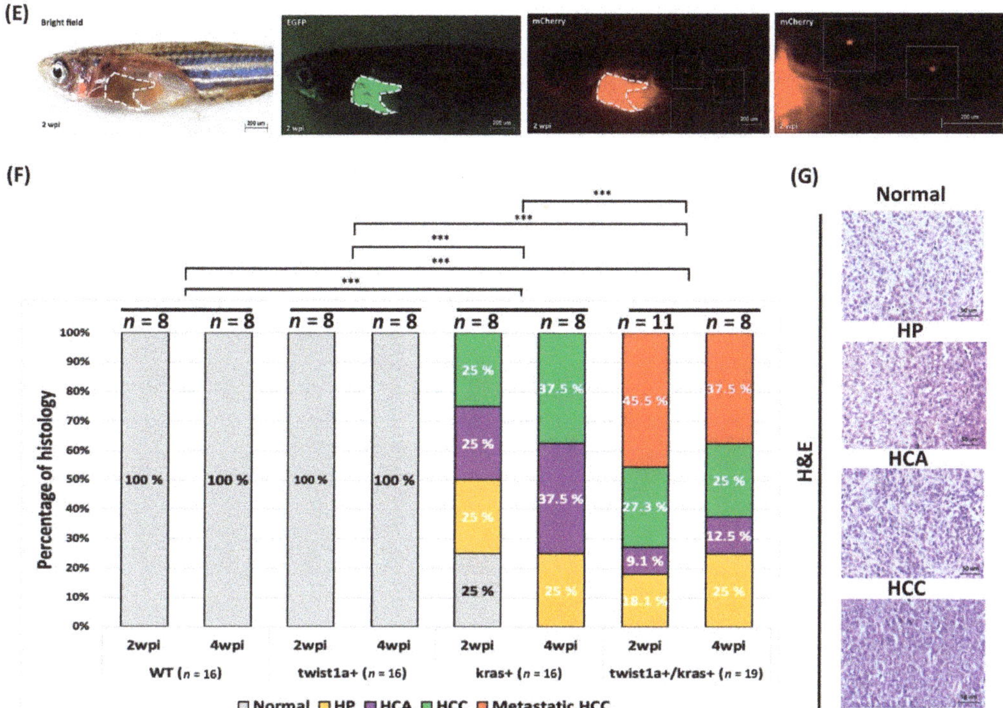

Figure 1. Induction of tumor metastasis in *twist1a+/kras+* transgenic zebrafish with Dox treatment. All zebrafish were treated with 20μg/mL Dox and 1μg/mL 4-OHT at 4 mpf, and samples were taken at 2 and 4 wpi. (**A**) Representative images of *wild-type*, *twist1a+*, *kras+*, and *twist1a+/kras+* transgenic zebrafish. The left column shows the external appearance, the middle column shows the internal organs of the abdomen with the liver outlined (white dotted line), and the right column shows H&E staining of liver tissues at 4 wpi. Compared with the wild-type group, (**B**) the body lengths of *kras+* and/or *twist1a+/kras+* transgenic zebrafish differed significantly at 2 and 4 wpi, whereas (**C**) the body weights of transgenic zebrafish did not differ at 2 and 4 wpi. (**D**) Kaplan-Meier survival curves showing the percentage of survival at 4 wpi. (**E**) Fluorescence analysis presenting evidence of metastatic HCC at 2 wpi in *twist1a+/kras+* zebrafish (white dotted line: primary and metastatic liver tumors). (**F**) Histological analysis revealed that wild-type, *twist1a+*, *kras+*, and *twist1a+/kras+* transgenic zebrafish developed HCC or metastatic HCC at 4 wpi. (**G**) Representative images of normal, HP, HCA, and HCC using histological analysis. Scale bars: 50 or 200 μm. Student's *t*-test or one-way ANOVA were used to assess differences between variables: * $p < 0.05$, *** $p < 0.001$.

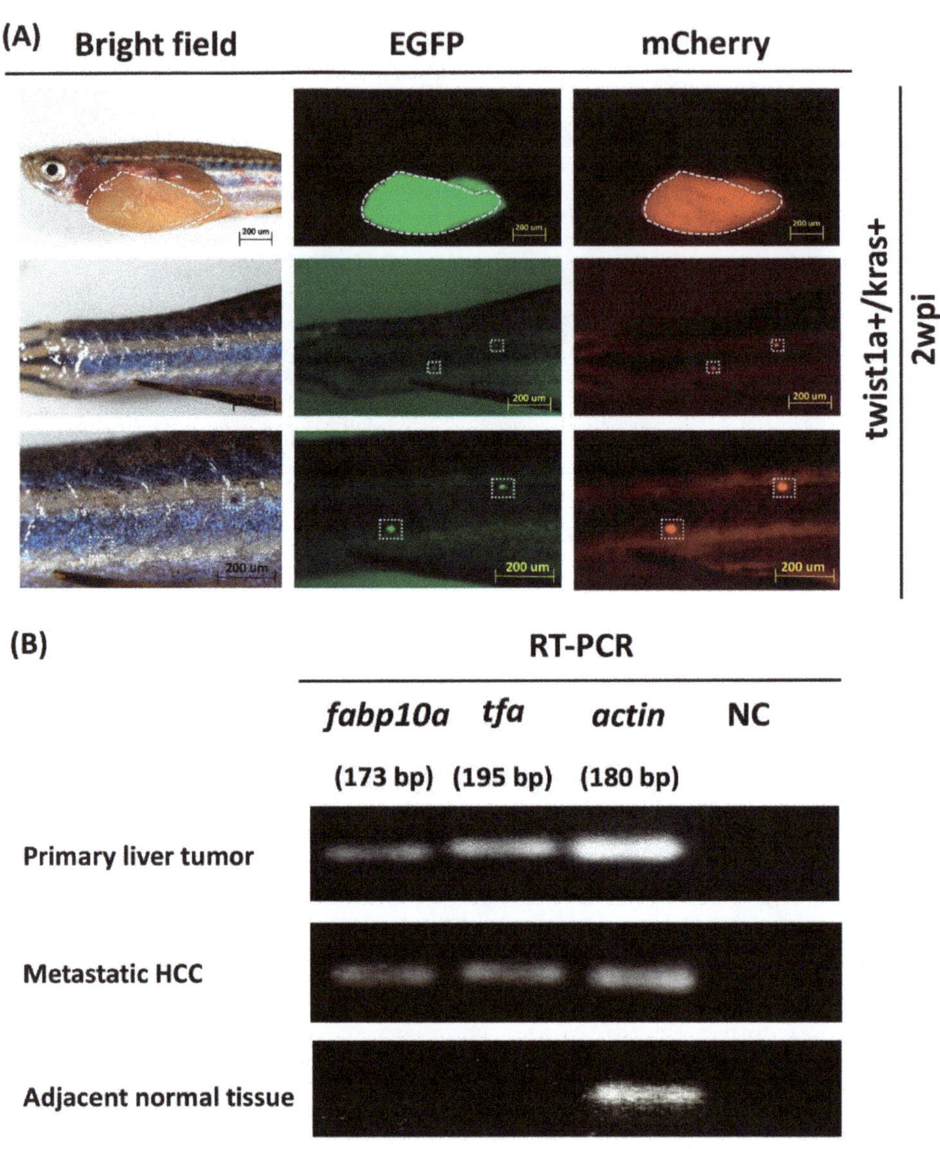

Figure 2. Liver markers in primary and metastatic liver tumor tissues from *twist1a+/kras+* doubl
transgenic zebrafish. (**A**) Immunofluorescence was used to visualize mCherry or EGFP-labele
metastatic liver tumors in *twist1a+/kras+* zebrafish (white dotted line: primary and metastatic live
tumors). (**B**) Semi-quantitative RT-PCR data showing expression levels of *fabp10a* and *tfa* in primar
liver tumors, metastatic HCC tissues, and adjacent normal muscle tissues. *Actin* and non-templat
samples were respectively used as positive and negative controls. Scale bar: 200 μm.

3.3. Co-Expression of twist1a/kras Significantly Increased Apoptosis, and twist1a Activated the EMT Pathway through E-Cadherin and Vimentin

To further compare the severity of liver tumorigenesis and metastasis between *kras*
and *twist1a+/kras+* zebrafish, the main hallmarks of cell proliferation and cell apoptosis, i.e
PCNA and caspase-3 staining, were examined in fish livers. After induction with 20 μg/m

Dox and 1 μg/mL 4-OHT of transgenic gene expression, *kras+* and *twist1a+/kras+* zebrafish showed a significant increase in the proliferation of liver cells and cell apoptosis compared with WT control zebrafish (Figure 3). At 4 wpi, we also found a dramatic increase in the percentage of *twist1a+/kras+* zebrafish undergoing liver cell apoptosis compared with *kras+* zebrafish (Figure 3A,B). Consistent with histological observations, the percentage of proliferating liver cells increased more rapidly in *kras+* zebrafish and *twist1a+/kras+* than in wild-type and *twist1a+* control zebrafish from 2 and 4 wpi (Figure 1F); however, at 4 wpi, the percentage of liver cells undergoing apoptosis was greater in *twist1a+/kras+* zebrafish than in *kras+* zebrafish (Figure 3A,B). These observations were consistent with findings from our previous studies [32,36,45].

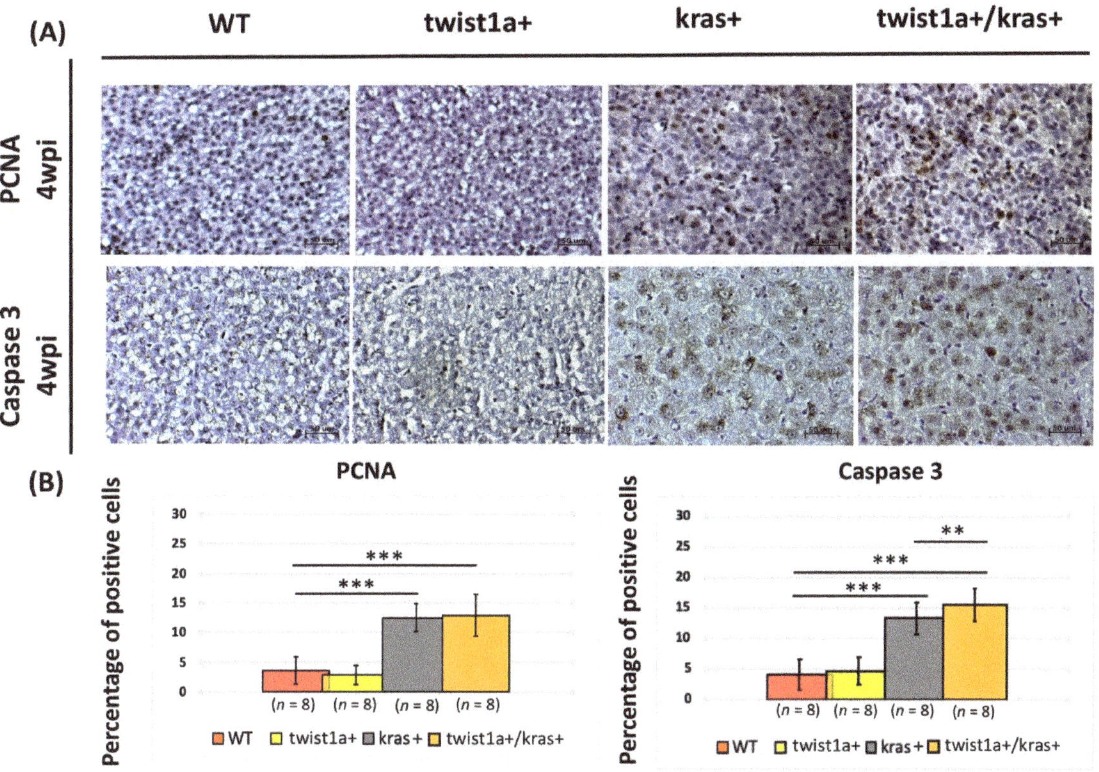

Figure 3. Main hallmarks of cell proliferation and cell apoptosis in liver tissues from *twist1a+/kras+* double transgenic zebrafish. (**A**) Immunohistochemical staining micrograph showing proliferating and apoptotic cells in liver cross-sections from wild-type, *twist1a+*, *kras+*, and *twist1a+/kras* zebrafish. (**B**) Quantification of the percentage of positive cells for the all fields following PCNA and caspase-3 staining at 4 wpi using ImageJ. Scale bar: 50 μm. Student's *t*-tests were used to assess differences between variables: ** $p < 0.01$, *** $p < 0.001$.

In order to identify a potential mechanism and pathway involved in liver tumorigenesis or metastasis, we further compared the severity of liver tumor occurrence and metastasis between *kras+* and *twist1a+/kras+* zebrafish. For this, the main EMT hallmarks during cancer metastasis, E-cadherin and Vimentin, were examined by immunohistochemical staining of the liver. Immunohistochemical staining results revealed that, following induction with 20 μg/mL Dox and 1 μg/mL 4-OHT, *kras+* zebrafish liver tissue had a decrease in E-cadherin and a corresponding increase in Vimentin at 4 wpi compared with wild-type zebrafish. However, the *kras+* zebrafish liver tissue also showed an increase in E-cadherin and corresponding decrease in Vimentin compared with *twist1a+/kras+* zebrafish (Figure 4A). Quantification of the percentage of positive cells for E-cadherin and Vimentin supported these findings (Figure 4B). These observations suggest that co-expression of *twist1a+* and *kras+* could trigger crosstalk along the EMT pathway and could contribute to the liver tumorigenesis or metastasis in *kras+* and *twist1a+/kras+* zebrafish. These observations are also consistent with findings from our previous studies [33,40].

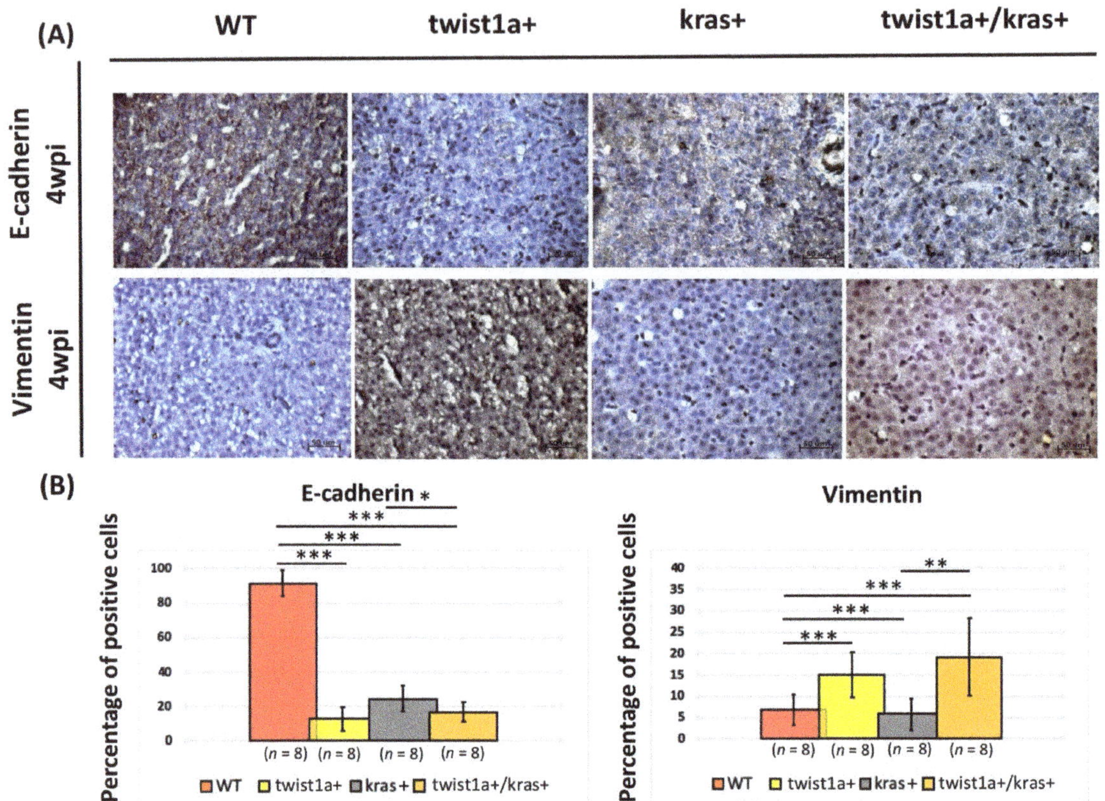

Figure 4. Main EMT hallmarks E-cadherin and Vimentin in liver tissues from *twist1a+/kras+* double transgenic zebrafish. (**A**) Immunohistochemical staining micrograph showing E-cadherin and Vimentin-positive cells in liver cross-sections of wild-type, *twist1a+*, *kras+*, and *twist1a+/kras+* zebrafish. (**B**) Quantification of the percentage of E-cadherin and Vimentin-positive cells for all fields following staining at 4 wpi using ImageJ. Scale bar: 50 μm. Student's *t*-tests were used to assess differences between variables: * $p < 0.05$, ** $p < 0.01$, *** $p < 0.001$.

3.4. Exposure to LPS Increased Liver Size in kras+ Transgenic Zebrafish Larvae

Before examining the long-term effects of LPS treatment, we examined the short-term effects of LPS treatment using *kras+* transgenic larvae and non-transgenic wild-type sibling larvae. For this, four-day-old *kras+* transgenic zebrafish larvae were treated with 20 μg/mL Dox alone or with 20 μg/mL Dox + 40 ng/mL LPS for 3 days. Wild-type (*kras−*) zebrafish larvae treated with 20 μg/mL Dox without exposure to LPS served as controls. All larvae in each group were imaged, and sizes of the livers were quantified (Figure 5A). Exposure to LPS significantly increases liver size in *kras+/LPS* larvae, compared with *kras+* and *kras−* control larvae (Figure 5B).

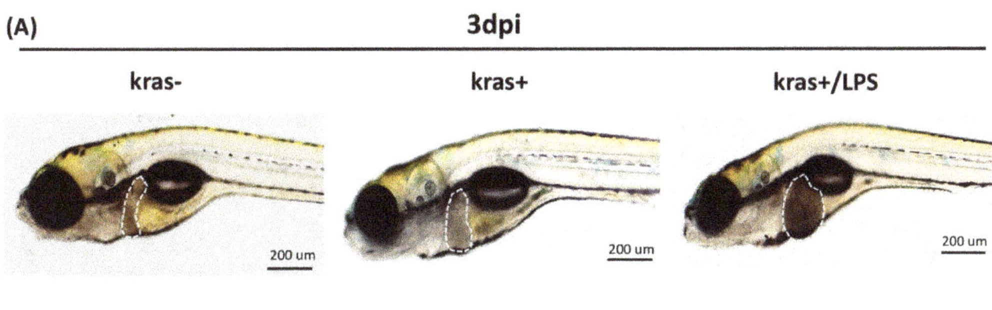

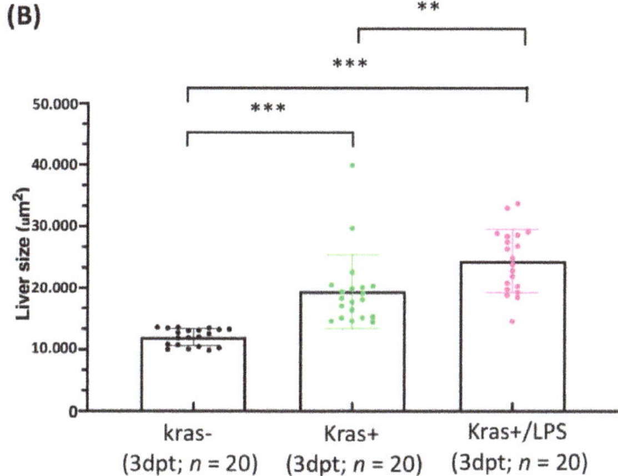

Figure 5. LPS increased liver size in *kras+* transgenic zebrafish larvae. (**A**) Representative images and (**B**) quantification of liver size in wild-type (*kras−*) control, *kras+*, and *kras+*/LPS zebrafish at 3 dpi (white dotted frame: liver; black dots; green dots; pink dots: the number of zebrafish larvae, respectively). Scale bar: 200 μm. Student's *t*-tests were used to assess differences between variables: ** $p < 0.01$, *** $p < 0.001$.

3.5. Liver Tumor Phenotypes Induced by Sustained Expression of kras and Exposure to LPS in Adult Transgenic Zebrafish

After determining that short-term exposure to LPS can increase liver size, we next sought to evaluate the long-term effects of LPS exposure. For this, five-month-old adult *kras+* transgenic zebrafish and their non-transgenic wild-type sibling zebrafish were treated with 10 μg/mL Dox alone or with 10 μg/mL Dox + 40 ng/mL LPS. Wild-type (*kras−*) adult zebrafish treated with 10 μg/mL Dox without exposure to LPS served as controls. The tumor status of all zebrafish in each group was examined at 2 wpi. H&E staining revealed that, following exposure to LPS, *kras+*/LPS zebrafish exhibited enlarged abdomens and obvious signs of liver overgrowth compared with *kras+* and wild-type (*kras−*) control zebrafish (Figure 6A). H&E staining also showed that liver tumors in both *kras+* and *kras+*/LPS zebrafish ranged from normal liver morphology to HCC and included the following classes: normal, HP, HCA, and HCC (Figure 6B). The classification of liver neoplasm types was based on established criteria as previously studied [41–44]. A significant increase in mortality was also observed in *kras+* and *kras+*/LPS transgenic zebrafish compared with wild-type (*kras−*) control zebrafish (Figure 6C). Histological analysis of *kras+* zebrafish revealed normal, HP, HCA, and HCC (2 wpi: 4/20; 3/20; 3/20; 10/20, respectively), whereas *kras+*/LPS zebrafish presented more severe evidence of HCC (2 wpi: 15/15). All wild-type (*kras−*) control zebrafish exhibited normal liver morphology (2 wpi: 20/20) (Figure 6D).

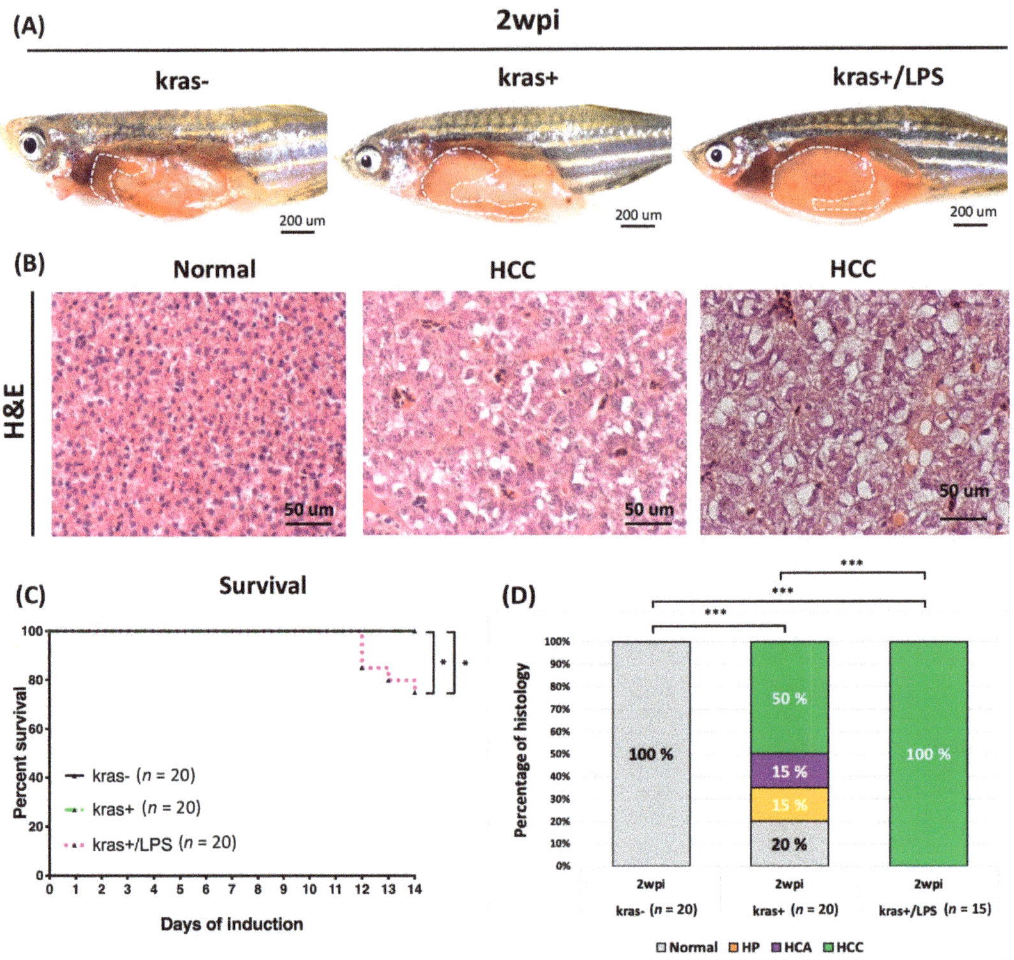

Figure 6. LPS promoted HCC progression in adult *kras+* transgenic zebrafish. The wild-type (*kras−* control, *kras+*, and *kras+/LPS* transgenic zebrafish were treated at 4 mpf with 10μg/mL Dox alone or with 10μg/mL Dox + 40ng/mL LPS, and samples were taken at 2 wpi. (**A**) The upper row shows the internal organs of the abdomen, and (**B**) the lower row displays H&E staining of liver sections (white dotted frame: liver). (**C**) Kaplan-Meier survival curves reveal the percentage of survival at 2 wpi. (**D**) Histological analysis shows that *kras+* and *kras+/LPS* transgenic zebrafish developed HCC at 2 wpi. Scale bar: 50 or 200 μm. Student's *t*-test or one-way ANOVA were used to assess differences between variables: * $p < 0.05$, *** $p < 0.001$.

3.6. LPS Exposure Exacerbated Liver Tumor Metastasis as Well as Hepatocyte-Specific Expression of twist1a and kras in Double Transgenic Zebrafish

After our aforementioned research results demonstrated that induction of *kras+* combined with LPS exposure increased liver size and the incidence of liver tumors, we next investigated how LPS exposure affected liver tumor metastasis in *twist1a+/kras+* double transgenic, adult-stage zebrafish. For this, three-month-old *twist1a+/kras+* zebrafish were treated with 10 μg/mL Dox and 1 μg/mL 4-OHT and exposed to 40 ng/mL LPS. Long-term treatment samples were collected and investigated at 4 weeks (Supplementary Figure S1B). HCC and metastatic HCC were examined using H&E or immunofluore cence analysis (Figure 7A,B). No significant differences in body lengths or body weights were found between *twist1a+/kras+* and *twist1a+/kras+/LPS* groups (Figure 7B,C), nor were there any differences in terms of mortality at *twist1a+/kras+* and *twist1a+/kras+/LPS* groups (Figure 7D). Histological analysis of *twist1a+/kras+* zebrafish revealed the presence of HP, HCA, HCC, and metastatic HCC in some fish (4 wpi: 1/9; 2/9; 4/9; 2/9, respectively), whereas the *twist1a+/kras+/LPS* zebrafish presented more severe metastatic HCC (4 wpi: 3/19; 1/19; 4/19; 11/19, respectively). Thus, the metastatic HCC status of *twist1a+/kras+/LPS* transgenic zebrafish was more severe following long-term LPS exposure (Figure 7E).

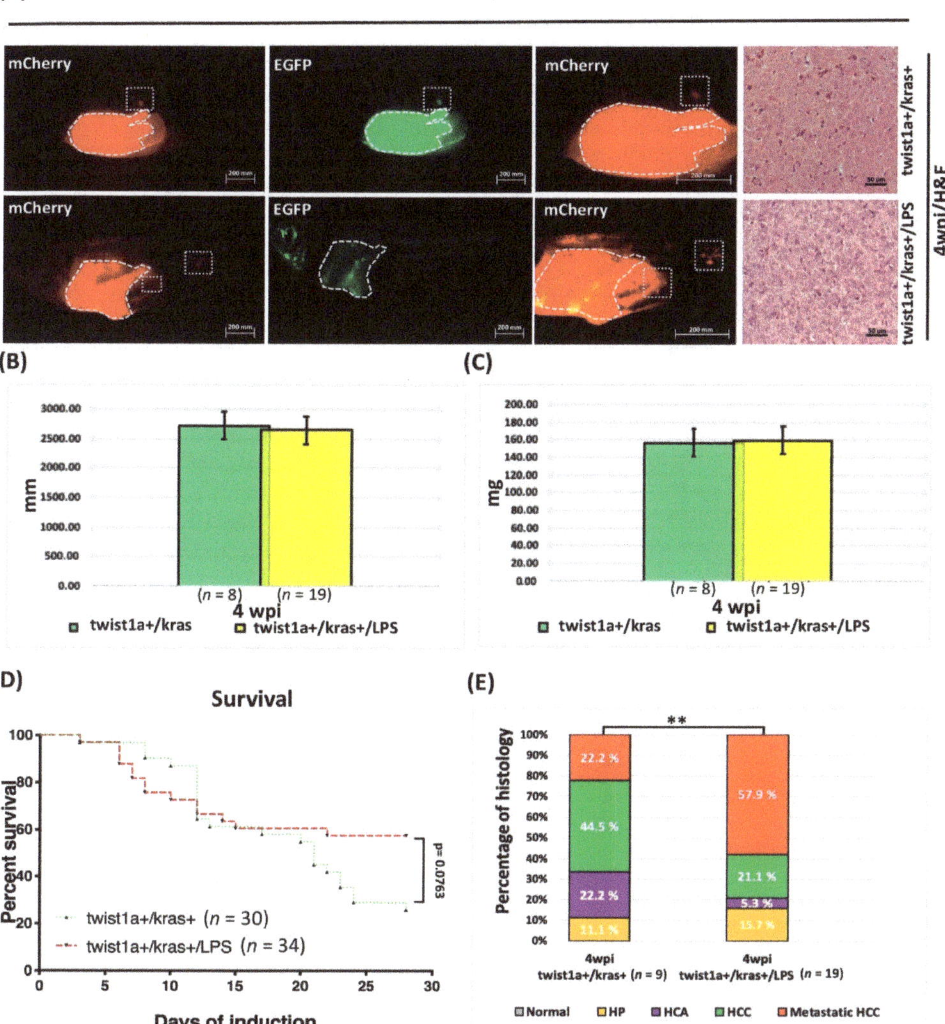

Figure 7. LPS enhanced HCC progression in adult *twist1a+/kras+* transgenic zebrafish. *Twist1a+/kras+* transgenic zebrafish were treated at 3 mpf with Dox and 4-OHT or with Dox, 4-OHT and LPS exposure. Samples were taken at 4 wpi. (**A**) Immunofluorescence analysis of mCherry and EGFP-labeled metastatic liver tumors and H&E staining of liver tissues from *twist1a+/kras+* and *twist1a+/kras+/LPS* zebrafish at 4 wpi (white dotted frame: primary and metastatic liver tumors). (**B**,**C**) No significant

differences in body lengths or weights were found between *twist1a+/kras+* and *twist1a+/kras+/LPS* transgenic zebrafish at 2 and 4 wpi. (**D**) Kaplan-Meier survival curves show the percentage of survival up to 4 wpi. (**E**) Histological analysis revealed that *twist1a+/kras+* and *twist1a+/kras+/LPS* transgenic zebrafish developed HCC or metastatic HCC at 4 wpi. Scale bar: 50 or 200 μm. Student's *t*-test or one-way ANOVA were used to assess differences between variables: ** $p < 0.01$.

4. Discussion

Recent evidence points to an increasing incidence of HCC among men in countries such as Japan, Italy, France, Switzerland, the United Kingdom, and the United States [46–48]. The process from clinical diagnosis to treatment primarily involves medical imaging, surgery, regional tumor treatment, and biotherapy. Although medical research has progressed remarkably with regard to HCC, effective treatments for patients are still lacking. Due to tumor recurrence and metastasis, the relative 5-year survival rate of liver cancer patients remains low [47,48], highlighting that further investigations into the mechanisms governing liver tumor metastasis should remain a top priority.

TWIST1 is involved in biological processes required for normal growth and development and regulates the expression of many specific genes [49]. However, *TWIST1* has known roles in the carcinogenesis of tumor cells. For example, *TWIST1* plays an important role in the vascular invasion and lung metastasis of tumor cells [11]. Primary tumor cells undergo EMT and, in the process of tumor metastasis, travel through the circulatory system to distant organs. The occurrence of EMT in tumor cells and the stimulation of tumor metastasis are promoted by *TWIST1*. In addition, *TWIST1* inhibits apoptosis and senescence and promotes the immortalization of cells [50]. However, the mechanisms by which *TWIST1* affect the metastasis of tumor cells remains unclear, although mechanisms are believed to differ depending on tumor type. One previous study that employed a *Kras*-induced lung cancer transgenic mouse model found that *Twist1* inhibited cell senescence, thereby accelerating and maintaining the tumorigenic effects of mutant *Kras genes* [51]. In the present study, our results showed that *twist1a+/kras+* double transgenic zebrafish developed spontaneous metastatic tumour. The mortality of *twist1a+/kras+* zebrafish was also significantly higher compared to that of *kras+* zebrafish (Figure 1) despite the significant increase in liver cell apoptosis in *twist1a+/kras+* zebrafish. However, the percentage of apoptotic liver cells was exacerbated in *twist1a+/kras+* zebrafish compared with *kras+* zebrafish (Figure 3). These results are also consistent with the previous results of *twist1a+/xmrk+* zebrafish compared with *xmrk+* zebrafish [33]. Thus, the increase in apoptosis of *twist1a* during the liver tumor metastasis at *kras+* or *xmrk+* zebrafish is different from the decrease in apoptosis found in rhabdomyosarcoma [52], which means that it has multiple functions. These observations suggest that cell apoptosis or cell proliferation were key factors in the tumorigenesis or metastasis of liver tumors in *kras+* or *twist1a+/kras+* zebrafish, which is consistent with findings from our previous study on *twist1a+/xmrk+* transgenic zebrafish [33].

TWIST1 acts as an EMT regulator and promotes tumor progression through distinct mechanisms. The upregulation of *TWIST1* in HCC cell lines promotes cell proliferation and migration [12]. *TWIST1* also regulates downstream genes such as *E-cadherin* and *vimentin* to promote EMT [53], wherein *E-cadherin* is the first confirmed gene target of *TWIST1*expression. *TWIST1* inhibits *E-cadherin* expression by binding to the E-cadherin promoter, resulting in the downregulation of *E-cadherin* and consequent attenuation of cell-cell adhesion as well as the enhancement of cell migration and invasion. An increase in angiogenesis related to tumor progression is also promoted by *TWIST1*, which acts by increasing the production of vascular endothelial growth factors [54]. Moreover, knockout of *TWIST1* has been found to significantly reduce the number of Vimentin-positive breast tumor cells, which indicates that *Twist1* expression is positively associated with *Vimentin* expression. In specific mouse models, *TWIST1* has also been shown to promote EMT in HCC by regulating vimentin via cullin2 circular RNA [12,55]. The results indicate that E-cadherin and Vimentin proteins are significantly increased or decreased, respectively, in *kras+* zebrafish compared with *twist1a+/kras+* zebrafish (Figure 4), which is consistent with findings from our previous studies [33,40].

The tumor microenvironment, which is composed of stromal cells, endothelial cells, immune cells, inflammatory cells, cytokines, and extracellular matrix, plays a major role in the initiation and development of HCC [56]. LPS induces inflammation in zebrafish by activating (1) TLR4/myeloid differentiation primary response 88 (MyD88)/nuclear factor kappa-light-chain-enhancer of activated B cells (NF-κB) as well as (2) TLR4/MyD88/mitogen-activated protein kinases (MAPKs) signaling pathways [57]. LPS is a ligand for TLR4 and functions by mediating specific effects of bacterial products. TLR4 is also expressed in different types of cells in the liver, including tumor, hepatic stellate, and kupffer cells [58–60]. In a mouse model of HCC, LPS was found to promote angiogenesis by stimulating the activation of hepatic stellate cells via the TLR4 pathway [27]. LPS has also been found to be related to the up-regulation of matrix metalloproteinases (MMPs) expression and activation by other pro-inflammatory signals, and zebrafish is a very suitable in vivo model for studying the regulation and activation of MMPs [61–63]. The upregulation of vascular endothelial growth factor (*VEGF*) expression by LPS has also been shown to induce angiogenesis in HCC cells through a STAT3-dependent pathway both in vitro and in vivo [25]. Another previous study found that LPS induced hepatic stellate cells to secrete a variety of pro-angiogenic factors, including *VEGF*, platelet-derived growth factor (*PDGF*), and angiopoietin-1 (*Ang-1*) [27]. Our results, obtained after exposing zebrafish larvae (Figure 5) and adult-stage zebrafish (Figure 6) to LPS, are consistent with those of previous studies that employed zebrafish models of *kras*-induced liver or gut tumors (i.e., these studies also reported that LPS treatment accelerated tumor progression) [32,33,36,64]. In addition, co-expression of *twist1a+/kras+* in zebrafish that had been exposed to LPS was found to exacerbate metastasis compared with *twist1a+/kras+* zebrafish that had not been exposed to LPS, indicating that LPS could activate liver tumor progression and metastasis in *kras* mutants in vivo by cooperating with the *twist1a* gene. On the other hand, we also noticed that the *twist1a+/kras+* group has a higher survival tendency than the *twist1a+/kras+/LPS* group. However, there is no statistically significant difference. We speculate that expanding the number of zebrafish will reduce this trend (Figure 7). Together, these results reveal the *twist1a+* and *kras+* genes have a cooperative relationship in chronic inflammation, which may contribute to interactions within the immune system that exacerbate the development of tumor metastasis.

5. Conclusions

In conclusion, our results indicate that *TWIST1* may be an effective target gene in treating HCC metastasis. This is the first in-vivo demonstration that *twist1a* plays an important role in the maintenance and acceleration of liver tumor metastasis in *kras+* adult-stage zebrafish. We also determined that the inflammatory agent LPS plays a significant role in *twist1a+/kras+* double transgenic zebrafish, which could exacerbate HCC metastasis.

Supplementary Materials: The following supporting information can be downloaded at: https://www.mdpi.com/article/10.3390/biomedicines10010095/s1, Figure S1: Experimental design and long-term treatment samples were collected weekly for investigation.

Author Contributions: Conceptualization, J.-W.L., L.-I.L. and Z.G.; methodology, J.-W.L., L.-I.L. and Y.S.; software, J.-W.L. and Y.S.; validation, J.-W.L., L.-I.L. and Y.S.; formal analysis, J.-W.L. and Y.S.; investigation, J.-W.L., L.-I.L., Y.S. and Z.G.; resources and data curation, J.-W.L., L.-I.L., Y.S., D.L. and Z.G.; writing—original draft preparation, J.-W.L.; writing—review and editing, J.-W.L., L.-I.L. and Z.G.; visualization, J.-W.L., Y.S. and D.L.; supervision, J.-W.L., L.-I.L. and Z.G.; project administration, J.-W.L., L.-I.L. and Z.G. All authors have read and agreed to the published version of the manuscript.

Funding: This work was supported by grants from Ministry of Education of Singapore (R154000B88112 and R154000B70114) at Singapore, National Taiwan University Hospital (UN109-062) at Taiwan, and Shenzhen-Hong Kong Institute of Brain Science-Shenzhen Fundamental Research Institutions (2021SHIBS0002) at China.

Institutional Review Board Statement: Not applicable.

Informed Consent Statement: Not applicable.

Data Availability Statement: Data are contained within the article.

Conflicts of Interest: The authors declare no conflict of interest.

References

1. Yang, J.D.; Hainaut, P.; Gores, G.J.; Amadou, A.; Plymoth, A.; Roberts, L.R. A global view of hepatocellular carcinoma: Trends, risk, prevention and management. *Nat. Rev. Gastroenterol. Hepatol.* **2019**, *16*, 589–604. [CrossRef]
2. Malkowski, P.; Pacholczyk, M.; Lagiewska, B.; Adadynski, L.; Wasiak, D.; Kwiatkowski, A.; Chmura, A.; Czerwinski, J. Hepatocellular carcinoma–Epidemiology and treatment. *Przegl. Epidemiol.* **2006**, *60*, 731–740.
3. Yang, J.D.; Roberts, L.R. Hepatocellular carcinoma: A global view. *Nat. Rev. Gastroenterol. Hepatol.* **2010**, *7*, 448–458. [CrossRef]
4. Forner, A.; Reig, M.; Bruix, J. Hepatocellular carcinoma. *Lancet* **2018**, *391*, 1301–1314. [CrossRef]
5. Bruix, J.; Gores, G.J.; Mazzaferro, V. Hepatocellular carcinoma: Clinical frontiers and perspectives. *Gut* **2014**, *63*, 844–855. [CrossRef]
6. Zhang, X.; Li, J.; Shen, F.; Lau, W.Y. Significance of presence of microvascular invasion in specimens obtained after surgical treatment of hepatocellular carcinoma. *J. Gastroenterol. Hepatol.* **2018**, *33*, 347–354. [CrossRef]
7. Yin, L.C.; Xiao, G.; Zhou, R.; Huang, X.P.; Li, N.L.; Tan, C.L.; Xie, F.J.; Weng, J.; Liu, L.X. MicroRNA-361-5p Inhibits Tumorigenesis and the EMT of HCC by Targeting Twist1. *BioMed Res. Int.* **2020**, *2020*, 8891876. [CrossRef]
8. Norozi, F.; Ahmadzadeh, A.; Shahjahani, M.; Shahrabi, S.; Saki, N. Twist as a new prognostic marker in hematological malignancies. *Clin. Transl. Oncol.* **2016**, *18*, 113–124. [CrossRef]
9. Kang, Y.; Massague, J. Epithelial-mesenchymal transitions: Twist in development and metastasis. *Cell* **2004**, *118*, 277–279. [CrossRef]
10. Georgakopoulos-Soares, I.; Chartoumpekis, D.V.; Kyriazopoulou, V.; Zaravinos, A. EMT Factors and Metabolic Pathways in Cancer. *Front. Oncol.* **2020**, *10*, 499. [CrossRef]
11. Yang, J.; Mani, S.A.; Donaher, J.L.; Ramaswamy, S.; Itzykson, R.A.; Come, C.; Savagner, P.; Gitelman, I.; Richardson, A.; Weinberg, R.A. Twist, a master regulator of morphogenesis, plays an essential role in tumor metastasis. *Cell* **2004**, *117*, 927–939. [CrossRef]
12. Meng, J.; Chen, S.; Han, J.X.; Qian, B.; Wang, X.R.; Zhong, W.L.; Qin, Y.; Zhang, H.; Gao, W.F.; Lei, Y.Y.; et al. Twist1 Regulates Vimentin through Cul2 Circular RNA to Promote EMT in Hepatocellular Carcinoma. *Cancer Res.* **2018**, *78*, 4150–4162. [CrossRef]
13. Zhu, Y.; Qu, C.; Hong, X.; Jia, Y.; Lin, M.; Luo, Y.; Lin, F.; Xie, X.; Xie, X.; Huang, J.; et al. Trabid inhibits hepatocellular carcinoma growth and metastasis by cleaving RNF8-induced K63 ubiquitination of Twist1. *Cell Death Differ.* **2019**, *26*, 306–320. [CrossRef]
14. Yin, L.C.; Luo, Z.C.; Gao, Y.X.; Li, Y.; Peng, Q.; Gao, Y. Twist Expression in Circulating Hepatocellular Carcinoma Cells Predicts Metastasis and Prognoses. *BioMed Res. Int.* **2018**, *2018*, 3789613. [CrossRef]
15. Yang, M.H.; Chen, C.L.; Chau, G.Y.; Chiou, S.H.; Su, C.W.; Chou, T.Y.; Peng, W.L.; Wu, J.C. Comprehensive analysis of the independent effect of twist and snail in promoting metastasis of hepatocellular carcinoma. *Hepatology* **2009**, *50*, 1464–1474. [CrossRef]
16. Ye, H.; Zhang, C.; Wang, B.J.; Tan, X.H.; Zhang, W.P.; Teng, Y.; Yang, X. Synergistic function of Kras mutation and HBx in initiation and progression of hepatocellular carcinoma in mice. *Oncogene* **2014**, *33*, 5133–5138. [CrossRef]
17. Adjei, A.A. Blocking oncogenic Ras signaling for cancer therapy. *J. Natl. Cancer Inst.* **2001**, *93*, 1062–1074. [CrossRef]
18. Kim, J.H.; Kim, H.Y.; Lee, Y.K.; Yoon, Y.S.; Xu, W.G.; Yoon, J.K.; Choi, S.E.; Ko, Y.G.; Kim, M.J.; Lee, S.J.; et al. Involvement of mitophagy in oncogenic K-Ras-induced transformation: Overcoming a cellular energy deficit from glucose deficiency. *Autophagy* **2011**, *7*, 1187–1198. [CrossRef]
19. Turhal, N.S.; Savas, B.; Coskun, O.; Bas, E.; Karabulut, B.; Nart, D.; Korkmaz, T.; Yavuzer, D.; Demir, G.; Dogusoy, G.; et al. Prevalence of K-Ras mutations in hepatocellular carcinoma: A Turkish Oncology Group pilot study. *Mol. Clin. Oncol.* **2015**, *4*, 1275–1279. [CrossRef]
20. Refolo, M.G.; Messa, C.; Guerra, V.; Carr, B.I.; D'Alessandro, R. Inflammatory Mechanisms of HCC Development. *Cancers* **2020**, *12*, 641. [CrossRef]
21. Tao, Y.; Li, Y.; Liu, X.; Deng, Q.; Yu, Y.; Yang, Z. Nonsteroidal anti-inflammatory drugs, especially aspirin, are linked to lower risk and better survival of hepatocellular carcinoma: A meta-analysis. *Cancer Manag. Res.* **2018**, *10*, 2695–2709. [CrossRef]
22. Yang, J.; Li, M.; Zheng, Q.C. Emerging role of Toll-like receptor 4 in hepatocellular carcinoma. *J. Hepatocell. Carcinoma* **2015**, *2*, 11–17.
23. Aggarwal, B.B.; Kunnumakkara, A.B.; Harikumar, K.B.; Gupta, S.R.; Tharakan, S.T.; Koca, C.; Dey, S.; Sung, B. Signal transducer and activator of transcription-3, inflammation, and cancer: How intimate is the relationship? *Ann. N. Y. Acad. Sci.* **2009**, *1171*, 59–76. [CrossRef]
24. Kortylewski, M.; Kujawski, M.; Herrmann, A.; Yang, C.; Wang, L.; Liu, Y.; Salcedo, R.; Yu, H. Toll-like receptor 9 activation of signal transducer and activator of transcription 3 constrains its agonist-based immunotherapy. *Cancer Res.* **2009**, *69*, 2497–2505. [CrossRef]

25. Wang, Z.; Yan, M.; Li, J.; Long, J.; Li, Y.; Zhang, H. Dual functions of STAT3 in LPS-induced angiogenesis of hepatocellular carcinoma. *Biochim. Biophys. Acta Mol. Cell Res.* **2019**, *1866*, 566–574. [CrossRef]
26. Bertani, B.; Ruiz, N. Function and Biogenesis of Lipopolysaccharides. *EcoSal Plus* **2018**, *8*, 1–19. [CrossRef]
27. Lu, Y.; Xu, J.; Chen, S.; Zhou, Z.; Lin, N. Lipopolysaccharide promotes angiogenesis in mice model of HCC by stimulating hepatic stellate cell activation via TLR4 pathway. *Acta Biochim. Biophys. Sin.* **2017**, *49*, 1029–1034. [CrossRef]
28. Jing, Y.Y.; Han, Z.P.; Sun, K.; Zhang, S.S.; Hou, J.; Liu, Y.; Li, R.; Gao, L.; Zhao, X.; Zhao, Q.D.; et al. Toll-like receptor 4 signaling promotes epithelial-mesenchymal transition in human hepatocellular carcinoma induced by lipopolysaccharide. *BMC Med.* **2012**, *10*, 98. [CrossRef]
29. Fausto, N. Mouse liver tumorigenesis: Models, mechanisms, and relevance to human disease. *Semin. Liver Dis.* **1999**, *19*, 243–252. [CrossRef]
30. Lewis, B.C.; Klimstra, D.S.; Socci, N.D.; Xu, S.; Koutcher, J.A.; Varmus, H.E. The absence of p53 promotes metastasis in a novel somatic mouse model for hepatocellular carcinoma. *Mol. Cell. Biol.* **2005**, *25*, 1228–1237. [CrossRef]
31. Nakayama, J.; Lu, J.W.; Makinoshima, H.; Gong, Z. A Novel Zebrafish Model of Metastasis Identifies the HSD11beta1 Inhibitor Adrenosterone as a Suppressor of Epithelial-Mesenchymal Transition and Metastatic Dissemination. *Mol. Cancer Res.* **2020**, *18*, 477–487. [CrossRef]
32. Li, Z.; Huang, X.; Zhan, H.; Zeng, Z.; Li, C.; Spitsbergen, J.M.; Meierjohann, S.; Schartl, M.; Gong, Z. Inducible and repressable oncogene-addicted hepatocellular carcinoma in Tet-on xmrk transgenic zebrafish. *J. Hepatol.* **2012**, *56*, 419–425. [CrossRef]
33. Lu, J.W.; Sun, Y.; Fong, P.A.; Lin, L.I.; Liu, D.; Gong, Z. Exacerbation of Liver Tumor Metastasis in twist1a+/xmrk+ Double Transgenic Zebrafish Following Lipopolysaccharide or Dextran Sulphate Sodium Exposure. *Pharmaceuticals* **2021**, *14*, 867. [CrossRef]
34. Lu, J.W.; Ho, Y.J.; Yang, Y.J.; Liao, H.A.; Ciou, S.C.; Lin, L.I.; Ou, D.L. Zebrafish as a disease model for studying human hepatocellular carcinoma. *World J. Gastroenterol.* **2015**, *21*, 12042–12058. [CrossRef]
35. Lu, J.W.; Hou, H.A.; Hsieh, M.S.; Tien, H.F.; Lin, L.I. Overexpression of FLT3-ITD driven by spi-1 results in expanded myelopoiesis with leukemic phenotype in zebrafish. *Leukemia* **2016**, *30*, 2098–2101. [CrossRef]
36. Lu, J.W.; Sun, Y.; Fong, P.A.; Lin, L.I.; Liu, D.; Gong, Z. Lipopolysaccharides Enhance Epithelial Hyperplasia and Tubular Adenoma in Intestine-Specific Expression of kras(V)(12) in Transgenic Zebrafish. *Biomedicines* **2021**, *9*, 974. [CrossRef]
37. Chew, T.W.; Liu, X.J.; Liu, L.; Spitsbergen, J.M.; Gong, Z.; Low, B.C. Crosstalk of Ras and Rho: Activation of RhoA abates *Kras*-induced liver tumorigenesis in transgenic zebrafish models. *Oncogene* **2014**, *33*, 2717–2727. [CrossRef]
38. Hou, H.A.; Lu, J.W.; Lin, T.Y.; Tsai, C.H.; Chou, W.C.; Lin, C.C.; Kuo, Y.Y.; Liu, C.Y.; Tseng, M.H.; Chiang, Y.C.; et al. Clinico-biological significance of suppressor of cytokine signaling 1 expression in acute myeloid leukemia. *Blood Cancer J.* **2017**, *7*, e588. [CrossRef]
39. Lu, J.W.; Hsieh, M.S.; Hou, H.A.; Chen, C.Y.; Tien, H.F.; Lin, L.I. Overexpression of SOX4 correlates with poor prognosis of acute myeloid leukemia and is leukemogenic in zebrafish. *Blood Cancer J.* **2017**, *7*, e593. [CrossRef]
40. Lu, J.W.; Raghuram, D.; Fong, P.A.; Gong, Z. Inducible Intestine-Specific Expression of kras(V12) Triggers Intestinal Tumorigenesis in Transgenic Zebrafish. *Neoplasia* **2018**, *20*, 1187–1197. [CrossRef]
41. Schlageter, M.; Terracciano, L.M.; D'Angelo, S.; Sorrentino, P. Histopathology of hepatocellular carcinoma. *World J. Gastroenterol.* **2014**, *20*, 15955–15964. [CrossRef]
42. Spitsbergen, J.M.; Tsai, H.W.; Reddy, A.; Miller, T.; Arbogast, D.; Hendricks, J.D.; Bailey, G.S. Neoplasia in zebrafish (*Danio rerio*) treated with 7,12-dimethylbenz[a]anthracene by two exposure routes at different developmental stages. *Toxicol. Pathol.* **2000**, *28*, 705–715. [CrossRef]
43. Spitsbergen, J.M.; Tsai, H.W.; Reddy, A.; Miller, T.; Arbogast, D.; Hendricks, J.D.; Bailey, G.S. Neoplasia in zebrafish (*Danio rerio*) treated with *N*-methyl-*N*'-nitro-*N*-nitrosoguanidine by three exposure routes at different developmental stages. *Toxicol. Pathol.* **2000**, *28*, 716–725. [CrossRef]
44. Lu, J.W.; Yang, W.Y.; Tsai, S.M.; Lin, Y.M.; Chang, P.H.; Chen, J.R.; Wang, H.D.; Wu, J.L.; Jin, S.L.; Yuh, C.H. Liver-specific expressions of HBx and src in the p53 mutant trigger hepatocarcinogenesis in zebrafish. *PLoS ONE* **2013**, *8*, e76951. [CrossRef]
45. Yan, C.; Yang, Q.; Huo, X.; Li, H.; Zhou, L.; Gong, Z. Chemical inhibition reveals differential requirements of signaling pathways in kras(V12)- and Myc-induced liver tumors in transgenic zebrafish. *Sci. Rep.* **2017**, *7*, 45796. [CrossRef]
46. Taylor-Robinson, S.D.; Foster, G.R.; Arora, S.; Hargreaves, S.; Thomas, H.C. Increase in primary liver cancer in the UK, 1979–1994. *Lancet* **1997**, *350*, 1142–1143. [CrossRef]
47. Landis, S.H.; Murray, T.; Bolden, S.; Wingo, P.A. Cancer statistics, 1998. *CA Cancer J. Clin.* **1998**, *48*, 6–29. [CrossRef]
48. Tang, Z.Y. Hepatocellular carcinoma-Cause, treatment and metastasis. *World J. Gastroenterol.* **2001**, *7*, 445–454. [CrossRef]
49. Puisieux, A.; Valsesia-Wittmann, S.; Ansieau, S. A twist for survival and cancer progression. *Br. J. Cancer* **2006**, *94*, 13–17. [CrossRef]
50. Smit, M.A.; Peeper, D.S. Deregulating EMT and senescence: Double impact by a single twist. *Cancer Cell* **2008**, *14*, 5–7. [CrossRef]
51. Tran, P.T.; Shroff, E.H.; Burns, T.F.; Thiyagarajan, S.; Das, S.T.; Zabuawala, T.; Chen, J.; Cho, Y.J.; Luong, R.; Tamayo, P.; et al. *Twist1* suppresses senescence programs and thereby accelerates and maintains mutant *Kras*-induced lung tumorigenesis. *PLoS Genet.* **2012**, *8*, e1002650. [CrossRef]
52. Maestro, R.; Dei Tos, A.P.; Hamamori, Y.; Krasnokutsky, S.; Sartorelli, V.; Kedes, L.; Doglioni, C.; Beach, D.H.; Hannon, G.J. Twist is a potential oncogene that inhibits apoptosis. *Genes Dev.* **1999**, *13*, 2207–2217. [CrossRef]

53. Tam, W.L.; Weinberg, R.A. The epigenetics of epithelial-mesenchymal plasticity in cancer. *Nat. Med.* **2013**, *19*, 1438–1449. [CrossRef]
54. Mironchik, Y.; Winnard, P.T., Jr.; Vesuna, F.; Kato, Y.; Wildes, F.; Pathak, A.P.; Kominsky, S.; Artemov, D.; Bhujwalla, Z.; Van Diest, P.; et al. Twist overexpression induces in vivo angiogenesis and correlates with chromosomal instability in breast cancer. *Cancer Res.* **2005**, *65*, 10801–10809. [CrossRef]
55. Xu, Y.; Lee, D.K.; Feng, Z.; Xu, Y.; Bu, W.; Li, Y.; Liao, L.; Xu, J. Breast tumor cell-specific knockout of Twist1 inhibits cancer cell plasticity, dissemination, and lung metastasis in mice. *Proc. Natl. Acad. Sci. USA* **2017**, *114*, 11494–11499. [CrossRef]
56. Tian, Z.; Hou, X.; Liu, W.; Han, Z.; Wei, L. Macrophages and hepatocellular carcinoma. *Cell Biosci.* **2019**, *9*, 79. [CrossRef]
57. Zhang, Y.; Takagi, N.; Yuan, B.; Zhou, Y.; Si, N.; Wang, H.; Yang, J.; Wei, X.; Zhao, H.; Bian, B. The protection of indolealkylamines from LPS-induced inflammation in zebrafish. *J. Ethnopharmacol.* **2019**, *243*, 112122. [CrossRef]
58. Zhu, Q.; Zou, L.; Jagavelu, K.; Simonetto, D.A.; Huebert, R.C.; Jiang, Z.D.; DuPont, H.L.; Shah, V.H. Intestinal decontamination inhibits TLR4 dependent fibronectin-mediated cross-talk between stellate cells and endothelial cells in liver fibrosis in mice. *J. Hepatol.* **2012**, *56*, 893–899. [CrossRef]
59. Ouyang, Y.; Guo, J.; Lin, C.; Lin, J.; Cao, Y.; Zhang, Y.; Wu, Y.; Chen, S.; Wang, J.; Chen, L.; et al. Transcriptomic analysis of the effects of Toll-like receptor 4 and its ligands on the gene expression network of hepatic stellate cells. *Fibrogenesis Tissue Repair* **2016**, *9*, 2. [CrossRef]
60. Stedman, C.A. Current prospects for interferon-free treatment of hepatitis C in 2012. *J. Gastroenterol. Hepatol.* **2013**, *28*, 38–45. [CrossRef]
61. Wyatt, R.A.; Keow, J.Y.; Harris, N.D.; Hache, C.A.; Li, D.H.; Crawford, B.D. The zebrafish embryo: A powerful model system for investigating matrix remodeling. *Zebrafish* **2009**, *6*, 347–354. [CrossRef]
62. Jeffrey, E.J.; Crawford, B.D. The epitope-mediated MMP activation assay: Detection and quantification of the activation of Mmp2 in vivo in the zebrafish embryo. *Histochem. Cell Biol.* **2018**, *149*, 277–286. [CrossRef]
63. Wyatt, R.A.; Crawford, B.D. Post-translational activation of Mmp2 correlates with patterns of active collagen degradation during the development of the zebrafish tail. *Dev. Biol.* **2021**, *477*, 155–163. [CrossRef]
64. Yang, Q.; Salim, L.; Yan, C.; Gong, Z. Rapid Analysis of Effects of Environmental Toxicants on Tumorigenesis and Inflammation Using a Transgenic Zebrafish Model for Liver Cancer. *Mar. Biotechnol.* **2019**, *21*, 396–405. [CrossRef]

Article

An Orthotopic Model of Uveal Melanoma in Zebrafish Embryo: A Novel Platform for Drug Evaluation

Chiara Tobia [1], Daniela Coltrini [1], Roberto Ronca [1], Alessandra Loda [1], Jessica Guerra [1], Elisa Scalvini [1], Francesco Semeraro [2] and Sara Rezzola [1,*]

[1] Department of Molecular and Translational Medicine, University of Brescia, 25123 Brescia, Italy; chiara.tobia@unibs.it (C.T.); daniela.coltrini@unibs.it (D.C.); roberto.ronca@unibs.it (R.R.); a.loda025@unibs.it (A.L.); jessica.guerra@unibs.it (J.G.); scalvini.elisa@yahoo.it (E.S.)

[2] Eye Clinic, Department of Medical and Surgical Specialties, Radiological Sciences and Public Health, University of Brescia, 25123 Brescia, Italy; francesco.semeraro@unibs.it

* Correspondence: sara.rezzola@unibs.it

Abstract: Uveal melanoma is a highly metastatic tumor, representing the most common primary intraocular malignancy in adults. Tumor cell xenografts in zebrafish embryos may provide the opportunity to study in vivo different aspects of the neoplastic disease and its response to therapy. Here, we established an orthotopic model of uveal melanoma in zebrafish by injecting highly metastatic murine B16-BL6 and B16-LS9 melanoma cells, human A375M melanoma cells, and human 92.1 uveal melanoma cells into the eye of zebrafish embryos in the proximity of the developing choroidal vasculature. Immunohistochemical and immunofluorescence analyses showed that melanoma cells proliferate during the first four days after injection and move towards the eye surface. Moreover, bioluminescence analysis of luciferase-expressing human 92.1 uveal melanoma cells allowed the quantitative assessment of the antitumor activity exerted by the canonical chemotherapeutic drugs paclitaxel, panobinostat, and everolimus after their injection into the grafted eye. Altogether, our data demonstrate that the zebrafish embryo eye is a permissive environment for the growth of invasive cutaneous and uveal melanoma cells. In addition, we have established a new luciferase-based in vivo orthotopic model that allows the quantification of human uveal melanoma cells engrafted in the zebrafish embryo eye, and which may represent a suitable tool for the screening of novel drug candidates for uveal melanoma therapy.

Keywords: uveal melanoma; zebrafish; orthotopic tumor; xenograft; luciferase

Citation: Tobia, C.; Coltrini, D.; Ronca, R.; Loda, A.; Guerra, J.; Scalvini, E.; Semeraro, F.; Rezzola, S. An Orthotopic Model of Uveal Melanoma in Zebrafish Embryo: A Novel Platform for Drug Evaluation. *Biomedicines* **2021**, *9*, 1873. https://doi.org/10.3390/biomedicines9121873

Academic Editors: James A. Marrs and Swapnalee Sarmah

Received: 17 September 2021
Accepted: 7 December 2021
Published: 10 December 2021

Publisher's Note: MDPI stays neutral with regard to jurisdictional claims in published maps and institutional affiliations.

1. Introduction

The zebrafish embryo has been successfully employed as a platform for modeling human diseases and for large-scale screening of new drugs [1–4]. Ease of manipulation, relatively low costs of maintenance, and optical transparency, combined with the opportunity to perform high-quality imaging, led to an extensive use of this model in cancer research. In this regard, mammalian tumor cell grafting in zebrafish embryos can be achieved in different anatomical sites, giving opportunity to study various aspects of the disease, such as tumor progression, angiogenesis, cancer cell spreading, and metastasis formation. Tumor cells have been successfully implanted in the perivitelline space, yolk ball, blood stream, pericardial cavity, eye, and brain (see [5–9] and references therein).

One of the major drawbacks of the use of the zebrafish embryo as a model in oncology is the quantification of tumor xenograft growth in the different anatomical sites, generally performed by measuring the fluorescence signal generated by engrafted fluorescent tumor cells [10,11]. This approach has also been used for the study of cancer growth following ocular transplantation of fluorescent tumor cells in zebrafish embryos [7–9]. However, the presence of the lens and the cup-like structure of the eye make difficult the acquisition of high-quality fluorescence images, which may lead to misleading results. This calls for alternative rapid and reliable quantification methods to be exploited for high throughput analysis.

Uveal melanoma represents the most common primary intraocular malignancy in adults. Classified as a rare neoplasm, its occurrence increases with age and its incidence is over 20 million/year. Despite the results obtained in terms of primary tumor management, the 5-year mortality rate of uveal melanoma patients (ranging from 26 to 32%) has not changed over the years [12–15]. Indeed, almost 50% of uveal melanoma patients develop metastatic disease through haematogenous dissemination [16], leading to an approximately 5–7-month median survival [13,14] which is rarely improved by chemotherapy [17]. At present, no drugs have been approved for the treatment of metastatic uveal melanoma patients and new therapeutic strategies are eagerly required. Nevertheless, despite the urgent need for an in vivo platform for the rapid screening of novel drug candidates, an orthotopic uveal melanoma model has not yet been implemented with zebrafish embryos.

Here, we propose a luciferase-based quantification method to demonstrate that transplantation of uveal melanoma cells into the eye of zebrafish embryos represents a useful in vivo orthotopic model suitable for the screening of novel drug candidates for uveal melanoma therapy.

2. Materials and Methods

2.1. Reagents

All reagents were of analytical grade. Dulbecco's modified Eagle medium (DMEM), RPMI 1640 medium, fetal bovine serum (FBS), non-essential amino acid (NEAA), and MEM vitamin solutions were obtained from GIBCO Life Technologies (Grand Island, NY, USA). Penicillin, streptomycin, sodium pyruvate, PTU, tricaine, bovine serum albumin (BSA), diaminobenzydine (DAB), and mouse anti-mouse vimentin antibody (Vim 13.2 clone) were from Sigma-Aldrich (St. Louis, MO, USA). Paclitaxel, panobinostat, and everolimus were from MedChemExpress (Monmouth Junction, NJ, USA). The Annexin-V/propidium iodide double staining kit was from Immunostep Biotec (Salamanca, Spain). The ONE Glo™ Luciferase Assay System was from Promega (Milan, Italy). Rat anti-mouse Ki-67 antibody (TEC-3) was from Dako (Santa Clara, CA, USA). Rabbit anti-human cleaved caspase 3 (Asp175) was from Cell Signaling (Danvers, MA, USA). Biotinylated anti-mouse IgM, anti-rat, and rabbit antibodies were from Abcam (Cambridge, UK). Biotin Avidin system Vectastain ABC reagent was from Vector Laboratories (Burlingame, CA, USA).

2.2. Cell Cultures

Murine melanoma B16-BL6 cells were grown in DMEM plus 10% FBS and 1% penicillin/streptomycin, and were stably transfected with DsRed fluorescent protein, thus generating B16-BL6-DsRed$^+$ cells [2]. Murine melanoma B16-LS9 cells [18] were kindly provided by Dr. L. Morbidelli (University of Siena, Siena, Italy) and were grown in DMEM plus 10% FBS, 1% penicillin/streptomycin. Luciferase-transfected B16-LS9 cells (B16-LS9-luc cells) were generated as previously described [19]. Human melanoma A375M cells were obtained from Dr. R. Giavazzi (Istituto Ricerche Farmacologiche Mario Negri, Bergamo, Italy) and were grown in DMEM plus 20% FBS, 1% NEAA, 2% MEM vitamin solution, 1% sodium pyruvate, and 1% penicillin/streptomycin. Human uveal melanoma 92.1 cells [20] were obtained from Dr. M. Jager (Leiden University, Leiden, The Netherlands) and were maintained in RPMI 1640 medium plus 10% FBS, 1% penicillin/streptomycin. A375M and 92.1 cells were infected with a lentivirus harboring the RFP/luciferase cDNA, thus generating stable A375M-RFP$^+$/luc$^+$ and 92.1-RFP$^+$/luc$^+$ cells that express both the red fluorescent RFP protein and the bioluminescent firefly luciferase. For eye injection, cells were suspended in PBS (final concentration equal to 100,000 cells/µL).

2.3. Cell Proliferation Assay

Cells were seeded on 48-well plates at 1.0×10^4 cells/cm^2 or at 1.5×10^4 cells/cm^2 for B16-LS9-luc$^+$ and 92.1-RFP$^+$/luc$^+$ cells, respectively. After 24 h, cells were treated with increasing concentrations of the different anticancer drugs. After a further 48 h or 72 h incubation, cells were trypsinized and viable cell counting was performed with the MACSQuant® Analyzer (Miltenyi Biotec, Bergisch Gladabach, Germany), as reported [21].

2.4. Apoptosis Assay

92.1-RFP$^+$/luc$^+$ cells were seeded on 6-well plates at 1.0×10^4 cells/cm^2. After 24 h, cells were treated with 140 nM paclitaxel, 20 nM panobinostat, and 60 nM everolimus. After 72 h of treatment, apoptotic cell death was assessed by Annexin-V/propidium iodide double staining according to the manufacturer's instructions, and cytofluorimetric analysis was performed using the MACSQuant® Analyzer.

2.5. Zebrafish Maintenance and Cell Transplantation

The transgenic tg(*kdrl*:EGFP) zebrafish line was maintained in the facility of the University of Brescia at 28 °C under standard conditions [22], and embryos were staged by h post-fertilization (hpf), as described [23]. To prevent pigmentation, embryo fish water was added with 0.2 mM 1-phenyl-2-thiourea (PTU) starting from 24 hpf. For cell injection and in vivo observation, embryos were anesthetized using 0.16 mg/mL tricaine. For cell engrafting, 48 hpf embryos were microinjected in the eye with tumor cells using a borosilicate needle and an Eppendorf FemtoJet microinjector equipped with an InjectMan NI2 manipulator. A single eye was injected with tumor cells in each zebrafish embryo. When indicated, 2.0 nL of a solution containing the anticancer drug under testing was injected in the same eye. After tumor cell injection, zebrafish embryos were selected under a fluorescence microscope to ensure that tumor cells were located only within the eyeballs and then grown at 33 °C.

2.6. Fluorescence and Light Sheet Microscopy

Live embryos were photographed at 1 h (t$_0$, 48 hpf), 1 day (t$_1$) and 4 days (t$_4$) post implantation on agarose-coated dishes using an AxioZoom V16 fluorescence stereomicroscope (Zeiss, Oberkochen, Germany, EU) equipped with a digital Axiocam 506 color camera (Zeiss). The mean area of the tumor was manually measured using FIJI software [24]. Light sheet microscopy experiments were performed using a Light Sheet Z.1 microscope (Zeiss). For this purpose, t$_0$, t$_1$, and t$_4$ embryos were embedded in a low melting agarose cylinder (1% low melting agarose:fish water, 1:1) and immersed in the observation chamber filled with fish water and anesthetic. Maximum intensity projections were obtained using the Zen software (Zeiss) and 3D reconstructions were made after z-stack processing with Arivis software (Zeiss).

To detect apoptotic cells, 48 hpf embryos were microinjected in the eye with 2.0 nL of a solution containing the anticancer drug under testing. After injection, zebrafish embryos were grown at 33 °C for 4 days. At t$_4$, live embryos were soaked in fish water containing 2 μg/mL acridine orange and incubated at 28 °C for 20 min. After 8 washes for 5 min each with fish water, embryos were anesthetized and analyzed immediately with a fluorescence stereomicroscope (Zeiss).

2.7. Immunohistochemical Analysis

After tumor cell injection, zebrafish embryos were formalin-fixed, paraffin-embedded, and sections of grafted eyes were analyzed at t$_0$ and t$_4$ by hematoxylin and eosin (H&E) or immunohistochemical staining [25]. Briefly, sections were de-waxed, rehydrated, and endogenous peroxidase activity blocked with 0.3% H$_2$O$_2$ in methanol. Antigen retrieval was performed using a thermostatic bath (Labochema, Vilnius, Lithuania), in 10 mM citrate buffer (pH 6.0). Sections were then washed in TBS (pH 7.4) and incubated overnight with a mouse monoclonal (IgM isotype) anti-mouse vimentin antibody (1:200) or with a rat anti-mouse Ki-67 antibody (1:100) or with a rabbit anti-human cleaved caspase 3 (Asp175) (1:100) diluted in TBS plus 1% BSA, 0.1% Triton x-100, and 0.1% Tween, followed by 1 h incubation with biotinylated anti-mouse IgM, anti-rat, or anti-rabbit antibody (1:200), respectively. Signal was revealed using Biotin Avidin system Vectastain ABC reagent followed by DAB as chromogen and hematoxylin as counterstain. Images were taken using an Axio Imager A2 microscope equipped with a digital AxioCam MRc5 camera (Zeiss).

2.8. Luciferase-Based Quantification Method

At different time points after intraocular grafting of luc^+ cells, enucleated eyes or anesthetized embryos were singularly placed in a well of a white polystyrene 96-well plate (Sigma-Aldrich). Embryo medium was removed and replaced with 50 μL of lysis buffer (80 mM Na_2HPO_4, 9.3 mM NaH_2PO_4, 2% TritonX100, 1.0 mM DTT in MilliQ water) and 50 μL of ONE-Glo™ Reagent. The luminescence was measured using an EnSight® Multimode Plate Reader (PerkinElmer, Milan, Italy) and expressed as relative luminescence units (RLUs).

To generate the calibration curve, a fixed number of B16-LS9-luc^+ cells (ranging from 0 to 1000 cells) was added to non-injected embryos and then the bioluminescence signal quantified as described above.

2.9. Statistical Analysis

Statistical analysis was performed with GraphPad Prism 8 (San Diego, CA, USA) using a Student's *t*-test for 2 groups of samples or one-way analysis of variance followed by Tukey's multiple comparison post hoc test for more than 2 groups. Differences were considered significant when *p*-values < 0.05.

3. Results and Discussion

3.1. Zebrafish Embryo Eye Is a Permissive Environment for the Growth of Engrafted Melanoma Cells

To evaluate whether the zebrafish embryo eye represents a microenvironment suitable for the grafting of melanoma cells, we first assessed the behavior of the well-characterized model of invasive murine melanoma represented by B16-BL6-DsRed$^+$ cells [2], which were injected into the eye of zebrafish embryos at 48 hpf. At this stage, the embryo eye consists of the retina (mainly composed of neuronal cells that will progressively organize in stratified retinal layers [26]), the hyaloid, and the ciliary vascular systems [27]. On this basis, B16-BL6-DsRed$^+$ cells were orthotopically injected in the posterior side of the developing eye of tg(*kdrl*:EGFP) embryos (100 cells/embryo) and monitored for the following 4 days by light sheet fluorescence microscopy. One hour after implantation (t_0), maximum intensity projection of the z-stacks and 3D reconstructions confirmed that DsRed$^+$ cells were present at the bottom of the eye in the proximity of the developing choroidal vasculature (Figure 1A,A'). One day post implantation (t_1), cells relocate towards the eye surface, interacting with the surrounding vasculature (Figure 1B,B'). At 4 days post implantation (t_4), DsRed$^+$ cells invaded the lens surface and grew without exerting a significant impact on the anatomical architecture of the eye (Figure 1C,C'). To confirm these observations, paraffin sections of tumor cell-grafted eyes were analyzed at t_0 and t_4 by H&E staining and by Ki-67 and vimentin immunostaining. As shown in Figure 2, implanted B16-BL6-DsRed$^+$ cells were able to proliferate, as demonstrated by the presence of Ki-67$^+$ cells, without affecting the physiological development of the retina. Moreover, cells moved towards the eye surface and invaded the lens (Figure 2B). Notably, preliminary observations suggest that the displacement of melanoma cells observed at t_4 is in part a consequence of the invasive properties of cancer cells and in part due to the remodeling of the eye that occurs during embryo development, which plays a not negligible role in tumor cell localization within the eye (data not shown).

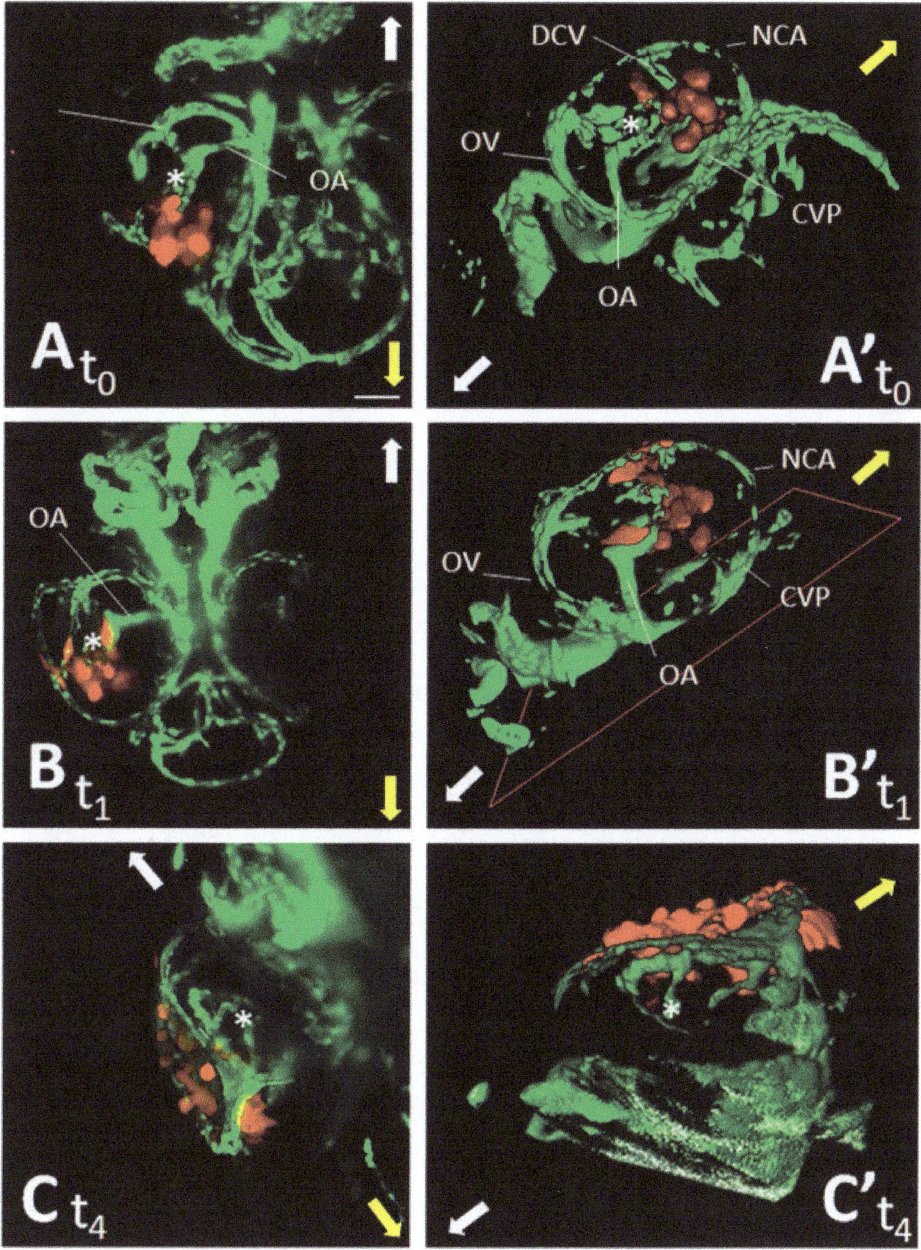

Figure 1. Zebrafish embryo eye is a microenvironment suitable for cell grafting. Murine melanoma B16-BL6-DsRed$^+$ cells (100 cells/embryo) were orthotopically injected in the posterior side of the developing eye of transgenic tg(*kdrl*:EGFP) zebrafish embryos at 48 hpf. Maximum intensity projection of the z-stacks (**A–C**) and 3D reconstructions (**A′–C′**) of B16-BL6-DsRed$^+$ cells performed at 1 h (t$_0$) (**A**,**A′**), 1 day (t$_1$) (**B**,**B′**), and 4 days (t$_4$) (**C**,**C′**) post implantation. (**A**,**B**) ventral view; (**C**) dorsal view. Asterisk indicates the hyaloid artery. Arrows indicate embryo orientation: white arrow, posterior side; yellow arrow, anterior side. CVP, choroidal vascular plexus; DCV, dorsal ciliary vein; NCA, nasal ciliary artery; OA, optic artery; OV, optic vein. Scale bar: 50 μm.

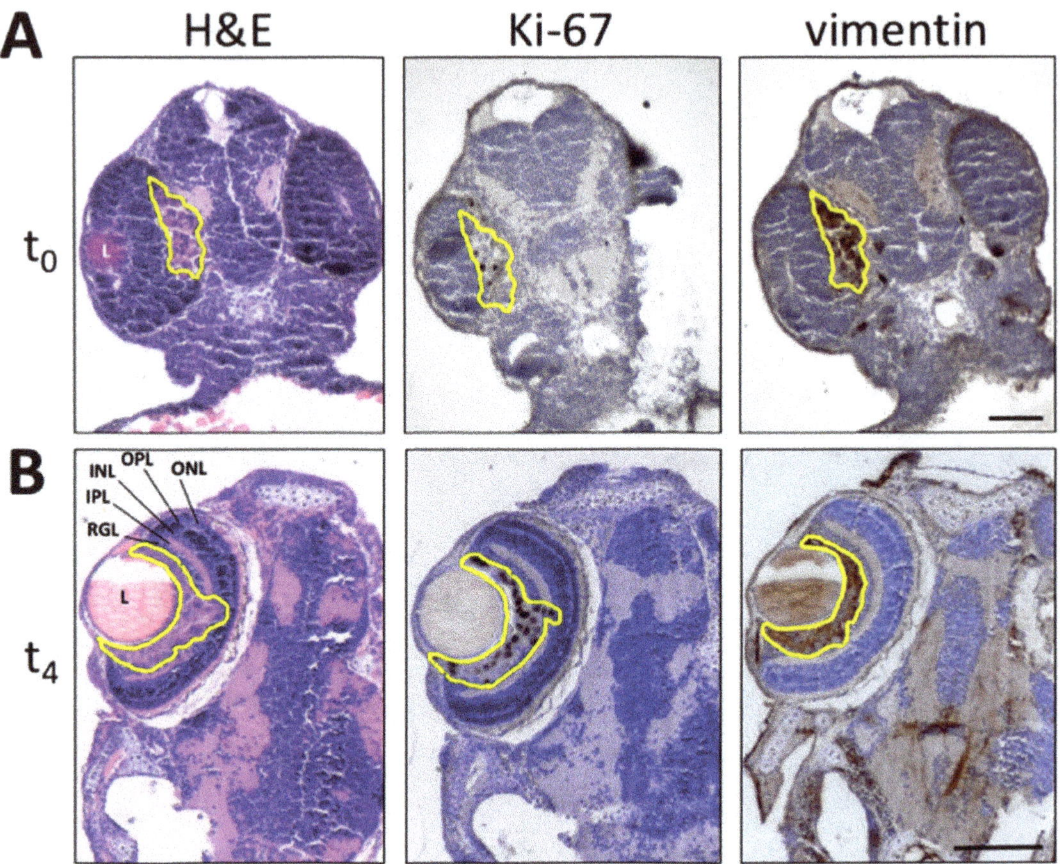

Figure 2. Histological analysis of melanoma B16-BL6-DsRed⁺ xenografts. Paraffin sections of B16-BL6-DsRed⁺ cells grafted into zebrafish embryo eyes obtained at 1 h (t_0) (**A**) or 4 days (t_4) post implantation (**B**) are stained by H&E (**left panel**) whereas Ki-67 (**central panel**) and vimentin (**right panel**) immunoreactivity is shown in brown. Tumor area is highlighted in yellow. L, lens; INL, inner nuclear layer; IPL, inner plexiform layer; ONL, outer nuclear layer; OPL, outer plexiform layer; RGL, retinal ganglion cell layer. Scale bars: 50 μm.

3.2. Quantification of Melanoma Xenograft Growth in the Zebrafish Embryo Eye

To obtain a reliable and reproducible quantification of melanoma cell growth in zebrafish embryo eyes, we performed a first set of experiments exploiting the fluorescence signal of B16-BL6-DsRed⁺ cells. For this purpose, we attempted to measure fluorescent tumor areas in engrafted embryos at t_0, t_1, and t_4 after injection. As anticipated, even though the analysis of digitalized images demonstrated an increase of DsRed⁺ tumor area at t_4 when compared to the other time points (Supplementary Materials, Figure S1), the results suffered significant drawbacks. Indeed, although extended depth of focus of the z-stacks provided a good quality lateral view of the xenografts at t_0 and t_1, the acquisition of images required to cover the entire thickness of the tumor was difficult at t_4 and was affected by the position of tumor cells that were close to the lens or deeply immersed in the eye.

In addition, the three-dimensional structure of the embryo eye and the presence of the lens, which may generate distorted images, made problematic the choice of the best angle for image acquisition. In this context, the optical accessibility of the zebrafish eye is further limited by the presence of pigmented cells, including neural crest-derived chromatophores (i.e., melanophores, iridophores, and xanthophores) and the retinal pigment epithelium [28]. Moreover, the blockade of zebrafish pigmentation by the addition of PTU in the fish water [22] has no effect on iridophores and on their nonspecific fluorescence signal [29], which impairs the reproducibility of the quantification technique.

On the other hand, it has been shown that the use of transparent *crystal* zebrafish mutants does not completely avoid refraction of the light due to the presence of the lens and of residual xanthophores present in the mutant eyes [29]. Finally, even though high-quality images may be obtained by confocal microscopy [11], acquisition and analysis procedures are time consuming and not suitable for high-throughput analysis.

To overcome these limitations, we developed an alternative quantification method exploiting the bioluminescence signal generated by tumor cells transduced with firefly luciferase. To this end, we took advantage of a red fluorescent and luciferase express-ing human melanoma cell line (A375M-RFP$^+$/luc$^+$) available in our laboratory. A375M-RFP$^+$/luc$^+$ cells were grafted in the eye of 48 hpf-old zebrafish embryos at 50, 100, and 200 cells/injection. Then, injected and not injected contralateral eyes were enucleated 1 h after grafting. As shown in the Supplementary Materials, Figure S2, analysis of grafted eyes indicates that the bioluminescence signal increases in a cell dose-dependent manner, being distinct from the basal levels measured in the contralateral control eyes. Similar results were obtained by measuring the bioluminescence signal generated by the lysates of the whole embryos engrafted with A375M-RFP$^+$/luc$^+$ cells (data not shown), thus avoiding the technically difficulty and the time-consuming eye enucleation procedure and confirming the reliability of this quantification method.

To assess whether this procedure allowed a quantitative evaluation of the growth of grafted tumors, A375M-RFP$^+$/luc$^+$ cells (100 cells/embryo) were injected in the eye of tg(*kdrl*:EGFP) embryos at 48 hpf. Then, injected embryos were analyzed at t_0 and t_4 by light sheet fluorescence microscopy or by evaluation of the bioluminescence of the lysates of the whole animals. As shown in Figure 3A, A375M-RFP$^+$/luc$^+$ cells were clearly visible 1 h after grafting in the embryo eye. At 4 days post implantation, grafted cells were alive and had moved from the injection site toward the lens surface, as already observed for B16-BL6-DsRed$^+$ cells. In parallel, a significant increase of the A375M-RFP$^+$/luc$^+$ cell-related bioluminescence signal was measured at t_4 when compared to t_0, thus confirming the capacity of this protocol to monitor the relative growth of tumor grafts (Figure 3B).

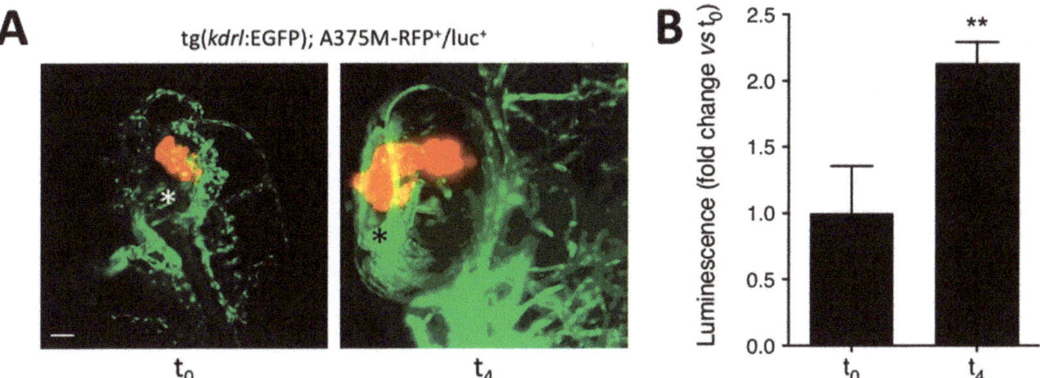

Figure 3. Luciferase-based quantification of the growth of human melanoma A375M-RFP$^+$/luc$^+$ xenografts. Human melanoma A375M-RFP$^+$/luc$^+$ cells (100 cells/embryo) were injected into the posterior side of the developing eye of trans-genic tg(*kdrl*:EGFP) zebrafish embryos at 48 hpf. (**A**) Maximum intensity projection of the z-stacks of A375M-RFP$^+$/luc$^+$ cells performed at 1 h (t_0) and 4 days (t_4) post implantation. T_0, lateral view, anterior to the top; t_4, dorsal view, anterior to the top. Asterisk indicates the superficial ocular vasculature. Scale bar: 50 µm. (**B**) Evaluation of A375M-RFP$^+$/luc$^+$ bioluminescence signal in the lysates of the whole embryos at t_0 and t_4. Data are the mean $\pm$ SEM ($n = 8$). ** $p < 0.01$ vs. t_0, Student's *t*-test.

3.3. Orthotopic Ocular Grafting in the Zebrafish Embryo as a Model for Uveal Melanoma Treatment

Given the promising capacity of luciferase-expressing melanoma cells to grow and to be quantified after grafting in zebrafish eyes, we decided to extend this assay to a well-established murine melanoma model suitable for investigating the mechanisms responsible for uveal melanoma liver tropism [30–32], immunologic and angiogenic aspects [33], and drug response [34–37]. On this basis, B16-LS9-luc$^+$ cells were injected in the zebrafish embryo eye, grafts were analyzed at t_0 and t_4, and immunohistochemical analysis of cell grafts showed that tumor cells proliferate, as already observed for B16-BL6 tumors (Figure 4A). In addition, bioluminescence quantification performed at different time points after injection (t_0, t_1, t_2, t_3, and t_4) showed that, after a slight decrease in cell growth at t_1, B16-LS9-luc$^+$ cells proliferate rapidly, their cell number increasing up to four times at t_3/t_4 when compared to t_0 (Figure 4B and Supplementary Materials, Figure S3).

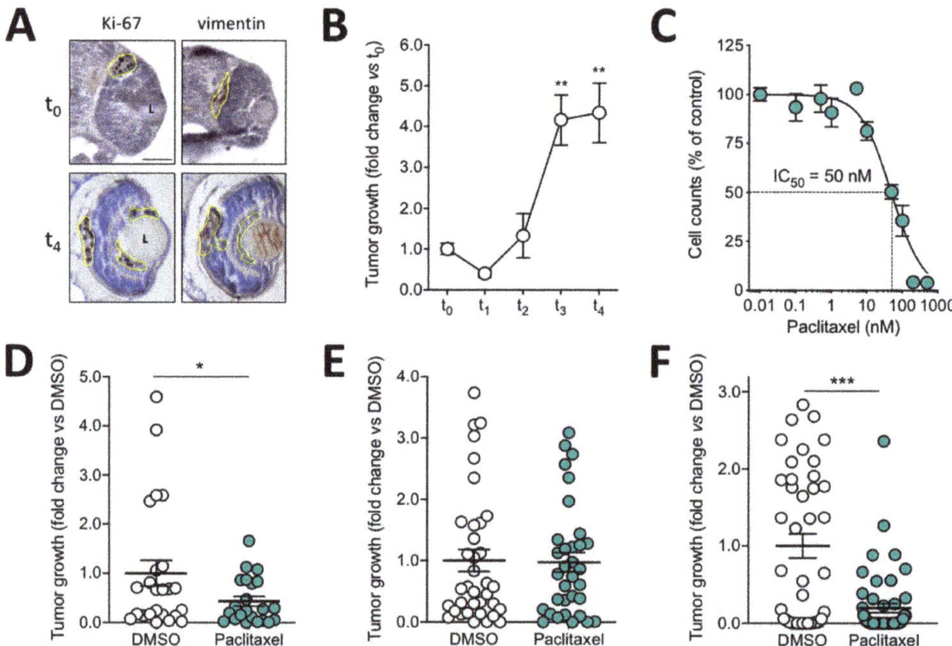

Figure 4. Effect of paclitaxel on the growth of murine melanoma B16-LS9-luc$^+$ xenografts. (**A**) Immunohistochemical analysis of zebrafish embryo eyes at 1 h (t_0) and 4 days (t_4) after orthotopic injection of B16-LS9-luc$^+$ cells. Ki-67 (**left panel**) and vimentin (**right panel**) are detected in brown. Tumor area is highlighted in yellow. L, lens. Scale bar: 50 μm. (**B**) B16-LS9-luc$^+$ bioluminescence signal was evaluated 1 h (t_0), 1 day (t_1), 2 days (t_2), 3 days (t_3), and 4 days (t_4) post implantation in the lysates of the whole embryos. Data are the mean ± SEM of five independent experiments. ** $p < 0.01$ vs. t_0 and t_1, ANOVA. (**C**) Effect of paclitaxel on the proliferation of B16-LS9-luc$^+$ cells in vitro. Viable cells were counted after 72 h of incubation with increasing concentrations of the drug. Data are the mean ± SEM of three independent experiments. (**D**) B16-LS9-luc$^+$ cells were cultured for 24 h in vitro in the absence or in the presence of 0.5 μM paclitaxel or with the corresponding volume of DMSO and then grafted in the zebrafish eye. Tumor growth was evaluated at t_4 by measuring the cell luminescence signal in the lysates of the whole embryos. Data are the mean ± SEM ($n = 20$). * $p < 0.05$ vs. DMSO, Student's *t*-test. (**E**) After injection of B16-LS9-luc$^+$ cells into the zebrafish eye, embryos were incubated at t_0 with 10 μM paclitaxel or with the corresponding volume of DMSO, both dissolved in fish water. Tumor growth was evaluated at t_4 by measuring the cell luminescence signal. Data are the mean ± SEM ($n = 35$). (**F**) After B16-LS9-luc$^+$ cell grafting into the zebrafish eye, 0.4 pmoles/embryo of paclitaxel or of the corresponding volume of DMSO were injected in the same eye. Tumor growth was evaluated at t_4 by measuring the cell luminescence signal. Data are the mean ± SEM ($n = 45$). In (**D–F**), each dot represents one embryo. *** $p < 0.0001$ vs. DMSO, Student's *t*-test.

In order to assess the response of tumor cells grafted in the embryo eye to anticancer drugs, preliminary experiments were carried out in which B16-LS9-luc$^+$ cells were treated in vitro for 72 h with increasing concentrations of the microtubule-disrupting agent paclitaxel [38]. As shown in Figure 4C, the compound inhibits the growth of B16-LS9-luc$^+$ cells with an ID$_{50}$ equal to 50 nM. On this basis, three different routes of in vivo administration of the drug were attempted in engrafted zebrafish embryos: (i) 24 h in vitro pretreatment of B16-LS9-luc$^+$ cells with 0.5 μM paclitaxel, followed by their injection in the zebrafish eye; (ii) injection of B16-LS9-luc$^+$ cells in the embryo eye, followed by incubation of engrafted embryos with 10 μM paclitaxel dissolved in fish water—an experimental procedure frequently used to test compounds in zebrafish [39]; (iii) engraftment of cells in the zebrafish embryo eye, followed by injection of the drug at 0.4 pmoles/embryo in the same eye. At the end of each protocol, the growth of B16-LS9-luc$^+$ grafts was assessed by bioluminescence-based quantification of luc$^+$ tumor cells performed at t_4.

As anticipated, pretreatment with paclitaxel resulted in a significant inhibition of the growth of the tumor grafts (Figure 4D). No inhibition of the growth of B16-LS9-luc$^+$ grafts was observed when engrafted embryos were treated with paclitaxel dissolved in the fish water, possibly as a consequence of the limited entry of the drug in the eye compartment (Figure 4E). Interestingly, a significant inhibition of B16-LS9-luc$^+$ tumor growth occurred when paclitaxel was directly injected in the embryo eye after cell grafting (Figure 4F).

Based on these observations, we decided to extend this experimental model by setting up an orthotopic experimental protocol in which human 92.1-RFP$^+$/luc$^+$ uveal melanoma cells were grafted (100 cells/embryo) in zebrafish embryo eyes at 48 hpf, followed by injection in the same eyes with 0.4 pmoles of different canonical chemotherapeutic drugs (i.e., paclitaxel [38], the histone deacetylase inhibitor panobinostat [40], the mTOR inhibitor everolimus [41], or vehicle). As shown in Figure 5A, all drugs inhibited the growth of uveal melanoma 92.1-RFP$^+$/luc$^+$ cells in vitro, with ID$_{50}$ values ranging between 10 nM and 67 nM. Accordingly, treatment of engrafted uveal melanoma cells by eye injection of paclitaxel, panobinostat, or everolimus caused a significant inhibition of tumor growth when assessed by measurement of bioluminescence (Figure 5B). Notably, panobinostat exerts a pro-apoptotic effect on cancer cells, both in vitro and in vivo (Supplementary Materials, Figure S4A,B). Moreover, no significant toxic or pro-apoptotic effect was observed in the zebrafish embryo eye tissue after the injection of the three drugs (Supplementary Materials, Figure S4C).

In addition, light sheet fluorescence microscopy confirmed the efficacy of drug treatment, uveal melanoma cells remaining confined at the bottom of the eye in the proximity of the choroidal vasculature (Figure 5C). These results are in line with previous observations about the efficacy of these drugs on uveal melanoma growth in in vitro and in vivo experimental models [42,43]. Relevant to this point, it must be pointed out that phase 2 clinical trials designed to evaluate the clinical benefits of paclitaxel or everolimus administration showed only a limited efficacy in uveal melanoma metastatic patients [44,45], whereas no data are available about the effect of panobinostat. Further studies will be required to assess the efficacy of histone deacetylase inhibitors in uveal melanoma.

In this paper, we describe the first orthotopic model of uveal melanoma in zebrafish, previous models of uveal melanoma being limited to the injection of cancer cells into the yolk sac of embryos [46–49]. Even though orthotopic models are usually less rapid and more technically challenging with respect to the heterotopic implants, these approaches are more tissue-specific and allow a more realistic recapitulation of the natural microenvironment in which the tumor originated. Altogether, our data extend previous observations about the possibility of engrafting tumor cells, including retinoblastoma and conjunctival melanoma cells, in zebrafish embryo eyes, thus generating orthotopic models of different ocular neoplasms [7–9]. In addition, it should be considered that the eye represents a metastatic site for various tumor types, including cutaneous melanoma, breast, and lung cancer, with choroidal metastases occurring in approximately 8% of human malignancies [50]. Thus, tumor cell grafting in the zebrafish embryo eye may be exploited as a useful orthotopic model to investigate novel therapeutic approaches not only for primary tumors but for eye metastases as well. Relevant to this point, our work focuses on providing a simple and reliable strategy for the accurate quantification of engrafted tumor cells by exploiting the bioluminescent signal of firefly luciferase-expressing cells. Indeed, the presence of the lens and the cup-like structure of the eye make difficult the acquisition of high-quality fluorescent images and may lead to misleading results. Moreover, the autofluorescent properties of zebrafish embryos and mammalian cells increase the non-specific background and decrease the sensitivity of signal detection. On the other hand, bioluminescence displays a higher detection capacity and allows for greater sensitivity because of the enzymatic nature of the bioluminescent reporter and the absence of the endogenous bioluminescence of cellular components.

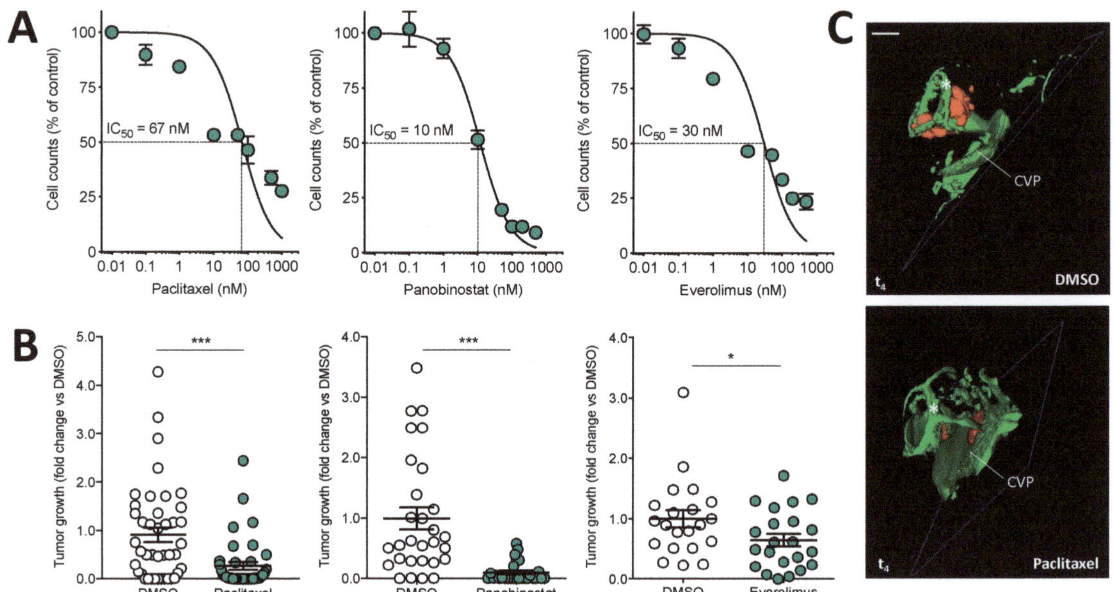

Figure 5. Effect of anticancer drugs on the growth of human uveal melanoma 92.1-RFP$^+$/luc$^+$ xenografts. (**A**) Effect of paclitaxel, panobinostat, and everolimus treatments on the proliferation of 92.1-RFP$^+$/luc$^+$ cells in vitro. Viable cells were counted after 48 h of incubation with increasing concentrations of paclitaxel or panobinostat or after 72 h of incubation with everolimus. Data are the mean ± SEM of two independent experiments. (**B**) After 92.1-RFP$^+$/luc$^+$ cell grafting into the zebrafish eye, 0.4 pmoles/embryo of paclitaxel, panobinostat, everolimus or the corresponding volume of DMSO were injected in the same eye. Tumor growth was evaluated at t_4 by measuring the cell luminescence signal in the lysates of the whole embryos. Data are the mean ± SEM of two independent experiments. Each dot represents one embryo. * $p < 0.05$ and *** $p < 0.001$ vs. DMSO, Student's t-test. (**C**) 3D reconstruction of the eye region of 92.1-RFP$^+$/luc$^+$ xenografts evaluated 4 days post implantation in the absence or in the presence of paclitaxel injection. Scale bar: 50 μm. Asterisk indicates the superficial ocular vasculature; CVP, choroidal vascular plexus.

The luminescence-based method herein described allows for a precise quantification without relying on any image analysis software and it provides a simple and quick in vivo evaluation of the efficacy of anticancer drugs after intraocular delivery. In this context, this approach may be exploited for high-throughput analysis and may have relevant implications for the evaluation of new low molecular weight compounds for the treatment of uveal melanoma and other primary ocular neoplasms and metastatic tumors endowed with ocular tropism.

4. Conclusions

In this paper, we described an orthotopic model of uveal melanoma in which tumor cells are grafted in the eye of zebrafish embryos in the proximity of the developing choroidal vasculature. In the following 3–4 days, grafted cells proliferate and move towards the eye surface, thus demonstrating that the zebrafish embryo eye is a permissive environment for the growth of UM cells. In addition, the use of firefly luciferase bioluminescent murine and human tumor cells allowed the assessment of the antitumor activity of candidate drugs when injected into the grafted eyes. In conclusion, we have established a new quantification method based on the ocular implantation of bioluminescent uveal melanoma cells in zebrafish embryos that may represent a useful in vivo orthotopic model suitable for the screening of novel drug candidates for uveal melanoma therapy.

Supplementary Materials: The following are available online at https://www.mdpi.com/article/10.3390/biomedicines9121873/s1, Figure S1: Fluorescence-based quantification of the growth of murine melanoma B16-BL6-DsRed$^+$ xenografts, Figure S2: Luciferase-based quantification of human melanoma A375M-RFP$^+$/luc$^+$ xenografts, Figure S3: Luciferase-based quantification of murine melanoma B16-LS9-luc$^+$ xenografts, Figure S4: Everolimus, paclitaxel, and panobinostat effects on cell apoptosis.

Author Contributions: Conceptualization, C.T. and S.R.; methodology, C.T. and S.R.; investigation, C.T., D.C., R.R., A.L., J.G., E.S. and S.R.; formal analysis: C.T. and S.R.; visualization: S.R.; writing—original draft preparation, C.T. and S.R.; writing—review and editing, R.R. and S.R.; supervision, F.S. and S.R.; project administration, S.R.; funding acquisition, R.R. and F.S. All authors have read and agreed to the published version of the manuscript.

Funding: This work was supported by Associazione Italiana Ricerca sul Cancro (AIRC) IG grant n° 23151 to R.R.; S.R. was supported by Fondazione Umberto Veronesi and Associazione Garda Vita R. Tosoni fellowships.

Institutional Review Board Statement: Not applicable.

Informed Consent Statement: Not applicable.

Data Availability Statement: Data are contained within the article or the Supplementary Materials.

Acknowledgments: The authors wish to thank M. Presta for his valuable suggestions and helpful discussion.

Conflicts of Interest: The authors declare no conflict of interest.

References

1. Santoriello, C.; Zon, L.I. Hooked! Modeling human disease in zebrafish. *J. Clin. Investig.* **2012**, *122*, 2337–2343. [CrossRef]
2. Tobia, C.; Gariano, G.; De Sena, G.; Presta, M. Zebrafish embryo as a tool to study tumor/endothelial cell cross-talk. *Biochim. Biophys. Acta* **2013**, *1832*, 1371–1377. [CrossRef]
3. Rezzola, S.; Paganini, G.; Semeraro, F.; Presta, M.; Tobia, C. Zebrafish (*Danio rerio*) embryo as a platform for the identification of novel angiogenesis inhibitors of retinal vascular diseases. *Biochim. Biophys. Acta* **2016**, *1862*, 1291–1296. [CrossRef] [PubMed]
4. Lee, H.-C.; Lin, C.-Y.; Tsai, H.-J. Zebrafish, an in vivo platform to screen drugs and proteins for biomedical use. *Pharmaceuticals* **2021**, *14*, 500. [CrossRef]
5. Barriuso, J.; Nagaraju, R.; Hurlstone, A. Zebrafish: A new companion for translational research in oncology. *Clin. Cancer Res.* **2015**, *21*, 969–975. [CrossRef]
6. Letrado, P.; De Miguel, I.; Lamberto, I.; Díez-Martínez, R.; Oyarzabal, J. Zebrafish: Speeding up the cancer drug discovery process. *Cancer Res.* **2018**, *78*, 6048–6058. [CrossRef]
7. Jo, D.H.; Son, D.; Na, Y.; Jang, M.; Choi, J.-H.; Kim, J.H.; Yu, Y.S.; Seok, S.H.; Kim, J.H. Orthotopic transplantation of retinoblastoma cells into vitreous cavity of zebrafish for screening of anticancer drugs. *Mol. Cancer* **2013**, *12*, 71–79. [CrossRef] [PubMed]
8. Chen, X.; Wang, J.; Cao, Z.; Hosaka, K.; Jensen, L.; Yang, H.; Sun, Y.; Zhuang, R.; Liu, Y.; Cao, Y. Invasiveness and metastasis of retinoblastoma in an orthotopic zebrafish tumor model. *Sci. Rep.* **2015**, *5*, srep10351. [CrossRef]
9. Chen, Q.; Ramu, V.; Aydar, Y.; Groenewoud, A.; Zhou, X.-Q.; Jager, M.J.; Cole, H.; Cameron, C.G.; McFarland, S.A.; Bonnet, S.; et al. TLD1433 photosensitizer inhibits conjunctival melanoma cells in zebrafish ectopic and orthotopic tumour models. *Cancers* **2020**, *12*, 587. [CrossRef] [PubMed]
10. Zhang, B.; Shimada, Y.; Kuroyanagi, J.; Umemoto, N.; Nishimura, Y.; Tanaka, T. Quantitative phenotyping-based in vivo chemical screening in a zebrafish model of leukemia stem cell xenotransplantation. *PLoS ONE* **2014**, *9*, e85439. [CrossRef]

11. Hill, D.; Chen, L.; Snaar-Jagalska, E.; Chaudhry, B. Embryonic zebrafish xenograft assay of human cancer metastasis. *F1000Research* **2018**, *7*, 1682. [CrossRef]
12. Jovanovic, P.; Mihajlovic, M.; Djordjevic-Jocic, J.; Vlajkovic, S.; Cekic, S.; Stefanovic, V. Ocular melanoma: An overview of the current status. *Int. J. Clin. Exp. Pathol.* **2013**, *6*, 1230–1244. [PubMed]
13. Yonekawa, Y.; Kim, I.K. Epidemiology and management of uveal melanoma. *Hematol. Oncol. Clin. N. Am.* **2012**, *26*, 1169–1184. [CrossRef]
14. Mahendraraj, K.; Lau, C.S.; Lee, I.; Chamberlain, R.S. Trends in incidence, survival, and management of uveal melanoma: A population-based study of 7516 patients from the Surveillance, Epidemiology, and End Results database (1973–2012). *Clin Ophthalmol.* **2016**, *10*, 2113–2119. [CrossRef]
15. Vivet-Noguer, R.; Tarin, M.; Roman-Roman, S.; Alsafadi, S. Emerging therapeutic opportunities based on current knowledge of uveal melanoma biology. *Cancers* **2019**, *11*, 1019. [CrossRef]
16. Bedikian, A.Y. Metastatic uveal melanoma therapy. *Int. Ophthalmol. Clin.* **2006**, *46*, 151–166. [CrossRef]
17. Croce, M.; Ferrini, S.; Pfeffer, U.; Gangemi, R. Targeted therapy of uveal melanoma: Recent failures and new perspectives. *Cancers* **2019**, *11*, 846. [CrossRef]
18. Diaz, C.E.; Rusciano, D.; Dithmar, S.; Grossniklaus, H.E. B16LS9 melanoma cells spread to the liver from the murine ocular posterior compartment (PC). *Curr. Eye Res.* **1999**, *18*, 125–129. [CrossRef]
19. Ronca, R.; Giacomini, A.; Di Salle, E.; Coltrini, D.; Pagano, K.; Ragona, L.; Matarazzo, S.; Rezzola, S.; Maiolo, D.; Torella, R.; et al. Long-pentraxin 3 derivative as a small-molecule FGF trap for cancer therapy. *Cancer Cell* **2015**, *28*, 225–239. [CrossRef] [PubMed]
20. De Waard-Siebinga, I.; Blom, D.-J.R.; Griffioen, M.; Schrier, P.I.; Hoogendoorn, E.; Beverstock, G.; Danen, E.H.J.; Jager, M.J Establishment and characterization of an uveal-melanoma cell line. *Int. J. Cancer* **1995**, *62*, 155–161. [CrossRef] [PubMed]
21. Rezzola, S.; Guerra, J.; Chandran, A.M.K.; Loda, A.; Cancarini, A.; Sacristani, P.; Semeraro, F.; Presta, M. VEGF-independent activation of Müller cells by the vitreous from proliferative diabetic retinopathy patients. *Int. J. Mol. Sci.* **2021**, *22*, 2179. [CrossRef]
22. Westerfield, M. *The Zebrafish Book. A Guide for the Laboratory Use of Zebrafish (Danio rerio)*, 4th ed.; University of Oregon Press Eugene, OR, USA, 2000.
23. Kimmel, C.B.; Ballard, W.W.; Kimmel, S.R.; Ullmann, B.; Schilling, T.F. Stages of embryonic development of the zebrafish. *Dev Dyn.* **1995**, *203*, 253–310. [CrossRef]
24. Schindelin, J.; Arganda-Carreras, I.; Frise, E.; Kaynig, V.; Longair, M.; Pietzsch, T.; Preibisch, S.; Rueden, C.; Saalfeld, S. Schmid, B.; et al. Fiji: An open-source platform for biological-image analysis. *Nat. Methods* **2012**, *9*, 676–682. [CrossRef]
25. Sabaliauskas, N.A.; Foutz, C.A.; Mest, J.R.; Budgeon, L.R.; Sidor, A.T.; Gershenson, J.A.; Joshi, S.B.; Cheng, K.C. High-throughput zebrafish histology. *Methods* **2006**, *39*, 246–254. [CrossRef] [PubMed]
26. Malicki, J.; Neuhauss, S.C.; Schier, A.F.; Solnica-Krezel, L.; Stemple, D.L.; Stainier, D.Y.; Abdelilah, S.; Zwartkruis, F.; Rangini, Z. Driever, W. Mutations affecting development of the zebrafish retina. *Development* **1996**, *123*, 263–273. [CrossRef]
27. Hashiura, T.; Kimura, E.; Fujisawa, S.; Oikawa, S.; Nonaka, S.; Kurosaka, D.; Hitomi, J. Live imaging of primary ocular vasculature formation in zebrafish. *PLoS ONE* **2017**, *12*, e0176456. [CrossRef]
28. Singh, A.; Nüsslein-Volhard, C. Zebrafish stripes as a model for vertebrate colour pattern formation. *Curr. Biol.* **2015**, *25*, R81–R92. [CrossRef]
29. Antinucci, P.; Hindges, R. A crystal-clear zebrafish for in vivo imaging. *Sci. Rep.* **2016**, *6*, 29490. [CrossRef] [PubMed]
30. Rusciano, D.; Lorenzoni, P.; Burger, M. Regulation of c-met expression in B16 murine melanoma cells by melanocyte stimulating hormone. *J. Cell Sci.* **1999**, *112 Pt 5*, 623–630. [CrossRef]
31. Elia, G.; Ren, Y.; Lorenzoni, P.; Zarnegar, R.; Burger, M.M.; Rusciano, D. Mechanisms regulating c-met overexpression in liver-metastatic B16-LS9 melanoma cells. *J. Cell. Biochem.* **2001**, *81*, 477–487. [CrossRef]
32. Jones, N.M.; Yang, H.; Zhang, Q.; Morales-Tirado, V.M.; Grossniklaus, H.E. Natural killer cells and pigment epithelial-derived factor control the infiltrative and nodular growth of hepatic metastases in an Orthotopic murine model of ocular melanoma. *BMC Cancer* **2019**, *19*, 484. [CrossRef]
33. Stei, M.M.; Loeffler, K.U.; Holz, F.G.; Herwig-Carl, M. Animal models of uveal melanoma: Methods, applicability, and limitations *BioMed Res. Int.* **2016**, *2016*, 4521807. [CrossRef]
34. Yang, W.; Li, H.; Mayhew, E.; Mellon, J.; Chen, P.W.; Niederkorn, J.Y. NKT cell exacerbation of liver metastases arising from melanomas transplanted into either the eyes or spleens of mice. *Investig. Opthalmol. Vis. Sci.* **2011**, *52*, 3094–3102. [CrossRef] [PubMed]
35. Yang, H.; Brackett, C.M.; Morales-Tirado, V.M.; Li, Z.; Zhang, Q.; Wilson, M.W.; Benjamin, C.; Harris, W.; Waller, E.K. Gudkov, A.; et al. The toll-like receptor 5 agonist entolimod suppresses hepatic metastases in a murine model of ocular melanoma via an NK cell-dependent mechanism. *Oncotarget* **2016**, *7*, 2936–2950. [CrossRef]
36. Ashur-Fabian, O.; Zloto, O.; Fabian, I.; Tsarfaty, G.; Ellis, M.; Steinberg, D.M.; Hercbergs, A.; Davis, P.J.; Fabian, I.D. Tetrac delayed the onset of ocular melanoma in an orthotopic mouse model. *Front. Endocrinol.* **2018**, *9*, 775. [CrossRef] [PubMed]
37. Rezzola, S.; Ronca, R.; Loda, A.; Nawaz, M.I.; Tobia, C.; Paganini, G.; Maccarinelli, F.; Giacomini, A.; Semeraro, F.; Mor, M.; et al. The autocrine FGF/FGFR system in both skin and uveal melanoma: FGF trapping as a possible therapeutic approach. *Cancers* **2019**, *11*, 1305. [CrossRef] [PubMed]
38. Owinsky, E.R.K.R.; Onehower, R.O.C.D. Paclitaxel (taxol). *N. Engl. J. Med.* **1995**, *332*, 1004–1014. [CrossRef]

9. Cassar, S.; Adatto, I.; Freeman, J.L.; Gamse, J.T.; Iturria, I.; Lawrence, C.; Muriana, A.; Peterson, R.T.; Van Cruchten, S.; Zon, L.I. Use of zebrafish in drug discovery toxicology. *Chem. Res. Toxicol.* **2020**, *33*, 95–118. [CrossRef]

10. Scuto, A.; Kirschbaum, M.; Kowolik, C.; Kretzner, L.; Juhasz, A.; Atadja, P.; Pullarkat, V.; Bhatia, R.; Forman, S.; Yen, Y.; et al. The novel histone deacetylase inhibitor, LBH589, induces expression of DNA damage response genes and apoptosis in Ph– acute lymphoblastic leukemia cells. *Blood* **2008**, *111*, 5093–5100. [CrossRef]

11. O'Reilly, T.; McSheehy, P.M. Biomarker development for the clinical activity of the mTOR inhibitor everolimus (RAD001): Processes, limitations, and further proposals. *Transl. Oncol.* **2010**, *3*, 65–79. [CrossRef]

12. Faião-Flores, F.; Emmons, M.F.; Durante, M.A.; Kinose, F.; Saha, B.; Fang, B.; Koomen, J.M.; Chellappan, S.P.; Maria-Engler, S.; Rix, U.; et al. HDAC inhibition enhances the in vivo efficacy of MEK inhibitor therapy in uveal melanoma. *Clin. Cancer Res.* **2019**, *25*, 5686–5701. [CrossRef] [PubMed]

13. Amirouchene-Angelozzi, N.; Frisch-Dit-Leitz, E.; Carita, G.; Dahmani, A.; Raymondie, C.; Liot, G.; Gentien, D.; Némati, F.; Decaudin, D.; Roman-Roman, S.; et al. The mTOR inhibitor Everolimus synergizes with the PI3K inhibitor GDC0941 to enhance anti-tumor efficacy in uveal melanoma. *Oncotarget* **2016**, *7*, 23633–23646. [CrossRef]

14. Shoushtari, A.N.; Ong, L.T.; Schoder, H.; Singh-Kandah, S.; Abbate, K.T.; Postow, M.A.; Callahan, M.K.; Wolchok, J.; Chapman, P.B.; Panageas, K.S.; et al. A phase 2 trial of everolimus and pasireotide long-acting release in patients with metastatic uveal melanoma. *Melanoma Res.* **2016**, *26*, 272–277. [CrossRef] [PubMed]

15. Homsi, J.; Bedikian, A.Y.; Papadopoulos, N.E.; Kim, K.B.; Hwu, W.-J.; Mahoney, S.L.; Hwu, P. Phase 2 open-label study of weekly docosahexaenoic acid–paclitaxel in patients with metastatic uveal melanoma. *Melanoma Res.* **2010**, *20*, 507–510. [CrossRef]

16. Van der Ent, W.; Burrello, C.; Teunisse, A.F.A.S.; Ksander, B.R.; Van Der Velden, P.A.; Jager, M.J.; Jochemsen, A.G.; Snaar-Jagalska, B.E. Modeling of human uveal melanoma in zebrafish xenograft embryos. *Investig. Opthalmol. Vis. Sci.* **2014**, *55*, 6612–6622. [CrossRef]

17. Fornabaio, G.; Barnhill, R.L.; Lugassy, C.; Bentolila, L.A.; Cassoux, N.; Roman-Roman, S.; Alsafadi, S.; Del Bene, F. Angiotropism and extravascular migratory metastasis in cutaneous and uveal melanoma progression in a zebrafish model. *Sci. Rep.* **2018**, *8*, 10448. [CrossRef]

18. Van der Ent, W.; Burrello, C.; De Lange, M.J.; Van Der Velden, P.A.; Jochemsen, A.G.; Jager, M.J.; Snaar-Jagalska, B.E. Embryonic zebrafish: Different phenotypes after injection of human uveal melanoma cells. *Ocul. Oncol. Pathol.* **2015**, *1*, 170–181. [CrossRef] [PubMed]

19. Yu, L.; Zhou, D.; Zhang, G.; Ren, Z.; Luo, X.; Liu, P.; Plouffe, S.W.; Meng, Z.; Moroishi, T.; Li, Y.; et al. Co-occurrence of BAP1 and SF3B1 mutations in uveal melanoma induces cellular senescence. *Mol. Oncol.* **2021**. [CrossRef] [PubMed]

20. Arepalli, S.; Kaliki, S.; Shields, C.L. Choroidal metastases: Origin, features, and therapy. *Indian J. Ophthalmol.* **2015**, *63*, 122–127. [CrossRef]

 *biomedicines*

Article

Knockout of *mafba* Causes Inner-Ear Developmental Defects in Zebrafish via the Impairment of Proliferation and Differentiation of Ionocyte Progenitor Cells

Xiang Chen [†], Yuwen Huang [†], Pan Gao, Yuexia Lv, Danna Jia, Kui Sun, Yunqiao Han, Hualei Hu, Zhaohui Tang, Xiang Ren * and Mugen Liu *

Key Laboratory of Molecular Biophysics of Ministry of Education, College of Life Science and Technology, Huazhong University of Science and Technology, 1037 Luoyu Road, Wuhan 430074, China; xiangchenhust816@outlook.com (X.C.); yuwenhuang_1994@foxmail.com (Y.H.); gaopan924989055@163.com (P.G.); lyuexia0614@163.com (Y.L.); jiadanna@foxmail.com (D.J.); shymcg@163.com (K.S.); 18064097079@163.com (Y.H.); d202080697@hust.edu.cn (H.H.); zh_tang@hust.edu.cn (Z.T.)
* Correspondence: renxiang@hust.edu.cn (X.R.); lium@hust.edu.cn (M.L.)
† These authors contributed equally to this work.

Abstract: Zebrafish is an excellent model for exploring the development of the inner ear. Its inner ear has similar functions to that of humans, specifically in the maintenance of hearing and balance. Mafb is a component of the Maf transcription factor family. It participates in multiple biological processes, but its role in inner-ear development remains poorly understood. In this study, we constructed *mafba* knockout (*mafba*$^{-/-}$) zebrafish model using CRISPR/Cas9 technology. The *mafba*$^{-/-}$ mutant inner ear displayed severe impairments, such as enlarged otocysts, smaller or absent otoliths, and insensitivity to sound stimulation. The proliferation of p63$^+$ epidermal stem cells and dlc$^+$ ionocyte progenitors was inhibited in *mafba*$^{-/-}$ mutants. Moreover, the results showed that *mafba* deletion induces the apoptosis of differentiated K$^+$-ATPase-rich (NR) cells and H$^+$-ATPase-rich (HR) cells. The activation of p53 apoptosis and G0/G1 cell cycle arrest resulted from DNA damage in the inner-ear region, providing a mechanism to account for the inner ear deficiencies. The loss of homeostasis resulting from disorders of ionocyte progenitors resulted in structural defects in the inner ear and consequently, loss of hearing. In conclusion, the present study elucidated the function of ionic channel homeostasis and inner-ear development using a zebrafish Mafba model and clarified the possible physiological roles.

Keywords: zebrafish; *mafba*; cell proliferation; cell differentiation; ion homeostasis; inner-ear development

Citation: Chen, X.; Huang, Y.; Gao, P.; Lv, Y.; Jia, D.; Sun, K.; Han, Y.; Hu, H.; Tang, Z.; Ren, X.; et al. Knockout of *mafba* Causes Inner-Ear Developmental Defects in Zebrafish via the Impairment of Proliferation and Differentiation of Ionocyte Progenitor Cells. *Biomedicines* **2021**, *9*, 1699. https://doi.org/10.3390/biomedicines9111699

Academic Editors: Jun Lu, James A. Marrs and Swapnalee Sarmah

Received: 11 August 2021
Accepted: 28 October 2021
Published: 16 November 2021

Publisher's Note: MDPI stays neutral with regard to jurisdictional claims in published maps and institutional affiliations.

1. Introduction

Approximately 360 million people are born with congenital and progressive hearing defects worldwide [1]. Genetic defects account for approximately 50% of hearing loss [2], most of which is caused by mutations in genes associated with inner-ear development [3]. The vertebrate inner ear is a conserved sensory organ responsible for vestibular and auditory functions [4]. It plays an important role in detecting body acceleration and keeping balance [5,6]. The mammalian ear is composed of inner, middle, and external components, while the zebrafish, belonging to the teleost fish, only exhibits the inner ear. Although there is considerable structural diversity of the inner ears among different species, the molecular basis of their development is relatively conserved [7]. Zebrafish has become an excellent vertebrate model for exploring the mechanism of inner-ear development and related diseases in recent years [8,9]. There is an urgent need to enhance our understanding of the molecular mechanisms of inner-ear development and to pragmatize biological strategies for restoring auditory functions.

The zebrafish otic vesicles consist of sensory hair cells and other non-sensory epithelial cells [10]. The barrier functions of the inner ear epithelial cells are essential to maintain homoeostasis in the otic lumen and the endolymph, which guides the development of hair cells, semicircular canals, and otoliths [11,12]. Ions and other components are secreted into the otic lumen and endolymph through ion channels of otic epithelial cells' membranes to maintain the steady state of endolymph homeostasis [13], which is also crucial for otolith formation [8,9,14,15]. Both the otic epithelium and the inner-ear epidermis regulate osmotic homeostasis through transporting ions and acid-base equivalents. In zebrafish embryos, the $p63^+$ epidermal stem cells become ionocytes or keratinocytes, determined by DeltaC (dlc)-Notch-mediated lateral inhibition [16,17]. The differentiated ionocytes develop mainly in the skin of embryos and maintain body fluid ionic homeostasis. Researchers have identified five types of ionocytes in zebrafish embryos, including NR cells, HR cells, SLC26-expressing cells, Na^+-Cl^- cotransporter-expressing (NCC) cells, and K^+-secreting (KS) cells [18]. These cells regulate osmotic homeostasis by utilizing various ion transporters [19,20]. Intriguingly, the specification and differentiation of ionocytes is controlled by a positive feedback loop between *foxi3a* and *foxi3b* [17,21,22]. It activates some important downstream transcription factors, such as *glial cell missing* 2 (*gcm2*), to determine which ionocyte progenitors differentiate into HR or NR cells following the upregulated expression of *foxi3a* or *foxi3b*, respectively [23]. The down-regulated expression of Na^+-K^+-Cl^- cotransporter *nkcc1* (*slc12a2*) and a few Na^+/K^+-ATPase channel genes results in a collapse of the otic vesicle in zebrafish embryos and the loss of endolymphatic fluid [24,25]. Consequently, the distribution of ion channels in the otic epithelial cells and inner-ear epidermis is essential for the homeostasis maintenance and inner-ear development.

Members of the MAF family are divided into two subfamilies. Large Maf family members (Mafa, Mafb, c-Maf, and Nrl) contain a C-terminal basic leucine-zipper domain (b-Zip), which mediates dimerization and DNA binding, and an N-terminal transactivation domain (TAD), which regulates target gene transcription. A small Maf family (Mafk, Mafg, and Maff) contains only a b-Zip domain [26]. The transcription factor *mafb* plays crucial roles in epidermal keratinocyte differentiation in the mammalian epidermis [27]. Numerous studies have identified the role of Mafb in the developmental of various tissues and in cellular differentiation. However, there is, as yet, no report on the association between ionocytes and *mafb*. For instance, *mafb* deprivation impairs the differentiation of podocyte, osteoclast, monocyte, and epidermal cells [27–30], the development of pancreatic islet α/β-cells [31], and the parathyroid gland [32] in mammals. Losses or decreases in the expression of mouse *mafb* may cause atopic dermatitis (AD) and psoriasis vulgaris [27]. In addition, *mafb* mutations in humans and mice lead to multicentric carpotarsal osteolysis (MCTO) [33], Duane retraction syndrome (DRS), aberrant extraocular muscle innervation, and inner-ear defects [34]. Prior studies have shown that *kreisler*/*mafb* mutant mice display inner-ear defects, which may be affected by abnormal adjacent segmented hindbrain development and signal transduction [35]. Silencing of the *valentino*/*mafba* in zebrafish impairs the signal regulation of *fgf3* in the hindbrain, which leads to the abnormal development of the hindbrain and otic vesicle [36]. In addition, a previous study showed that *mafb* deficiency promotes tumorigenesis in clinical colorectal cancer (CRC) and causes cell cycle arrest in the G0/G1 phase [37]. The balance of *mafb* and *cFos* or *cJun* heterodimer complexes controls the apoptosis-survival fate, which is required for triggering p53-dependent apoptosis in response to DNA damage [38]. These studies reveal the role of Mafb in regulating cell survival and various developmental processes. However, the specific mechanism of how *mafb* mutants cause inner-ear defects is poorly understood. Here, we suggest that the proliferation and differentiation of ionocytes are directly influenced by the transcription factor *mafba*, and then ion homeostasis, which affects the inner ear.

We constructed a zebrafish *mafba* (the homologous gene of mammalian *mafb*) knockout model using CRISPR/Cas9 technology. Loss of Mafba impairs inner-ear development in zebrafish embryos. The inner-ear defect in *mafba*$^{-/-}$ embryos results from a loss of ionic homeostasis, due to reduced populations of ionocyte progenitors and the apoptosis of

HR and NR ionocytes in the otic epithelium. We further confirmed that *mafba* deletion induces apoptosis in HR and NR cells, resulting from G0/G1 cell cycle arrest, which is associated with DNA damage. Our findings provide a novel insight into Mafba's role in the maintenance of ionic channel homeostasis and inner-ear development.

2. Materials and Methods

2.1. Zebrafish Maintenance

Zebrafish (*Danio rerio*) larvae and adults were raised in 28.5 °C-constant temperature recirculating water under a 14-h light/10-h dark cycle [39]. We collected and maintained embryos in E3 medium for 72 h until the larvae hatched. If needed, we added 0.003% phenylthiourea to the E3 medium at 12 h after fertilization to inhibit pigmentation of embryos. We anaesthetized the embryos and larvae with 0.02% tricaine before further analysis. The specific developmental stages of zebrafish are defined as hours postfertilization (hpf) and days postfertilization (dpf). All zebrafish experimental procedures were performed in accordance with guidelines approved by the Experimental Animal Ethics Committee of Huazhong University of Science and Technology.

2.2. Generation of Mafba Mutant by CRISPR/Cas9 Technology

The target sites were designed using a web tool (http://chopchop.cbu.uib.no/, accessed on 30 August 2019). The target sequences of *mafba* we used are listed in Table S1. The gRNA and Cas9 mRNA were synthesized with a TranscriptAid T7 High Yield Transcription Kit (Thermo Scientific, Waltham, MA, USA) and mMESSAGE mMACHINE T7 transcription synthesis kit (Invitrogen, Carlsbad, CA, USA), respectively. A mixture of 600 pg Cas9 mRNA and 200 pg gRNA were co-injected into the one-cell-stage, wild-type embryos. At 3 dpf, 18 injected zebrafish embryos were collected to identify F0 founders by genotyping in order to evaluate CRISPR/Cas9 system target efficiency. A 464-bases pair (bp) DNA product containing the *mafba* targeting site was amplified by PCR (primers used are listed in Table S1) and sequenced. The F1 generation was obtained through the outcrossing of F0 mutant zebrafish to wild-type zebrafish. The F2 generation of mutant zebrafish were obtained and raised to adults using the same method as for the F1 generation described above. The F2 generation *mafba*$^{+/-}$ zebrafish were crossed to produce the sibling (*mafba*$^{+/+}$) *mafba*$^{+/-}$, and *mafba*$^{-/-}$ mutant embryos.

2.3. Whole-Mount In Situ Hybridization

Whole-mount in situ hybridization (WISH) for zebrafish embryos was executed as previously described [40]. The sequences of all probes and primers are listed in Table S1. The images were acquired using an optical microscope (10 × lens, BX53, Olympus, Tokyo, Japan). After imaging, we identified the genotypes by extracting DNA from the embryos. The numbers at the bottom right of each picture indicate the number of embryos with a similar staining pattern across all embryos examined.

2.4. Whole-Mount Immunofluorescence Assay

Whole-mount immunofluorescence was performed as described [41,42]. The embryos were fixed with 4% paraformaldehyde at 4 °C overnight. After three washes with 0.5% PBST (PBS + 0.5% Triton X-100) for 10 min each, embryos were permeabilized in solution (PBS + 2% Triton X-100) to dissolve the otolith at room temperature for 8 h (time until the otolith is completely dissolved may be longer). Then, we washed the embryos with 0.5% PBST twice and blocked with them 10% goat serum in 0.5% PBST for 2 h. Next, the embryos were incubated with primary antibodies (diluted in 10% goat serum in 0.5% PBST) overnight at 4 °C. The following primary antibodies were used: p63 (Ab735, Abcam; 1:50, Cambridge, UK), phosphorylated histone H3 (AF3358, Affinity; 1:200), p63 (Ab124762, Abcam; 1:100), Dlc (Ab7336, Abcam; 1:100), Atp1a1a.4 (p09572, DHSB; 1:50) Atp6v1a (GTX129736, GeneTex; 1:100), γH2AX (9178s, CST; 1:200), Myo7a (20720, Proteintech; 1:200), Acetylated-α-Tubulin (sc-23950, Santa Cruz Biotechnology; 1:100), Mafba

(GTX128295, GenxTex; 1:100), and Alexa Fluor 594 Phalloidin (A12381, Thermo Scientific; 1:200). After 0.5% PBST washes, the embryos were treated with the corresponding secondary antibody (Alexa-Fluor 405-nm, Alexa-Fluor 488-nm, or 594-nm secondary antibody, 1:500, diluted in 10% goat serum of 0.5% PBST) at room temperature in the dark for 3 h. For the double immunofluorescence labeling, the samples continued treatment in accordance with the standard immunofluorescence procedure described above. We washed the embryos with PBST twice and stored them in PBS at 4 °C. The samples images were captured using a confocal microscope (FV31S-SW, Olympus) with a 10 × lens (NA = 0.40), fully automatic and continuously adjustable in the range of 50–800 nm, independent spectral fluorescence detector, and fluorescence detection channel. The spectral windows were blue (359–461 nm), green (500–580 nm), and red (610–700 nm), respectively. Histogram normalizations and gamma adjustments were adjusted before the image acquisition, and we did not make adjustments at the later stage. Image stitching and projections were carried out through the ImageJ v1.8.0 software [43,44].

2.5. EDU and TUNEL Assay

Zebrafish embryos were treated with 2 mM EDU (5-ethynyl-2′- deoxyuridine) for 30 min at 4 °C in egg-water. The embryos were transferred to fresh E3 medium for 2 h and fixed in 4% paraformaldehyde at 4 °C overnight. We treated the zebrafish embryos with Cell-Light™ EdU Cell Proliferation Detection (C10310-1, Ribobio, Guangzhou, China), according to the operating protocol, to visualize the proliferating cells. For the triple labeling of EDU, p63, and dlc, the samples were continuously treated in accordance with the standard immunofluorescence procedure and assessed via a confocal microscope (FV31S-SW, Olympus), as mentioned in Section 2.4.

TUNEL staining was performed as previously described [45]. The apoptosis cells were labeled with the TUNEL BrightRed Apoptosis Detection Kit (Vazyme Biotech, NanJing, China). For double labeling of the TUNEL and atp1a1a.4 and the TUNEL and atp6v1a, as well as triple labeling the TUNEL, p63, and dlc, the samples were treated in accordance with the standard immunofluorescence procedure, and assessed by a confocal microscope (FV31S-SW, Olympus), as mentioned in Section 2.4.

2.6. Western Blot

Zebrafish embryos were identified and collected for protein extraction at 4 dpf. Approximately 25 heads of the same genotypes, including the inner-ear region, were lysed with RIPA lysis buffer containing cOmplete protease inhibitor cocktail (04693132001, Roche, Basel, Switzerland), while the tails of embryos were used for the genotyping assay. Western blot was performed as described previously [46]. The following primary antibodies were used in this study: anti-Mafb (PA5-40756, Thermo Scientific; 1:500), anti-p53 (3121, Daian; 1:1000), anti-γH2AX (9178s, CST; 1:1000), anti-GAPDH (60004, Proteintech; 1:1000), anti-CDK2 (R1309, HUABIO; 1:1000), anti-CDK6 (ER40101, HUABIO; 1:1000), anti-Cyclin D1 (0427, HUABIO; 1:1000), anti-Cyclin E1 (ER1906, HUABIO; 1:1000), anti-PCNA (10205, Proteintech; 1:2000), anti-Phospho-ATR (2853, CST; 1:1000), anti-Phospho-ATM (13050, CST; 1:1000), and anti-Rb1 (10048, Proteintech; 1:1000). The images were captured with a ChemiDocTM XRS$^+$ system (BIO-RAD).

2.7. Bioinformatics

All relevant Maf factor structures and sequences are available at Genome Browser (http://genome.ucsc.edu/cgi-bin/hgBlat, accessed on 12 July 2020) and Ensembl (http://asia.ensembl.org/, accessed on 12 July 2020). SWISS-MODEL (https://swissmodel.expasy.org/, accessed on 12 July 2020) provided the fully automated protein structure homology modeling [47]. The amino acid sequences of Maf family factors from the other species mentioned above were downloaded from the NCBI GenBank database (http://www.ncbi.nlm.nih.gov/protein, accessed on 12 July 2020) and used for the multiple sequence alignments performed by ClustalW software.

2.8. Comet Assay

Heads from 30 zebrafish embryos were cut off and placed in 1.5 mL of ice-cold PBS (PBS + 20 mM EDTA). The cell suspensions were prepared by mincing the tissues. Alkaline Comet Assay was carried out in accordance with the manufacturer's procedure using the Comet Assay Kit (Abcam, Ab238544) [48]. DNA damage in single-strand breaks (SSB) and double-strand breaks (DSB) can be detected by alkaline electrophoresis. Images were captured by fluorescence microscopy (Nikon Ti_i80).

2.9. Startle Response Tests

The sound-evoked C-start escape response was examined in 96-well plastic plates and recorded with a high-speed camera (Olympus, E-M10, 1000 fps) under infrared light illumination. Pure tone stimulations (10 ms, 500 Hz) with two different intensities (20 dB and 40 dB) were supplied through a plastic board mounted on a voice box (HiVi, D1080MKII). Positive C-startle response was confirmed if the response occurred within 20 ms after the application of stimuli. For each group, 32 larvae were tested. Each larva was tested 10 times and each larva's C-startle times was computed. The possibility of measuring a C-startle response in a group of larvae was defined as the average percentage of C-startle [49].

2.10. Quantitative RT-PCR Analysis

Zebrafish embryos at 4 dpf were collected for total RNA extraction. The tails were dissected and used for genotyping. The residual head parts of approximately 30 were cut off and blended together to extract RNA by using the TRIzol Reagent (Vazyme Biotech, NanJing, China). They were then reverse transcribed into cDNA using HiScript® II Q RT SuperMix (+gDNA wiper) for qPCR (Vazyme Biotech, Nanjing, China). The qPCR was carried out with AceQ™ qPCR SYBR Green Master Mix (Vazyme Biotech, Nanjing, China) in a StepOnePlus™ real time PCR machine, and analyzed with Graphpad 6.1 software. The relative mRNA expression levels of the selected genes were calculated using the $2^{-\Delta\Delta C}$ method [50]. The β-actin was set as an internal control. All qPCR primers used in this research are shown in Supplementary Table S1.

2.11. Flow Cytometry Analysis for Cell Cycle

For 4-dpf fish, 100 embryos were collected and washed once in ice-cold PBS. We removed all the solution into a 50-mL Falcon tube, then passed it through a 100-μm nylon filter (BD Falcon, Cat. 352360, Franklin Lakes, NJ, USA). We aspirated the solution and resuspended the pellet in 1 mL of Dispase II (Gibco, 1U/mL in PBS, Shanghai, China) to break up tissues. The solution was transferred into a 1.5-mL tube and incubated at 37 °C for 20 min. Then 1 mL of washing buffer (Hanks buffered saline solution containing 20% FBS, 5 mM $CaCl_2$, and 0.5 u/mL DNAse) was added to the samples. The samples were treated with 0.2 mL of PBT (100 μg/mL RNase A and 50 00B5g/mL PI in PBS). Setting 200 μL PBS as the control group, we commenced dark treatment at room temperature for 30 min. We resuspended the cells in 1 mL PBS (with 2–3% FBS), which was then passed through a 40-μm nylon filter (BD Falcon, Cat. 352340, Franklin Lakes, NJ, USA), and then we counted the cells. The labeled cells were detected for PI staining using a Beckman Coulter CytoFlex 5, and the percentages of cells in the G0/G1, S, and G2/M phases in each sample were determined via gating using the Flowjo 10.0 software (Beckman Coulter, Brea, CA, USA).

2.12. Statistical Analysis

Statistical analyses were performed using GraphPad Prism 6.0 software. The experiments were repeated at least three times. The numbers of samples measure in every experiment are discussed in the Section 2 and in the figure legends. The two-tailed Student's *t*-test was used for determining significance. The results are presented as the mean ± SD. The level of significance was set to $p < 0.05$, and $p > 0.05$, $p < 0.05$, $p < 0.01$, $p < 0.001$, and $p < 0.0001$ were labeled as ns, *, **, ***, and ****, respectively.

3. Results

3.1. Generation of mafba$^{-/-}$ Mutant Zebrafish with CRISPR/Cas9 Technology

There are two *mafba* transcripts (*mafba*-201(NM_131015.3) and *mafba*-202 (AB006322.1)) listed in the ZFIN database. Each contains only one exon, encoding 397 and 356 amino acids, respectively. Both can be detected in zebrafish embryos. Zebrafish Mafba proteins are 73%/72% identical at the amino acid level to the human/mouse Mafb protein, respectively. These are significantly higher than the sequence similarities among zebrafish Mafba, Mafbb, and Mafa proteins (Figure S1A). The functional domains of the Maf family members are highly conserved among zebrafish and mammalian Mafb proteins, including N-terminal transcriptional activation domain (TAD), the C-terminal basic domain (BR), and the leucine zipper domain (LZ) (Figure S1A,C), which are the functional domains of Maf family members [26]. All of the tertiary structures consist of three α-helices and two β-sheets sharing many similarities (Figure S1B). In vertebrates, small Maf evolves into three members (MafF, MafG, and MafK), while large Maf evolves into four members (MafA, MafB, C-MAF, and Nrl). Some of the genes duplicate because of an extra copy of the genome in the teleost [51]. Zebrafish have five large Maf genes and four small Maf genes. Due to an extra copy of the genome in the teleost fish, gene duplication may occur in both MafB and MafG [52].

To elucidate the physiological role of Mafba in vivo, we generated *mafba* knockout zebrafish using CRISPR/Cas9 technology. The target sites were designed in the exon of *mafba*-201 (NM_131015.3, NP_571090.2), in a sequence common to both transcripts. The detailed design of the *mafba* knockout is shown in Figure 1A. A 7-bp deletion mutation (c. 298_304delACTCCTA) would generate a premature termination codon and encode a truncated Mafba protein (p. Ser100Argfs*141). The del7 mutation may abolish the Mafba protein's function, and so this zebrafish line was selected for our research (Figure 1B).

To further ensure the 7-bp deletion mutation, we verified the expression of *mafba* on the mRNA and protein levels. Firstly, semi-quantitative RT-PCR analysis showed that the expression levels of both transcripts (*mafba*-201, *mafba*-202) are significantly reduced in *mafba* mutant zebrafish (Figure S2). Secondly, qRT-PCR analysis showed that *mafba* mRNA expression decreased by 25% in *mafba*$^{+/-}$ embryos and decreased more significantly in *mafba*$^{-/-}$ mutant embryos at 4 days postfertilization (dpf) (Figure 1C). This suggests that the truncated *mafba* mRNAs are subject to nonsense-mediated decay. Thirdly, qRT-PCR analysis showed that the mRNA expressions of Maf homologues, such as large Maf (*mafba*, *mafbb*, *mafa*, *c-maf*, and *nrl*) and small Maf (*maff*, *mafk*, *mafga*, and *mafgb*), decrease significantly in the *mafba* mutants (Figure S3). We speculated that *mafba* mutants may have systemic non-specific effects that reduce the expressions of many genes, otherwise affecting the viability of some tissues. Additionally, the temporal and spatial expression of *mafba* was determined in sibling (Figure S4A–J) and *mafba*$^{-/-}$ mutants (Figure S4K–T) via whole-mount in situ hybridization (WISH). *Mafba* displayed a dynamic expression pattern throughout embryo development (Figure S4A–J). It emerged in bud-stage embryos (Figure S4C) and was then expressed throughout the embryos (Figure S4D–J). Noticeably, at 36 hpf, *mafba* was broadly expressed in the rhombomeres 5 and 6 [36], the inner ear, the eye, and the kidney (Figure S4F,G). Moreover, the expression pattern seen at 2 dpf was maintained until 5 dpf (Figure S4H–J). Semi-quantitative RT-PCR analysis showed that the expression of *mafba* was initiated at early developmental stages (6-somites stage) and then stabilized from 12 h postfertilization (hpf) to 6 dpf (Figure S4U). This is similar to the temporal expression pattern of siblings, shown in Figure S4A–J. WISH results showed that *mafba* can be detected in *mafba*$^{-/-}$ mutants, and that it displayed strong expression at the bud stage and 6-somites stage (Figure S4M,N). However, from 24 hpf onwards, *mafba* mRNA levels were dramatically decreased in *mafba*$^{-/-}$ embryos and were almost undetectable from 3 dpf (Figure S4O–T). Western blot analysis also showed that the Mafba protein's expression was markedly decreased in the mutant line at 4 dpf, but did not completely disappear, likely due to the maternal deposit (Figure 1D). In addition, immunostaining showed that Mafba was localized in the rhombomere (r) and inner-ear region of the sibling at 2 dpf. However, almost no Mafba signals were detected in *mafba*$^{-/-}$ embryos (Figure S4V), suggesting its potential correlation with hindbrain and inner-ear development.

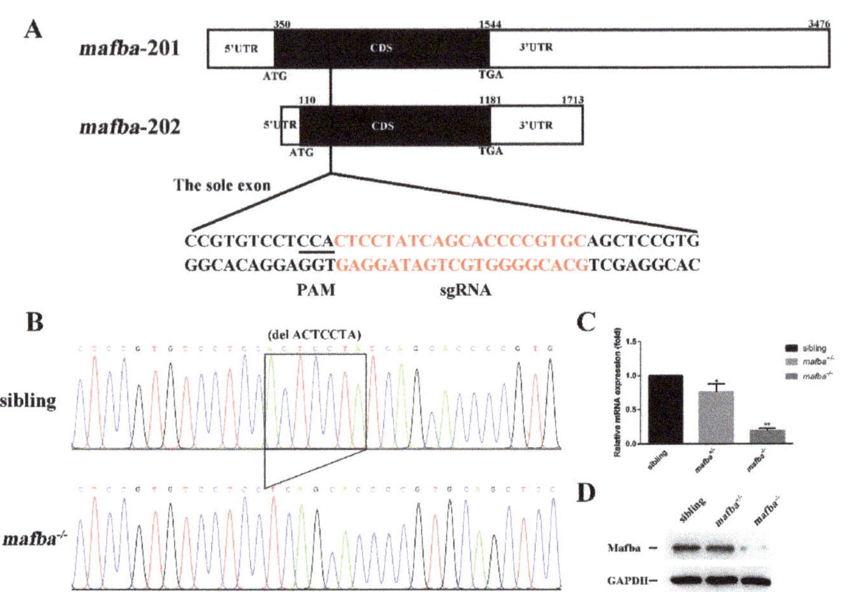

Figure 1. CRISPR/Cas9-mediated mutagenesis of *mafba*. (**A**) The sole exon of two *mafba* transcripts and the CRISPR/Cas9 target site are shown. The 5′UTR, CDS, and 3′UTR regions of the two *mafba* transcripts are shown in detail. (**B**) DNA sequencing identifies the c.298_304delACTCCTA *mafba* mutation lines. The 7-bp deletion is indicated with a black frame. (**C**) Relative expressions of *mafba* determined by qRT-PCR analysis in sibling, *mafba*$^{+/-}$, and *mafba*$^{-/-}$ embryos at 4 dpf. The β-actin served as the endogenous control. Data are represented as mean ± SD; *, $p < 0.05$; **, $p < 0.01$ (**D**) Western blot analysis shows the similar protein expression of Mafba between siblings and *mafba*$^{+/-}$ embryos, but a significant decrease in *mafba*$^{-/-}$ mutants at 4 dpf. GAPDH was used as the internal control.

3.2. Depletion of Zebrafish mafba Results in Inner-Ear Defects and Hearing Loss

To determine whether *mafba* is essential for inner-ear development in zebrafish, the phenotype of *mafba*$^{-/-}$ embryos was examined and compared with that of sibling embryos. The *mafba*$^{-/-}$ mutants were indistinguishable in gross morphology from the sibling and *mafba*$^{+/-}$ embryos in the inner ear at 4 dpf. Morphological defects of the inner ear abruptly occurred in the *mafba*$^{-/-}$ embryos at 5 dpf (Figure 2A,B). At 5 dpf, otoliths in the mutants showed varying degrees of size reduction. The otic defects of 5 dpf *mafba*$^{-/-}$ mutants included larger otocysts, thinner otic epithelia, and smaller or even eliminated otoliths (Figure 2A,C). According to the otolith size, we categorized the *mafba*$^{-/-}$ mutants into four groups: normal otoliths (accounting for 10.3%, $n = 92$), with otoliths comparable in size to the sibling embryos (on average 4662 ± 238 μm^2); small otoliths (accounting for 24.4%), at 35–75% of the average otolith size of the siblings; tiny otoliths (30.1%), 10–34% of the average otolith size of the siblings; and absent otoliths (35.2%), at less than 10% of the otolith size of the siblings (Figure 2B). Less than 4% of the otic vesicles in sibling embryos ($n = 117$) could be categorized into the small group. Both the otic vesicle lumen and otolith sizes of zebrafish embryos were measured, and their areas were compared. The results further suggested the otic defect in mutants (Figure 2C). The inner-ear structures of *mafba*$^{-/-}$ mutants were almost destroyed. The mutation caused defects in inner-ears in a recessive mode.

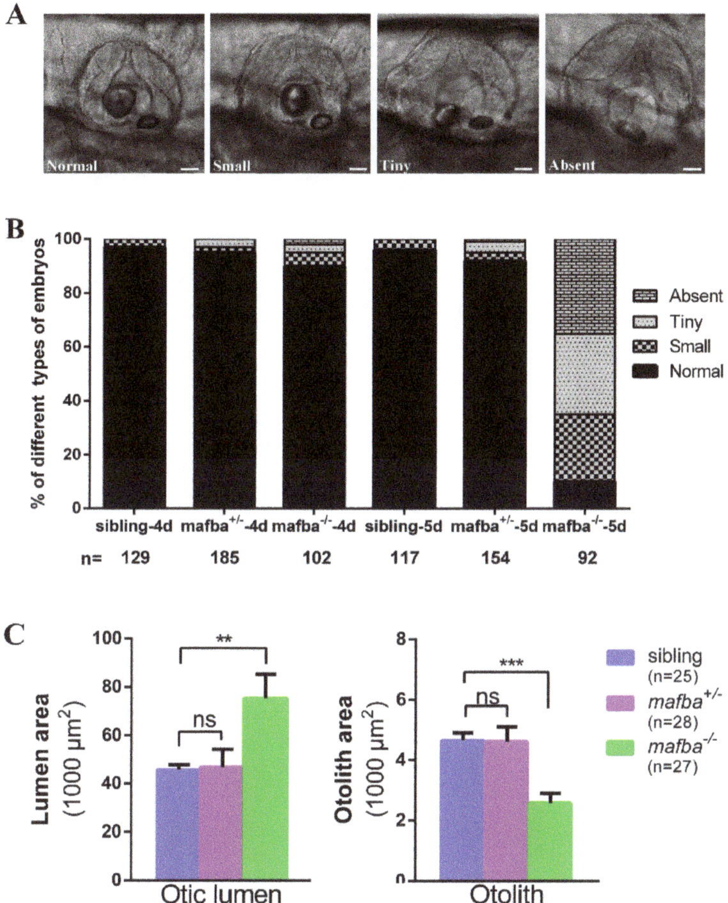

Figure 2. Deletion of *mafba* leads to inner-ear morphological defects. (**A**) The variable otolith sizes of *mafba*$^{-/-}$ mutant embryos at 5 dpf. According to the otolith sizes, mutant embryos were classified into four groups: normal, small, tiny, and absent. Scale bars: 40 μm. (**B**) Percentages of embryos in sibling and *mafba*$^{-/-}$ group at 4 dpf and 5 dpf; *n*, the number of observed embryos. (**C**) Statistical analysis of the otic lumen area and otolith area in different types of embryos at 5 dpf. Individual embryos were randomly picked from each type for statistical analysis. The otic lumen and otolith areas were measured with SPOT Advanced software (version 4.6) in the focal plane representing the maximal area. Data are represented as mean ± SD; ns, $p > 0.05$; **, $p < 0.01$; ***, $p < 0.001$.

The inner ear plays crucial roles in zebrafish hearing and balance maintenance [53]. Due to the severe defects in the inner ears of *mafba*$^{-/-}$ mutants, we further investigated hearing abilities of the mutants. We assessed the fast escape reflex (named the C-shaped startle response) using near-field pure tone stimulation at two different levels of sound intensity [8]. The sibling and *mafba*$^{+/-}$ group larvae responded within 9 ms, but the *mafba*$^{-/-}$ larvae showed either a delayed response or no response at all at 5 dpf (Figure 3A).

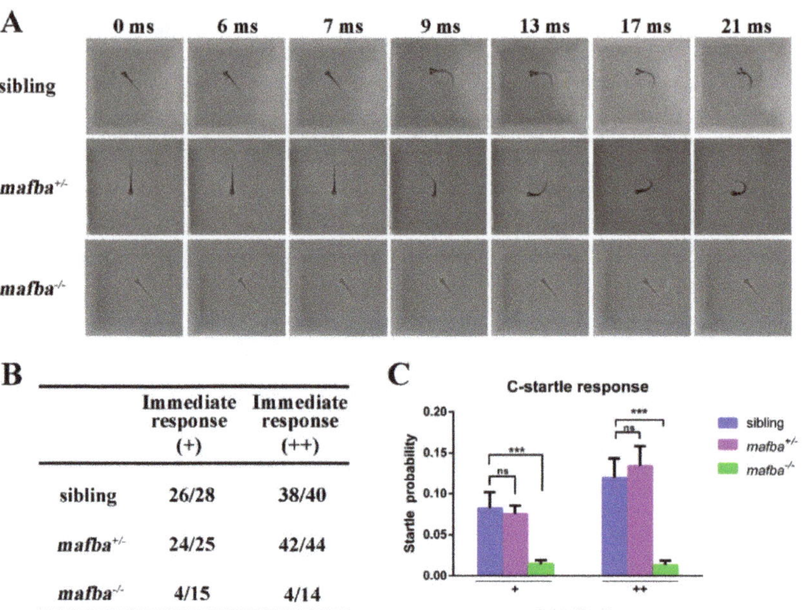

Figure 3. Hearing disability test for *mafba*$^{-/-}$ mutants at 5 dpf. (**A**) C-startle escape response of embryos of different genotypes in response to the sound stimuli at 500-Hz frequency and 10-ms duration. Typical C-startle escape response initiated within 9 ms after the sound stimulation. (**B**) Statistical data show the proportion of immediate responses in three genotypes at two different sound intensities. Data represent the number of immediate responses versus the number of total responses. '+' and '++' mean 20 dB and 40 dB, respectively. (**C**) The average C-startle response probability. For each group, 32 larvae were tested. Data are mean $\pm$ SD; ns, $p > 0.05$; ***, $p < 0.001$.

For three genotypes, the numbers of immediate responses and delayed responses were recorded and are shown in Figure 3B. As described previously [54], we eliminated delayed responses and identified the percentage of successful C-shaped startle responses, and the result was that the probabilities of a C-startle response in sibling and *mafba*$^{+/-}$ embryos were indistinguishable and markedly higher than those of *mafba*$^{-/-}$ mutant (Figure 3C). These data support the idea that *mafba*$^{-/-}$ mutant zebrafish embryos display the hearing loss.

The survival rates of *mafba*$^{-/-}$ mutants are comparable with those of sibling and *mafba*$^{+/-}$ embryos up to 5 dpf. Most of the *mafba*$^{-/-}$ larvae showed inner-ear defects at 5 dpf. The survival rate of *mafba*$^{-/-}$ mutants decreased to 65% under normal feeding conditions, while the survival rate of *mafba*$^{+/-}$ embryos at 6 dpf was 92% (data not shown). This reduction may have been caused by the dietary deficiency resulting from the defective hearing and balance system [55].

3.3. Deficiency of mafba Does Not Affect Otic Patterning or Hair Cell Development

The early development of the inner ear appears be normal in *mafba*$^{-/-}$ mutants. The induction of the otic placode appeared as expected at 18 hpf. In order to comprehend the molecular function of *mafba* in ontogenesis, we assessed the expression levels of several marker genes involved in the inner-ear patterning structure. The expressions of patterning markers for medial otic vesicle (*pax2a*), dorsal otic vesicle (*dlx3b*), and ventral otic vesicle (*foxj1b*) showed no difference between the sibling and *mafba*$^{-/-}$ mutant embryos (Figure S6A–F). The expressions of the semicircular canal projections' marker *udg* (Figure S7G,H), the semicircular canal sensory cristae marker *bmp4* (Figure S7A,B), the utricular and saccular maculae marker *cahz* (Figure S7C,D), and the endolymphatic duct marker *foxi1* (Figure S7E,F) were also the same in the *mafba*$^{-/-}$ embryos and the sibling. In addition, live bright-field images of siblings (Figure S5A,C,E) and *mafba*$^{-/-}$ mutants (Figure S5B,D,F) showed that the inner-ear sizes were comparable at 54 hpf and the semi-circular canal projections were forming normally. There was no obvious difference between the mutant and the sibling at 78 hpf in the otocysts, semicircular canal, utricle, and saccule. The organogenesis of the inner ear appeared to be normal until 102 hpf, suggesting the correct specification of early otic tissue.

The formation and mineralization of the otolith is a complex process that requires the orderly regulation and participation of many developmental processes, such as the secretion of matrix proteins [8], hair cell development, and endolymphatic fluid home-ostasis [20], which are necessary for zebrafish otolith formation and growth. The WISH results showed that the expressions' levels of the matrix protein genes *otomp* and *starmaker* (*stm*) at 2 dpf and 5 dpf in the sibling group were in accordance with those in *mafba*$^{-/-}$ embryos (Figure S8A–H). Then, we tested the otic hair cells, which play essential roles in the development of the inner ear. There was no significant difference between the *mafba*$^{-/-}$ and sibling embryos as stained by the hair cell marker *atp1b2b* at 24 hpf (Figure S9A). Immunostaining using the anti-Myo7a antibody indicated that the number of hair cells in *mafba*$^{-/-}$ and sibling embryos showed no difference in the utricular and saccular maculae at 36 and 48 hpf (Figure S9B,C). Phalloidin staining for stereociliary bundles and kinocilium growth labeled by Acetylated-α-Tubulin both suggested that the differentiation of the sensory cristae and maculae was normal in *mafba*$^{-/-}$ mutant embryos at 5 dpf (Figure S9D). In conclusion, these results implied that the patterning and structural specification of the otic vesicles, the differentiation and maturation of hair cells, and the growth of ciliary bundles were normal in the *mafba*-deficient embryos.

3.4. Knockout of mafba Suppresses the Proliferation of p63+ Epidermal Stem Cells and Reduces dlc+ Ionocyte Progenitor Cell Number

Mouse mafb plays an important role in regulating epidermal differentiation and homeostasis [56]. After confirming that matrix proteins and hair cell development were normal in the *mafba*$^{-/-}$ inner ear, the balance of ionocytes' homeostasis was explored. Epidermal stem cells were marked with p63 and are known to produce both keratinocytes (dlc$^-$p63$^+$ cells) and ionocytes (dlc$^+$p63$^+$ cells). Whether *mafba* is expressed in the epidermis and affects the development of ionocytes in zebrafish embryos needs further investigation. The p63 expression was observed in the ventral ectoderm of 90% epiboly and bud-stage embryos [18], while the expression of Mafba occurred in the bud stage and 5-somites stage. The p63$^+$Mafba$^+$ colocalized cell number was high in the dorsal ectoderm region at the bud stage and 5-somites stage (Figure S10A). In addition, colocalization of Mafba and dlc was observed at the bud stage and 5-somites stages in the epidermal ionocyte region. Almost half of the dlc$^+$ cells were positively stained with Mafba (Figure S10B). The colocalization of p63$^+$Mafba$^+$ and dlc$^+$Mafba$^+$ suggests that Mafba participates in the development of epidermal stem cells and ionocytes.

Mafb is widely known for regulating epidermal keratinocyte differentiation and epidermal homeostasis in mammals [27]. We speculated that zebrafish Mafba also modulates p63$^+$ epidermal stem cell proliferation and dlc$^+$ ionocyte progenitor differentiation. The p63$^+$ epidermal stem cell number was reduced by 14.4% in *mafba*$^{-/-}$ mutants compared to the sibling group at the bud stage (Figure 4A(g,m)), and we also saw a 19.6% reduction in the proportion of p63$^+$EDU$^+$ epidermal stem cells in *mafba*$^{-/-}$ embryos. These results suggest that Mafba is essential to maintaining the proliferation rate of p63$^+$ epidermal stem cells. The reduction in p63$^+$ epidermal stem cell led to a decreased number of dlc$^-$p63$^+$ keratinocytes and dlc$^+$p63$^+$ ionocytes in *mafba*$^{-/-}$ mutant embryos at the bud stage (Figure 4A(g,h,j,m)). The percentage of EDU$^+$ signals colocalized with dlc$^-$p63$^+$ and dlc$^+$p63$^+$ was significantly decreased in *mafba*$^{-/-}$ mutant embryos at the bud stage (Figure 4A(k,l,n)). A previous study suggested that *foxi3a* and *foxi3b* are the main regulators of epidermal ionocytes' specification in zebrafish embryos [20]. At the bud stage, a prominent reduction in ionocyte progenitors was detected by dlc labeling (Figure 4C). A significant decrease in the densities of ionocyte progenitors expressing either *foxi3a* or *foxi3b* was observed in *mafba*$^{-/-}$ embryos at the 5-somites stage (Figure 4D). Mafba is, thus, essential to maintaining both the proliferation of p63$^+$ epidermal stem cells and the cell numbers of dlc$^-$p63$^+$ keratinocytes and dlc$^+$p63$^+$ ionocytes.

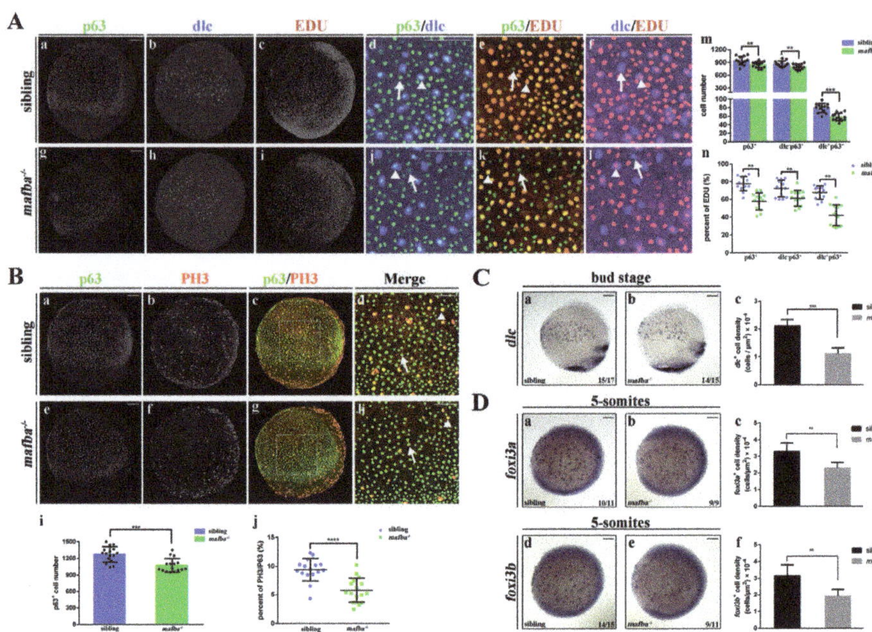

Figure 4. Knockout of *mafba* reduced the proliferation of epidermal stem cells and the *dlc*$^+$ ionocyte progenitor cell number. (**A**) Examples of p63 with dlc or p63 with EDU or dlc with EDU colocalized (arrowhead) or non-colocalized (arrow) cells between sibling and *mafba*$^{-/-}$ groups are shown in (**a–l**). The quantitative analyses of the p63$^+$ (marker for epidermal stem cells), dlc$^-$p63$^+$ (marker for keratinocyte precursors), and dlc$^+$p63$^+$ (marker for ionocyte precursors) cells in each group at the bud stage are shown in (**m**). The quantitative analyses of p63$^+$, dlc$^-$p63$^{+,}$ and dlc$^+$p63$^+$ cells colocalized with EDU (S-phase cells) in sibling and *mafba*$^{-/-}$ embryos at the bud stage are shown in (**n**). The *n* = 13 for each panel. Scale bars: 100 μm. (**B**) Double staining of p63 and pH3 (M-phase cells) in the siblings and *mafba*$^{-/-}$ group at the bud stage. Examples of p63 and pH3 colocalized (arrowhead) or non-colocalized (arrow) cells are shown. The quantitative analyses of p63$^+$ cell and p63$^+$ colocalized with pH3-positive cells of sibling and *mafba*$^{-/-}$ embryos at the bud stage are shown in (**i**) and (**j**), respectively. The *n* = 16 for each panel. Scale bars: 100 μm. (**C**) The *dlc*$^+$ ionocyte progenitors' cell density was reduced in the *mafba*$^{-/-}$ group (**b**) as compared to the sibling group (**a**) at the bud stage. The quantitative analysis of *dlc*$^+$ ionocyte progenitors' cell densities are shown in (**c**). (**D**) The cell density of *foxi3a*$^+$ and *foxi3b*$^+$ ionocyte progenitors are compared between the sibling (**a,d**) and *mafba*$^{-/-}$ group (**b,e**) at the 5-somites stage, respectively. Cell densities of *foxi3a*$^+$ and *foxi3b*$^+$ ionocyte progenitors are quantified in (**c**) and (**f**), respectively. Scale bars: 100 μm. Data are expressed as mean ± SD; **, *p* < 0.01; ***, *p* < 0.001; ****, *p* < 0.0001.

The significant reduction in p63$^+$ epidermal stem cells in *mafba*$^{-/-}$ embryos suggests that these could be arrested in the cell cycle. To determine the cell cycle status of p63$^+$ epidermal stem cells in *mafba*$^{-/-}$ embryos, EdU incorporation (labeling cells in S-phase) combined with the immunofluorescence of phospho-histone3 (labeling cells in M-phase) was performed at the bud stage. The p63$^+$ epidermal stem cell number is also decreased in *mafba*$^{-/-}$ embryos (Figure 4B(e,i)) in accordance with the results shown in Figure 4A(m).

Double immunostaining showed that the proportion of p63$^+$pPH$^+$ cells decreased significantly, by 3.6%, in *mafba*$^{-/-}$ mutant embryos (Figure 4B(g,h,j)). We thus speculated that the epidermal stem cells lacking Mafba may be arrested in G0/G1, and this may result in the increased apoptosis of these ionocyte progenitors. However, the TUNEL assays showed only a slight increase in apoptotic signal in the p63$^+$ cells (by 1.73‰) and dlc$^+$p63$^+$ cells (by 6.71‰) in *mafba*$^{-/-}$ mutants compared to the sibling group, while no significant change was observed in dlc$^-$p63$^+$ cells (Figure 5). In summary, we confirmed a reduction in the proliferation of p63$^+$ epidermal stem cells and the cell numbers of dlc$^-$p63$^+$ keratinocytes and dlc$^+$p63$^+$ ionocytes as well as the differentiation of the epidermal ionocytes' progenitor in *mafba*$^{-/-}$ embryos.

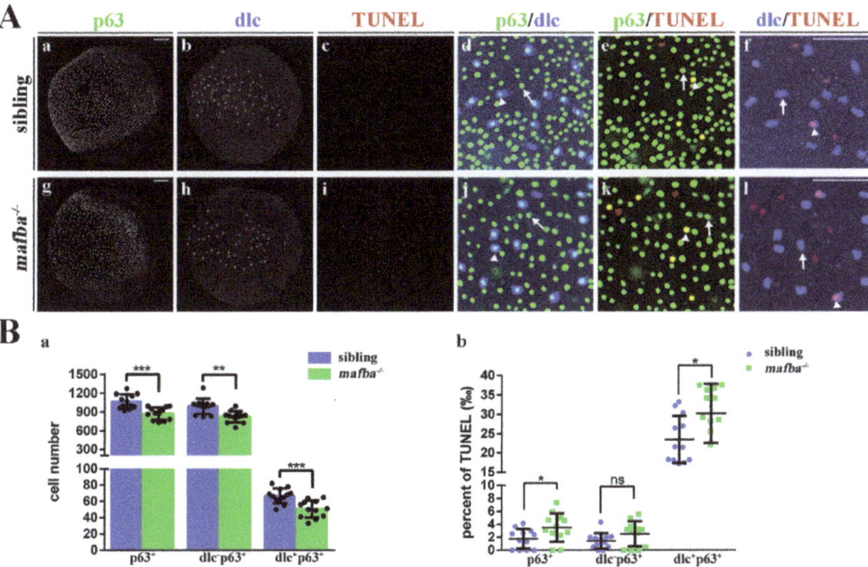

Figure 5. Increased apoptosis in the epidermal stem cells and *dlc*$^+$ ionocyte progenitor cells of *mafba*$^{-/-}$ zebrafish. (**A**) Examples of p63 with dlc, p63 with TUNEL, or dlc with TUNEL colocalized (arrowhead) or non-colocalized (arrow) cells are shown. (**B**) The quantitative analyses of p63$^+$ cells, dlc$^-$p63$^+$ cells, and dlc$^+$p63$^+$ cells of sibling and *mafba*$^{-/-}$ embryos at the bud stage are shown in (**a**). The quantitative analysis of p63$^+$, dlc$^-$p63$^+$, and dlc$^+$p63$^+$ cell numbers colocalized with TUNEL in sibling and *mafba*$^{-/-}$ embryos at the bud stage are shown in (**b**). The *n* = 12 for each panel. Scale bars: 100 μm. Data are mean ± SD; ns, *p* > 0.05; *, *p* < 0.05; **, *p* < 0.01; ***, *p* < 0.001.

3.5. Apoptosis of the Differentiated Epidermal Ionocytes Is Increased in mafba$^{-/-}$ Embryos

A previous study suggested that *foxi3a* and *foxi3b* control ionocyte progenitor specification into NR and HR cells through a positive feedback loop [24]. Ionocyte progenitors' regulators, such as *dlc*, *foxi3a*, and *foxi3b*, were downregulated in *mafba*$^{-/-}$ mutant embryos. We, thus, wished to assess whether the formation of NR and HR cells is affected by deprivation of *mafba* expression. We performed WISH using relevant markers (*atp1a1a.1*, *atp1a1a.4*, and *atp1b1a* for NR cells, *atp6v1aa* for HR cells) to evaluate the differentiated epidermal ionocytes in the inner ear. At 24 hpf, the expressions of *atp1a1a.1*, *atp1a1a.4*, *atp1b1a*, and *atp6v1aa* in the *mafba*$^{-/-}$ embryos reduced markedly compared to those in the siblings (Figure 6A). Similarly, a significant reduction in atp1a1a.4 and atp6v1aa expression in the inner ear was observed in *mafba*$^{-/-}$ mutants from 2 dpf to 5 dpf (Figure 6B,C), respectively. Considering the dramatic decline in atp1a1a.4- and atp6v1aa-expressing cells in the inner ear at 5 dpf, we suspected that the apoptosis may also occur in *mafba* mutants. To validate this hypothesis, we measured the apoptosis of atp1a1a.4- and atp6v1aa-expression cells in *mafba*$^{-/-}$ embryos using TUNEL assays. There was a significant increase in apoptotic atp1a1a.4 and atp6v1aa cells' signals in the inner-ear region in *mafba*$^{-/-}$ embryos compared with those in the siblings at 4 dpf and 5 dpf (Figure 6B(j),C(j)). The percentage of apoptotic cells with atp1a1a.4 and atp6v1aa at 5 dpf was notably higher than that at 4 dpf. The defective inner ear might result from the decreased ionocyte progenitors and increased apoptosis in NR and HR cells. In summary, Mafba is essential to the proliferation of the epidermal stem cells and ionocyte progenitors, which, in turn, determine the numbers of NR and HR cells and maintain ion homeostasis in zebrafish inner ear. The differentiated ionocyte progenitors provide a stable environment for the development of statoacoustic organs such as otoliths and otocysts, which are essential to the hearing and balance system (Figure 6D). However, more research is needed to reveal the specific regulatory mechanism of inner-ear developmental defects.

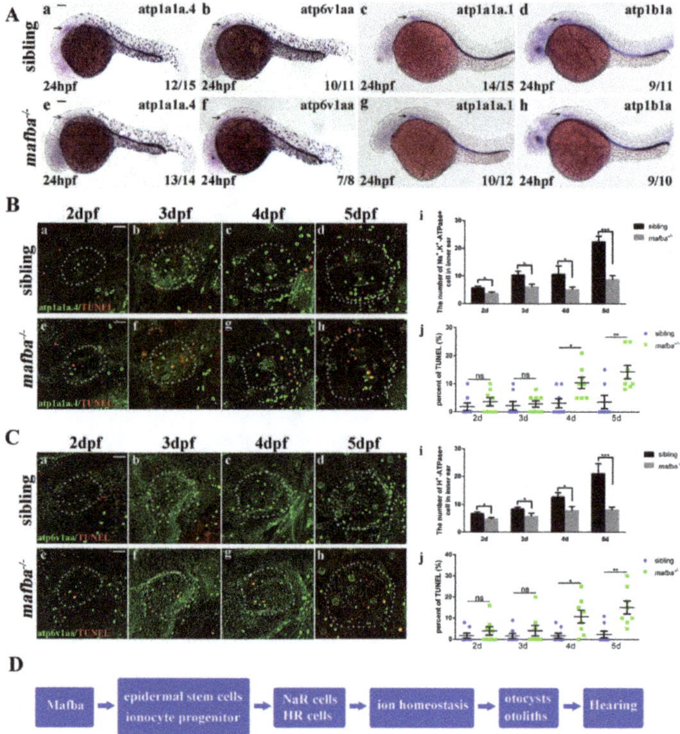

Figure 6. *Mafba* deprivation impaired ionocyte progenitor cell differentiation and triggered the apoptosis of NR and HR cells. (**A**) The in situ hybridization staining of markers for NR cells (*atp1a1a.4, atp1a1a.1, atp1b1a*) and for HR cells (*atp6v1aa*) at 24 hpf. The reductions in differentiated NR and HR cells in the inner-ear (arrow) region are shown. Scale bars: 100 µm. (**B**) Double immunostaining of atp1a1a.4 and TUNEL from 2 dpf to 5 dpf. The *n* = 8 for each panel. Scale bars: 20 µm (**a–h**). (**i**) The quantitative analysis of NR cells of the inner ear (the white, dotted, circled area) from 2 dpf to 5 dpf between sibling and *mafba*$^{-/-}$ mutants. (**j**) The percentage of the apoptosis in the NR cell from 2 dpf to 5 dpf indicated increased apoptosis at 4 dpf and 5 dpf in *mafba*$^{-/-}$ embryos' inner-ear regions. (**C**) Double immunostaining of atp6v1aa and TUNEL from 2 dpf to 5 dpf. The *n* = 8 for each panel. Scale bars: 20 µm (**a–h**). (**i**) The quantitative analysis of HR cells of the inner-ear (the white, dotted, circled area) from 2 dpf to 5 dpf between sibling and *mafba*$^{-/-}$ mutants. (**j**) The quantification of the apoptosis percentage of NR cells from 2 dpf to 5 dpf indicates an increase at 4 dpf and 5 dpf in *mafba*$^{-/-}$ embryos' inner-ear regions. Data are represented as mean ± SD; ns, $p > 0.05$; *, $p < 0.05$; **, $p < 0.01$; ***, $p < 0.001$. (**D**) A model of Mafba functions in zebrafish inner-ear development and hearing.

3.6. Deprivation of mafba Activates p53 Apoptosis Pathway and Arrests the Cell Cycle in the G0/G1 Phase

In order to determine how Mafba deprivation triggers apoptosis, we quantified the expression levels of several apoptosis-related genes, including *caspase8*, *caspase10*, and *p53*, and its target genes in zebrafish embryos. The expressions of *p53* and its target genes *baxa*, *mdm2*, and *puma*, along with apoptosis-related genes *caspase10* and *caspase8*, showed significant increases in *mafba$^{-/-}$* mutants at 4 dpf. In addition, the expressions of *mdm4* and *bcl2a*, typical inhibitors of *p53*, decreased significantly (Figure 7A). Based on the observation that the proliferation of epidermal stem cells was suppressed, the p53 pathway was activated, and the apoptosis of NR and HR cells was increased, we hypothesize that G0/G1 cell cycle arrest may also occur in *mafba* mutants. A previous study revealed that mouse Mafb promotes cell proliferation with detectable changes in cell cycle factors [57]. Cell cycle distribution was examined via flow cytometry. The proportions of G0/G1 phase cells increased significantly, while S and G2/M phase cells decreased in the *mafba$^{-/-}$* mutant at 4 dpf (Figure 7B). The knockout of *mafba* led to G0/G1 cell cycle arrest, suggesting it affects cell cycle factors.

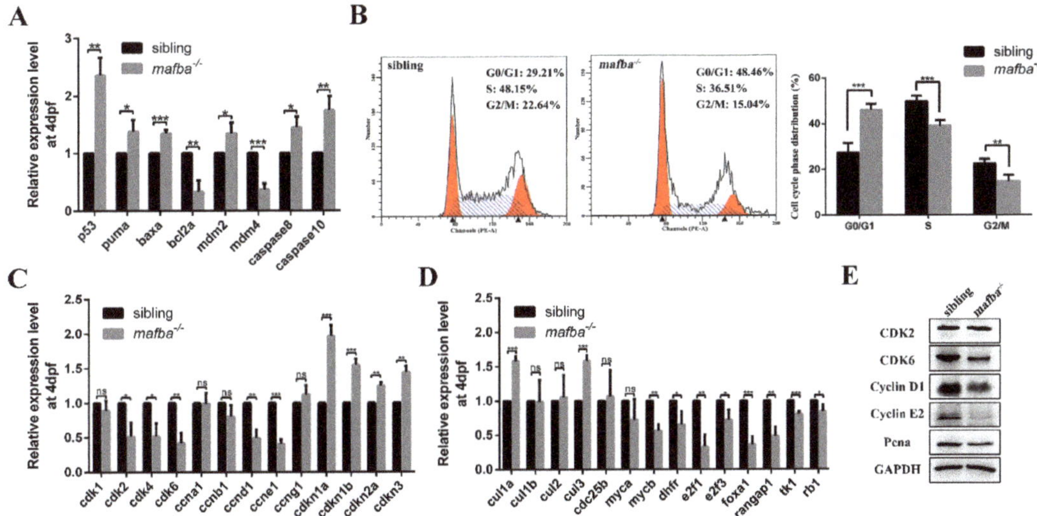

Figure 7. *Mafba* deficiency causes G0/G1 cell cycle arrest and activates the p53 pathway. (**A**) The upregulation of p53 pathway genes in *mafba* mutants at 4 dpf was detected by qRT-PCR. (**B**) The cell cycle phase distribution was performed by flow cytometry at 4 dpf. The proportion of sibling and *mafba$^{-/-}$* embryos in the G0/G1, S, and G2/M phase cell numbers. The β-actin served as endogenous control. (**C**) Quantitative RT-PCR analysis of the primarily relevant cell cycle factors in sibling and *mafba$^{-/-}$* mutant embryos at 4 dpf. (**D**) The key cell cycle factors controlling G1/S phase transition were analyzed via RT-qPCR at 4 dpf. (**E**) Western blot analysis of CDK2, CDK6, Cyclin D1, Cyclin E1, and Pcna at 4 dpf. GAPDH was used to normalize the protein. Data are mean ± SD; ns, $p > 0.05$; *, $p < 0.05$; **, $p < 0.01$; ***, $p < 0.001$ compared to the sibling control (by ANOVA followed by Dunnett's multiple comparison).

A number of cell cycle factors that control G1/S phase transition were analyzed via RT-qPCR. The expressions of *cdk2*, *cdk4*, *cdk6*, *cyclin d1* (*ccnd1*), and *cyclin e1* (*ccne1*) were downregulated. This agrees with the results that the protein levels of CDK6, Cyclin D1, and Cyclin E1 were lower in *mafba$^{-/-}$* mutants than in sibling embryos at 4 dpf (Figure 7C,E). Meanwhile, the cycle regulators of the G1/S checkpoint, such as *cnkn1a* (*p21*), *cdkn1b* (*p27*), *cdkn3*, *cul1a*, and *cul3*, were upregulated, while *mycb*, *dhfr*, *e2f1*, *e2f3*, *foxa1*, *rangap1*, *tk1*, and *rb1* were downregulated in *mafba$^{-/-}$* mutants at 4 dpf (Figure 7C,D). These results indicated that cell cycle arrest occurred between the G1 and S phases in the mutants. The reduced protein levels of Pcna also confirmed the arrest of G1/S phase transition. These results suggest that Mafba is essential for cell proliferation via the regulation of cell cycle factors levels.

3.7. Accumulation of DNA Damages in mafba$^{-/-}$ Inner-Ear

The apoptosis of the differentiated ionocytes may have resulted from the activation of the p53 pathway and/or cell cycle arrest at 4 dpf in *mafba*$^{-/-}$ inner ear. DNA damage may also occur in *mafba*$^{-/-}$ mutants, which would explain the apoptosis. The immunostaining of γH2AX indicated the DNA damage in sibling and *mafba*$^{-/-}$ inner ears at 3, 4, and 5 dpf (Figure 8A,B). The γH2AX-positive cells were observed in *mafba*$^{-/-}$ mutant inner ears at 3 dpf. The numbers of γH2AX-labeled cells in *mafba*$^{-/-}$ embryos inner ears increased dramatically at 4 and 5 dpf, compared to those in the sibling. DNA single/double-strand break levels were assessed directly via an alkaline comet assay [58]. We also detected more DNA breaking signals in *mafba*$^{-/-}$ mutants at 4 and 5 dpf (Figure 8C,D). Meanwhile Western blot analysis also showed that the expression levels of γH2AX, P-ATM, P-ATR and p53 accumulated gradually in *mafba*$^{-/-}$ embryos (Figure 8E). These results suggest that Mafba is essential for preventing DNA damage and maintaining genomic stability in differentiated ionocytes.

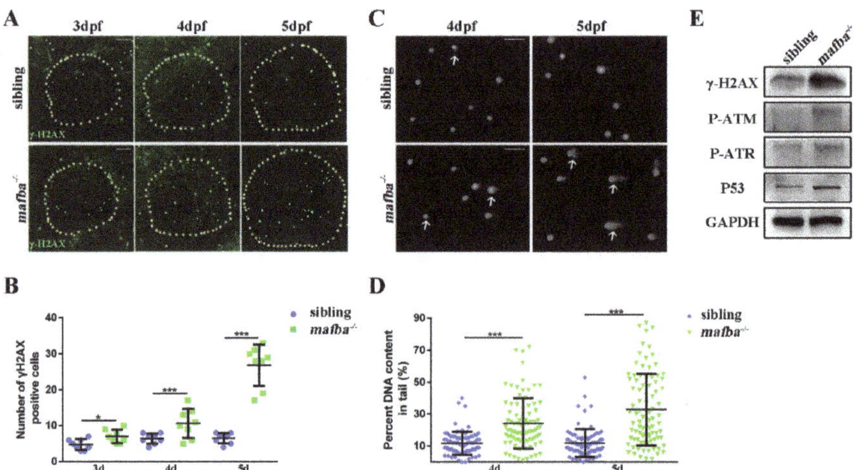

Figure 8. Analysis of the accumulation of DNA damage in *mafba*$^{-/-}$ mutants' inner-ear. (**A**) Whole-mount immunofluorescence analysis using the anti-γH2AX antibody in siblings and *mafba*$^{-/-}$ inner-ear at 3, 4, and 5 dpf. The white, dotted circle represents the inner ear area. Scale bars: 20 μm. (**B**) Statistical analysis of the γH2AX-positive cells is shown in (**A**). (**C**) The alkaline comet assay showed increased DNA damage in *mafba*$^{-/-}$ embryos at 4 and 5 dpf. White arrows show DNA-damaged cells with single- or double-strand breaks. Scale bars: 10 μm. (**D**) Quantitative analysis of 88 cells from 10 embryos in siblings and *mafba*$^{-/-}$ group are shown. (**E**) Western blot analysis of γH2AX, P-ATM, P-ATR, and p53 in siblings and *mafba*$^{-/-}$ zebrafish at 4 dpf. GAPDH was used as the normalized protein control. Data are represented as mean ± SD; *, $p < 0.05$; ***, $p < 0.001$.

4. Discussion

In the present study, we characterized a zebrafish *mafba* knockout line generated using CRISPR/Cas9 technology. The inner ears of mutants showed specific defects, including enlarged otocysts, small or absent otoliths, and insensitivity to sound stimulation, whereas matrix protein expression and hair cells' development appeared to be normal. Genetically *mafba* plays a positive role in epidermal stem cells' proliferation and can also promote the differentiation of the ionocyte progenitor in zebrafish inner ears. It also takes part in the differentiation of NR and HR cells, which are necessary for the ion homeostasis and epidermis differentiation.

The mouse *kreisler/mafb* is a member of the Maf transcription factor family. When mutated, it is identified as the causative gene of physiological defects resulting in segmentation abnormalities in the caudal hindbrain and defective inner-ear development [59]. In various vertebrates (including zebrafish), there are two paralogous othologues of mammalian *mafb*, i.e., *mafba* and *mafbb* [60,61]. Progressive syndromic deafness caused by heterozygous loss-of-function *Mafb* mutations was identified in a large family, though severity and age of onset differed among individuals. The mutation was located in the DNA-binding domain [33,34]. Researchers have been unable to characterize this inner-ear defect in mice because mutations in the mafb locus are lethal in the early postnatal period [62]. The locus of the mutation in our *mafba* zebrafish line is similar to that of the 0819 mutation in human MAFB [34], which explains why *mafba*$^{+/-}$ mutants have no effect on inner-ear defects. A mouse model with conditional knockout *Mafb* in the otic epithelial cells may help to illustrate the function of *mafba* in mammalian inner-ear development. Previous studies have shown that Mafb not only promoted the differentiation of otic epithelial cells, but also acted in the gene expression program that positively regulates epidermal keratinocyte differentiation [27]. Our research revealed that zebrafish *mafba*$^{-/-}$ mutant embryos exhibit inner-ear defects, with variable expressivity between individuals. The mutants with the more severe phenotypes possess bigger otocysts and smaller otoliths. Therefore, the mutant zebrafish *mafba*$^{-/-}$ line is the optimum animal model for progressive hearing loss in humans caused by *mafb* mutation, and can be applied to further study the mechanisms underlying *mafb*-associated inner-ear developmental defects.

Mafba plays crucial roles in various tissues and developmental stages. Based on a previous study and our research, zebrafish *mafba* is expressed in the otic vesicles, hindbrain, renal system, visual system, and other tissues during embryogenesis [36,63]. Developmental abnormalities in the inner ear were first detected in *mafba*$^{-/-}$ mutants at 5 dpf. Human progressive hearing loss caused by Mafb mutation is syndromic with other detectable defects, such as atopic dermatitis (AD), psoriasis vulgaris [27], aberrant extraocular muscle innervation, MCTO, and focal segmental glomerulosclerosis (FSGS) with DRS [33,34]. Zebrafish Mafba and human Mafb share a highly conserved amino acid sequence. The similarity in the inner-ear phenotype between zebrafish and human *mafb*-deficient individuals implies that *mafba* may be essential to inner-ear development and the maintenance of hearing capacity. In addition, *mafba*$^{-/-}$ mutants need to be examined at later stages of development to determine whether other tissues/organs are affected.

It is widely known that apical junctional complex proteins (AJC), extracellular matrix (ECM), and ion channels in the otic epithelial/epiderma cells act as barriers and are important to homeostasis maintenance, which is essential to normal primary inner-ear formation and subsequent development. The adhesion class G protein-coupled receptor (Gpr126) acts through a cAMP-mediated pathway to control the outgrowth and adhesion of canal projections in the zebrafish inner ear via the regulation of ECM gene expression [64]. The secretion of ECM molecules drives the growth of the epithelial projections and regulates the morphogenesis of the semicircular canals [65]. Regulated fluid secretion is crucial for many organs. The loss of chloride channel cystic fibrosis transmembrane conductance regulator (CFTR)-mediated fluid secretion impairs Kupffer's vesicle (KV) lumen expansion, leading to defects in organ laterality [66]. Interestingly, a previous study showed that fluid accumulation increases hydrostatic pressure in the lumen. Consequently, the stress passes into the epithelium and expands the otic vesicle. In turn, this pressure inhibits fluid transport into the lumen. This negative feedback between pressure and transport allows the otic vesicle to adjust its growth rate, resulting in otic vesicle size variations [67]. A previous study showed that mouse *Kreisler/Mafb* and zebrafish *valentino/mafba* mutants display early inner-ear defects that are related to abnormalities in the hindbrain development [36,68]. Facial motor neurons will fail to migrate through r5/r6 and complete caudal migration in zebrafish embryos with *mafba* deficiencies. [69]. The expression of genes in r5/r6 is regulated by RA, FGF, *hnf1ba*, and *valentino* (*mafba*); any losses of these factors will abolish r5/r6 gene expression [70]. Mafba is also essential to abducens' motor neuron development in zebrafish [71]. It has also been shown that the expression of *fgf3* in the hindbrain rhombomere is essential to the dorsal otic vesicle pattern in *valentino mutant zebrafish*, and its primary role is to maintain and focus the expression of the dorsal inner-ear gene that is induced by WNT signals [72]. Both of these reports imply that the otic vesicle pattern is regulated or influenced by hindbrain development. Interestingly, other researchers have provided evidence that *fgf3* is not sufficient to guide otic regionalization [35]. However both of these studies focused on early inner-ear defects caused by cascading effects of the external environment, such as abnormal hindbrain development. Whether there is any direct internal mechanism responsible for the abnormal inner-ear structure is yet to be determined.

The transcription factor *Mafb* plays a crucial role in epidermal keratinocyte differentiation in the mammalian epidermis by regulating downstream key transcription factors, such as *grhl3, znf750, klf4*, and *prdm1* [73,74] The ectopic expression of Mafb in basal keratinocyte results in excessive proliferation and disturbs epidermal homeostasis [56]. The terrestrial vertebrate epidermis is a stratified epithelium consisting of distinct layers of keratinocytes which build a functional osmotic barrier that prevents dehydration. In contrast, the fish epidermis is constituted of keratinocytes and ionocytes, which transport ions and acid-base substances to maintain the homeostasis of bodily fluids [75]. Most, if not all, studies of Mafb have focused on the homeostasis of epidermal keratinocytes and pay no attention to the ionocytes. In zebrafish embryos, epidermal stem cells specify and differentiate into ionocytes and keratinocytes during the bud stage, which is controlled by the Dlc-Notch mediated lateral inhibition [18,75]. Our data revealed that Mafba helps to maintain the ionocyte progenitor population by regulating epidermal stem cell proliferation, resulting in fewer stem cells and a decreasing number of differentiated ionocyte progenitors, and then decreases NR and HR cells in the inner-ear region. Finally, the loss of homeostasis could impede the volume control of the otocysts and the growth of otoliths. Collectively these defects lead to hearing and balance problems (Figure 6D). Previous studies have shown that *klf4*, which serves as a downstream regulator of *mafba*, maintains the ionocyte progenitor population via *dlc*-mediated lateral inhibition [18]. Taken together, our result clarify the novel and key role of *mafba* in the maintenance of ionocytes and ion homeostasis during zebrafish embryogenesis.

Interestingly, unlike in mice [37], the apoptosis of NR and HR cells increases markedly in *mafba*$^{-/-}$ mutant inner ears compared with those of sibling embryos from 4 dpf. The discrepancy between mice and zebrafish phenotypes may stem from differences between species or it might also be attributed to diversities between ionocytes and other cells. The reduction in NR and HR cells was due to cell apoptosis, and this led to inner-ear ion homeostasis imbalances, which eventually resulted in the *mafba*$^{-/-}$ mutant inner-ear defects at 5 dpf. Follow-up studies identified that the activation of p53-mediated apoptosis and cell cycle arrest at the G0/G1 phase in *mafba*$^{-/-}$ embryos at 4 dpf and DNA damage appeared in the inner-ear area from 3 dpf to 5 dpf. DNA damage is a relatively common event that leads to the activation of DNA damage checkpoints, such as ATM and ATR. These signal transducers activate p53 and deactivate cyclin-dependent kinases to inhibit cell cycle progression [76]. These results imply that DNA damage is the major factor causing p53 activation in response to zebrafish *mafba* deletion. Our results show that *mafba* depletion leads to a reduction in CDK6 protein level and transcriptional activity accompanied with an accumulation of *p53* mRNA and protein expression, which may be the mechanism of cell cycle arrest. Previous research has also shown that *mafb* deprivation destroys direct downstream regulator cyclin-dependent kinase 6 (CDK6) transcription and impedes clinical colorectal cancer (CRC) cell proliferation, as a result of the cell cycle arrest at the G0/G1 phase [37]. The specific deletion of *mafb* suppresses nasopharyngeal carcinoma cell (NCC) proliferation [77]. Overexpression of *mafb* enhances cell foci formation and increases cyclin B1 and D2 expression [40]. These studies suggest that DNA damage, p53 pathway activation, and the impairment of cell proliferation are responsible for the ion channel homeostasis imbalances in *mafba*$^{-/-}$ mutant inner ears. Aside from the ion channels, junction proteins are important for the homeostasis maintenance and inner-ear development [8,9]. Therefore, it is expected that *mafba* will exert its function on inner-ear development through the regulation of the junction proteins' factor. Further studies may focus on identifying the downstream targets among these candidates in order to clarify the specific mechanism of zebrafish *mafba*$^{-/-}$ mutant inner-ear defects.

In summary, we established a *mafba* knockout zebrafish line displaying auditory impairment for the first time. We elucidated the roles of Mafba in maintaining ion channel homeostasis in the zebrafish inner ear via the control of the proliferation of the ionocyte progenitor and the populations of differentiated NR and HR cells. This study provides novel insights into the inner-ear pathogenic mechanisms of Mafba and offers an ideal model for identifying new therapeutic interventions for inner-ear defects.

Supplementary Materials: The following are available online at https://www.mdpi.com/article/10.3390/biomedicines9111699/s1. Figure S1: The domain features and homologous comparison of the transcription factor Mafb among mammals and zebrafish. Figure S2: The *mafba* mRNA levels were detected in 4-dpf sibling and *mafba* mutant zebrafish by qRT-PCR. Figure S3. Relative expressions of large maf (*mafba*, *mafbb*, *mafa*, *c-maf*, and *nrl*) and small maf (*maff*, *mafk*, *mafga*, and *mafgb*) were determined by qRT-PCR analysis in sibling, *mafba*$^{+/-}$, and *mafba*$^{-/-}$ embryos at 4 dpf. Figure S4: Expression of the zebrafish *mafba* gene during embryonic development in siblings and its corresponding homozygous mutants. Figure S5: Early patterning of the inner-ear appears normal in *mafba*$^{-/-}$ mutants. Figure S6: Early patterning of the otic vesicle appears normal in *mafba*$^{-/-}$ mutants. Figure S7: The expression of otic specification-related markers was normal in *mafba*$^{-/-}$ mutant embryos. Figure S8: The expression of two otic matrix protein markers was normal in *mafba*$^{-/-}$ mutant embryos at 2 and 5 dpf. Figure S9: Deficiency of *mafba* does not affect hair cell development. Figure S10: Mafba protein expression pattern and colocalization of Mafba/p63 and Mafba/dlc during late-gastrulation and early-somite stages. Table S1: Primers used for Cas9 target design, WISH probes construction, genotype identification, and qRT-PCR.

Author Contributions: Conceptualization, X.C., Z.T., X.R., and M.L.; methodology, X.C.; software, Y.H. (Yuwen Huang); validation, Y.H. (Yuwen Huang); formal analysis, X.C.; investigation, X.C. and Y.H. (Yuwen Huang); resources, X.R. and M.L.; data curation, X.C.; writing—original draft preparation, X.C.; writing—review and editing, X.R., and M.L.; visualization, Y.L. and H.H.; supervision, P.G., D.J., and Y.H.(Yunqiao Han); project administration, K.S.; funding acquisition, Z.T. and M.L. All authors have read and agreed to the published version of the manuscript.

Funding: This research was funded by the Ministry of Science and Technology of China (2018YFA0801000) and the National Natural Science Foundation of China (82071010, 81801260, 81800870, 31801041, and 81870691).

Institutional Review Board Statement: The study was conducted according to the guidelines of the Declaration of Helsinki and approved by the Experimental Animal Ethics Committee of Huazhong University of Science and Technology (Approval Code: 2019s907 Approval Date: 25 February 2019).

Informed Consent Statement: Not applicable.

Data Availability Statement: The data presented in this study are available on request from the corresponding author.

Acknowledgments: The authors thank Fei Liu at the Institute of Hydrobiology, Chinese Academy of Sciences, for technical support. We appreciate Stephen Archacki's help correcting the grammar errors and polishing our manuscript.

Conflicts of Interest: The authors declare no conflict of interest.

References

1. Mittal, R.; Nguyen, D.; Patel, A.P.; Debs, L.H.; Mittal, J.; Yan, D.; Eshraghi, A.A.; Van De Water, T.R.; Liu, X.Z. Recent Advancements in the Regeneration of Auditory Hair Cells and Hearing Restoration. *Front. Mol. Neurosci.* **2017**, *10*, 236. [CrossRef]
2. Ganapathy, A.; Pandey, N.; Srisailapathy, C.R.S.; Jalvi, R.; Malhotra, V.; Venkatappa, M.; Chatterjee, A.; Sharma, M.; Santhanam, R.; Chadha, S.; et al. Non-Syndromic Hearing Impairment in India: High Allelic Heterogeneity among Mutations in TMPRSS3, TMC1, USHIC, CDH23 and TMIE. *PLoS ONE* **2014**, *9*, e84773. [CrossRef]

3. Nance, W.E.; Lim, B.G.; Dodson, K.M. Importance of congenital cytomegalovirus infections as a cause for pre-lingual hearing loss. *J. Clin. Virol.* **2006**, *35*, 221–225. [CrossRef]

4. Fritzsch, B.; Pauley, S.; Beisel, K.W. Cells, molecules and morphogenesis: The making of the vertebrate ear. *Brain Res.* **2006**, *1091*, 151–171. [CrossRef]

5. Jarvis, B.L.; Johnston, M.; Sulik, K. Congenital Malformations of the External Middle, and Inner Ear Produced by Isotretinoin Exposure in Mouse Embryos. *Otolaryngol. Head Neck Surg.* **1990**, *102*, 391–401. [CrossRef] [PubMed]

6. Schmitt, A. Accelerating our understanding of the inner ear of sauropodomorpha: First global, statistical analysis of semicircular canals of diplodocid and macronarian dinosaurs and its implication for neck-posture. In Proceedings of the SVP 74th Annual Meeting, Berlin, Germany, 6 November 2014.

7. Whitfield, T.T.; Mburu, P.; Hardisty-Hughes, R.E.; Brown, S.D. Models of congenital deafness: Mouse and zebrafish. *Drug Discov. Today Dis. Model.* **2005**, *2*, 85–92. [CrossRef]

8. Han, Y.; Mu, Y.; Li, X.; Xu, P.; Tong, J.; Liu, Z.; Ma, T.; Zeng, G.; Yang, S.; Du, J.; et al. Grhl2 deficiency impairs otic development and hearing ability in a zebrafish model of the progressive dominant hearing loss DFNA28. *Hum. Mol. Genet.* **2011**, *20*, 3213–3226. [CrossRef]

9. Li, X.; Song, G.; Zhao, Y.; Zhao, F.; Liu, C.; Liu, D.; Li, Q.; Cui, Z. Claudin7b is required for the formation and function of inner ear in zebrafish. *J. Cell. Physiol.* **2017**, *233*, 3195–3206. [CrossRef]

10. Abbas, L.; Whitfield, T.T. The zebrafish inner ear. *Fish Physiol.* **2010**, *29*, 123–171. [CrossRef]

11. Whitfield, T.T.; Riley, B.; Chiang, M.-Y.; Phillips, B. Development of the zebrafish inner ear. *Dev. Dyn.* **2002**, *223*, 427–458. [CrossRef]

12. Whitfield, T.T.; Granato, M.; van Eeden, F.J.; Schach, U.; Brand, M.; Furutani-Seiki, M.; Haffter, P.; Hammerschmidt, M.; Heisenberg, C.P.; Jiang, Y.J.; et al. Mutations affecting development of the zebrafish inner ear and lateral line. *Development* **1996**, *123*, 241–256. [CrossRef]

13. Trune, D.R. Ion homeostasis in the ear: Mechanisms, maladies, and management. *Curr. Opin. Otolaryngol. Head Neck Surg.* **2010**, *18*, 413–419. [CrossRef]

14. Ellertsdottir, E.; Ganz, J.; Dürr, K.; Loges, N.; Biemar, F.; Seifert, F.; Ettl, A.-K.; Kramer-Zucker, A.K.; Nitschke, R.; Driever, W. A mutation in the zebrafish Na,K-ATPase subunitatp1a1a.1provides genetic evidence that the sodium potassium pump contributes to left-right asymmetry downstream or in parallel to nodal flow. *Dev. Dyn.* **2006**, *235*, 1794–1808. [CrossRef] [PubMed]

15. Blasiole, B.; Canfield, V.A.; Vollrath, M.A.; Huss, D.; Mohideen, M.-A.P.; Dickman, J.D.; Cheng, K.; Fekete, D.M.; Levenson, R. Separate Na,K-ATPase genes are required for otolith formation and semicircular canal development in zebrafish. *Dev. Biol.* **2006**, *294*, 148–160. [CrossRef]

16. Jänicke, M.; Carney, T.J.; Hammerschmidt, M. Foxi3 transcription factors and Notch signaling control the formation of skin ionocytes from epidermal precursors of the zebrafish embryo. *Dev. Biol.* **2007**, *307*, 258–271. [CrossRef] [PubMed]

17. Hsiao, C.-D.; You, M.-S.; Guh, Y.-J.; Ma, M.; Jiang, Y.-J.; Hwang, P.-P. A Positive Regulatory Loop between foxi3a and foxi3b Is Essential for Specification and Differentiation of Zebrafish Epidermal Ionocytes. *PLoS ONE* **2007**, *2*, e302. [CrossRef]

18. Chen, Y.-C.; Liao, B.-K.; Lu, Y.-F.; Liu, Y.-H.; Hsieh, F.-C.; Hwang, P.-P.; Hwang, S.-P.L. Zebrafish Klf4 maintains the ionocyte progenitor population by regulating epidermal stem cell proliferation and lateral inhibition. *PLoS Genet.* **2019**, *15*, e1008058. [CrossRef]

19. Evans, D.H. Teleost fish osmoregulation: What have we learned since August Krogh, Homer Smith, and Ancel Keys. *Am. J. Physiol. Integr. Comp. Physiol.* **2008**, *295*, R704–R713. [CrossRef] [PubMed]

20. Hwang, P.-P.; Chou, M.-Y. Zebrafish as an animal model to study ion homeostasis. *Pflüg. Arch.* **2013**, *465*, 1233–1247. [CrossRef]

21. Hunter, G.L.; Hadjivasiliou, Z.; Bonin, H.; He, L.; Perrimon, N.; Charras, G.; Baum, B. Coordinated control of Notch-Delta signalling and cell cycle progression drives lateral inhibition mediated tissue patterning. *Development* **2016**, *143*, 2305–2310. [CrossRef]

22. Cruz, S.A.; Chao, P.-L.; Hwang, P.-P. Cortisol promotes differentiation of epidermal ionocytes through Foxi3 transcription factors in zebrafish (Danio rerio). *Comp. Biochem. Physiol. Part A Mol. Integr. Physiol.* **2013**, *164*, 249–257. [CrossRef]

23. Chang, W.J.; Horng, J.L.; Yan, J.J.; Hsiao, C.D.; Hwang, P.P. The transcription factor, glial cell missing 2, is involved in differentiation and functional regulation of H+-ATPase-rich cells in zebrafish (Danio rerio). *Am. J. Physiol. Regul. Integr. Comp. Physiol.* **2009**, *296*, R1192–R1201. [CrossRef] [PubMed]

24. Abbas, L.; Whitfield, T.T. Nkcc1 (Slc12a2) is required for the regulation of endolymph volume in the otic vesicle and swim bladder volume in the zebrafish larva. *Development* **2009**, *136*, 2837–2848. [CrossRef] [PubMed]

25. Cruz, S.; Shiao, J.-C.; Liao, B.-K.; Huang, C.-J.; Hwang, P.-P. Plasma membrane calcium ATPase required for semicircular canal formation and otolith growth in the zebrafish inner ear. *J. Exp. Biol.* **2009**, *212*, 639–647. [CrossRef]

26. Kataoka, K.; Noda, M.; Nishizawa, M. Maf nuclear oncoprotein recognizes sequences related to an AP-1 site and forms heterodimers with both Fos and Jun. *Mol. Cell Biol.* **1994**, *14*, 700–712. [CrossRef] [PubMed]

27. Moriguchi, T.; Hamada, M.; Morito, N.; Terunuma, T.; Hasegawa, K.; Zhang, C.; Yokomizo, T.; Esaki, R.; Kuroda, E.; Yoh, K.; et al. MafB Is Essential for Renal Development and F4/80 Expression in Macrophages. *Mol. Cell. Biol.* **2006**, *26*, 5715–5727. [CrossRef] [PubMed]

28. Aziz, A.; Vanhille, L.; Mohideen, P.; Kelly, L.M.; Otto, C.; Bakri, Y.; Mossadegh, N.; Sarrazin, S.; Sieweke, M.H. Development of Macrophages with Altered Actin Organization in the Absence of MafB. *Mol. Cell. Biol.* **2006**, *26*, 6808–6818. [CrossRef]

9.   Kim, K.; Kim, J.H.; Lee, J.; Jin, H.M.; Kook, H.; Kim, K.K.; Lee, S.Y.; Kim, N. MafB negatively regulates RANKL-mediated osteoclast differentiation. *Blood* **2006**, *109*, 3253–3259. [CrossRef]

10.   Miyai, M.; Hamada, M.; Moriguchi, T.; Hiruma, J.; Kamitani-Kawamoto, A.; Watanabe, H.; Hara-Chikuma, M.; Takahashi, K.; Takahashi, S.; Kataoka, K. Transcription Factor MafB Coordinates Epidermal Keratinocyte Differentiation. *J. Investig. Dermatol.* **2016**, *136*, 1848–1857. [CrossRef]

11.   Artner, I.; Le Lay, J.; Hang, Y.; Elghazi, L.; Schisler, J.; Henderson, E.; Sosa-Pineda, B.; Stein, R. MafB: An activator of the glucagon gene expressed in developing islet alpha- and beta-cells. *Diabetes* **2006**, *55*, 297–304. [CrossRef] [PubMed]

12.   Kamitani-Kawamoto, A.; Hamada, M.; Moriguchi, T.; Miyai, M.; Saji, F.; Hatamura, I.; Nishikawa, K.; Takayanagi, H.; Hitoshi, S.; Ikenaka, K.; et al. MafB interacts with Gcm2 and regulates parathyroid hormone expression and parathyroid development. *J. Bone Miner. Res.* **2011**, *26*, 2463–2472. [CrossRef]

13.   Sato, Y.; Tsukaguchi, H.; Morita, H.; Higasa, K.; Tran, M.T.N.; Hamada, M.; Usui, T.; Morito, N.; Horita, S.; Hayashi, T.; et al. A mutation in transcription factor MAFB causes Focal Segmental Glomerulosclerosis with Duane Retraction Syndrome. *Kidney Int.* **2018**, *94*, 396–407. [CrossRef]

14.   Park, J.G.; Tischfield, M.A.; Nugent, A.A.; Cheng, L.; Di Gioia, S.A.; Chan, W.-M.; Maconachie, G.; Bosley, T.M.; Summers, C.G.; Hunter, D.; et al. Loss of MAFB Function in Humans and Mice Causes Duane Syndrome, Aberrant Extraocular Muscle Innervation, and Inner-Ear Defects. *Am. J. Hum. Genet.* **2016**, *98*, 1220–1227. [CrossRef] [PubMed]

15.   Vázquez-Echeverría, C.; Dominguez-Frutos, E.; Charnay, P.; Schimmang, T.; Pujades, C. Analysis of mouse kreisler mutants reveals new roles of hindbrain-derived signals in the establishment of the otic neurogenic domain. *Dev. Biol.* **2008**, *322*, 167–178. [CrossRef]

16.   Kwak, S.J.; Phillips, B.T.; Heck, R.; Riley, B.B. An expanded domain of fgf3 expression in the hindbrain of zebrafish valentino mutants results in mis-patterning of the otic vesicle. *Development* **2002**, *129*, 5279–5287. [CrossRef] [PubMed]

17.   Yang, L.-S.; Zhang, X.-J.; Xie, Y.-Y.; Sun, X.-J.; Zhao, R.; Huang, Q.-H. SUMOylated MAFB promotes colorectal cancer tumorigenesis. *Oncotarget* **2016**, *7*, 83488–83501. [CrossRef] [PubMed]

18.   Suda, N.; Itoh, T.; Nakato, R.; Shirakawa, D.; Bando, M.; Katou, Y.; Kataoka, K.; Shirahige, K.; Tickle, C.; Tanaka, M. Dimeric combinations of MafB, cFos and cJun control the apoptosis-survival balance in limb morphogenesis. *Development* **2014**, *141*, 2885–2894. [CrossRef]

19.   Gao, M.; Huang, Y.; Wang, L.; Huang, M.; Liu, F.; Liao, S.; Yu, S.; Lu, Z.; Han, S.; Hu, X.; et al. HSF4 regulates lens fiber cell differentiation by activating p53 and its downstream regulators. *Cell Death Dis.* **2017**, *8*, e3082. [CrossRef]

20.   Thisse, C.; Thisse, B. High-resolution in situ hybridization to whole-mount zebrafish embryos. *Nat. Protoc.* **2007**, *3*, 59–69. [CrossRef]

21.   Malicki, J.; Avanesov, A.; Li, J.; Yuan, S.; Sun, Z. Analysis of Cilia Structure and Function in Zebrafish. *Method Cell Biol.* **2011**, *101*, 39–74. [CrossRef]

22.   Hu, X.; Lu, Z.; Yu, S.; Reilly, J.; Liu, F.; Jia, D.; Qin, Y.; Han, S.; Liu, X.; Qu, Z.; et al. CERKL regulates autophagy via the NAD-dependent deacetylase SIRT1. *Autophagy* **2018**, *15*, 453–465. [CrossRef] [PubMed]

23.   Preibisch, S.; Saalfeld, S.; Tomancak, P. Globally optimal stitching of tiled 3D microscopic image acquisitions. *Bioinformatics* **2009**, *25*, 1463–1465. [CrossRef] [PubMed]

24.   Schindelin, J.; Rueden, C.T.; Hiner, M.C.; Eliceiri, K.W. The ImageJ ecosystem: An open platform for biomedical image analysis. *Mol. Reprod. Dev.* **2015**, *82*, 518–529. [CrossRef]

25.   Yu, S.; Jiang, T.; Jia, D.; Han, Y.; Liu, F.; Huang, Y.; Qu, Z.; Zhao, Y.; Tu, J.; Lv, Y.; et al. BCAS2 is essential for hematopoietic stem and progenitor cell maintenance during zebrafish embryogenesis. *Blood* **2019**, *133*, 805–815. [CrossRef]

26.   Li, J.; Liu, F.; Lv, Y.; Sun, K.; Zhao, Y.; Reilly, J.; Zhang, Y.; Tu, J.; Yu, S.; Liu, X.; et al. Prpf31 is essential for the survival and differentiation of retinal progenitor cells by modulating alternative splicing. *Nucleic Acids Res.* **2021**, *49*, 2027–2043. [CrossRef] [PubMed]

27.   Liu, F.; Qin, Y.; Yu, S.; Soares, D.; Yang, L.; Weng, J.; Li, C.; Gao, M.; Lu, Z.; Hu, X.; et al. Pathogenic mutations in retinitis pigmentosa 2 predominantly result in loss of RP2 protein stability in humans and zebrafish. *J. Biol. Chem.* **2017**, *292*, 6225–6239. [CrossRef]

28.   Końca, K.; Lankoff, A.; Banasik, A.; Lisowska, H.; Kuszewski, T.; Gozdz, S.; Koza, Z.; Wojcik, A. A cross-platform public domain PC image-analysis program for the comet assay. *Mutat. Res. Toxicol. Environ. Mutagen.* **2002**, *534*, 15–20. [CrossRef]

29.   Han, S.; Liu, X.; Xie, S.; Gao, M.; Liu, F.; Yu, S.; Sun, P.; Wang, C.; Archacki, S.; Lu, Z.; et al. Knockout of ush2a gene in zebrafish causes hearing impairment and late onset rod-cone dystrophy. *Qual. Life Res.* **2018**, *137*, 779–794. [CrossRef] [PubMed]

30.   Zhu, K.-C.; Wu, M.; Zhang, D.-C.; Guo, H.-Y.; Zhang, N.; Guo, L.; Liu, B.-S.; Jiang, S.-G. Toll-Like Receptor 5 of Golden Pompano *Trachinotus ovatus* (Linnaeus 1758): Characterization, Promoter Activity and Functional Analysis. *Int. J. Mol. Sci.* **2020**, *21*, 5916. [CrossRef]

31.   Yang, Y.; Hongyan, L.; Shicui, Z. Phylogenetic analysis of genes in the Maf family. *Ludong Univ. J. Nat. Sci. Ed.* **2014**, *30*, 330–335. [CrossRef]

32.   Kajihara, M.; Kawauchi, S.; Kobayashi, M.; Ogino, H.; Takahashi, S.; Yasuda, K. Isolation, characterization, and expression analysis of zebrafish large Mafs. *J. Biochem.* **2001**, *129*, 139–146. [CrossRef]

33.   Nicolson, T. The Genetics of Hearing and Balance in Zebrafish. *Annu. Rev. Genet.* **2005**, *39*, 9–22. [CrossRef] [PubMed]

54. Baxendale, S.; Whitfield, T.T. Zebrafish Inner Ear Development and Function. In *Development of Auditory and Vestibular System* 4th ed.; Romand, R., Varela-Nieto, I., Eds.; Academic Press: Amsterdam, The Netherlands, 2014; pp. 63–105. [CrossRef]
55. Riley, B.B.; Moorman, S.J. Development of utricular otoliths, but not saccular otoliths, is necessary for vestibular function and survival in zebrafish. *J. Neurobiol.* **2015**, *43*, 329–337. [CrossRef]
56. Miyai, M.; Tsunekage, Y.; Saito, M.; Kohno, K.; Takahashi, K.; Kataoka, K. Ectopic expression of the transcription factor Maf in basal keratinocytes induces hyperproliferation and perturbs epidermal homeostasis. *Exp. Dermatol.* **2017**, *26*, 1039–1041. [CrossRef] [PubMed]
57. Lu, J.; Hamze, Z.; Bonnavion, R.; Herath, N.; Pouponnot, C.; Assade, F.; Fontaniere, S.; Bertolino, P.; Cordierbussat, M.; Zhang, C.X. Reexpression of oncoprotein MafB in proliferative β-cells and Men1 insulinomas in mouse. *Oncogene* **2011**, *31*, 3647–3654. [CrossRef] [PubMed]
58. Karbaschi, M.; Ji, Y.; Abdulwahed, A.M.S.; Alohaly, A.; Bedoya, J.F.; Burke, S.L.; Boulos, T.M.; Tempest, H.G.; Cooke, M. Evaluation of the Major Steps in the Conventional Protocol for the Alkaline Comet Assay. *Int. J. Mol. Sci.* **2019**, *20*, 6072. [CrossRef]
59. Cordes, S.P.; Barsh, G.S. The mouse segmentation gene kr encodes a novel basic domain-leucine zipper transcription factor. *Cell* **1994**, *79*, 1025–1034. [CrossRef]
60. Schvarzstein, M.; Kirn, A.; Haffter, P.; Cordes, S.P. Expression of Zkrml2, a homologue of the Krml1/val segmentation gene during embryonic patterning of the zebrafish (Danio rerio). *Mech. Dev.* **1999**, *80*, 223–226. [CrossRef]
61. Katzenback, B.A.; Karpman, M.; Belosevic, M. Distribution and expression analysis of transcription factors in tissues and progenitor cell populations of the goldfish (Carassius auratus L.) in response to growth factors and pathogens. *Mol. Immunol.* **2011**, *48*, 1224–1235. [CrossRef]
62. Sadl, V.S.; Jin, F.; Yu, J.; Cui, S.; Holmyard, D.; Quaggin, S.E.; Barsh, G.S.; Cordes, S.P. The Mouse Kreisler (Krml1/MafB) Segmentation Gene Is Required for Differentiation of Glomerular Visceral Epithelial Cells. *Dev. Biol.* **2002**, *249*, 16–29. [CrossRef]
63. Koltowska, K.; Paterson, S.; Bower, N.I.; Baillie, G.J.; Lagendijk, A.K.; Astin, J.W.; Chen, H.; Francois, M.; Crosier, P.S.; Taft, R. mafba is a downstream transcriptional effector of Vegfc signaling essential for embryonic lymphangiogenesis in zebrafish. *Genes Dev.* **2015**, *29*, 1618–1630. [CrossRef] [PubMed]
64. Geng, F.-S.; Abbas, L.; Baxendale, S.; Holdsworth, C.J.; Swanson, A.G.; Slanchev, K.; Hammerschmidt, M.; Topczewski, J.; Whitfield, T.T. Semicircular canal morphogenesis in the zebrafish inner ear requires the function of gpr126 (lauscher), an adhesion class G protein-coupled receptor gene. *Development* **2013**, *140*, 4362–4374. [CrossRef] [PubMed]
65. Clément, A.; Blanco-Sánchez, B.; Peirce, J.L.; Westerfield, M. Cog4 is required for protrusion and extension of the epithelium in the developing semicircular canals. *Mech. Dev.* **2018**, *155*, 1–7. [CrossRef]
66. Navis, A.; Marjoram, L.; Bagnat, M. Cftr controls lumen expansion and function of Kupffer's vesicle in zebrafish. *Development* **2013**, *140*, 1703–1712. [CrossRef] [PubMed]
67. Mosaliganti, K.R.; Swinburne, I.A.; Chan, C.U.; Obholzer, N.D.; A Green, A.; Tanksale, S.; Mahadevan, L.; Megason, S.G. Size control of the inner ear via hydraulic feedback. *eLife* **2019**, *8*. [CrossRef]
68. Choo, D.; Ward, J.; Reece, A.; Dou, H.; Lin, Z.; Greinwald, J. Molecular mechanisms underlying inner ear patterning defects in kreisler mutants. *Dev. Biol.* **2006**, *289*, 308–317. [CrossRef] [PubMed]
69. Zannino, D.A.; Sagerström, C.G.; Appel, B. olig2-expressing hindbrain cells are required for migrating facial motor neurons. *Dev. Dyn.* **2011**, *241*, 315–326. [CrossRef] [PubMed]
70. Ghosh, P.; Maurer, J.M.; Sagerström, C.G. Analysis of novel caudal hindbrain genes reveals different regulatory logic for gene expression in rhombomere 4 versus 5/6 in embryonic zebrafish. *Neural Dev.* **2018**, *13*, 1–24. [CrossRef]
71. Asakawa, K.; Kawakami, K. Protocadherin-Mediated Cell Repulsion Controls the Central Topography and Efferent Projections of the Abducens Nucleus. *Cell Rep.* **2018**, *24*, 1562–1572. [CrossRef] [PubMed]
72. Hatch, E.P.; Noyes, C.A.; Wang, X.; Wright, T.J.; Mansour, S.L. Fgf3 is required for dorsal patterning and morphogenesis of the inner ear epithelium. *Development* **2007**, *134*, 3615–3625. [CrossRef]
73. Lopez-Pajares, V.; Qu, K.; Zhang, J.; Webster, D.E.; Barajas, B.C.; Siprashvili, Z.; Zarnegar, B.J.; Boxer, L.; Rios, E.J.; Tao, S.; et al. A LncRNA-MAF:MAFB Transcription Factor Network Regulates Epidermal Differentiation. *Dev. Cell* **2015**, *32*, 693–706. [CrossRef] [PubMed]
74. Labott, A.T.; Lopez-Pajares, V. Epidermal differentiation gene regulatory networks controlled by MAF and MAFB. *Cell Cycle* **2016**, *15*, 1405–1409. [CrossRef] [PubMed]
75. Chang, W.-J.; Hwang, P.-P. Development of zebrafish epidermis. *Birth Defects Res. Part C Embryo Today Rev.* **2011**, *93*, 205–214. [CrossRef]
76. Sancar, A.; Lindsey-Boltz, L.A.; Ünsal-Kaçmaz, K.; Linn, S. Molecular Mechanisms of Mammalian DNA Repair and the DNA Damage Checkpoints. *Annu. Rev. Biochem.* **2004**, *73*, 39–85. [CrossRef] [PubMed]
77. Yang, W.; Lan, X.; Li, D.; Li, T.; Lu, S. MiR-223 targeting MAFB suppresses proliferation and migration of nasopharyngeal carcinoma cells. *BMC Cancer* **2015**, *15*, 461. [CrossRef]

biomedicines

Article

Functions of *SMC2* in the Development of Zebrafish Liver

Xixi Li [1], Guili Song [2], Yasong Zhao [2,3], Jing Ren [1], Qing Li [2] and Zongbin Cui [1,2,*]

[1] Guangdong Provincial Key Laboratory of Microbial Culture Collection and Application, State Key Laboratory of Applied Microbiology Southern China, Institute of Microbiology, Guangdong Academy of Sciences, Guangzhou 510070, China; lixixi@ihb.ac.cn (X.L.); renjing@ihb.ac.cn (J.R.)

[2] Institute of Hydrobiology, Chinese Academy of Sciences, Wuhan 430072, China; guilisong@ihb.ac.cn (G.S.); yszhao@ihb.ac.cn (Y.Z.); qli@ihb.ac.cn (Q.L.)

[3] College of Advanced Agricultural Sciences, University of Chinese Academy of Sciences, Beijing 100049, China

* Correspondence: cuizb@gdim.cn; Tel.: +86-020-87137656

Abstract: *SMC2* (structural maintenance of chromosomes 2) is the core subunit of condensins, which play a central role in chromosome organization and segregation. However, the functions of *SMC2* in embryonic development remain poorly understood, due to the embryonic lethality of homozygous $SMC2^{-/-}$ mice. Herein, we explored the roles of *SMC2* in the liver development of zebrafish. The depletion of *SMC2*, with the CRISPR/Cas9-dependent gene knockout approach, led to a small liver phenotype. The specification of hepatoblasts was unaffected. Mechanistically, extensive apoptosis occurred in the liver of *SMC2* mutants, which was mainly associated with the activation of the *p53*-dependent apoptotic pathway. Moreover, an aberrant activation of a series of apoptotic pathways in *SMC2* mutants was involved in the defective chromosome segregation and subsequent DNA damage. Therefore, our findings demonstrate that *SMC2* is necessary for zebrafish liver development.

Keywords: zebrafish; *SMC2*; liver development; DNA damage; apoptosis

Citation: Li, X.; Song, G.; Zhao, Y.; Ren, J.; Li, Q.; Cui, Z. Functions of *SMC2* in the Development of Zebrafish Liver. *Biomedicines* **2021**, *9*, 240. https://doi.org/10.3390/biomedicines9091240

Academic Editor: James A. Marrs

Received: 17 August 2021
Accepted: 10 September 2021
Published: 16 September 2021

Publisher's Note: MDPI stays neutral with regard to jurisdictional claims in published maps and institutional affiliations.

1. Introduction

Chromosomes undergo essential changes in morphology, to control the proper expression of genes, and these changes are partially mediated by the structural maintenance of chromosome (*SMC*) proteins [1]. *SMC* proteins are evolutionarily conserved from bacteria to human, and function in chromosome condensation, sister-chromatid cohesion, DNA repair and recombination, and gene dosage compensation in somatic and meiotic cells [2]. *SMC* proteins were initially found in Saccharomyces cerevisiae, and later in all tested eukaryotes [3]. Bacteria contain a single gene that encodes a single *SMC* protein to form homodimers [4]. In eukaryotes, at least six members of the *SMC* protein family are found in individual organisms. The primary structure of *SMC* proteins consists of the following five distinct domains: two nucleotide-binding motifs, Walker A and Walker B motifs that are located in the highly conserved N-terminal and C-terminal domains, respectively, and the central domain, which is composed of a moderately conserved "hinge" sequence that is flanked by two long coiled-coil motifs [5].

As members of the *SMC* family, *SMC2* and *SMC4* form a heterodimer that is the catalytic subunit of both condensins I and II complexes, which play roles in mitotic and meiotic chromosome condensation and rigidity, interphase ribosomal DNA compaction, and removal of cohesion during mitosis and meiosis [6–8]. Previous knockout or knockdown studies of *SMC2* revealed its importance for mitotic and meiotic chromosome condensation and segregation in Drosophila melanogaster [9], Caenorhabditis elegans [10], S. cerevisiae [10], and other species [3]. In mice, the knockout of *SMC2* led to embryonic lethality [11]. Thus, the functions of *SMC2* in the embryonic development of vertebrates remain largely unknown.

The liver is a visceral organ in vertebrates that has many important functions in metabolism, secretion, detoxification, and homeostasis. The advantages of high fecundity, transparent embryos, and small size make zebrafish a powerful model for specialized mutagenesis screens, to identify genes whose counterparts can regulate liver development in humans. Liver organogenesis begins with the establishment of a population of cells gaining hepatic competency within the ventral foregut endoderm, instructed by Foxa and Gata factors. Thereafter, mesodermal signals, including Fgfs, Bmps, Wnt2b, and retinoic acid, induce the specification of hepatoblasts, which then migrate and proliferate to form a discrete liver bud. Finally, hepatoblasts in the liver bud undergo rapid proliferation and differentiation, giving rise to bile duct cells and functional hepatocytes [12–15].

In zebrafish, there are the following three main stages of hepatogenesis: (1) specification (as part of endoderm patterning); (2) differentiation (the budding phase); and (3) hepatic outgrowth, accompanied by morphogenesis [16]. Hepatoblast specification is thought to occur at approximately 22 h post-fertilization, as marked by the localized endodermal expression of *hhex* and *prox1* [14], which are two transcription factor genes that are also expressed in mice [17] and chicks [18]. The budding phase occurs from 24 to 50 hpf. At the subsequent growth stage, the liver undergoes dramatic changes in its size, shape, and placement, because of rapid cell proliferation.

In this study, we addressed an in vivo role of *SMC2* in the liver development of zebrafish. The mutation of *SMC2*, with a CRISPR/Cas9-mediated approach, led to a small liver phenotype, due to elevated apoptosis in the liver and decreased cell proliferation. We then found that extensive apoptosis occurred within the defective liver, due to the activation of intrinsic apoptotic signaling pathways, especially the p53-dependent apoptotic pathway. We further demonstrated that the aberrant activation of the apoptotic pathways was closely associated with DNA damage.

2. Materials and Methods

2.1. Zebrafish Husbandry and Ethics

Zebrafish AB strain was used and maintained under standard conditions in this study. The $p53^{M241K/M214K}$ and $Tg(fabp10a:dsRed;ela3l:EGFP)$ lines were previously described [19,20]. All zebrafish studies were conducted according to standard animal guidelines and approved by the Institutional Animal Care and Use Committee of the Institute of Hydrobiology, Chinese Academy of Sciences (approval ID: Keshuizhuan 0829).

2.2. Generation of SMC2 Mutant Zebrafish Lines

The *SMC2* mutant lines were generated with the CRISPR/Cas9 system following previous methods [21]. Briefly, the target site sequence of *SMC2* is ATCACTGGACT-GAACGGCAG, which is located in the second exon. The gRNAs were synthesized in vitro with T7 RNA polymerase (ThermoFisher, Waltham, MA, USA). The Cas9 mRNA was synthesized using the mMESSAGE mMACHINE T7 kit (Invitrogen, Carlsbad, CA, USA). A total of 400 pg Cas9-mRNA and 50 pg *SMC2*-gRNA were co-injected into zebrafish embryos at one-cell stage. The *SMC2* mutations were examined by PCR, and the amplified fragments were sequenced using the following primer pair: 5′-TGGTTGAACTGAAAGCAACG-3′ and 5′-CTTCCAGTTGTTTGCATCTCG-3′.

Because $SMC2^{-/-}$ died at about 7 days post-fertilization (dpf), the $SMC2^{+/c504}$ ($SMC2^{+/-}$) fish were used to cross with Tg $(fabp10a:dsRed; ela3l:EGFP)$ fish. To obtain the ($SMC2^{+/-}$; DsRed) adult zebrafish, the fluorescence microscope was used to identify DsRed-expressing fish, followed by the genotype of $SMC2^{+/-}$ fish with the primer pair SMC2-gF/SMC2-gR (Table S1). The ($SMC2^{+/-}$; DsRed) adult zebrafish were inbred to obtain the ($SMC2^{-/-}$; DsRed) embryos by phenotypic observation and fluorescence microscope since the $SMC2^{-/-}$ mutants exhibited abnormal brain and eyes at about 28 h post-fertilization (hpf).

To obtain the ($SMC2^{-/-}$; $p53^{-/-}$; DsRed) fish, the ($SMC2^{+/-}$; DsRed) adult zebrafish were crossed with $p53^{-/-}$ fish. Fluorescence microscope was used to identify DsRed-expressing fish, followed by the genotype of $SMC2^{+/-}$ and $p53^{+/-}$ with two primer pairs SMC2-gF/SMC2-gR and tp53-gF/tp53-gR, thus obtaining the ($SMC2^{+/-}$; $p53^{+/-}$; DsRed) adult fish. The ($SMC2^{+/-}$; $p53^{+/-}$; DsRed) adult fish were inbred to obtain the ($SMC2^{-/-}$;

p53⁻/⁻; DsRed) embryos by the observation of dsRed fluorescence and abnormal phenotypes, followed by tail-PCR of 96-hpf embryos with primer pair tp53-gF/tp53-gR (Table S1).

2.3. Quantitative Real-Time PCR (qPCR)

Total RNA was extracted from 50 embryos at indicated stages with TRIZOL reagent (Invitrogen, cat#15596026) according to the manufacturer's instructions. A first strand cDNA synthesis kit (ThermoFisher, Waltham, MA, USA) was used to synthesize cDNAs. SYBR Green real-time PCR master mix (Bio-Rad) was used for qPCR. The qPCR primers used in this study are listed in Table S1.

2.4. Whole-Mount In Situ Hybridization (WISH)

Zebrafish embryos at desired stages were fixed in 4% paraformaldehyde (PFA) overnight before processing for WISH analysis as described [22]. Digoxigenin-UTP-labeled antisense RNA probes for *fabp10a, cp, insulin, typsin, fabp2, hhex, prox1, foxa1, foxa3* and *gata4* were generated with an in vitro transcription method using T7 RNA polymerase (ThermoFisher, Waltham, MA, USA). The primer sequences for these genes are listed in Table S1.

2.5. TUNEL Assay

Embryos at 96 hpf were fixed with 4% formaldehyde for 4 h at room temperature and embedded in OCT compound overnight. The embryos were then sectioned at 10 μm thickness using a Leica cryostat. TUNEL assays were performed with in situ cell death detection kit, Fluorescein (Roche, Wilmington, MA, USA), following the manufacture's instruction.

2.6. Immunofluorescence Staining

Sectioned samples were fixed in 4% paraformaldehyde for 20 min and washed five times of 5 min each in PBS. After completely removing paraformaldehyde, samples were permeabilized with 0.5% Triton X-100 (Sigma-Aldrich) in PBS for 15 min at room temperature and blocked in 1% bovine serum albumin for 1 h. Incubation with the primary antibody occurred at 4 °C overnight, followed by overnight incubation with the secondary antibody at 4 °C. Primary antibodies against γ-H2AX (1:400; 9718T; Cell Signaling Technology) were used. The fluorescence-conjugated second antibody FITC goat anti-rabbit IgG (H + L) (BOSTER, Wuhan, China) was applied to sample at a dilution of 1:50. Sections were counterstained with DAPI for 30 min at room temperature and then mounted with an antifade agent. The samples were imaged under an SP8 confocal microscope (Leica).

2.7. EdU Labeling

Embryos at 96 hpf were microinjected into the yolk cell with 10 mM EdU. After two hours, the embryos were fixed with 4% formaldehyde for 4 h at room temperature and embedded in OCT compound overnight. The embryos were then sectioned at 10 μm thickness using a Leica cryostat. EdU staining was performed according to the manufacture's instructions (Beyotime, Shanghai, China).

2.8. Synthesis of Capped SMC2 mRNA

The coding sequence of zebrafish *SMC2* was cloned into the vector pSBRNAX and linearized for in vitro transcription. mRNAs were synthesized using the mMESSAGE mMACHINE T7 kit (Invitrogen, Carlsbad, CA, USA). One-cell stage embryos were injected.

2.9. Fluorescence-Activated Cell Sorting

Sib and *SMC2⁻/⁻* mutant embryos at 96 hpf were used. The embryos were transferred into 1.5 mL centrifuge tubes and washed three times in PBS. Then 1 mL of 0.25% trypsin was added to each centrifuge tube. Embryos were passed through the syringe to generate cell suspensions, and the cell suspensions were passed through a 40 μm filter. Cells were washed in ice-cold PBS and incubated for 30 min in 1 mL DAPI. Samples were filtered

again over a 40 μm filter, washed with the ice-cold PBS and finally resuspended in PBS fo
sorting by flow cytometry (BD Biosciences, Franklin Lakes, NJ, USA).

2.10. Statistical Analysis

The data are presented as mean ± standard deviation. Statistical differences betweer
two sets of data were analyzed using two-tailed paired Student's *t*-test, and a value o
$p < 0.05$ was considered to indicate significance.

3. Results

3.1. Conservation of SMC2-Containing Complexes among Vertebrates, and the Expression Patterns of SMC2 during Embryonic Development of Zebrafish

To illuminate the potential conserved functions of *SMC2* among vertebrates, we
first compared the similarity of *SMC2*-containing complexes among zebrafish, mice, and
humans. The zebrafish genome contains a single copy of each gene encoding subunits o
condensin I and II complexes, and the amino acid identity between zebrafish and their humar
counterparts ranged from 36.5 to 74.2% (Table S2), indicating that the two *SMC2*-containing
complexes are highly conserved among vertebrates. Zebrafish *SMC2* encodes a proteir
consisting of 1199 amino acids. Amino acid sequence alignment indicates that zebrafish *SMC.*
is the most conserved subunit, sharing 74.2% identity with human *SMC2* (Table S2).

Next, we detected the spatiotemporal expression patterns of *SMC2* with RT-PCF
and WISH. Transcripts of *SMC2* can be detected in developing embryos at 0 to 120 hp
(Figure 1A), weakly detected in embryos at the one-cell stage (Figure 1B(a)), and ubic
uitously expressed in embryos at two-cells, shield, and 12-hpf stages (Figure 1B(b–d)
indicating the maternal origin of *SMC2* transcripts. The *SMC2* transcripts were expressed ir
diencephalon (di), mesencephalon (me), eye (e), and endoderm (en), at 48 hpf (Figure 1B(f)
Later, its expression was found in the forebrain ventricular zone (fvz), branchial arche
(ba), midbrain–hindbrain boundary (mhb), liver (lv), and intestine (i), at 72 and 96 hp
(Figure 1B(g–i)). These results indicate that *SMC2* functions in early embryonic develop
ment and the formation of multiple organs in zebrafish.

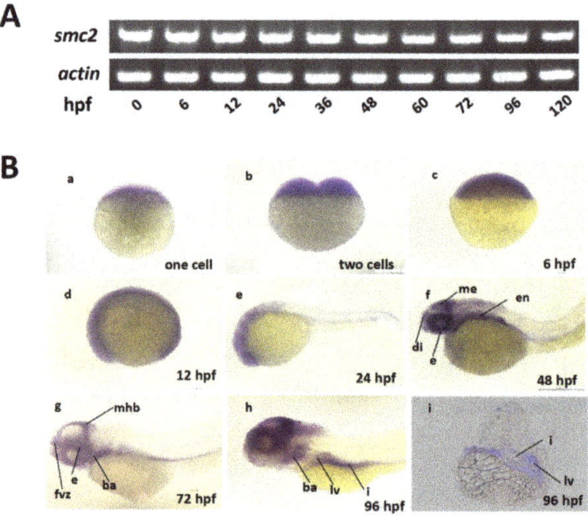

Figure 1. Expression of *SMC2* mRNA during zebrafish embryogenesis. (**A**) Expression levels of *SMC.*
at different stages were analyzed with RT-PCR and the expression of *β-actin* served as the contro
(**B**) Detection of *SMC2* transcripts during embryogenesis with WISH. hpf, hours post-fertilizatior
me, mesencephalon; di, diencephalon; fvz, forebrain ventricular zone; ba, branchial arches; mhl
midbrain–hindbrain boundary; i, intestine; en, endoderm; e, eye; lv, liver.

3.2. Knockout of SMC2 in Zebrafish

To investigate the function of *SMC2*, we generated *SMC2* mutant zebrafish with the CRISPR/Cas9 system. Two mutant alleles were obtained from different P0 founders, by targeting exon 1 of the *SMC2* gene. One contains a 31-bp deletion (named *SMC2$^{c504/c504}$*), and the other has a 15-bp deletion and 1-bp insertion (named *SMC2$^{c505/c505}$*) (Figure 2A). Both of the two mutants led to frame-shift mutations of the open reading frame, and premature stop codons that can abolish all functions of *SMC2* (Figure 2B). The relative mRNA expression levels were significantly reduced in homozygotes of *SMC2$^{c504/c504}$* and *SMC2$^{c505/c505}$* embryos (Figure 2C), likely through a nonsense-mediated decay mechanism [23].

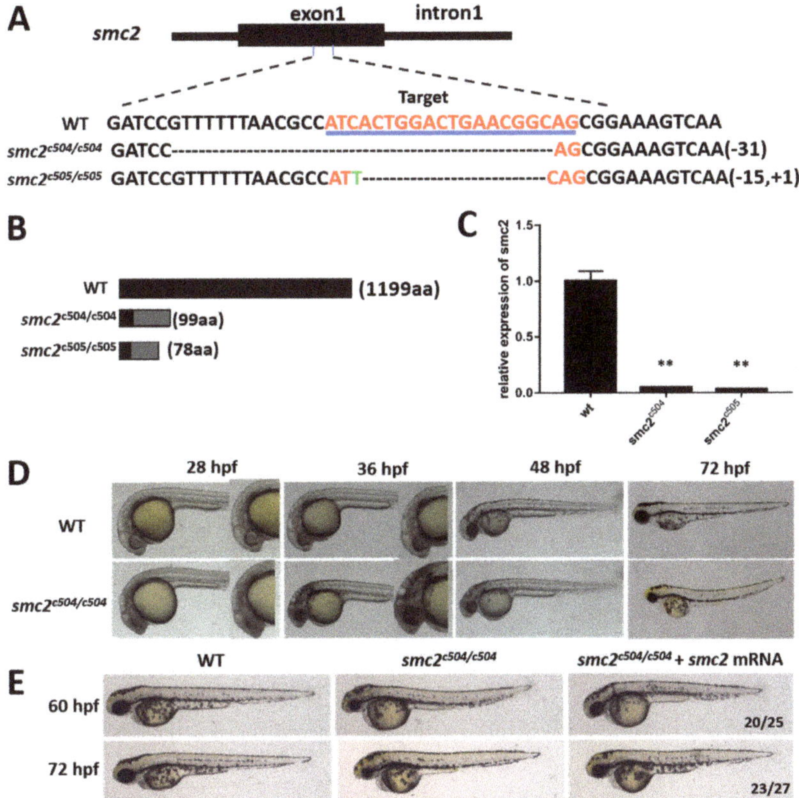

Figure 2. (**A**) Top panel: the schematic diagram shows the structure of the target region in the *SMC2* gene. Black box, exon; solid line, introns. The bottom panel shows a comparison of the genomic DNA sequences among WT and two mutant alleles, *SMC2$^{c504/504}$* mutants carrying a 31-bp deletion and *SMC2$^{c505/505}$* mutants carrying a 15-bp deletion and 1-bp insertion. The target sequence is indicated in red and underlined. (**B**) Schematic structures of WT and truncated *SMC2* proteins. (**C**) The relative mRNA levels of *SMC2* in WT and homozygous mutants were assayed by qPCR at 4 dpf. The results are expressed as the mean ± SD of three independent experiments (**, $p < 0.01$; *t*-test). (**D**) Lateral views showing the morphology of WT and *SMC2$^{-/-}$* mutants at 28, 36, 48 and 72 hpf. (**E**) Morphology of WT and *SMC2$^{-/-}$* embryos injected with or without 200 pg *SMC2*-mRNAs at indicated stages. The smaller eyes and smaller head were significantly reduced in *SMC2*-mRNA-injected mutant embryos. The ratios at the bottom right corners indicate the number of embryos with indicated phenotypes vs. total number of observed embryos.

Heterozygous $SMC2^{c504/+}$ or $SMC2^{c505/+}$ fish that showed no discernable phenotypes are viable, and can develop into fertile adults. However, homozygous mutants obtained from a cross between either $SMC2^{c504/c504}$ or $SMC2^{c505/c505}$ homozygotes exhibited a dark phenotype in the head, due to extensive cell death after 28 hpf, and this extended to the whole brain and spinal cord afterwards (Figure 2D). The homozygous mutant embryos displayed small eyes and a small head at 48 and 72 hpf (Figure 2D), and died at about 7 dpf. Since the $SMC2^{c504/c504}$ and $SMC2^{c505/c505}$ larvae showed exactly the same phenotypes, $SMC2^{c504/c504}$ mutants ($SMC2^{-/-}$ mutants hereafter) were used for further analysis. To determine whether the phenotypes of mutant embryos resulted from $SMC2$ depletion, synthesized $SMC2$ mRNA was injected into F2 $SMC2^{-/-}$ larvae. The injection of $SMC2$ mRNA did not cause any obvious morphological defects in WT embryos, but markedly rescued the small eyes and small head phenotypes of $SMC2^{-/-}$ larvae at 60 hpf and 72 hpf (Figure 2E). These findings suggest a crucial role of $SMC2$ during zebrafish embryogenesis.

3.3. Loss of SMC2 Led to a Small Liver Phenotype in $SMC2^{-/-}$ Mutants

The existence of $SMC2$ transcripts in the endoderm at 24 hpf, and liver at 96 hpf, suggests the involvement of $SMC2$ in the liver development of zebrafish. We evaluated liver development in homozygous $SMC2$ mutants, using the hepatocyte marker gene *fabp10a* as a probe of WISH. As shown in Figure 3A, the expression of *fabp10a* in $SMC2$ mutants at 72 and 96 hpf severely reduced. We also examined the liver phenotype of $SMC2$ mutants in the *Tg (fabp10a:dsRed; ela3l:EGFP)* line that expresses DsRed specifically in differentiated hepatocytes. Markedly reduced sizes of livers were shown in $SMC2$ mutant larvae in comparison with those in WT larvae, at 72 and 96 hpf (Figure 3B).

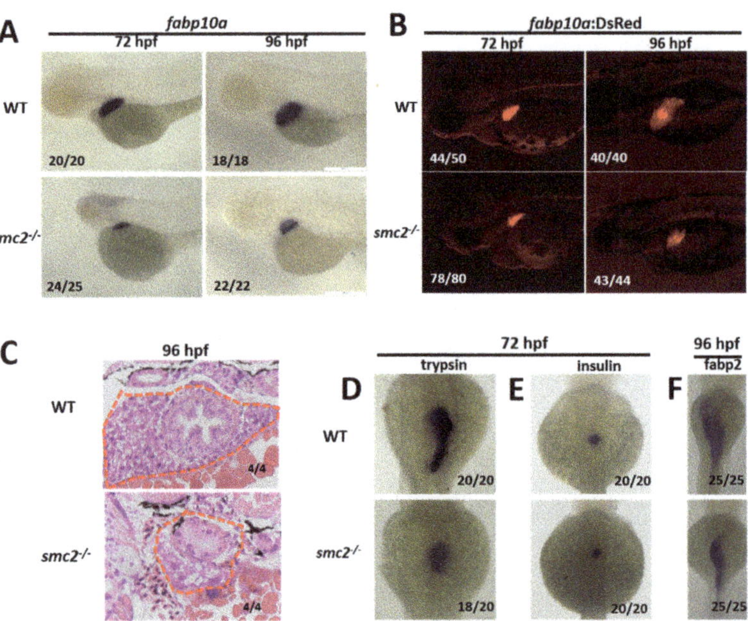

Figure 3. Loss of $SMC2$ confers a small liver phenotype. (**A**) WT and $SMC2^{-/-}$ mutant embryos were stained with RNA probes of *fabp10a*, a marker of hepatocytes at 72 and 96 hpf. (**B**) Liver size in $SMC2^{-/-}$ mutants is smaller than that in the WT at 72 and 96 hpf, under the *Tg (fabp10a:dsRed, ela3l:EGFP)* transgenic background. (**C**) Livers from WT and $SMC2^{-/-}$ mutants were analyzed with hematoxylin and eosin staining. (**D–F**) WISH using the exocrine pancreas marker *trypsin* (**D**), endocrine pancreas marker *insulin* (**E**), and intestinal maker *fabp2* (**F**) as RNA probes.

Next, the livers in WT and $SMC2^{-/-}$ mutants were analyzed with hematoxylin and eosin staining, and the small liver in $SMC2^{-/-}$ mutants at 96 hpf was clearly shown (Figure 3C). The development of an exocrine pancreas and islet was examined by checking the expression of *trypsin* and *insulin* with WISH at 72 hpf, respectively. We found that the exocrine pancreas in $SMC2^{-/-}$ mutants significantly reduced in size (Figure 3D), while the islet was slightly affected by the loss of $SMC2$ (Figure 3E). Moreover, the gut formation was detected with the expression of the intestinal marker gene *fabp2*, and a small gut phenotype was exhibited in $SMC2^{-/-}$ mutants at 96 hpf (Figure 3F).

Taken together, these data demonstrate that $SMC2$ is required for the digestive system development in zebrafish.

3.4. SMC2 Is Required for Liver Expansion

Liver development in zebrafish begins with the specification of hepatoblasts to form a liver bud at about 30 hpf, and these progenitor cells are later expanded and differentiated into either hepatocytes or bile duct cells [24]. We first examined whether the specification of hepatoblasts was affected by the loss of $SMC2$. As shown in Figure 4, the expression patterns of endodermal marker genes, including *foxa1*, *foxa3*, and *gata4*, which are required for the establishment of competent hepatic cells, were similar in the liver *primordioum* of WT and $SMC2^{-/-}$ mutants at 30 and 34 hpf. We also detected the expression of *prox1* and *hhex*, which are the earliest markers for definitive hepatoblasts [25,26]. The expression of both *prox1* and *hhex* was not affected in $SMC2^{-/-}$ mutants at 30 and 34 hpf (Figure 4). However, the expression of *foxa1*, *foxa3*, *gata4*, *prox1*, and *hhex*, as well as hepatic marker (*cp*), reduced in the liver region of $SMC2^{-/-}$ mutants at 48 hpf (Figure 4 and Figure S1). These data indicate that the expansion of the liver bud, but not the specification of hepatoblasts, was affected by the loss of $SMC2$.

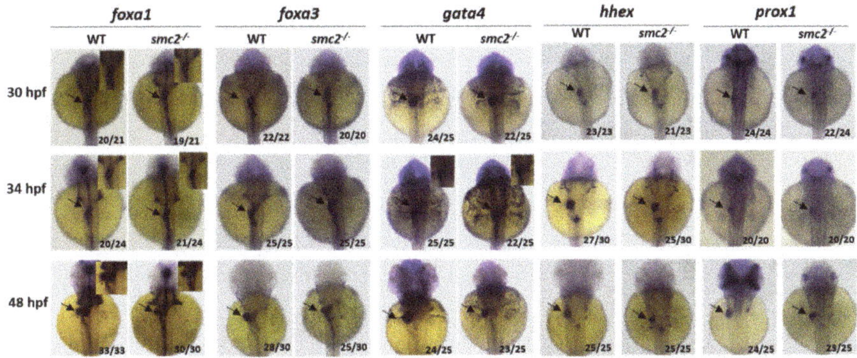

Figure 4. $SMC2$ is required for liver expansion. WT and $SMC2^{-/-}$ mutant embryos were analyzed with markers for liver specification and liver bud expansion at 30, 34 and 48 hpf. WISH probes used include pan-endodermal markers *gata4*, *foxa1*, *foxa3*, and hepatic markers *prox1* and *hhex*. Black arrowhead: liver.

3.5. Hepatocellular Apoptosis Increased in SMC2$^{-/-}$ Mutants

Since the specification of hepatoblasts was not disturbed in $SMC2^{-/-}$ mutants, the reduction in liver size may have resulted from increased cell death or a decreased cell proliferation rate [27]. TUNEL assays and EdU (5-ethynyl-2-deoxyuridine) detection were performed in the liver region of zebrafish larvae at 96 hpf. The results of the TUNEL assays indicated that the liver cells of $SMC2^{-/-}$ mutants underwent active apoptosis, whereas no apoptotic cells were observed in the same region of sibling WT larvae (Figure 5A,B). Moreover, extensive EdU$^+$ signals and a high proportion of dividing cells were observed in the liver region of WT larvae, but not in the liver of $SMC2^{-/-}$ larvae, suggesting that a loss of $SMC2$ also reduced the proliferation of liver cells (Figure 5C,D).

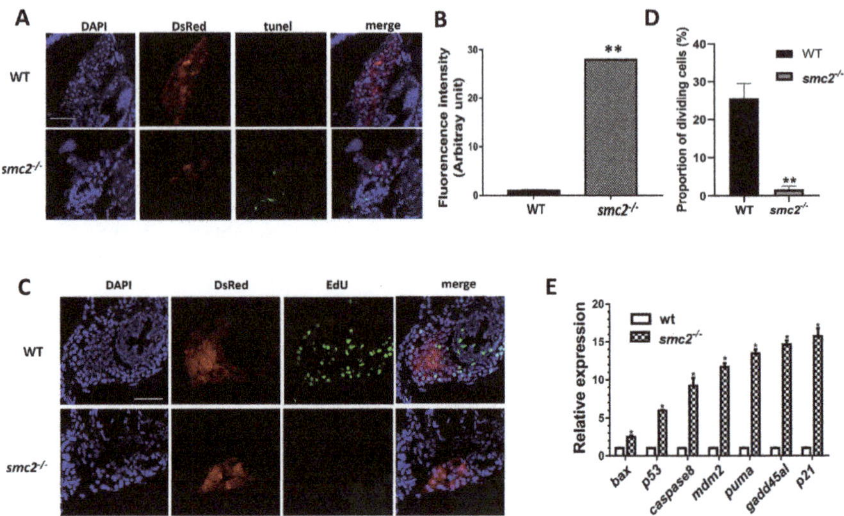

Figure 5. Hepatocellular apoptosis increased in $SMC2^{-/-}$ mutants. (**A**) TUNEL analysis of apoptotic cells in the liver of WT and homozygous mutants under the *Tg (fabp10a:dsRed; ela3l: EGFP)* transgenic background at 96 hpf. Scale bar, 50 µm. (**B**) Quantitative analysis of the apoptotic cells in the liver. Fluorescence intensities of three WT and three mutant embryos across the liver were determined using the ImageJ software. **, $p < 0.01$. (**C**) Frozen sections were stained with EdU staining buffer and nuclei counterstained with DAPI (blue) under the *Tg (fabp10a:dsRed; ela3l: EGFP)* transgenic background at 96 hpf. Scale bar, 50 µm. (**D**) The proportions of EdU-positive cells vs. DAPI-positive cells in the liver of three WT and three *SMC2* mutant embryos were determined using the ImageJ software. **, $p < 0.01$. (**E**) The mRNA levels of genes involved in apoptotic pathways were analyzed with qPCR. Expression levels were normalized to WT. The data expressed as mean ± SD were representatives of three independent experiments containing 40 embryos per sample. *, $p < 0.01$.

We then employed qPCR to detect the expression of a set of genes related to apoptosis during embryogenesis. We found that a number of genes, including *bax, p53, caspase 8, mdm2, puma, gadd45al*, and *p21*, were highly up-regulated in $SMC2^{-/-}$ mutants (Figure 5E). These observations indicate that the loss of *SMC2* has triggered intrinsic and extrinsic apoptotic pathways in developing zebrafish.

3.6. Activation of the p53-Driven Apoptotic Pathway Contributed to the Small Liver Phenotype in $SMC2^{-/-}$ Mutants

p53 is a key signal molecule of both intrinsic and extrinsic apoptotic pathways, which can regulate the expression of the genes hastening apoptosis and cell cycle arrest [28]. Abnormally elevated expression of *p53* in $SMC2^{-/-}$ mutants suggests the possibility that p53-dependent apoptosis might be a key cause for the small liver phenotype. To address whether *p53* deficiency could suppress the increased apoptosis in the liver, the $SMC2^{+/c504}$ ($SMC2^{+/-}$ hereafter) fish were crossed with $p53^{M214K/M214K}$ ($p53^{-/-}$ hereafter) fish [19], and then double heterozygotes ($SMC2^{+/c504}$; $p53^{+/M214K}$) were identified by genotyping. Double homozygote mutants ($SMC2^{-/-}$; $p53^{-/-}$) were subsequently obtained.

We then examined the liver phenotype of the double homozygous mutants that express DsRed from the intercross with the *Tg (fabp10a: dsRed; ela3l: EGFP)* line, and found that the small liver defects of $SMC2^{-/-}$ mutants at 72 hpf were partially rescued by the loss of p53 (Figure 6A). WISH assays using *fabp10a* as a probe also showed a partially recovered liver in double homozygotes mutants ($SMC2^{-/-}$; $p53^{-/-}$) at 96 hpf (Figure 6B). TUNEL assays showed that the number of TUNEL-positive cells in *SMC2* and *p53* double-deficient larvae at 96 hpf was obviously reduced when compared to that in $SMC2^{-/-}$ mutants

(Figure 6C). These data suggest that a loss of p53 could rescue the liver defects, mainly by a reduction in hepatocyte apoptosis in $SMC2^{-/-}$ mutants.

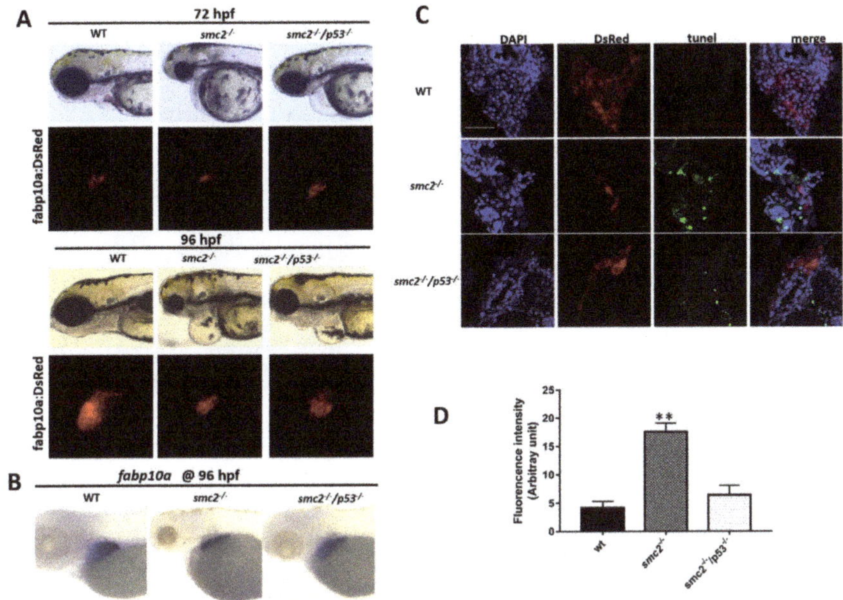

Figure 6. Activation of the p53-dependent apoptotic pathway contributed to the small liver phenotype in $SMC2^{-/-}$ mutants. (**A**) Phenotype comparison of WT, $SMC2^{-/-}$ and $SMC2^{-/-}/p53^{-/-}$ embryos under the *Tg (fabp10a:dsRed;ela3l:EGFP)* transgenic background at 72 hpf and 96 hpf. (**B**) WT embryos, $SMC2^{-/-}$ and $SMC2^{-/-}/p53^{-/-}$ embryos stained with the *fabp10a* probe at 96 hpf. (**C**) TUNEL analysis of apoptotic cells in the liver of $SMC2^{-/-}/p53^{-/-}$ mutants compared to WT and $SMC2^{-/-}$ mutants at 96 hpf under the *Tg (fabp10a:dsRed;ela3l:EGFP)* transgenic background. Scale bar, 50 µm. (**D**) Quantitative analysis of the apoptotic cells in the liver. Fluorescence intensities of three WT embryos, three $SMC2^{-/-}$ mutant embryos and three $SMC2^{-/-}/p53^{-/-}$ embryo across the liver were determined using the ImageJ software. **, $p < 0.01$.

3.7. Extensive Apoptosis Occurring in the Liver of SMC2 Mutants Is Attributable to DNA Damage

It has previously shown that simultaneous depletion of two condensins led to severe defects in chromosome assembly and segregation, which, in turn, caused DNA damage and triggered p53-induced apoptosis in cells [11]. Thus, we suspected that the DNA damage pathway that functions upstream of p53 signaling may be responsible for the small liver phenotype of $SMC2^{-/-}$ mutants. DNA content analysis with flow cytometry revealed that $SMC2^{-/-}$ mutants showed an increase, from 9.99% to 16.44%, in mitotic 4N and apoptotic sub-G1 populations (Figure 7A), and a significant increase in the number of cells with a big nucleus (DAPI staining), in comparison with the WT liver (Figure 7B,C). These data demonstrate that a loss of $SMC2$ resulted in the blockage of cell cycles at the M-phase, and defects in chromosome segregation.

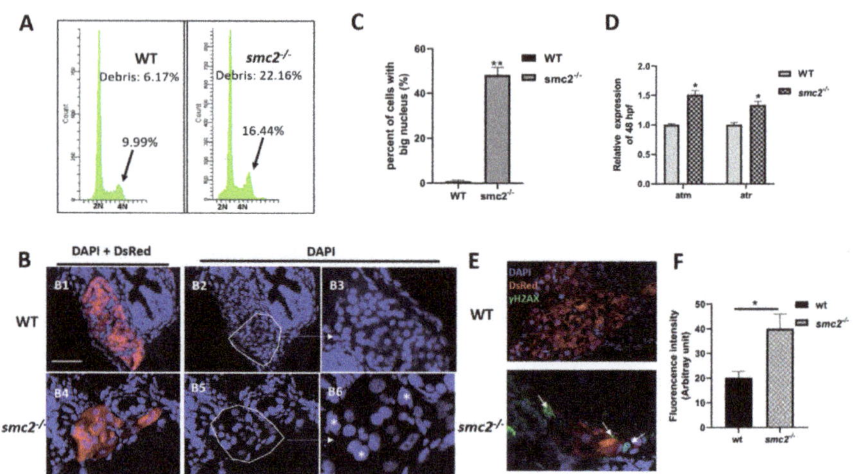

Figure 7. Extensive apoptosis occurring in liver of $SMC2^{-/-}$ mutants is attributable to DNA damage. (**A**) FACS analyses after DAPI staining of dissociated cells at 96 hpf from wild-type and $SMC2^{-/-}$ embryos. The $SMC2^{-/-}$ embryos have an accumulation of cells in the G2/M phase and more debris. (**B**) Frozen sections were stained with DAPI to visualize nuclei under the *Tg (fabp10a:dsRed; ela3l:EGFP)* transgenic background at 96 hpf. The $SMC2^{-/-}$ embryos have an increased number of cells with bigger nucleus compared with WT. Figure 7(B3,B6) are enlargement of the framed part in Figure 7(B2,B5), respectively. *, bigger nuclei. (**C**) Quantitative analysis of the cell numbers with big nucleus in the liver. Three WT and three mutant embryos across the liver were determined using the ImageJ software. **, $p < 0.01$. (**D**) Expression levels of genes (*atm* and *atr*) implicated in DNA damage response pathway were analyzed by qPCR in WT and $SMC2^{-/-}$ mutants at 96 hpf. Expression levels were normalized to WT embryos. Data expressed as mean ± SD were representative of three independent experiments. * $p < 0.01$. (**E**) Frozen sections were stained against γ-H2AX antibody which marks DNA double-stranded breaks and nuclei counterstained with DAPI (blue) under the *Tg (fabp10a:dsRed; ela3l:EGFP)* transgenic background at 96 hpf. The signal was labeled with white arrowhead. (**F**) Quantitative analysis of the apoptotic cells in the liver. Fluorescence intensities of three WT and three $SMC2^{-/-}$ mutant across the liver were determined using the ImageJ software. **, $p < 0.01$.

Defects in chromosome segregation provide a potential source of DNA damage. qPCR was performed to determine the expression levels of *atm* and *atr*, which are the critical components of the DNA damage response pathway. As expected, the qPCR results revealed increased expression levels of these two genes in the DNA damage response pathway in *SMC2* mutants, when compared to those in the WT controls (Figure 7D). Phosphorylated H2AX, referred to as γ-H2AX, can be detected within minutes after the induction of a DNA strand break [29]. Frozen sections were stained with γ-H2AX antibody and nuclei were counterstained with DAPI (blue), under the *Tg (fabp10a:dsRed; ela3l: EGFP)* transgenic background, at 96 hpf. The results revealed an increased number of γ-H2AX foci in the liver of $SMC2^{-/-}$ mutants when compared to those in the WT controls (Figure 7E,F).

Taken together, these findings indicate that loss of *SMC2* led to defects in chromosome segregation, followed by DNA damage and the activation of apoptotic pathways in *SMC2* expressing tissues, including the liver region of zebrafish.

4. Discussion

Previous studies have shown that condensins play a central role in chromosome organization and segregation [11], but the functions and mechanisms of condensins during the development of the liver in vertebrates remain to be explored. In this study, we demonstrated pivotal roles of zebrafish *SMC2* in the development of the liver. Using the CRISPR/Cas9 technology, we generated two zebrafish homozygous *SMC2* mutant lines, and found that a loss of *SMC2* led to a morphogenetic malformation in the liver. In comparison with WT larvae, homozygous *SMC2* mutants exhibited a small liver phenotype, suggesting a specific function of *SMC2* in the development of the liver. The liver region affected by a loss of *SMC2* matched the position where *SMC2* transcripts were expressed during zebrafish development. Therefore, the tissue-specific expression of *SMC2* in the liver region is required for the appropriate formation of the liver in zebrafish.

The liver morphogenesis process can be arbitrarily divided into the following two phases: budding and growth [13]. The budding phase occurs from 24 to 50 hpf, and at the subsequent growth stage, the liver undergoes dramatic changes in its size, shape, and placement [24]. During the dynamic process, *SMC2* transcripts were detectable in the endoderm and liver region, at and after 48 hpf. A loss of *SMC2* did not impair the specification of hepatoblasts, as characterized by the unperturbed expression of the early endodermal markers *foxa1*, *foxa3*, and *gata4*, as well as the earliest hepatoblast markers *prox1* and *hhex*, at 30 and 34 hpf. However, *SMC2* played indispensable roles in liver expansion, as evidenced by the restricted expression of these makers and *cp* at 48 hpf.

Condensins I and II are two large protein complexes that play a central role in chromosome organization and segregation [30]. Eukaryotic condensins I and II share the core *SMC2* and *SMC4* subunits, but differ in their auxiliary non-*SMC* components, called condensin-associated proteins (CAP-D2, CAP-G, and CAP-H for condensin I; CAP-D3, CAP-G2, and CAP-H2 for condensin II) [31]. Previous studies have shown that condensins I and II are both essential for early embryonic divisions in mice [11,32]. Simultaneous depletion of condensins I and II, from neuronal stem cells, caused severe defects in chromosome assembly and segregation, eventually leading to p53-induced apoptosis [11].

In this study, *SMC2* transcripts were detected in embryos at one-cell stage, indicating its maternal origin. The $SMC2^{-/-}$ mutant embryos displayed small eyes and small heads at 48 and 72 hpf, and died at about 7 dpf, suggesting that the existence of maternal *SMC2* transcripts would lead to mild mutant phenotypes of $SMC2^{-/-}$ embryos, and the depletion of maternal *SMC2*-mRNA with *SMC2* morpholinos may cause severe abnormal phenotypes. It is known that a heterodimer of *SMC2* and *SMC4* formed the core of eukaryotic condensins. We found that the abnormal phenotypes of $SMC2^{-/-}$ mutant embryos can be rescued by the injection of capped *SMC2*-mRNA, and the expression of *SMC4* in embryos at 36 hpf was not upregulated (data not shown). Thus, the mild abnormal phenotypes of $SMC2^{-/-}$ early developing embryos are caused by the reduction in *SMC2*, but not the redundant function of *SMC4*.

We found that *SMC2* plays critical roles in zebrafish liver morphogenetic processes, since *SMC2* knockout led to an obviously reduced size of the liver. Further evidence from this study indicates that the small liver was caused by increased cell death and reduced cell proliferation in $SMC2^{-/-}$ mutants. Moreover, increased cell death in the developing liver of *SMC2* mutants was caused by the significantly elevated expression of many genes associated with apoptotic pathways. Among these apoptotic pathways, p53-dependent apoptotic signaling appears to play a key role in the formation of small liver, due to the elevated expression level of *p53* and the partial rescue of the small liver by *p53* knockout in *SMC2* mutants.

p53 is thought to be a decision-making transcription factor that selectively activates genes to determine cellular outcomes [33]. Upon DNA damage, the p53 protein accumulates rapidly through a post-transcriptional mechanism(s), and is also activated as a transcription factor, leading to growth arrest or apoptosis [34,35]. Cells can respond to DNA damage by instigating robust DNA damage response pathways [36], which serve as cellular surveillance systems to sense the presence of damaged DNA, and elicit checkpoint activation and subsequent lesion repair in preventing the amplification or loss of genes or chromosomes [37]. Thus, the activation of p53 apoptotic signaling by the loss of *SMC2* is likely mediated by the abnormality of chromosome organization and segregation. Indeed, we found that *SMC2* mutants had a low fraction of cells in the G1 phase and an accumulation of cells in the G2/M phases. In comparison with WT, *SMC2* mutants exhibited an increased number of cells with a big nucleus, indicating that the knockout of *SMC2* resulted in defects in chromosome segregation. Moreover, we found that defects in the chromosome segregation of *SMC2* mutants led to DNA damage responses, as evidenced by the increased expression levels of *atm* and *atr*, and the significantly elevated expression of γ-H2AX. Thus, the small liver phenotype is mainly attributable to the extensive apoptosis caused by defective chromosome segregation and DNA damage in *SMC2* mutants.

5. Conclusions

The loss of *SMC2* led to a small liver phenotype in $SMC2^{-/-}$ mutants. The expansion of the liver bud, but not the specification of hepatoblasts, was affected by the loss of *SMC2*. Increased cell apoptosis and decreased cell proliferation are responsible for the small liver phenotype in $SMC2^{-/-}$ mutants. The p53-driven apoptotic pathway was activated in $SMC2^{-/-}$ mutants. Extensive apoptosis occurring in the liver of *SMC2* mutants is attributable to DNA damage.

Supplementary Materials: The following are available online at https://www.mdpi.com/article/10.3390/biomedicines9091240/s1, Figure S1: WISH of specific maker (*cp*) for hepatic cells; Table S1: the primers used in this study; Table S2: comparison of condensing I and condensing II subunits of human, mouse, and zebrafish.

Author Contributions: Conceptualization, Z.C. and X.L.; methodology, X.L., G.S. and Y.Z.; software, X.L. and J.R.; validation, X.L. and G.S.; formal analysis, X.L. and Q.L.; investigation, X.L., G.S. and Y.Z.; resources, Q.L. and Y.Z.; data curation, X.L.; writing—original draft preparation, X.L.; writing—review and editing, X.L. and Z.C.; visualization, X.L. and G.S.; supervision, Z.C.; project administration, Z.C.; funding acquisition, Z.C. All authors have read and agreed to the published version of the manuscript.

Funding: This research was funded by National Key R & D Program of China, 2018YFA0800503; Special Fund Project for Guangdong Academy of Sciences to Build Domestic First-class Research Institutions, 2021GDASYL-20210102003 and 2021GDASYL-20210103025; National Natural Science Foundation of China, 31772836 and 31571504; State Key Laboratory of Freshwater Ecology and Biotechnology, 2019FBZ05. The funders played no role in the design of the study and collection, analysis, and interpretation of data and preparation of the manuscript.

Institutional Review Board Statement: The study was conducted according to the guidelines of the Institutional Animal Care and Use Committee of Institute of Hydrobiology (protocol code E01F0501, most recent approval 16 September 2019).

Informed Consent Statement: Not applicable.

Data Availability Statement: Data are contained within the article.

Acknowledgments: We thank the Analysis and Testing Center at IHB for technical supports.

Conflicts of Interest: The authors declare that they have no conflict of interest.

References

1. Wood, A.J.; Severson, A.F.; Meyer, B.J. Condensin and cohesin complexity: The expanding repertoire of functions. *Nat. Rev. Genet.* **2010**, *11*, 391–404. [CrossRef]
2. Jessberger, R. The many functions of *SMC* proteins in chromosome dynamics. *Nat. Rev. Mol. Cell Biol.* **2002**, *3*, 767–778. [CrossRef]
3. Wang, H.; Liu, Y.; Yuan, J.; Zhang, J.; Han, F. The condensin subunits *SMC2* and *SMC4* interact for correct condensation and segregation of mitotic maize chromosomes. *Plant J. Cell Mol. Biol.* **2020**, *102*, 467–479. [CrossRef]
4. Nolivos, S.; Sherratt, D. The bacterial chromosome: Architecture and action of bacterial *SMC* and *SMC*-like complexes. *FEMS Microbiol. Rev.* **2014**, *38*, 380–392. [CrossRef] [PubMed]
5. Matityahu, A.; Onn, I. A new twist in the coil: Functions of the coiled-coil domain of structural maintenance of chromosome (SMC) proteins. *Curr. Genet.* **2018**, *64*, 109–116. [CrossRef]
6. Losada, A. Dynamic molecular linkers of the genome: The first decade of *SMC* proteins. *Genes Dev.* **2005**, *19*, 1269–1287. [CrossRef]
7. Tsang, C.K.; Wei, Y.; Zheng, X.F. Compacting DNA during the interphase: Condensin maintains rDNA integrity. *Cell Cycle* **2007**, *6*, 2213–2218. [CrossRef] [PubMed]
8. Fazzio, T.G.; Panning, B. Condensin complexes regulate mitotic progression and interphase chromatin structure in embryonic stem cells. *J. Cell Biol.* **2010**, *188*, 491–503. [CrossRef]

9. Hartl, T.A.; Sweeney, S.J.; Knepler, P.J.; Bosco, G. Condensin II resolves chromosomal associations to enable anaphase I segregation in Drosophila male meiosis. *PLoS Genet.* **2008**, *4*, e1000228. [CrossRef] [PubMed]
10. Hagstrom, K.A.; Holmes, V.F.; Cozzarelli, N.R.; Meyer, B.J.C. elegans condensin promotes mitotic chromosome architecture, centromere organization, and sister chromatid segregation during mitosis and meiosis. *Genes Dev.* **2002**, *16*, 729–742. [CrossRef]
11. Nishide, K.; Hirano, T. Overlapping and non-overlapping functions of condensins I and II in neural stem cell divisions. *PLoS Genet.* **2014**, *10*, e1004847. [CrossRef] [PubMed]
12. Zaret, K.S. Regulatory phases of early liver development: Paradigms of organogenesis. *Nat. Rev. Genet.* **2002**, *3*, 499–512. [CrossRef]
13. Field, H.A.; Ober, E.A.; Roeser, T.; Stainier, D.Y.R. Formation of the digestive system in zebrafish. I. Liver morphogenesis. *Dev. Biol.* **2003**, *253*, 279–290. [CrossRef]
14. Ober, E.A.; Verkade, H.; Field, H.A.; Stainier, D.Y. Mesodermal Wnt2b signalling positively regulates liver specification. *Nature* **2006**, *442*, 688–691. [CrossRef]
15. Shin, D.; Shin, C.H.; Tucker, J.; Ober, E.A.; Rentzsch, F.; Poss, K.D.; Hammerschmidt, M.; Mullins, M.C.; Stainier, D.Y. Bmp and Fgf signaling are essential for liver specification in zebrafish. *Development* **2007**, *134*, 2041–2050. [CrossRef]
16. Chu, J.; Sadler, K.C. New school in liver development: Lessons from zebrafish. *Hepatology* **2009**, *50*, 1656–1663. [CrossRef] [PubMed]
17. Burke, Z.; Oliver, G. Prox1 is an early specific marker for the developing liver and pancreas in the mammalian foregut endoderm. *Mech. Dev.* **2002**, *118*, 147–155. [CrossRef]
18. Zhang, W.; Yatskievych, T.A.; Baker, R.K.; Antin, P.B. Regulation of Hex gene expression and initial stages of avian hepatogenesis by Bmp and Fgf signaling. *Dev. Biol.* **2004**, *268*, 312–326. [CrossRef]
19. Berghmans, S.; Murphey, R.D.; Wienholds, E.; Neuberg, D.; Kutok, J.L.; Fletcher, C.D.; Morris, J.P.; Liu, T.X.; Schulte-Merker, S.; Kanki, J.P.; et al. tp53 mutant zebrafish develop malignant peripheral nerve sheath tumors. *Proc. Natl. Acad. Sci. USA* **2005**, *102*, 407–412. [CrossRef]
20. Korzh, S.; Pan, X.; Garcia-Lecea, M.; Winata, C.L.; Pan, X.; Wohland, T.; Korzh, V.; Gong, Z. Requirement of vasculogenesis and blood circulation in late stages of liver growth in zebrafish. *BMC Dev. Biol.* **2008**, *8*, 84. [CrossRef]
21. Chang, N.; Sun, C.; Gao, L.; Zhu, D.; Xu, X.; Zhu, X.; Xiong, J.W.; Xi, J.J. Genome editing with RNA-guided Cas9 nuclease in zebrafish embryos. *Cell Res.* **2013**, *23*, 465–472. [CrossRef] [PubMed]
22. Chitramuthu, B.P.; Bennett, H.P. High resolution whole mount in situ hybridization within zebrafish embryos to study gene expression and function. *J. Vis. Exp. JoVE* **2013**, *80*, e50644. [CrossRef]
23. Wittkopp, N.; Huntzinger, E.; Weiler, C.; Saulière, J.; Schmidt, S.; Sonawane, M.; Izaurralde, E. Nonsense-mediated mRNA decay effectors are essential for zebrafish embryonic development and survival. *Mol. Cell. Biol.* **2009**, *29*, 3517–3528. [CrossRef]
24. Tao, T.; Peng, J. Liver development in zebrafish (*Danio rerio*). *J. Genet. Genom.* **2009**, *36*, 325–334. [CrossRef]
25. Ober, E.A.; Field, H.A.; Stainier, D.Y.R. From endoderm formation to liver and pancreas development in zebrafish. *Mech. Dev.* **2003**, *120*, 5–18. [CrossRef]
26. Wallace, K.N.; Pack, M. Unique and conserved aspects of gut development in zebrafish. *Dev. Biol.* **2003**, *255*, 12–29. [CrossRef]
27. Poulain, M.; Ober, E.A. Interplay between Wnt2 and Wnt2bb controls multiple steps of early foregut-derived organ development. *Development* **2011**, *138*, 3557–3568. [CrossRef]
28. Yang, X.; Li, X.; Gu, Q.; Li, Q.; Cui, Z. Nucleoporin 62-Like Protein is Required for the Development of Pharyngeal Arches through Regulation of Wnt/beta-Catenin Signaling and Apoptotic Homeostasis in Zebrafish. *Cells* **2019**, *8*, 1038. [CrossRef]
29. Price, B.D.; D'Andrea, A.D. Chromatin remodeling at DNA double-strand breaks. *Cell* **2013**, *152*, 1344–1354. [CrossRef]
30. Hirano, T. Condensin-Based Chromosome Organization from Bacteria to Vertebrates. *Cell* **2016**, *164*, 847–857. [CrossRef] [PubMed]
31. Green, L.C.; Kalitsis, P.; Chang, T.M.; Cipetic, M.; Kim, J.H.; Marshall, O.; Turnbull, L.; Whitchurch, C.B.; Vagnarelli, P.; Samejima, K.; et al. Contrasting roles of condensin I and condensin II in mitotic chromosome formation. *J. Cell Sci.* **2012**, *125*, 1591–1604. [CrossRef] [PubMed]
32. Houlard, M.; Godwin, J.; Metson, J.; Lee, J.; Hirano, T.; Nasmyth, K. Condensin confers the longitudinal rigidity of chromosomes. *Nat. Cell Biol.* **2015**, *17*, 771–781. [CrossRef] [PubMed]
33. Hafner, A.; Bulyk, M.L.; Jambhekar, A.; Lahav, G. The multiple mechanisms that regulate p53 activity and cell fate. *Nat. Rev. Mol. Cell Biol.* **2019**, *20*, 199–210. [CrossRef]
34. Gottlieb, T.M.; Oren, M. P53 in growth control and neoplasia. *Biochim. Biophys. Acta* **1996**, *1287*, 77–102. [CrossRef]
35. Levine, A.J. p53, the cellular gatekeeper for growth and division. *Cell* **1997**, *88*, 323. [CrossRef]
36. Chatterjee, N.; Walker, G.C. Mechanisms of DNA damage, repair, and mutagenesis. *Environ. Mol. Mutagenesis* **2017**, *58*, 235–263. [CrossRef] [PubMed]
37. Zhou, B.-B.S.; Elledge, S.J. The DNA damage response: Putting checkpoints in perspective. *Nature* **2000**, *408*, 433–439. [CrossRef]

Article

Lipopolysaccharides Enhance Epithelial Hyperplasia and Tubular Adenoma in Intestine-Specific Expression of $kras^{V12}$ in Transgenic Zebrafish

Jeng-Wei Lu [1,2,*], Yuxi Sun [1,3], Pei-Shi Angelina Fong [1], Liang-In Lin [2,4], Dong Liu [3] and Zhiyuan Gong [1,*]

[1] Department of Biological Sciences, National University of Singapore, Singapore 117543, Singapore; e0437708@u.nus.edu (Y.S.); a0131047_angelina@u.nus.edu (P.-S.A.F.)
[2] Department of Clinical Laboratory Sciences and Medical Biotechnology, National Taiwan University, Taipei 10048, Taiwan; lilin@ntu.edu.tw
[3] Department of Biology, Southern University of Science and Technology, Shenzhen 518055, China; liud@sustech.edu.cn
[4] Department of Laboratory Medicine, National Taiwan University Hospital, Taipei 10048, Taiwan
[*] Correspondence: jengweilu@gmail.com (J.-W.L.); dbsgzy@nus.edu.sg (Z.G.); Tel.: +65-6516-2860 (Z.G.)

Citation: Lu, J.-W.; Sun, Y.; Fong, P.-S.A.; Lin, L.-I.; Liu, D.; Gong, Z. Lipopolysaccharides Enhance Epithelial Hyperplasia and Tubular Adenoma in Intestine-Specific Expression of $kras^{V12}$ in Transgenic Zebrafish. *Biomedicines* **2021**, *9*, 974. https://doi.org/10.3390/biomedicines9080974

Academic Editors: James A. Marrs and Swapnalee Sarmah

Received: 1 July 2021
Accepted: 4 August 2021
Published: 7 August 2021

Publisher's Note: MDPI stays neutral with regard to jurisdictional claims in published maps and institutional affiliations.

Abstract: Intestinal carcinogenesis is a multistep process that begins with epithelial hyperplasia followed by a transition to an adenoma and then to a carcinoma. Many etiological factors, including *KRAS* mutations and inflammation, have been implicated in oncogenesis. However, the potential synergistic effects between *KRAS* mutations and inflammation as well as the potential mechanisms by which they promote intestinal carcinogenesis remain unclear. Thus, the objective of this study was to investigate the synergistic effects of $kras^{V12}$, lipopolysaccharides (LPS), and/or dextran sulfate sodium (DSS) on inflammation, tumor progression, and intestinal disorders using transgenic adults and larvae of zebrafish. Histopathology and pathological staining were used to examine the intestines of $kras^{V12}$ transgenic zebrafish treated with LPS and/or DSS. LPS and/or DSS treatment enhanced intestinal inflammation in $kras^{V12}$ transgenic larvae with concomitant increases in the number of neutrophils and macrophages in the intestines. The expression of $kras^{V12}$, combined with LPS treatment, also enhanced epithelial hyperplasia and tubular adenoma, demonstrated by histopathological examinations and by increases in cell apoptosis, cell proliferation, and downstream signaling of phosphorylated AKT serine/threonine kinase 1 (AKT), extracellular-signal-regulated kinase (ERK), and histone. We also found that $kras^{V12}$ expression, combined with LPS treatment, significantly enhanced changes in intestinal morphology, specifically (1) decreases in goblet cell number, goblet cell size, villi height, and intervilli space, as well as (2) increases in villi width and smooth muscle thickness. Moreover, $kras^{V12}$ transgenic larvae cotreated with DSS and LPS exhibited exacerbated intestinal inflammation. Cotreatment with DSS and LPS in $kras^{V12}$-expressing transgenic adult zebrafish also enhanced epithelial hyperplasia and tubular adenoma, compared with wild-type fish that received the same cotreatment. In conclusion, our data suggest that $kras^{V12}$ expression, combined with LPS and/or DSS treatment, can enhance intestinal tumor progression by activating the phosphatidylinositol-3-kinase (PI3K)/AKT signaling pathway and may provide a valuable in vivo platform to investigate tumor initiation and antitumor drugs for gastrointestinal cancers.

Keywords: colorectal cancer; dextran sulfate sodium; lipopolysaccharides; $kras^{V12}$; intestinal tumor; transgenic zebrafish

1. Introduction

Colorectal cancer (CRC) ranks third in terms of global cancer incidence. This disease causes more than 600,000 deaths every year, and the number of affected individuals continues to increase around the world [1–3]. In recent years, the incidence and mortality of CRC have also continued to rise rapidly. Between 2007 and 2016, the high mortality rates for CRC were 35% (United States), 45% (Europe), and 47.8% (worldwide) [4–6]. Furthermore, the high recurrence and low survival rates of this disease seriously affect patient quality of life [7].

RAS proteins, which are GTPases that regulate the RAS signaling pathway and control cell proliferation and cell survival, are often mutated in human cancers [8]. CRC results from a multistep process of carcinogenesis that is caused by the accumulation of genetic mutations as well as changes in signal transduction pathways. Approximately 35% of CRC cases are caused by genetic mutations, with KRAS and NRAS gene mutations respectively accounting for 40% and 5% of these cases caused by genetic mutations [9,10]. Previous research reported that 85% of KRAS gene mutations occur in codons 12 and 13 of exon 2 [11]. During carcinogenesis, the activation of KRAS proteins triggers tumor initiation and accelerates tumor growth. KRAS mutations have been detected in both early- and late-stage CRC patients, which indicates that KRAS mutations likely occur in the early stages of tumor development [12].

CRC development is also related to the composition of the gut microbiota because it involves immune, structural, and metabolic processes [13,14]. Importantly, the destruction of gut microbiota and/or imbalance in gut microbiota composition may be related to CRC formation [15,16]. Toll-like receptor 4 (TLR4) can bind to lipopolysaccharides (LPS), which activate the nuclear transcription factor kappa B (NF-κB) signaling pathway, from Gram-negative bacteria [17]. This in turn induces other innate immune responses as well as proinflammatory gene expression and recruitment of the adaptive immune system [18]. Notably, LPS strongly stimulates innate immune signaling, thereby impairing intestinal homeostasis and normal host physiology. LPS thus plays a vital role in promoting the progression and metastasis of CRC [17,19,20]. In addition, exposure to the inflammatory agent dextran sulfate sodium (DSS) for 1 week in combination with a single treatment of azoxymethane can accelerate the induction of CRC in rodents [21].

Zebrafish are considered an excellent animal model for studying human gastrointestinal cancer [22,23]. In terms of histological morphology and the expression profiles of dysregulated genes, many zebrafish models of intestinal diseases and tumors are similar to human disease states. As a vertebrate, the zebrafish has a highly conserved anatomical structure and homologous organs with higher vertebrates, including humans, and most of the signal pathways that control apoptosis, proliferation, differentiation, and movement are also highly conserved between zebrafish and human. Furthermore, the high degree of homology and oncogenes and tumor suppressor genes also reveal that the carcinogenic mechanism between zebrafish and higher vertebrates is also highly conserved. Zebrafish studies also show the activation of carcinogenic signaling pathways, such as RAS, tumor protein p53 (Tp53), and Wnt/β-catenin pathways, which plays an important role in CRC. It has been found that the tumor biology, intestinal disorders caused by carcinogens, and the morphological pattern of tumors are highly similar between zebrafish and humans [23]. Intestinal tumors can be induced by 7,12-dimethylbenz(a)anthracene (DMBA). Moreover, in adenomatous polyposis coli (*apc*) mutated zebrafish [24], intestinal diseases and tumors can be driven by inducible *kras*^{V12} or by the continuous expression of *kras*^{G12} or *Helicobacter. pylori* virulence factor cagA (*cagA*) with wild-type (WT) or *tp53* mutated zebrafish [25–28]. The zebrafish xenograft model also provides an excellent platform for studying CRC tumor metastasis and drug screening [29–31]. Previous research has shown zebrafish patient-derived xenografts (zPDX) derived from surgically resected human CRC samples and treated with the same treatment administered to the donor patient. This study provides a proof-of-concept experiment that can compare the CRC patient and zPDX with chemotherapy and biotherapy response [30].

In this study, we report the potential synergy between intestine-specific overexpression of *kras*^{V12} [27] and LPS and/or DSS treatment in the development and progression of intestinal tumors in transgenic zebrafish. We used histopathology and pathological staining to examine the potential effects of *kras*^{V12} and LPS and/or DSS treatment on intestinal disease and intestinal tumors. In so doing, a further goal was to investigate the potential mechanisms that underlie these disorders.

2. Materials and Methods

2.1. Zebrafish Husbandry

Zebrafish embryos and larvae as well as adult zebrafish (*Danio rerio*) were maintained according to established protocols described in our previous studies [27,32–34]. Tg(ifabp:EGFP-kras^{V12}), Tg(lyz:DsRed), and Tg(mpeg1:mCherry) transgenic zebrafish embryos and larvae as well as adult zebrafish were maintained at 28 °C under continuous water flow and a 14 h light and 10 h dark cycle. The intestine-specific *kras*^{V12} transgenic zebrafish were generated as previously described [27], and wild-type (WT) zebrafish were used as a control. All experiments involving zebrafish were approved by the Institutional Animal Care and Use Committee (IACUC) of the National University of Singapore and National Taiwan University.

2.2. Mifepristone and Chemical Treatments of Zebrafish

Each larva treatment group included 20 larvae, which were maintained in six-well plates. Each well contained 1X E3 medium and mifepristone (catalog number: M8046; Sigma-Aldrich, St. Louis, MO, USA), DSS (catalog number: D8906; Sigma-Aldrich, St. Louis, MO, USA), and/or LPS (catalog number: L4391; Sigma-Aldrich, St. Louis, MO, USA). For the DSS and/or LPS treatment groups, larvae were treated with 0.05% DSS and/or 40 ng/mL of LPS for 2 or 3 days postinduction (dpi). To induce *kras*^{V12} expression, larvae were also treated with 4 μM mifepristone for 2 or 3 dpi. Larvae were incubated at 28 °C, and mortality was determined daily. The 1X E3 medium, fresh mifepristone, and chemicals were treated every other day.

Each adult zebrafish treatment group was maintained in a 5 L tank at room temperature, and all zebrafish were fed normally. At 4 weeks postinduction (wpi), samples were collected to investigate long-term treatment effects. For the DSS and/or LPS treatment groups, 4-month-old zebrafish were treated with 0.00625% DSS and/or 40 ng/mL of LPS for 4 wpi. To induce *kras*^{V12} expression, zebrafish were exposed to 2 μM mifepristone at 4 wpi. The mortality of adult zebrafish was determined daily, and water, fresh mifepristone, and chemicals were treated every other day.

2.3. Tissue Collection and Histopathology of Zebrafish Intestines

Control and transgenic zebrafish were euthanized at 5 months of age using 0.02% tricaine (catalog number: E10521; Sigma-Aldrich, St. Louis, MO, USA). Zebrafish intestines were then collected and fixed in 10% neutral buffered formalin solution (catalog number: HT501128; Sigma-Aldrich, St. Louis, MO, USA) overnight, embedded in paraffin, sectioned into 4 μm sections, and then mounted on poly-L-lysine-coated slides at different time points following mifepristone induction and chemical treatment. The slides were stored in slide boxes at room temperature.

Cytological analysis was also performed on the collected zebrafish intestines. After hematoxylin and eosin (H&E) staining was completed, intestinal histopathology was assessed via a single-blind evaluation of all samples. All intestine tissue evaluations were based on four consecutive sagittal serial sections, which composed the entire intestinal tract anterior to posterior. Specifically, tissue samples were evaluated for epithelial hyperplasia, dysplasia, and tumors according to previously described diagnostic criteria [26,27].

2.4. Immunofluorescence (IF) and Periodic Acid–Schiff (PAS) Staining

The 4 μm zebrafish sections were dewaxed using histoclear (catalog number: H2779; Sigma-Aldrich, St. Louis, MO, USA) and hydrated in an ethanol gradient and Milli-Q water for 10 min, respectively. For antigen retrieval, endogenous peroxidase activity was blocked by heating the slides at 100 °C for 20 min in 10 mM citric acid buffer (catalog number: C9999; Sigma-Aldrich, St. Louis, MO, USA). This was followed by blocking with

3% H_2O_2 for 15 min. The slides were then washed three times with 1X phosphate-buffered saline (catalog number: P3813; Sigma-Aldrich, St. Louis, MO, USA) with 0.1% Tween 20 (catalog number: 9005-64-5; Sigma-Aldrich, St. Louis, MO, USA) (PBST) for 5 min. Following this, slides were blocked again using 5% bovine serum albumin (BSA) (catalog number: A2153; Sigma-Aldrich, St. Louis, MO, USA) at room temperature for 30 min and then incubated with specific primary antibodies in a humidifying chamber at 4 °C overnight. After being washed with 1X PBST, the slides were incubated with conjugated fluorescent secondary antibodies and then incubated with 4′,6-diamidino-2-phenylindole (DAPI) (catalog number: D9542 Sigma-Aldrich, St. Louis, MO, USA) for 10 min. Finally, the slides were dehydrated, cleared, and mounted. To determine the specificity of primary antibodies for IF staining, we performed experiments using both appropriate positive (from a previously known positive case) and negative controls (slides not incubated with primary antibodies).

For PAS staining, tissues were also dewaxed using histoclear and hydrated in an ethanol gradient and Milli-Q water for 10 min, respectively. Following this, staining was performed using the Periodic Acid–Schiff Stain Kit (catalog number: 24200-1; Polysciences, Inc., Warrington, PA, USA) to detect goblet cells. Goblet cells were evaluated according to the number of villi. Finally, the slides were dehydrated, cleared, and mounted and then examined using light or fluorescent microscopy.

2.5. Statistical Analysis

Differences between experimental and control groups were analyzed by two-tailed Student's t-test and one-way ANOVA. Plot survival curves were derived using the Kaplan–Meier method, and log-rank tests were used to examine differences between experimental and control groups. *p*-Values of less than 0.05 were considered statistically significant.

3. Results

3.1. Effects of LPS or DSS Treatment on Intestinal Inflammation in kras^{V12} Transgenic Zebrafish Larvae

To examine the effects of intestinal inflammation, the following groups of zebrafish larvae from 4 to 6 or 7 dpf were treated with 40 ng/mL of LPS or 0.5% DSS in addition to treatment with 4 μM of mifepristone (to induce *kras^{V12}* expression): WT/lyz+/LPS, kras+/lyz+/LPS, WT/mpeg1+/LPS, kras+/mpeg1+/LPS, WT/lyz+/DSS, kras+/lyz+/DSS, WT/mpeg1+/DSS, kras+/mpeg1+/DSS. WT/lyz+, kras+/lyz+, WT/mpeg1+, and kras+/mpeg1+ zebrafish larvae without treatments were used as controls. All larvae from each group were imaged. Neutrophils and macrophages in intestines were respectively counted based on the fluorescence of DsRed or mCherry. We observed significant increases in neutrophils in the kras+/lyz+/LPS and kras+/lyz+/DSS larva groups compared with WT/lyz+, WT/lyz+ with LPS or DSS treatment, and kras+/lyz+ (Figure 1A–D) larva groups. We also observed significant increases in macrophages in the kras+/mpeg1+/LPS and kras+/mpeg1+/DSS larva groups compared with the kras+/mpeg1, WT/mpeg1+ with DSS treatment, and kras+/mpeg1+ (Figure 1E–H) larva groups.

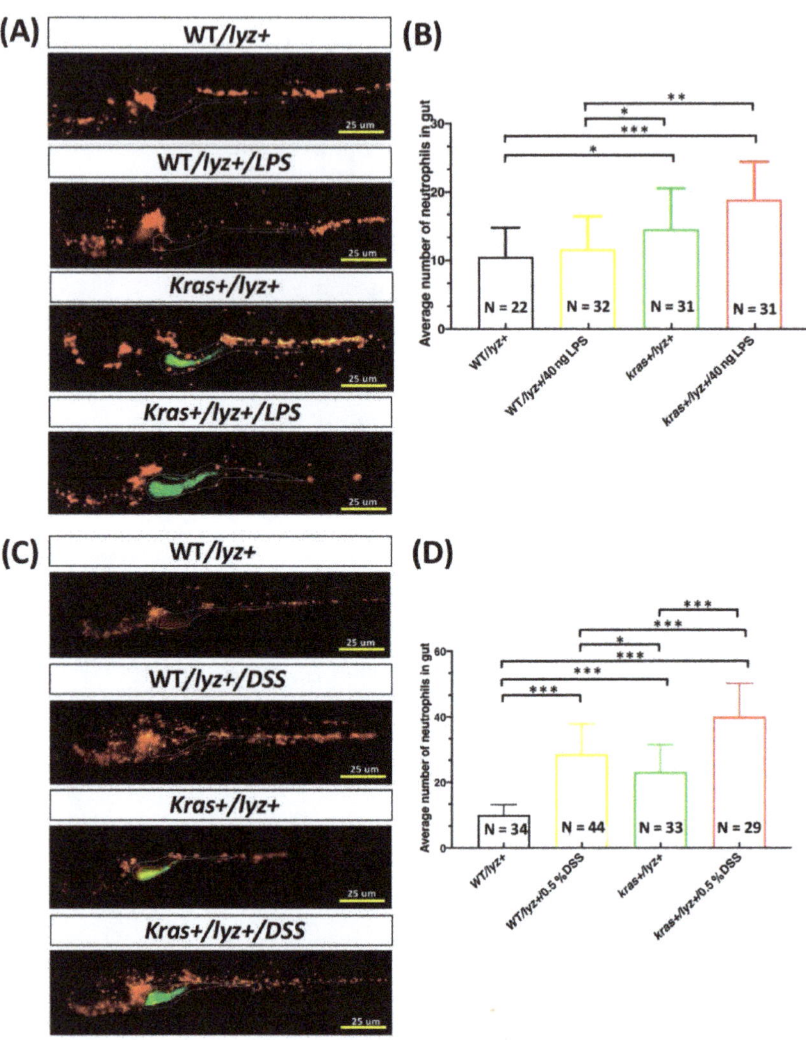

Figure 1. *Cont.*

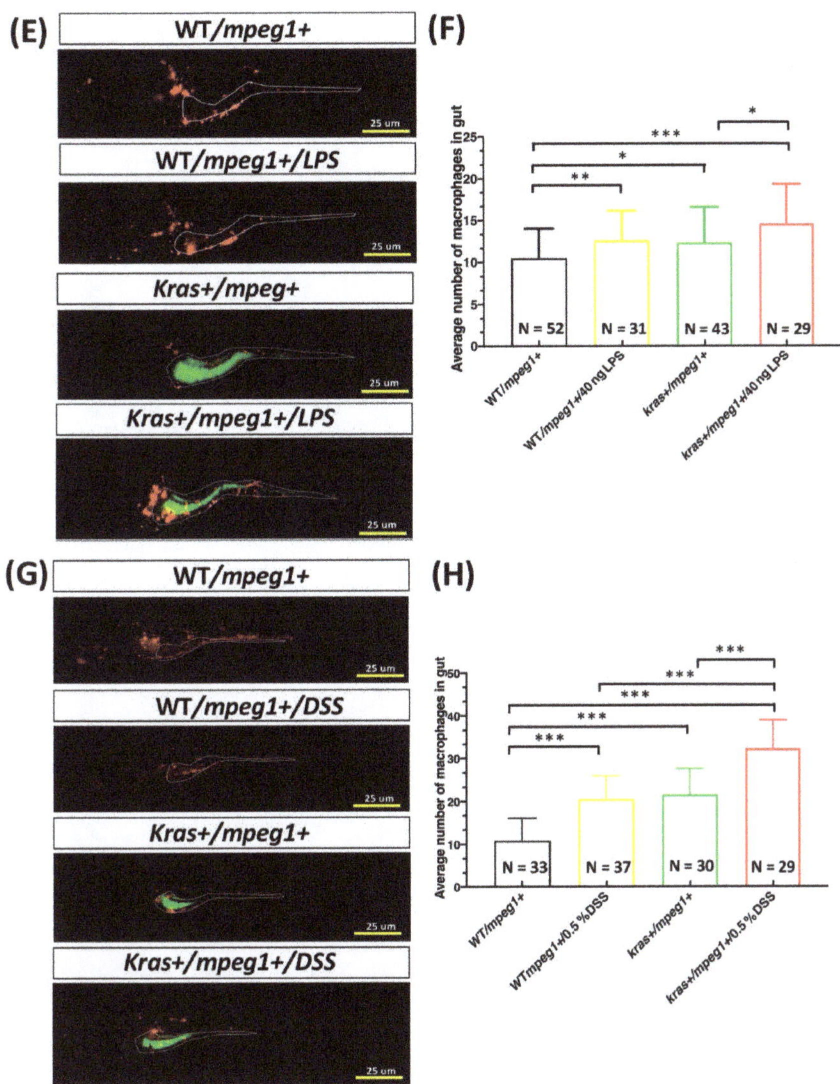

Figure 1. DSS or LPS enhances the increase in intestinal neutrophils and macrophages in kras+/lyz+ and kras+/mepg1+ zebrafish larvae. (**A,C,E,G**) Fluorescence of neutrophils and macrophages in the intestine. (**B,D,F,H**) Quantification of the number of positive cells as revealed by fluorescence of neutrophils or macrophages. Differences among the variables were assessed using Student's t-tests. Statistical significance: * $p < 0.05$, ** $p < 0.01$, *** $p < 0.001$. Scale bar: 25 µm.

3.2. Phenotype of Intestinal Tumors Induced by Sustained Expression of $kras^{V12}$ with LPS Treatment in Transgenic Zebrafish

Results of previous studies demonstrated that treating adult-stage zebrafish with the colitogenic agent DSS can enhance intestinal tumorigenesis in $kras^{V12}$-expressing transgenic zebrafish [27]. To examine whether LPS could also enhance tumorigenesis in $kras^{V12}$ adult transgenic zebrafish through the induction of inflammation, heterozygous $kras^{V12}$ transgenic zebrafish were cotreated with 2 µM of mifepristone and 40 ng/mL of LPS for 4 weeks at 4 months postfertilization (mpf). We found no significant difference in body length between $kras^{V12}$ transgenic zebrafish treated with LPS (kras+/LPS) and the control group (Figure 2A). However, LPS treatment did lead to significantly reduced body weights in WT/LPS and kras+/LPS adult zebrafish compared with WT, WT/LPS, and kras+ zebrafish (Figure 2B). Furthermore, by 4 wpi, four of the kras+ zebrafish and eight of the kras+/LPS zebrafish had died. During the same period, five WT/LPS and no WT control zebrafish died (Figure 2C).

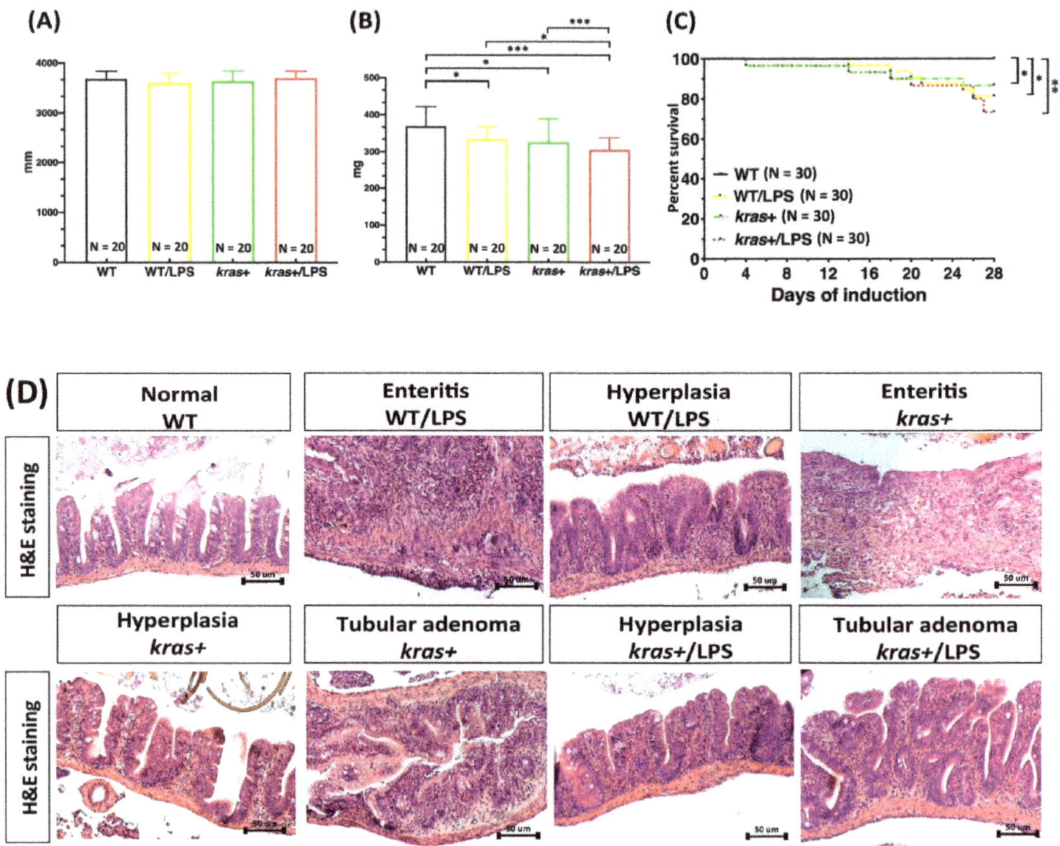

Figure 2. *Cont.*

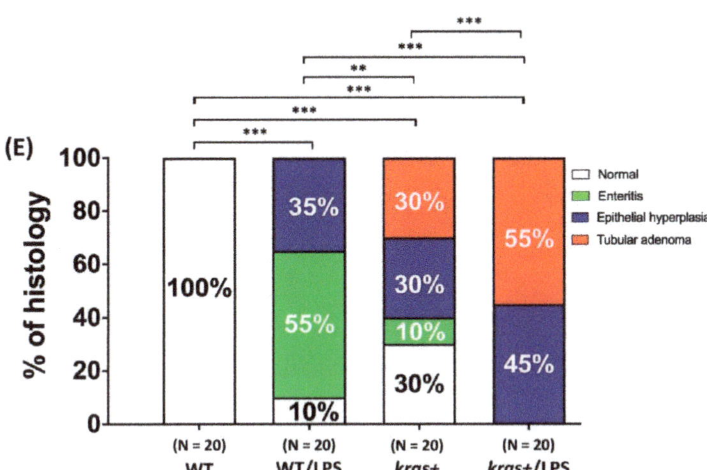

Figure 2. Synergistic effect of *kras*^{V12} expression and LPS treatment in intestinal tumorigenesis. Four-month-postfertilization WT and kras+ zebrafish were cotreated with 2 µM of mifepristone and 40 ng/mL of LPS for 4 weeks, and samples were then collected for gross observations and histological analyses. There were four experimental groups: WT, WT/LPS, kras+, and kras+/LPS. (**A,B**) Body length and body weight. (**C**) Survival curves. (**D**) Examples of normal intestines, enteritis, hyperplasia, and tubular adenoma as revealed by H&E-stained sections of the intestine. (**E**) Summary of intestinal histological abnormalities observed in the four experimental groups. These data were generated from results of a blinded histological analysis (WT, N = 20; WT/LPS, N = 20; kras+, N = 20; kras+/LPS, N = 20). Differences among the variables were assessed using Student's t-tests or one-way ANOVA. Statistical significance: * $p < 0.05$, ** $p < 0.01$, *** $p < 0.001$. Scale bar: 50 µm.

At 4 wpi, we also evaluated the entire intestinal tract of all adult zebrafish for enteritis, epithelial hyperplasia, and the presence of tumors (Figure 2D). Intestinal samples were collected from the WT, WT/LPS, kras+, and kras+/LPS groups, and histological examinations were then performed. While all WT zebrafish showed normal intestinal histology, LPS treatment respectively caused enteritis and hyperplasia in 55% and 35% of WT zebrafish. Conversely, kras+ zebrafish treated with mifepristone exhibited inflammation (10%), hyperplasia (30%), and tubular adenoma (30%). LPS treatment in kras+ zebrafish further increased the prevalence of abnormalities: 45% and 55% of adult kras+/LPS zebrafish respectively showed hyperplasia and tubular adenoma. These observations suggest that *kras*^{V12} expression combined with LPS treatment significantly enhanced intestinal tumors in adult zebrafish compared with untreated WT zebrafish, WT zebrafish treated with LPS, and kras+ zebrafish that did not undergo LPS treatment (Figure 2E).

3.3. Induction of kras^{V12} Expression with LPS Treatment Decreased the Number of Goblet Cells, Goblet Cell Size, Villi Height, and Intervilli Space and Increased Villi Width and Smooth Muscle Thickness in Fish Intestines

To examine the effects of LPS on the intestinal morphology of WT/LPS and kras+/LPS zebrafish, we examined the number and size of goblet cells; the height, intervilli space, and width of the villi; and the thickness of smooth muscles (Figure 3A). In a 4-week induction experiment, histology analysis revealed that the number (Figure 3B) and size (Figure 3C) of goblet cells were significantly reduced in WT/LPS and kras+LPS zebrafish compared with WT, WT/LPS, and kras+ zebrafish. Intestinal villi height (Figure 3D) and intervilli space (Figure 3F) were also significantly reduced in WT/LPS zebrafish compared with WT/WT/LPS and kras+ zebrafish. Conversely, intestinal villi width (Figure 3E) and smooth muscle thickness (Figure 3G) were significantly increased in WT/LPS and kras+/LPS zebrafish. These observations indicate that $kras^{V12}$ expression combined with LPS treatment significantly enhanced changes in intestinal morphology.

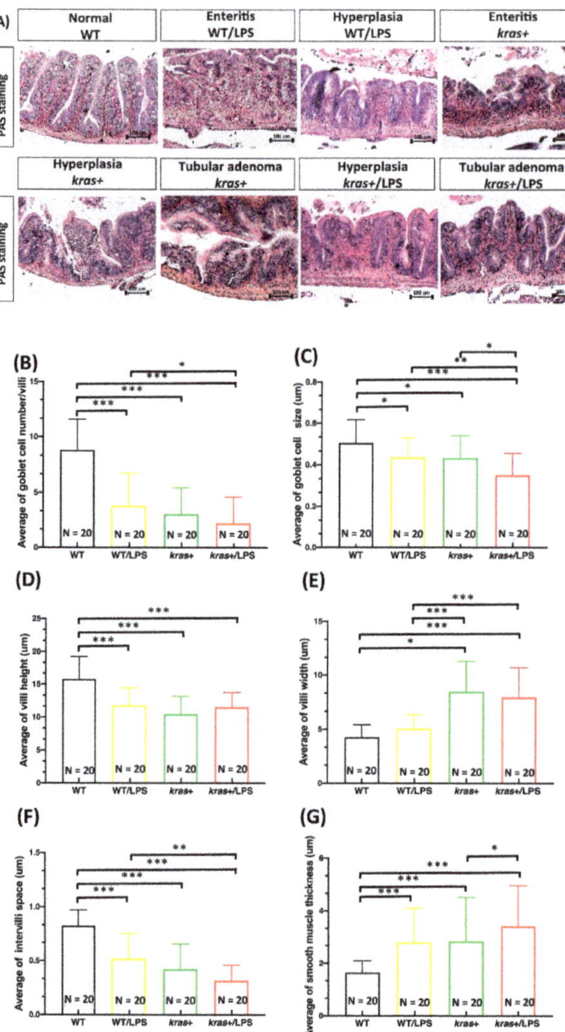

Figure 3. Expression of $kras^{V12}$ decreased goblet cell number, goblet cell size, villi height, and intervilli space and increased villi width and smooth muscle thickness in the intestine. At 4 mpf, WT and kras+ zebrafish were treated with 2 μM of mifepristone and 40 ng/mL of LPS for 4 weeks, and samples were then collected for gross observations and histological analysis. There were four experimental groups: WT, WT/LPS, kras+, and kras+/LPS (WT, N = 20; WT/LPS, N = 20; kras+, N = 20; kras+/LPS, N = 20). (**A**) PAS staining was carried out in intestinal sections. (**B–G**) Quantification of goblet cell number, goblet cell size of villi, villi height, intervilli space, villi width, and smooth muscle thickness in the intestine. Differences among the variables were assessed using Student's t-tests. Statistical significance: * $p < 0.05$, ** $p < 0.01$, *** $p < 0.001$. Scale bar: 100 μm.

3.4. Increases in Cell Apoptosis, Proliferation, and Downstream Signaling of Phosphorylated AKT and ERK Induced by Sustained Expression of kras^{V12} with LPS Treatment in Transgenic Zebrafish

We next aimed to clarify the effects of LPS treatment on intestinal cell apoptosis and proliferation in *kras*^{V12}-expressing zebrafish. For this, immunofluorescence staining was performed on intestine paraffin sections from WT, WT/LPS, kras+, and kras+/LPS zebrafish treated with 2 µM of mifepristone or 40 ng/mL of LPS. We found that *kras*^{V12}-expressing zebrafish treated with LPS showed significant increases in both caspase-3 (Figure 4A,B) and PCNA-labeled cells (Figure 4C,D) compared with WT/LPS and kras+ zebrafish. In addition, *kras*^{V12} expression combined with LPS treatment increased the expression of a mitotic marker of phosphorylated histone (*p*-histone) in intestinal epithelial cells compared with WT/LPS and kras+ zebrafish (Figure S1). Our previous results revealed that the expression of *kras*^{V12} in zebrafish intestines significantly stimulated AKT and ERK activities, leading to the upregulation of the MAPK/ERK pathway (the MAPK/ERK pathway is the main downstream effector of RAS in the development of intestinal tumors) [27]. Furthermore, immunofluorescence staining for phosphorylated AKT (*p*-AKT) and ERK (*p*-ERK) revealed that compared with WT/LPS and kras+ zebrafish, *kras*^{V12} expression combined with LPS treatment significantly increased *p*-AKT levels (Figure 5A,B) but not *p*-ERK levels (Figure 5C,D) in intestinal epithelial cells.

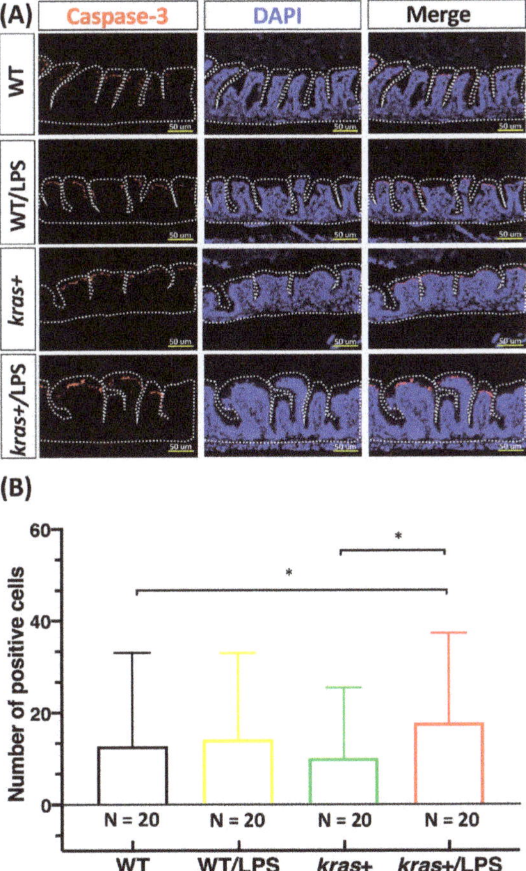

Figure 4. *Cont.*

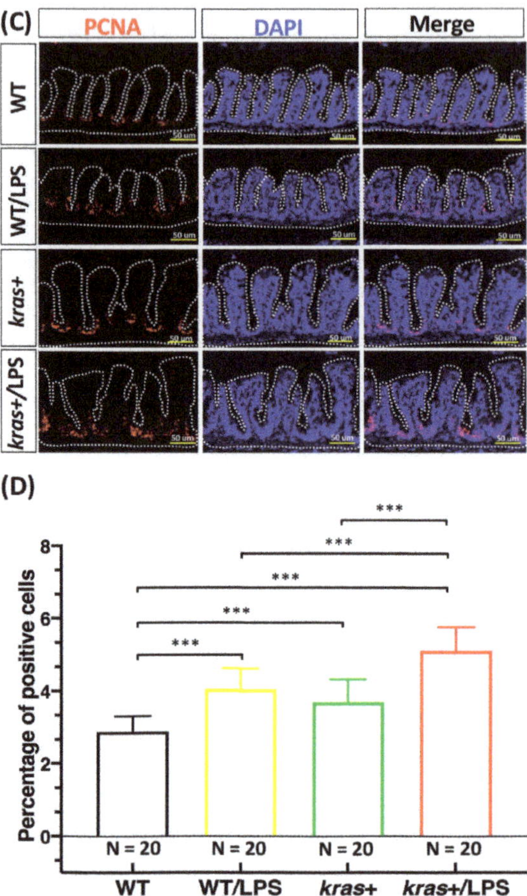

Figure 4. Expression of *kras*^{V12} with LPS treatment enhanced the increase in cell apoptosis and cell proliferation in the intestinal epithelium. (**A,C**) Immunofluorescence staining (red) was carried out in intestinal paraffin sections of WT (N = 20), WT/LPS (N = 20), kras+ (N = 20), and kras+/LPS (N = 20) zebrafish. (**B,D**) Immunofluorescence staining of caspase-3 showing (1) apoptosis and (2) PCNA as a marker for cell proliferation as well as (3) quantification of the number and percentage of positive cells. Differences among the variables were assessed using Student's t-tests. Statistical significance * $p < 0.05$, *** $p < 0.001$. Scale bar: 50 μm.

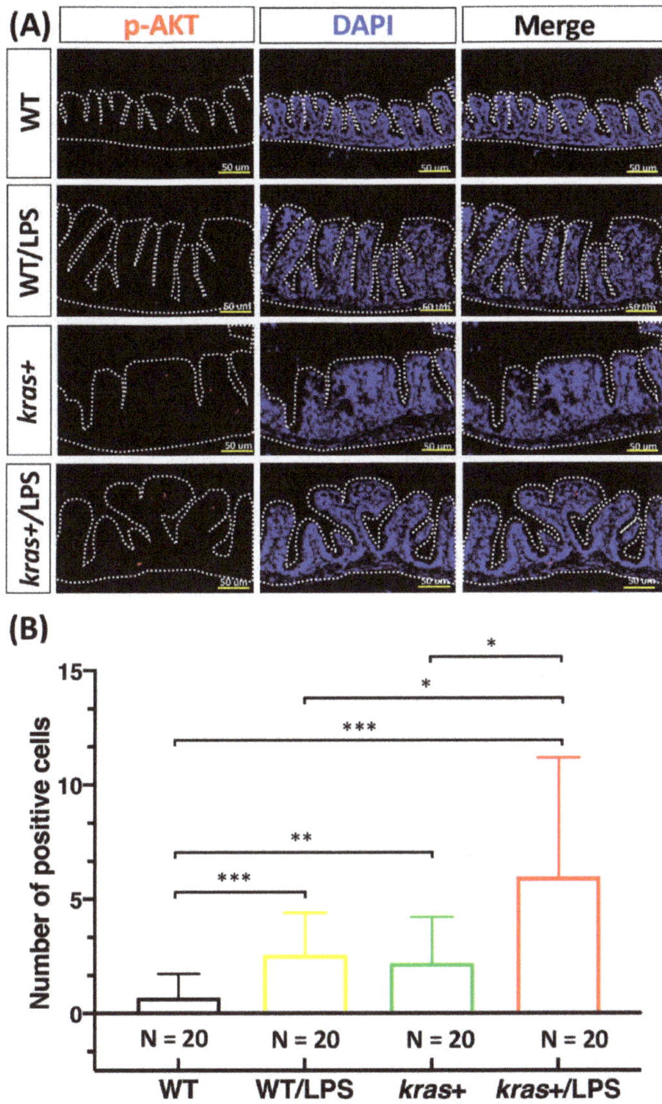

Figure 5. *Cont.*

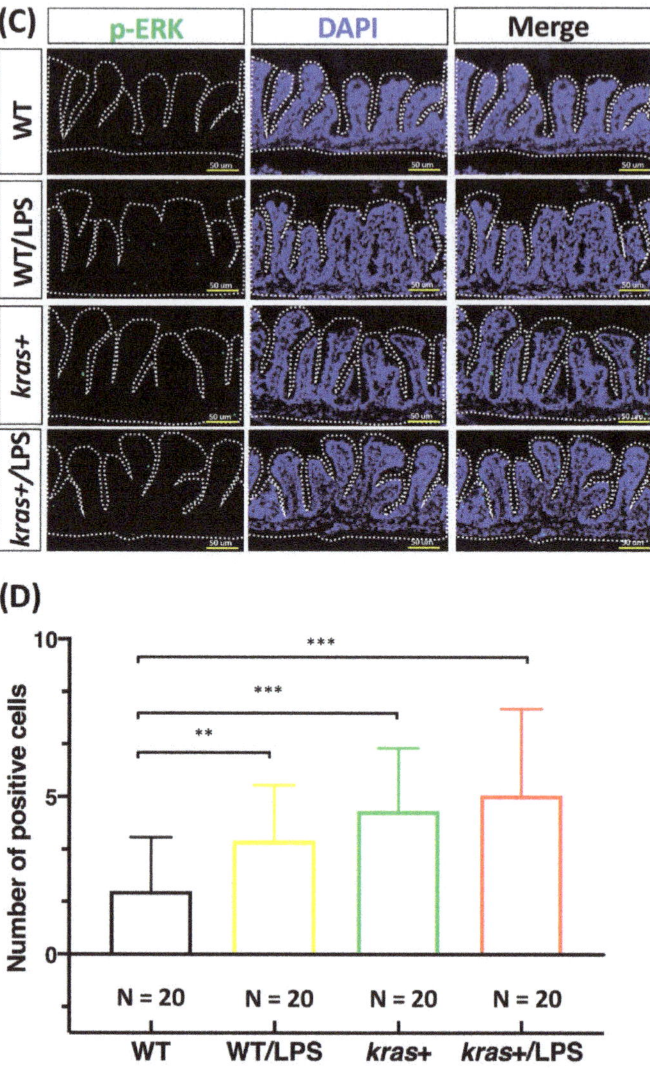

Figure 5. Expression of *kras*^{V12} with LPS treatment enhanced the increase in *p*-AKT in intestinal epithelial cells. (**A,C**) Immunofluorescence staining (red) was performed on intestinal paraffin sections of WT (N = 20), WT/LPS (N = 20), kras+ (N = 20), and kras+/LPS (N = 20) zebrafish. (**B,D**) Immunofluorescence staining of *p*-AKT and *p*-ERK as a marker of RAS signaling and quantification of the number of positive cells. Differences among the variables were assessed using Student's t-tests. Statistical significance: * $p < 0.05$, ** $p < 0.01$, *** $p < 0.001$. Scale bar: 50 µm.

3.5. Cotreatment with DSS and LPS Enhanced Intestinal Inflammation in kras^{V12} Transgenic Zebrafish Larvae

We further tested potential synergistic effects on intestinal inflammation by cotreating WT/lyz+, kras+/lyz, WT/mpeg1+, and kras+/mpeg1+ zebrafish larvae from 4 to 6 or 7 dpf with 0.5% DSS and 40 ng/mL of LPS, in addition to treatment with 4 µM of mifepristone (to induce *kras*^{V12} expression) (WT/lyz+, kras+/lyz+, WT/mpeg1+, kras+/mpeg1+ without treatments served as controls). All larvae from each group were imaged. Cotreatment with LPS and DSS revealed that both numbers of neutrophils and macrophages were significantly increased in kras+/lyz+/DSS/LPS and kras+/mpeg1+/DSS/LPS larvae compared with WT/lyz+/DSS/LPS (Figure 6A,B) and WT/mpeg1+/DSS/LPS larvae (Figure 6C,D). We also observed significant increases in neutrophils and macrophages in kras+/lyz+/DSS/LPS and kras+/mpeg1+/DSS/LPS larvae compared with kras+/lyz+/LPS, kras+/lyz+/DSS, kras+/lyz+/DSS, and kras+/mpeg1+/DSS larvae (Figure S2).

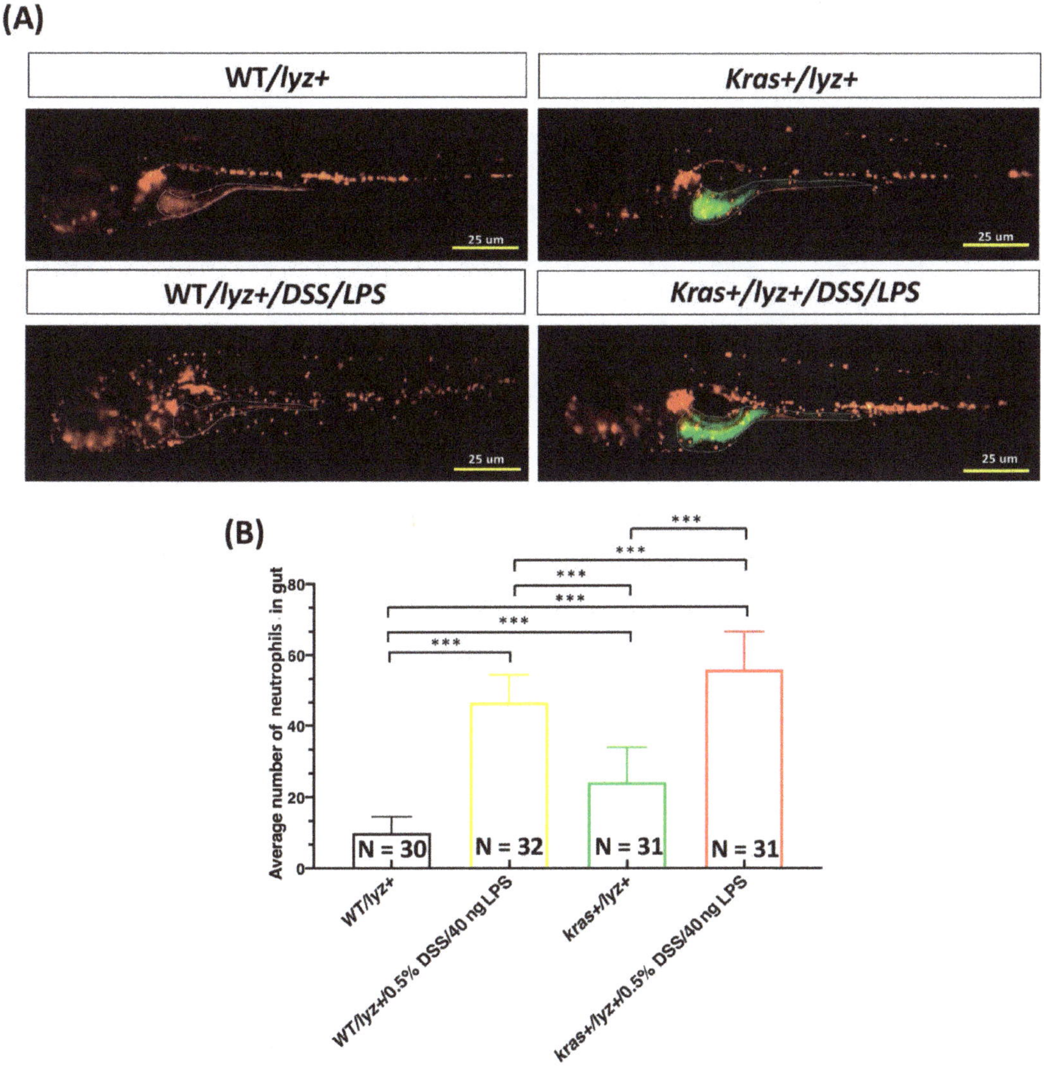

(A)

(B)

Figure 6. *Cont.*

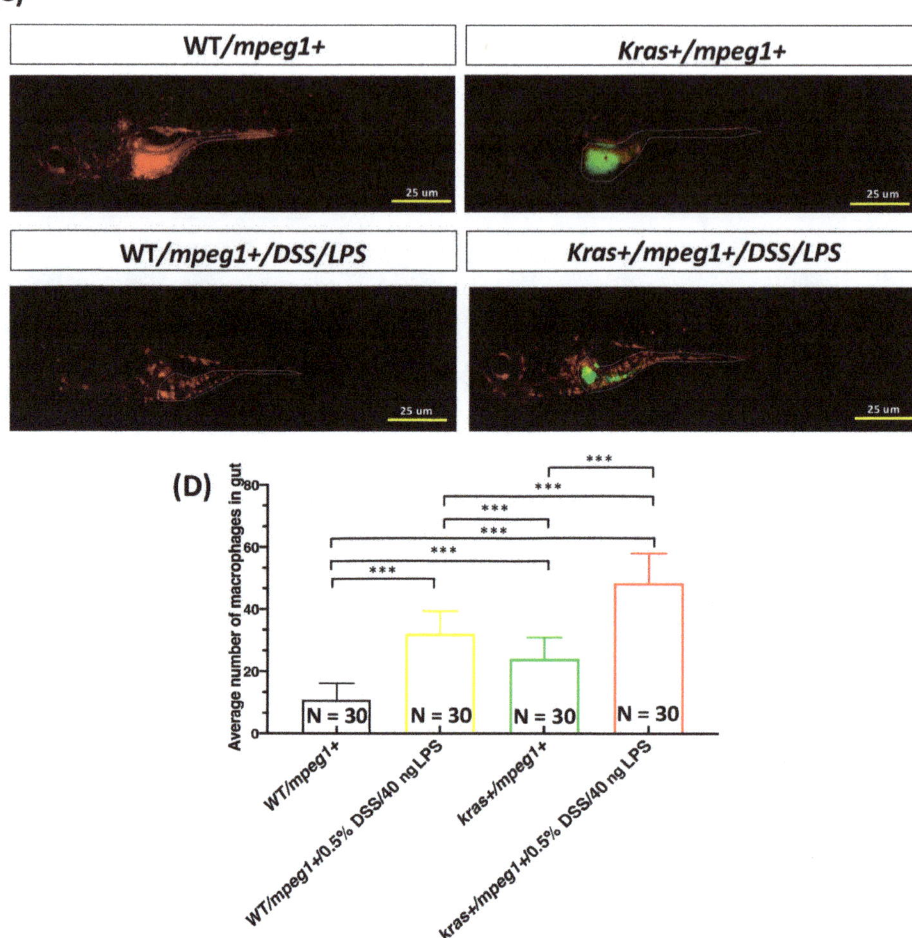

Figure 6. Cotreatment with DSS and LPS exacerbated the increased number of neutrophils and macrophages in the intestines of kras+/lyz+ and kras+/mepg1+ zebrafish. (**A,C**) Fluorescence of neutrophils (WT/lyz+, N = 30; WT/lyz+/DSS/LPS, N = 32; kras+/lyz+, N = 31; kras+/lyz+/DSS/LPS, N = 31) or macrophages (WT/mpeg1+, N = 30; WT/mepg1+/DSS/LPS, N = 30; kras+/mpeg1+, N = 30; kras+/mepg1+/DSS/LPS, N = 30) in the intestines. (**B,D**) Quantification of the number of positive cells as revealed by fluorescence of neutrophils or macrophages. Differences among the variables were assessed using Student's t-tests. Statistical significance: *** $p < 0.001$. Scale bar: 25 μm.

3.6. Phenotype of Intestinal Tumors Induced by Sustained Expression of kras^{V12} with DSS and LPS Cotreatment in Transgenic Zebrafish

We further confirmed the cotreatment of DSS/LPS in kras^{V12} transgenic zebrafish at the adult stage. For this, 4 mpf heterozygous kras^{V12} zebrafish were cotreated with 2 μM of mifepristone and DSS/LPS (0.0625%; 40 ng/mL) for 4 weeks. No significant differences were found in terms of body length between kras+ zebrafish treated with DSS/LPS and control group fish (Figure 7A). However, we observed significantly reduced body weights in adult WT/DSS/LPS and kras+/DSS/LPS zebrafish compared with WT or WT/DSS/LPS zebrafish (Figure 7B).

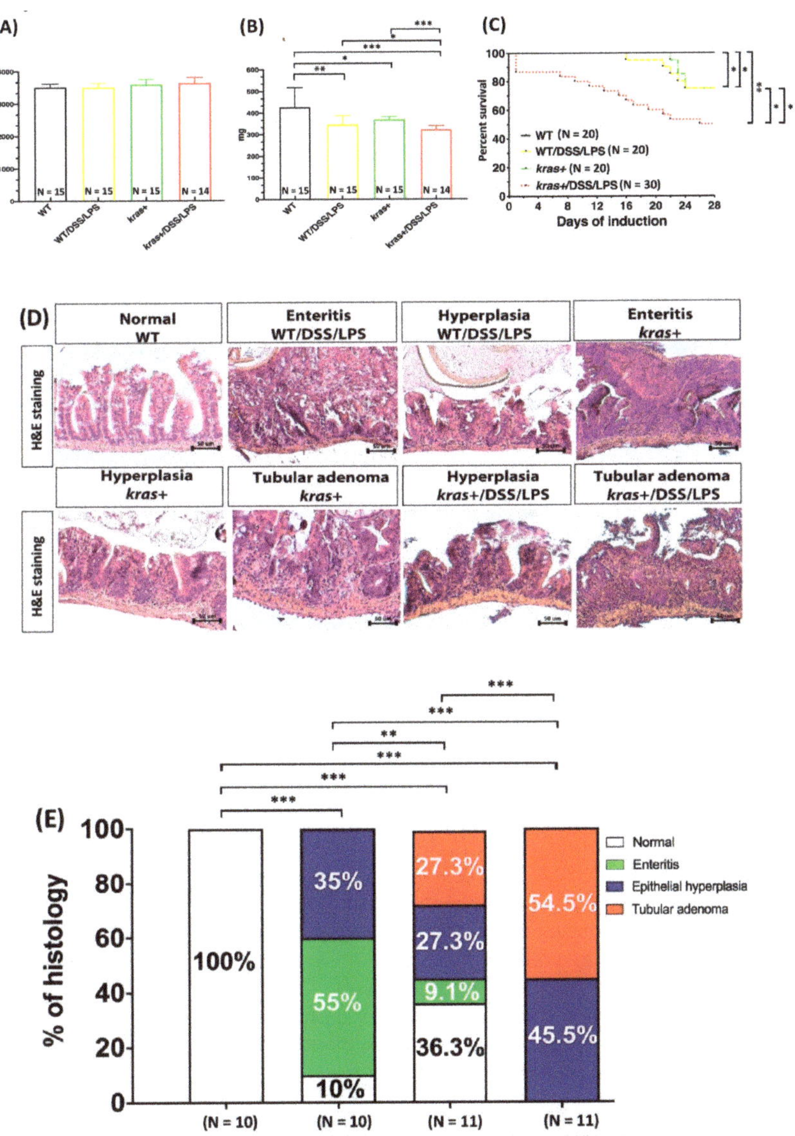

Figure 7. Synergistic effect of *kras*^{V12} expression and LPS/DSS on intestinal tumorigenesis. Four-month-postfertilization wild-type and kras+ zebrafish were cotreated with 2 μM of mifepristone, 40 ng/mL of LPS, and 0.0625% DSS for 4 weeks, and samples were then collected for gross observations and histological analyses. There were four experimental groups: WT, WT/DSS/LPS, kras+, and kras+/DSS/LPS. (**A,B**) Body length and body weight. (**C**) Survival curves. (**D**) Examples of normal intestines, enteritis, hyperplasia, and tubular adenoma as revealed by H&E staining of intestinal sections. (**E**) Summary of intestinal histological abnormalities observed in the four experimental groups. These data were generated as a result of a blinded histological analysis (WT, N = 10; WT/DSS/LPS, N = 10; kras+, N = 11; kras+ with DSS/LPS, N = 11). Differences among the variables were assessed using Student's t-tests or one-way ANOVA. Statistical significance: * $p < 0.05$, ** $p < 0.01$, *** $p < 0.001$. Scale bar: 50 μm.

By 4 wpi, 5 of the WT/DSS/LPS or kras+ zebrafish from each group and 13 of the kras+/DSS/LPS zebrafish had died, whereas no WT zebrafish died during the same period (Figure 7C). At 4 wpi, we evaluated the entire intestinal tract of adult zebrafish for enteritis, epithelial hyperplasia, and the presence of tumors (Figure 7D). For this, intestinal samples were collected from WT, WT/DSS/LPS, kras+, and kras+/DSS/LPS groups, and histological examinations were then performed. While all WT zebrafish showed normal intestinal histology, DSS/LPS treatment resulted in 55% and 35% of WT zebrafish respectively developing enteritis and hyperplasia. In kras+ zebrafish, mifepristone treatment led to inflammation, hyperplasia, and tubular adenoma in 9.1%, 27.3%, and 27.3% of these zebrafish, respectively. DSS/LPS treatment in kras+ zebrafish further increased the prevalence of abnormalities, whereby 45.5% and 54.5% of zebrafish respectively exhibited hyperplasia and tubular adenoma. These observations suggest that $kras^{V12}$ expression combined with DSS/LPS treatment significantly increased the prevalence of intestinal tumors in adult zebrafish compared with WT zebrafish treated with DSS/LPS (Figure 7E). We also confirmed the WT or kras+ cotreatment of DSS/LPS at the adult stage of zebrafish compared with WT/LPS or kras+/LPS. Results from WT/DSS/LPS and kras+/DSS/LPS transgenic zebrafish were not significantly different from those of WT/LPS and kras+/LPS control groups (Figure S3).

4. Discussion

The development of CRC is a multistep process that involves the progression of normal epithelium to an adenoma and then to an adenocarcinoma, where the adenocarcinoma may eventually metastasize to different organs [35]. The development a genetic model of colorectal tumorigenesis was introduced in 1990. In this model, APC, KRAS, TP53, and DCC were proposed as genes that promote CRC progression [36]. Since the introduction of this model, many potential molecular mechanisms of CRC have been investigated. There is a consensus that CRC development is related to LPS-induced systemic inflammation, and these events alter the signal transduction of TLR4, NF-κB, and transforming growth factor beta 1 (TGF-β1) pathways [17,37]. In mice, LPS has been found to contribute to tumor progression and hepatic metastasis of colon cancer cells [17,38,39]. Furthermore, the DSS induced colitis model is widely used because it has many similarities with human ulcerative colitis [40]. Furthermore, a huge advantage of transgenic zebrafish can be an exceptional platform for mimicking human intestinal disorder and establishing vertebrate models for drug screening. For intestinal disease research, high tumor incidence and convenient chemical treatment make the inducible transgenic zebrafish a reasonable platform for intestinal inflammation and tumor progression [23,27]. However, no previously published studies have reported that $kras^{V12}$ expression combined with LPS treatment can induce intestinal tumors in zebrafish. This is also the first study to report that LPS and/or DSS treatment promotes intestinal tumor progression in $kras^{V12}$-expressing transgenic zebrafish.

Tumor-associated macrophages and tumor-associated neutrophils are among the most abundant immune cells in the tumor microenvironment. In CRC, they play pivotal tumor-supporting roles [41,42]. In this study, a zebrafish model was used to study the effects of LPS and/or DSS treatment on intestinal inflammation in $kras^{V12}$ transgenic zebrafish larvae. Specifically, we were interested in (1) the effects that $kras^{V12}$ expression has on neutrophils and macrophages when combined with LPS and/or DSS treatment and (2) how these effects stimulate the immune system (Figure 1). We previously found that treating $kras^{V12}$ zebrafish larvae with LPS led to significant increases in neutrophil count and neutrophil density in the liver. These increases in neutrophils further led to an enlargement in liver size [43]. In adult zebrafish, the intestine-specific expression of $kras^{V12}$ with LPS treatment also led to an increase in hyperplasia and tubular adenoma (Figure 2D,E). In addition, intestinal morphology (Figure 3A) revealed that goblet cell number, goblet cell size, villi height, intervilli space, villi width, and smooth muscle thickness (Figure 3B–G) were also significantly altered in these $kras^{V12}$ zebrafish. Goblet cells in intestinal epithelium produce and secrete mucins (predominantly MUC2), which enter the intestinal lumen to form a mucus layer [44]. The mucus and mucins of goblet cells and intestinal epithelial cells compose the first line of defense for the gastrointestinal tract and interact with the immune system [45]. In clinical CRC samples from mice, SCF/c-KIT signaling has been shown to promote mucus secretion in colonic goblet cells as well as the development of mucinous colorectal adenocarcinoma [46]. CRC tumors and cell lines are characterized by an increased expression of goblet cell marker genes, and there is an association between the proportion of goblet cells in a tumor and the probability that the tumor is assigned as consensus molecular subtype 3 (CMS3) (CMS3 is a tumor subtype that is mutually exclusive from mucinous adenocarcinoma pathologies [47]). Changes in mucin gene expression and mucin glycan structure can occur in intestinal cancers, which leads to cancer progression [48]. Our results indicated that $kras^{V12}$ and/or LPS can critically interact with the immune system and may be involved in the development of intestinal carcinogenesis.

We analyzed intestinal tumor formation using histological and immunochemical methods. Immunochemical data showed an increase in active caspase-3 (Figure 4A,B), PCNA expression (Figure 4C,D), and downstream signaling of phosphorylated AKT (Figure 5A,B) in kras+/LPS zebrafish compared with kras+ zebrafish, which suggests that LPS is associated with intestinal tumor formation. LPS has been reported to cause rapid apoptosis and death in intestinal epithelial cells as well as loss of epithelial integrity, which is contingent on apoptosis functioning normally [49–51]. LPS treatment has also been found to increase (1) AKT phosphorylation at residue Ser473 and (2) increase liver metastasis of HT-29 cells, both in vitro and in vivo [17]. Our data indicate that $kras^{V12}$ and LPS can be linked through the AKT pathway. The AKT pathway has roles in apoptosis [52], cell proliferation, cell migration [53], angiogenesis [54], and metabolism [55]. Thus, this kras/LPS/AKT link may be the mechanism that underlies the development of intestinal carcinogenesis in $kras^{V12}$ transgenic zebrafish.

Recently, inflammation has been found to increase the risk of CRC [56–58]. DSS has been shown to induce inflammation of the colonic mucosa and has a tumor-promoting effect in mouse and zebrafish models [27,59]. Moreover, a DSS-induced inflammatory bowel disease (IBD)-like enterocolitis model has also been established in zebrafish [60]. Inflammatory stimuli induced by DSS treatment following initiation with AOM carcinogens are effective at rapidly inducing inflammation and intestinal tumors [61]. In mice, high-fat diets can also promote colon tumors associated with AOM/DSS-induced colitis [62]. In addition, DSS treatment has been found to initiate the development of small intestinal polyps or tumors at 2% and 4%, respectively, in mouse models [63,64]. Therefore, we further investigated whether cotreatment with DSS/LPS can exacerbate intestinal inflammation associated with neutrophils or macrophages in $kras^{V12}$ transgenic zebrafish larvae. In $kras^{V12}$ zebrafish larvae, DSS/LPS treatment led to significant increases in both of neutrophil and macrophage numbers in the intestines (Figure 6). In addition, LPS/DSS cotreatment significantly enhanced the increase in neutrophils and macrophages in the intestines in kras+/lyz+/LPS and kras+/lyz+/DSS as well as in kras+/mpeg1+/LPS and kras+/mpeg1+/DSS zebrafish during the larval stage (Figure S2). For adult zebrafish, $kras^{V12}$ expression combined with DSS/LPS cotreatment led to an increase in the prevalence of hyperplasia and tubular adenoma compared with WT/DSS/LPS adult zebrafish (Figure 7D,E). Therefore, the current research also strongly supports a relationship between chronic inflammation and intestinal tumorigenesis. However, we did not observe significant differences in intestinal tumorigenesis between WT/LPS and WT/LPS/DSS or between kras+/LPS and kras+/DSS/LPS adult-stage zebrafish (Figure S3). These results indicate that extending the treatment time may be necessary to strengthen the state of chronic inflammation [65,66].

5. Conclusions

In conclusion, our results (based on zebrafish treated with LPS alone or cotreated with LPS and DSS) provide evidence that LPS and/or DSS exacerbates the development and progression of intestinal tumors in $kras^{V12}$ transgenic zebrafish. These findings are an extension of our previous data [27], which showed that DSS increased $kras^{V12}$-induced intestinal tumors in zebrafish. The current study demonstrated that $kras^{V12}$ expression combined with LPS and/or DSS treatment also significantly exacerbated the development and progression of intestinal tumors, and this is achieved through the modulation of the PI3K–AKT signaling pathway. Therefore, further research is necessary to explore the effect of other inflammatory agents on CRC progression in humans.

Supplementary Materials: The following are available online at https://www.mdpi.com/article 10.3390/biomedicines9080974/s1: Figure S1: Expression of $kras^{V12}$ with LPS treatment enhance the increase in p-Histone in intestinal epithelial cells, Figure S2: LPS/DSS co-treatment significantl enhanced the increase number in neutrophils and macrophages in the intestine during the larval stag in kras+/lyz+/LPS and kras+/lyz+/DSS as well as in kras+/mpeg1+/LPS and kras+/mpeg1+/DS zebrafish, Figure S3: No significant synergistic effects on intestinal tumorigenesis between WT/LP and WT/LPS/DSS or between kras+/LPS and kras+ with DSS/LPS adult stage zebrafish, Table S Information of antibodies used in this study.

Author Contributions: Conceptualization, J.-W.L., L.-I.L. and Z.G.; methodology, J.-W.L., Y.S., P S.A.F. and L.-I.L.; software, J.-W.L., Y.S. and P.-S.A.F.; validation, J.-W.L. and P.-S.A.F.; formal analysis J.-W.L. and Y.S.; investigation, J.-W.L., Y.S., P.-S.A.F., L.-I.L., and Z.G.; resources and data curation J.-W.L., Y.S., P.-S.A.F., L.-I.L., D.L. and Z.G.; writing—original draft preparation, J.-W.L.; writing—review and editing, J.-W.L., L.-I.L. and Z.G.; visualization, J.-W.L. and Y.S.; supervision, J.-W.L., L.-I.L and Z.G.; project administration, J.-W.L., L.-I.L. and Z.G. All authors have read and agreed to the published version of the manuscript.

Funding: This work was supported by grants from the Ministry of Education of Singapore (R154000B and R154000B70114) in Singapore and the National Taiwan University Hospital (UN109-062) in Ta wan.

Institutional Review Board Statement: Not applicable.

Informed Consent Statement: Not applicable.

Data Availability Statement: Data are contained within the article.

Conflicts of Interest: The authors declare no conflict of interest.

References

1. Brenner, H.; Kloor, M.; Pox, C.P. Colorectal cancer. *Lancet* **2014**, *383*, 1490–1502. [CrossRef]
2. Kuipers, E.J.; Grady, W.M.; Lieberman, D.; Seufferlein, T.; Sung, J.J.; Boelens, P.G.; van de Velde, C.J.; Watanabe, T. Colorecta cancer. *Nat. Rev. Dis. Primers* **2015**, *1*, 15065. [CrossRef] [PubMed]
3. Mármol, I.; Sánchez-De-Diego, C.; Dieste, A.P.; Cerrada, E.; Yoldi, M.J.R. Colorectal Carcinoma: A General Overview and Futur Perspectives in Colorectal Cancer. *Int. J. Mol. Sci.* **2017**, *18*, 197. [CrossRef] [PubMed]
4. Bray, F.; Ferlay, J.; Soerjomataram, I.; Siegel, R.L.; Torre, L.A.; Jemal, A. Global cancer statistics 2018: GLOBOCAN estimates o incidence and mortality worldwide for 36 cancers in 185 countries. *CA Cancer J. Clin.* **2018**, *68*, 394–424. [CrossRef]
5. Siegel, R.L.; Miller, K.D.; Jemal, A. Cancer statistics, 2019. *CA Cancer J. Clin.* **2019**, *69*, 7–34. [CrossRef] [PubMed]
6. Zhang, Z.; Tao, Y.; Hua, Q.; Cai, J.; Ye, X.; Li, H. SNORA71A Promotes Colorectal Cancer Cell Proliferation, Migration, and Invasion. *BioMed Res. Int.* **2020**, *2020*, 1–11. [CrossRef]
7. Rawla, P.; Sunkara, T.; Barsouk, A. Epidemiology of colorectal cancer: Incidence, mortality, survival, and risk factors. *Prz Gastroenterol.* **2019**, *14*, 89–103. [CrossRef]
8. Jančík, S.; Drabek, J.; Radzioch, D.; Hajdúch, M. Clinical Relevance of KRAS in Human Cancers. *J. Biomed. Biotechnol.* **2010**, *2010* 1–13. [CrossRef]
9. Marley, A.R.; Nan, H. Epidemiology of colorectal cancer. *Int. J. Mol. Epidemiol. Genet.* **2016**, *7*, 105–114.
10. Zenonos, K.; Kyprianou, K. RAS signaling pathways, mutations and their role in colorectal cancer. *World J. Gastrointest. Onco* **2013**, *5*, 97–101. [CrossRef]
11. Vakiani, E.; Solit, D.B. KRAS and BRAF: Drug targets and predictive biomarkers. *J. Pathol.* **2010**, *223*, 220–230. [CrossRef [PubMed]
12. Saud, S.M.; Li, W.; Morris, N.L.; Matter, M.; Colburn, N.H.; Kim, Y.S.; Young, M.R. Resveratrol prevents tumorigenesis in mous model of Kras activated sporadic colorectal cancer by suppressing oncogenic Kras expression. *Carcinogenesis* **2014**, *35*, 2778–278 [CrossRef] [PubMed]
13. Gagnière, J.; Raisch, J.; Veziant, J.; Barnich, N.; Bonnet, R.; Buc, E.; Bringer, M.-A.; Pezet, D.; Bonnet, M. Gut microbiota imbalanc and colorectal cancer. *World J. Gastroenterol.* **2016**, *22*, 501–518. [CrossRef] [PubMed]
14. Brennan, C.A.; Garrett, W.S. Gut Microbiota, Inflammation, and Colorectal Cancer. *Annu. Rev. Microbiol.* **2016**, *70*, 395–41 [CrossRef]

15. Flemer, B.; Lynch, D.B.; Brown, J.M.R.; Jeffery, I.; Ryan, F.; Claesson, M.; O'Riordain, M.; Shanahan, F.; O'Toole, P.W. Tumour-associated and non-tumour-associated microbiota in colorectal cancer. *Gut* **2017**, *66*, 633–643. [CrossRef] [PubMed]

16. Song, M.; Chan, A.T. Environmental Factors, Gut Microbiota, and Colorectal Cancer Prevention. *Clin. Gastroenterol. Hepatol.* **2019**, *17*, 275–289. [CrossRef]

17. Hsu, R.Y.; Chan, C.H.; Spicer, J.D.; Rousseau, M.C.; Giannias, B.; Rousseau, S.; Ferri, L.E. LPS-induced TLR4 signaling in human colorectal cancer cells increases beta1 integrin-mediated cell adhesion and liver metastasis. *Cancer Res.* **2011**, *71*, 1989–1998. [CrossRef]

18. Pastorelli, L.; De Salvo, C.; Mercado, J.R.; Vecchi, M.; Pizarro, T.T. Central Role of the Gut Epithelial Barrier in the Pathogenesis of Chronic Intestinal Inflammation: Lessons Learned from Animal Models and Human Genetics. *Front. Immunol.* **2013**, *4*, 280. [CrossRef]

19. Yesudhas, D.; Gosu, V.; Anwar, M.A.; Choi, S. Multiple Roles of Toll-Like Receptor 4 in Colorectal Cancer. *Front. Immunol.* **2014**, *5*, 334. [CrossRef]

20. Simiantonaki, N.; Kurzik-Dumke, U.; Karyofylli, G.; Jayasinghe, C.; Michel-Schmidt, R.; Kirkpatrick, C.J. Reduced expression of TLR4 is associated with the metastatic status of human colorectal cancer. *Int. J. Mol. Med.* **2007**, *20*, 21–29. [CrossRef] [PubMed]

21. Fazio, V.M.; De Robertis, M.; Massi, E.; Poeta, M.L.; Carotti, S.; Morini, S.; Cecchetelli, L.; Signori, E. The AOM/DSS murine model for the study of colon carcinogenesis: From pathways to diagnosis and therapy studies. *J. Carcinog.* **2011**, *10*, 9. [CrossRef]

22. Lu, J.-W.; Ho, Y.-J.; Yang, Y.-J.; Liao, H.-A.; Ciou, S.-C.; Lin, L.-I.; Ou, D.-L. Zebrafish as a disease model for studying human hepatocellular carcinoma. *World J. Gastroenterol.* **2015**, *21*, 12042–12058. [CrossRef]

23. Lu, J.-W.; Ho, Y.-J.; Ciou, S.-C.; Gong, Z. Innovative Disease Model: Zebrafish as an In Vivo Platform for Intestinal Disorder and Tumors. *Biomedicines* **2017**, *5*, 58. [CrossRef] [PubMed]

24. Haramis, A.G.; Hurlstone, A.; Van Der Velden, Y.; Begthel, H.; Born, M.V.D.; Offerhaus, G.J.A.; Clevers, H.C. Adenomatous polyposis coli-deficient zebrafish are susceptible to digestive tract neoplasia. *EMBO Rep.* **2006**, *7*, 444–449. [CrossRef] [PubMed]

25. Le, X.; Langenau, D.M.; Keefe, M.D.; Kutok, J.L.; Neuberg, D.S.; Zon, L.I. Heat shock-inducible Cre/Lox approaches to induce diverse types of tumors and hyperplasia in transgenic zebrafish. *Proc. Natl. Acad. Sci. USA* **2007**, *104*, 9410–9415. [CrossRef] [PubMed]

26. Neal, J.; Peterson, T.S.; Kent, M.L.; Guillemin, K.H. pylori virulence factor CagA increases intestinal cell proliferation by Wnt pathway activation in a transgenic zebrafish model. *Dis. Model. Mech.* **2013**, *6*, 802–810. [CrossRef]

27. Lu, J.W.; Raghuram, D.; Fong, P.A.; Gong, Z. Inducible Intestine-Specific Expression of kras(V12) Triggers Intestinal Tumorigenesis in Transgenic Zebrafish. *Neoplasia* **2018**, *20*, 1187–1197. [CrossRef] [PubMed]

28. Enya, S.; Kawakami, K.; Suzuki, Y.; Kawaoka, S. A novel zebrafish intestinal tumor model reveals a role for cyp7a1-dependent tumor-liver crosstalk in causing adverse effects on the host. *Dis. Models Mech.* **2018**, *11*. [CrossRef]

29. Topi, G.; Satapathy, S.R.; Dash, P.; Fred Mehrabi, S.; Ehrnstrom, R.; Olsson, R.; Lydrup, M.L.; Sjolander, A. Tumour-suppressive effect of oestrogen receptor beta in colorectal cancer patients, colon cancer cells, and a zebrafish model. *J. Pathol.* **2020**, *251*, 297–309. [CrossRef]

30. Fior, R.; Póvoa, V.; Mendes, R.V.; Carvalho, T.; Gomes, A.; Figueiredo, N.; Ferreira, M.G. Single-cell functional and chemosensitive profiling of combinatorial colorectal therapy in zebrafish xenografts. *Proc. Natl. Acad. Sci. USA* **2017**, *114*, E8234–E8243. [CrossRef]

31. De Almeida, C.R.; Mendes, R.V.; Pezzarossa, A.; Gago, J.; Carvalho, C.; Alves, A.; Nunes, V.; Brito, M.J.; Cardoso, M.J.; Ribeiro, J.; et al. Zebrafish xenografts as a fast screening platform for bevacizumab cancer therapy. *Commun. Biol.* **2020**, *3*, 299. [CrossRef] [PubMed]

32. Li, H.; Lu, J.-W.; Huo, X.; Li, Y.; Li, Z.; Gong, Z. Effects of sex hormones on liver tumor progression and regression in Myc/xmrk double oncogene transgenic zebrafish. *Gen. Comp. Endocrinol.* **2019**, *277*, 112–121. [CrossRef] [PubMed]

33. Lu, J.-W.; Hsieh, M.-S.; Hou, H.-A.; Chen, C.-Y.; Tien, H.-F.; Lin, L.-I. Overexpression of SOX4 correlates with poor prognosis of acute myeloid leukemia and is leukemogenic in zebrafish. *Blood Cancer J.* **2017**, *7*, e593. [CrossRef] [PubMed]

34. Lu, J.-W.; Hou, H.-A.; Hsieh, M.-S.; Tien, H.-F.; Lin, L.-I. Overexpression of FLT3-ITD driven by spi-1 results in expanded myelopoiesis with leukemic phenotype in zebrafish. *Leukemia* **2016**, *30*, 2098–2101. [CrossRef] [PubMed]

35. Balch, C.; Ramapuram, J.B.; Tiwari, A.K. The Epigenomics of Embryonic Pathway Signaling in Colorectal Cancer. *Front. Pharmacol.* **2017**, *8*, 267. [CrossRef] [PubMed]

36. Fearon, E.R.; Vogelstein, B. A genetic model for colorectal tumorigenesis. *Cell* **1990**, *61*, 759–767. [CrossRef]

37. Zhao, L.; Yang, R.; Cheng, L.; Wang, M.; Jiang, Y.; Wang, S. LPS-Induced Epithelial-Mesenchymal Transition of Intrahepatic Biliary Epithelial Cells. *J. Surg. Res.* **2011**, *171*, 819–825. [CrossRef]

38. Zhu, G.; Huang, Q.; Huang, Y.; Zheng, W.; Hua, J.; Yang, S.; Zhuang, J.; Wang, J.; Ye, J. Lipopolysaccharide increases the release of VEGF-C that enhances cell motility and promotes lymphangiogenesis and lymphatic metastasis through the TLR4-NF-kappaB/JNK pathways in colorectal cancer. *Oncotarget* **2016**, *7*, 73711–73724. [CrossRef]

39. Liu, W.-T.; Jing, Y.-Y.; Yan, F.; Han, Z.-P.; Lai, F.-B.; Zeng, J.-X.; Yu, G.-F.; Fan, Q.-M.; Li, R.; Zhao, Q.-D.; et al. LPS-induced CXCR4-dependent migratory properties and a mesenchymal-like phenotype of colorectal cancer cells. *Cell Adhes. Migr.* **2016**, *11*, 13–23. [CrossRef]

40. Chassaing, B.; Aitken, J.D.; Malleshappa, M.; Vijay-Kumar, M. Dextran Sulfate Sodium (DSS)-Induced Colitis in Mice. *Curr. Protoc. Immunol.* **2014**, *104*, 15.25.1–15.25.14. [CrossRef] [PubMed]

41. Mizuno, R.; Kawada, K.; Itatani, Y.; Ogawa, R.; Kiyasu, Y.; Sakai, Y. The Role of Tumor-Associated Neutrophils in Colorectal Cancer. *Int. J. Mol. Sci.* **2019**, *20*, 529. [CrossRef] [PubMed]

42. Zhong, X.; Chen, B.; Yang, Z. The Role of Tumor-Associated Macrophages in Colorectal Carcinoma Progression. *Cell. Physiol. Biochem.* **2018**, *45*, 356–365. [CrossRef] [PubMed]

43. Yan, C.; Huo, X.; Wang, S.; Feng, Y.; Gong, Z. Stimulation of hepatocarcinogenesis by neutrophils upon induction of oncogenic kras expression in transgenic zebrafish. *J. Hepatol.* **2015**, *63*, 420–428. [CrossRef] [PubMed]

44. Tytgat, K.M.; Büller, H.A.; Opdam, F.J.; Kim, Y.S.; Einerhand, A.W.; Dekker, J. Biosynthesis of human colonic mucin: Muc2 is the prominent secretory mucin. *Gastroenterology* **1994**, *107*, 1352–1363. [CrossRef]

45. Pelaseyed, T.; Bergström, J.H.; Gustafsson, J.K.; Ermund, A.; Birchenough, G.M.H.; Schütte, A.; van der Post, S.; Svensson, F.; Rodríguez-Piñeiro, A.M.; Nyström, E.E.L.; et al. The mucus and mucins of the goblet cells and enterocytes provide the first defense line of the gastrointestinal tract and interact with the immune system. *Immunol. Rev.* **2014**, *260*, 8–20. [CrossRef]

46. Li, G.; Yang, S.; Shen, P.; Wu, B.; Sun, T.; Sun, H.; Ji, F.; Zhou, D. SCF/c-KIT signaling promotes mucus secretion of colonic goblet cells and development of mucinous colorectal adenocarcinoma. *Am. J. Cancer Res.* **2018**, *8*, 1064–1073.

47. Miller, S.A.; Ghobashi, A.H.; O'Hagan, H.M. Consensus molecular subtyping of colorectal cancers is influenced by goblet cell content. *Cancer Genet.* **2021**, *254–255*, 34–39. [CrossRef] [PubMed]

48. Kim, Y.S.; Ho, S.B. Intestinal Goblet Cells and Mucins in Health and Disease: Recent Insights and Progress. *Curr. Gastroenterol. Rep.* **2010**, *12*, 319–330. [CrossRef]

49. Yu, L.C.-H.; Flynn, A.N.; Turner, J.R.; Buret, A.G. SGLT-1-mediated glucose uptake protects intestinal epithelial cells against LPS-induced apoptosis and barrier defects: A novel cellular rescue mechanism? *FASEB J.* **2005**, *19*, 1822–1835. [CrossRef]

50. Forsythe, R.M.; Xu, D.; Lu, Q.; Deitch, E.A. Lipopolysaccharide-Induced Enterocyte-Derived Nitric Oxide Induces Intestinal Monolayer Permeability in an Autocrine Fashion. *Shock* **2002**, *17*, 180–184. [CrossRef]

51. Liu, C.; Li, A.; Weng, Y.B.; Duan, M.L.; Wang, B.E.; Zhang, S.W. Changes in intestinal mucosal immune barrier in rats with endotoxemia. *World J. Gastroenterol.* **2009**, *15*, 5843–5850. [CrossRef]

52. Franke, T.F.; Hornik, C.P.; Segev, L.; Shostak, G.A.; Sugimoto, C. PI3K/Akt and apoptosis: Size matters. *Oncogene* **2003**, *22*, 8983–8998. [CrossRef] [PubMed]

53. Lu, J.-W.; Liao, C.-Y.; Yang, W.-Y.; Lin, Y.-M.; Jin, S.-L.C.; Wang, H.-D.; Yuh, C.-H. Overexpression of Endothelin 1 Triggers Hepatocarcinogenesis in Zebrafish and Promotes Cell Proliferation and Migration through the AKT Pathway. *PLoS ONE* **2014**, *9*, e85318. [CrossRef]

54. Somanath, P.R.; Razorenova, O.V.; Chen, J.; Byzova, T.V. Akt1 in endothelial cell and angiogenesis. *Cell Cycle* **2006**, *5*, 512–518. [CrossRef]

55. Coloff, J.L.; Rathmell, J.C. Metabolic regulation of Akt: Roles reversed. *J. Cell Biol.* **2006**, *175*, 845–847. [CrossRef]

56. Munkholm, P. Review article: The incidence and prevalence of colorectal cancer in inflammatory bowel disease. *Aliment Pharmacol. Ther.* **2003**, *18*, 1–5. [CrossRef] [PubMed]

57. Van Der Woude, C.J.; Kleibeuker, J.H.; Jansen, P.L.M.; Moshage, H. Chronic inflammation, apoptosis and (pre-)malignant lesions in the gastro-intestinal tract. *Apoptosis* **2004**, *9*, 123–130. [CrossRef]

58. Seril, D.N.; Liao, J.; Yang, G.-Y.; Yang, C.S. Oxidative stress and ulcerative colitis-associated carcinogenesis: Studies in humans and animal models. *Carcinogenesis* **2003**, *24*, 353–362. [CrossRef]

59. Okayasu, I.; Hatakeyama, S.; Yamada, M.; Ohkusa, T.; Inagaki, Y.; Nakaya, R. A novel method in the induction of reliable experimental acute and chronic ulcerative colitis in mice. *Gastroenterology* **1990**, *98*, 694–702. [CrossRef]

60. Oehlers, S.; Flores, M.V.; Hall, C.; Okuda, K.S.; Sison, J.O.; Crosier, K.E.; Crosier, P.S. Chemically Induced Intestinal Damage Models in Zebrafish Larvae. *Zebrafish* **2013**, *10*, 184–193. [CrossRef] [PubMed]

61. Parang, B.; Barrett, C.W.; Williams, C.S. AOM/DSS Model of Colitis-Associated Cancer. *Methods Mol. Biol.* **2016**, *1422*, 297–307. [CrossRef] [PubMed]

62. Jin, B.-R.; Chung, K.-S.; Lee, M.; An, H.-J. High-Fat Diet Propelled AOM/DSS-Induced Colitis-Associated Colon Cancer Alleviated by Administration of Aster glehni via STAT3 Signaling Pathway. *Biology* **2020**, *9*, 24. [CrossRef] [PubMed]

63. Tanaka, T.; Kohno, H.; Suzuki, R.; Hata, K.; Sugie, S.; Niho, N.; Sakano, K.; Takahashi, M.; Wakabayashi, K. Dextran sodium sulfate strongly promotes colorectal carcinogenesis inApcMin/+ mice: Inflammatory stimuli by dextran sodium sulfate results in development of multiple colonic neoplasms. *Int. J. Cancer* **2005**, *118*, 25–34. [CrossRef]

64. Cooper, H.S.; Everley, L.; Chang, W.; Pfeiffer, G.; Lee, B.; Murthy, S.; Clapper, M.L. The role of mutant Apc in the development of dysplasia and cancer in the mouse model of dextran sulfate sodium-induced colitis. *Gastroenterology* **2001**, *121*, 1407–1416. [CrossRef] [PubMed]

65. Dubois, R.N. Role of Inflammation and Inflammatory Mediators in Colorectal Cancer. *Trans. Am. Clin. Clim. Assoc.* **2014**, *125*, 358–373.

66. Park, C.H.; Eun, C.S.; Han, D.S. Intestinal microbiota, chronic inflammation, and colorectal cancer. *Intest. Res.* **2018**, *16*, 338–345. [CrossRef]

 *biomedicines*

Article

Depletion of Alpha-Melanocyte-Stimulating Hormone Induces Insatiable Appetite and Gains in Energy Reserves and Body Weight in Zebrafish

Yang-Wen Hsieh [1,2,†], Yi-Wen Tsai [3,4,†], Hsin-Hung Lai [2], Chi-Yu Lai [1,2], Chiu-Ya Lin [1,2] and Guor Mour Her [2,*]

1 Department of Bioscience and Biotechnology, National Taiwan Ocean University, Keelung 202, Taiwan; hearhero@hotmail.com (Y.-W.H.); c.y.stephen.lai@gmail.com (C.-Y.L.); vista_jckey_1590@livemail.tw (C.-Y.L.)
2 Institute of Biopharmaceutical Sciences, National Yang Ming Chiao Tung University, Taipei 112, Taiwan; s232579@gmail.com
3 Department of Family Medicine, Chang Gung Memorial Hospital, Keelung 204, Taiwan; tsaiyiwen@gmail.com
4 College of Medicine, Chang Gung University, Taoyuan 333, Taiwan
* Correspondence: gmher@nycu.edu.tw; Tel.: +886-2-2826-7000 (ext. 67990)
† These authors contributed equally to this work.

Abstract: The functions of anorexigenic neurons secreting proopiomelanocortin (POMC)/alpha-melanocyte-stimulating hormone (α-MSH) of the melanocortin system in the hypothalamus in vertebrates are energy homeostasis, food intake, and body weight regulation. However, the mechanisms remain elusive. This article reports on zebrafish that have been genetically engineered to produce α-MSH mutants, α-MSH^{-7aa} and α-MSH^{-8aa}, selectively lacking 7 and 8 amino acids within the α-MSH region, but retaining most of the other normal melanocortin-signaling (Pomc-derived) peptides. The α-MSH mutants exhibited hyperphagic phenotypes leading to body weight gain, as observed in human patients and mammalian models. The actions of several genes regulating appetite in zebrafish are similar to those in mammals when analyzed using gene expression analysis. These include four selected orexigenic genes: Promelanin-concentrating hormone (*pmch*), agouti-related protein 2 (*agrp2*), neuropeptide Y (*npy*), and hypothalamic hypocretin/orexin (*hcrt*). We also study five selected anorexigenic genes: Brain-derived neurotrophic factor (*bdnf*), single-minded homolog 1-a (*sim1a*), corticotropin-releasing hormone b (*crhb*), thyrotropin-releasing hormone (*trh*), and prohormone convertase 2 (*pcsk2*). The orexigenic actions of α-MSH mutants are rescued completely after hindbrain ventricle injection with a synthetic analog of α-MSH and a melanocortin receptor agonist, Melanotan II. We evaluate the adverse effects of MSH depletion on energy balance using the Alamar Blue metabolic rate assay. Our results show that α-MSH is a key regulator of POMC signaling in appetite regulation and energy expenditure, suggesting that it might be a potential therapeutic target for treating human obesity.

Keywords: orexigen; obesogens; adipogenesis; hypothalamus; appetite

Citation: Hsieh, Y.-W.; Tsai, Y.-W.; Lai, H.-H.; Lai, C.-Y.; Lin, C.-Y.; Her, G.M. Depletion of Alpha-Melanocyte-Stimulating Hormone Induces Insatiable Appetite and Gains in Energy Reserves and Body Weight in Zebrafish. *Biomedicines* **2021**, *9*, 941. https://doi.org/10.3390/biomedicines9080941

Academic Editor: James A. Marrs

Received: 28 June 2021
Accepted: 27 July 2021
Published: 2 August 2021

Publisher's Note: MDPI stays neutral with regard to jurisdictional claims in published maps and institutional affiliations.

1. Introduction

The arcuate melanocortin system in vertebrates consists of anorexigenic neurons expressing proopiomelanocortin peptide (POMC), and orexigenic neurons expressing neuropeptide Y/agouti-related peptide (NPY/AgRP). The POMC-expressing neurons of the hypothalamic arcuate nucleus (ARC) participate in the control of food intake, energy homeostasis, body weight (BW) regulation, and other metabolic processes in the melanocortin system [1]. The POMC precursor peptide is processed into a series of biologically active components, including alpha-melanocyte-stimulating hormone (α-MSH), adrenocorticotrophic hormone (ACTH), β-MSH, and β-endorphin (β-END) by tissue-specific proteolysis [2,3]. α-MSH functions in anorexigenic responses by activating the melanocortin

4 receptor (MC4R), which is expressed on distinct second-order neurons, whereas AGRP, the orexigenic peptide, conveys as an antagonist of MC4R [4,5]. In mammals, loss of the genes encoding POMC [6] or MC4R [7,8] leads to severe obesity.

Melanocortin peptides are known to regulate feeding behavior in mammals [9,10]. The suppression of appetite in mammals mediated by α-MSH/β-MSH is attributed to MC4R [7]. In animal models, hyperphagia, an obese phenotype, and hyperinsulinemia were observed in the obese yellow mouse (Ay) model of interruption of α-MSH central signaling by the ubiquitous constitutive expression of the agouti gene [11]. Recently, genetically modified *Pomc*^{*tm1/tm1*} mice, which had a mutation in the *Pomc* gene, were unable to synthesize desacetyl-α-MSH and α-MSH. *Pomc*^{*tm1/tm1*} mice are hyperphagic and show an obese phenotype even when fed a regular chow diet [12,13]. It was recently reported in a model using Labrador retriever dogs that a 14-bp deletion in the gene encoding pro-POMC in these canines is associated with obesity [14]. The β-MSH [15–17] and β-endorphin [18,19] coding sequences are also functionally associated with adiposity and greater appetite [20]. In fact, novel mutations (Phe144Leu [21] and Arg145Cys [22]) located in the α-MSH domain of the *POMC* gene were observed to be associated with early-onset obesity.

Additionally, moderately linear growth, which is primarily regulated by growth hormone (GH) released by somatotrophs in the adenohypophysis of the pituitary gland, was also observed in both loss-of-function mutant *Pomc* or *Mc4r* rodents and human models [6,7,23–26]. Familial glucocorticoid deficiency (FGD), whose clinical features are enhanced longitudinal bone growth and advanced bone age, is an ACTH-insensitivity disorder characterized by the overproduction of ACTH [27–29].

According to previous studies, Pomc in zebrafish has highly conserved regions similar to ACTH, α-MSH, β-END, β-MSH, and possibly to N-POMC-derived peptides, but lacks γ-MSH compared with mammalian models [30]. However, among the two *pomc* family member genes, *pomca* and *pomcb*, observed in zebrafish [2], only *pomca*, expressed in the pituitary gland, is responsible for developing the interrenal organ of zebrafish [30,31]. In zebrafish, knockdown of the gene encoding Pomca resulted in a significant reduction in Acth immunoreactivity, and attenuated melanosome dispersal at five days postfertilization (dpf) following injection of a designed antisense morpholino oligonucleotide [32]. Shi et al. demonstrated that the increased somatic growth without obesity in *pomca* mutant zebrafish was associated with reduced anxiety-like behaviors but not with appetite or energy expenditure.

Most research on the biological functions of Pomc in zebrafish focusing on anxiety disorder has been based on the evaluation of changes in Acth expression levels. Here, we used zebrafish to test whether depletion of α-MSH signaling might disturb food intake and influence obesity, as the hypothalamic neural circuits involved in food intake are highly conserved in fish species. Our findings provide comprehensive information about the dynamic expression of genes controlling appetite and growth in nonfunctional α-MSH signaling in the Pomc neurons of zebrafish.

2. Materials and Methods

2.1. Fish Husbandry

All zebrafish were maintained in a controlled environment with a 14/10-h light-dark cycle at 28 °C. They were fed twice daily with brine shrimp and commercial fish food pellets. All animal experiments were conducted in accordance with the guidelines and approval of the Institutional Animal Care and Use Committee (IACUC) of the National Yang Ming Chiao Tung University, Taiwan.

2.2. TALEN Cloning and Targeted Mutagenesis

For each mutant target site in the *pomca* locus(acc. no. NM_181438.3), two 18 bp TALEN binding sites were selected (exon3: 5′-CCTACTCCATGGAGCAC-3′, 5′-CGGTCGG CCGCAAACGC-3′). A restriction enzyme (AgeI) site between each TALEN pair was used for genotyping by restriction fragment length polymorphism (RFLP) analysis. TALEN constructs were cloned using Golden Gate assembly (Cermak et al., 2011) and an accompanying plasmid kit from Addgene (Addgene Kit #1000000024). TALEN mRNAs were injected into 1-cell embryos to generate stable mutant lines. The final TALEN expression plasmids were linearized by digestion with the NotI restriction enzyme. TALEN mRNAs were transcribed using the mMESSAGE SP6 kit (Ambion, USA) and purified using the RNeasy Mini kit (QIAGEN, Hilden, Germany).

2.3. Quantitative Reverse Transcription Polymerase Chain Reaction (RT–qPCR)

Total RNA was extracted using TRIzol reagent (Thermo Fisher Scientific, USA) or the RNeasy Mini Kit (QIAGEN, Hilden, Germany) and reverse transcribed using the first-strand cDNA synthesis kit (K1691; Thermo Fisher Scientific). Real-time RT–qPCR was performed using the StepOne Real-Time PCR System (Applied Biosystems, Foster City, USA). The genes and their corresponding primer sequences are listed in Supplementary Table S1. These genes, including those encoding growth hormone (*gh*), growth hormone receptor b (*ghrb*), insulin-like growth factor 2b (*igf2b*), insulin-like growth factor binding protein 1b (*igfbp1b*), insulin-like growth factor binding protein 5b (igfbp5b), and insulin-like growth factor binding protein 6b (*igfbp6b*) have been validated in zebrafish [33].

2.4. In-Situ Hybridization (ISH)

The gene-specific probes were cloned by PCR into a pGEM®-T Easy TA cloning vector (Promega, Madison, WI USA) using the primers listed in Supplementary Table S2. Antisense probes were synthesized by in vitro transcription using the DIG RNA Labelling Kit (SP6/T7) (Roche Applied Science, Mannheim, Germany). ISH was performed as described previously [34].

2.5. Quantification of Food Intake

Our procedure was adapted from that described previously [35]. The larval fish food was fluorescently labeled paramecia prepared using the lipophilic tracer 4-(4-(Didecylamino)styryl)-*N*-Dethylpyridinium iodide (4-Di-10-ASP; Invitrogen, Carlsbad, CA, USA). We conducted feeding of 7 dpf zebrafish in 6-well plates to allow for free swimming. At 1.5 h after feeding, the larvae were anesthetized. After two washes to remove residual paramecia, groups of five larvae were transferred into a 96-well round-bottom black plate in an anesthetic solution. The intra-abdominal fluorescent signal was measured using the Synergy™ HT Multi-Detection Reader (BioTek Instruments, Winooski, VT) in fluorescence area scan mode 11×11 multipoint/well, 0.1 s/point using a fluorescein filter set (excitation wavelength, 485 nm; emission wavelength, 590 nm).

2.6. Hindbrain Ventricle Injection of Zebrafish Larvae

Briefly, anesthetized fish were placed on a 3% agarose plate with water containing 0.05% MS222. The skulls were impaled with a 0.53 mm diameter needle attached to a syringe in the midline at the telencephalon–diencephalon border. The fish were injected intracranially with 4.6 nL of sterile water, an α-MSH analog (100 µM in sterile water; M4135, Sigma-Aldrich), or Melanotan II (MTII, 100 µM in sterile water; M8693, Sigma-Aldrich) into the cranial cavity by a heat-pulled glass capillary micropipette attached to microinjection equipment. After administration, the cut heads were sampled, and at least 30 larvae were pooled to yield enough total RNA.

2.7. Histology and Immunohistochemistry (IHC)

Tissues were fixed using 4% paraformaldehyde solution in phosphate-buffered saline (PBS) overnight at 4 °C, washed in PBS, and equilibrated in 30% sucrose/PBS overnight at 4 °C. Then, they were embedded in an OCT compound and cut into 4 µm sections using a cryostat. IHC was performed as described previously [36,37]. Rabbit anti-α-MSH antibody (1:1000, RayBiotech 130-10355) was used to detect α-MSH immunoreactivity. Protein levels were detected using horseradish peroxidase (HRP)-conjugated secondary antibody (1:1000 Jackson Immuno Research AB_2313567).

2.8. Alamar Blue Metabolic Rate Assay

Our metabolic rate assay was adapted from an established method [38]. The assay buffer was supplemented with egg water containing 0.1% DMSO, 1% Alamar Blue (A50101 Thermo Fisher Scientific), and 4 mM sodium bicarbonate. Zebrafish were rinsed with 0.22 μm filtered egg water and pipetted into a 96-well plate (three specimens per well). Fluorescence of the plate was immediately read on the Infinite 200 PRO multimode plate reader (Tecan Group Ltd., Switzerland) with excitation at 530 nm and emission monitored at 590 nm (fluorescence area scan mode 3×3 multipoint/well, 10 times/point per reading). The plate was incubated in the dark at 28 °C for 24 h and then read again. Any wells containing dead fish were excluded from the analyses. The change in fluorescence from time 0 to 24 h was then calculated. The data were corrected by setting the average of the wild-type (WT) controls to 1.

2.9. Whole-Mount Oil Red O Staining

Zebrafish larvae were fixed in 4% paraformaldehyde solution in PBS overnight at 4 °C. Equal numbers of control and mutant larvae were transferred to 1.5 mL Eppendorf tubes and rinsed three times (5 min each) with PBS. After removing the PBS, the larvae were prestained with a mixture of 60% isopropyl alcohol for at least 1 h. Then, fresh 0.5% Oil Red O solution was added at 4 °C for 1 h, and larvae were washed in PBS. They were stored in 70% glycerol and imaged using a bright-field dissecting microscope (Stemi 305 Carl Zeiss AG, Oberkochen, Germany).

2.10. Growth Rate

The growth rate was recorded monthly starting at two months postfertilization (mpf) and ending at 12 mpf. Twenty fish per diet (mixed sex) were selected randomly. Body length (BL) was measured using a standard metric ruler and determined the distance from the snout to the caudal peduncle. BL (cm) and BW (mg) were used to calculate the body mass index (BMI).

2.11. Morphological and Morphometric Studies: Analysis of Zebrafish Fat Tissues and Adipocytes

These studies were carried out on histological sections according to Mon-Talbano et al. [39]

2.12. Statistical Analysis

All data are presented as the mean $\pm$ standard error of the mean (SEM). Statistical analysis was performed using analysis of variance (ANOVA) followed by Bonferroni post-hoc tests. All analyses were performed using GraphPad Prism software (v. 8.0; GraphPad, San Diego, CA, USA). Differences were considered statistically significant at $p < 0.05$.

3. Results

3.1. Generation of α-MSH Depletion Lines in Zebrafish

The TALEN-based genome editing technique was performed for the depletion of α-MSH. According to the sequence information, the targeted site of TALENs was located in the third exon of the *pomca* gene locus. Both arms of the designed binding arms were 18 bp long. The spacer between the two arms was 17 bp. The AgeI restriction digestion site within the space region was used for genotyping (Figure 1A).

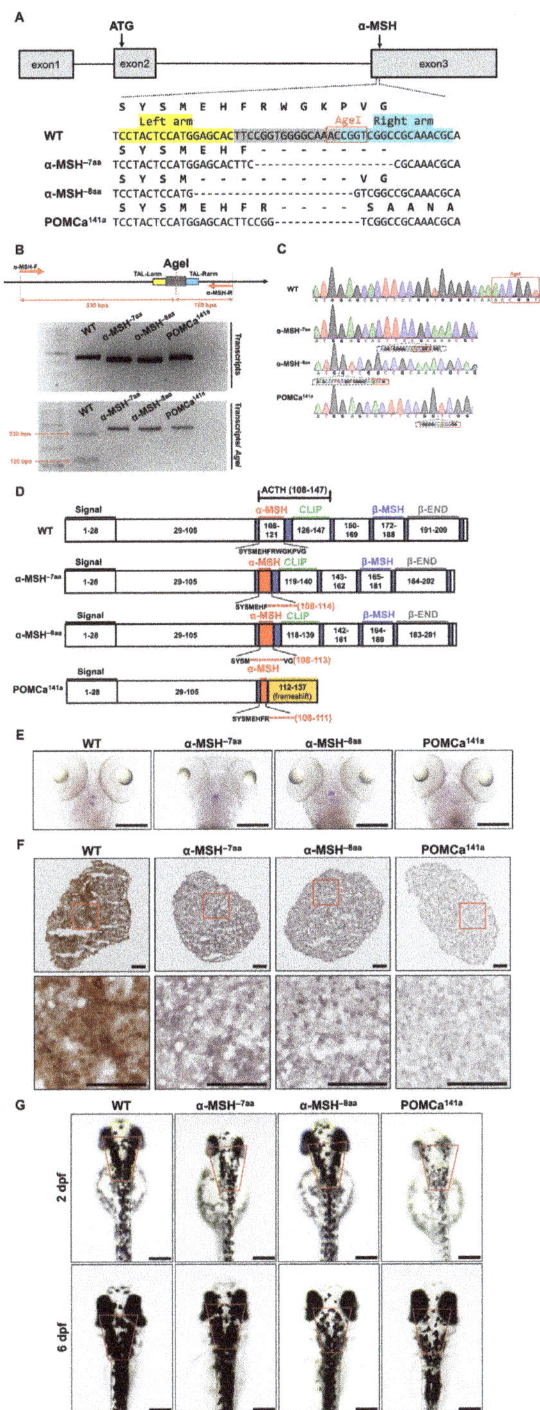

Figure 1. Generation of zebrafish *pomca* mutants using TALEN. (**A**) Schematics of the three *pomca* mutant alleles generated. The sequences in the third exon that were targeted by the TALEN pairs

are shown in gray boxes. The AgeI restriction enzyme recognition site for genotyping purposes i shown in a red box; dashed lines indicate deleted nucleotides. The left and right TALEN targeting sites are highlighted in yellow and sky blue, respectively. (**B**) Upper: Schematic illustration of the primers used for RT–qPCR detection of mutations. The specific primer is to the mutated site/schem of the locations of primers (red arrows) designed to detect a disruption in the spacer of the third exor Yellow box, left TALEN sites (TAL-L); sky blue box, right TALEN sites (TAL-R); grey box, spacer; re dashed line, AgeI restriction site. Lower: Results of RT–qPCR analyses on fin clips of heterozygous F fish containing one of the corresponding mutations (as indicated in A). (**C**) Chromographs illustratin, the sequences in the third exon of the *pomca* WT controls and the nucleotide deletion of *pomca* mutant The boxed sequence in red indicates the restriction enzyme AgeI cutting sites in WT controls that wer deleted in the mutant fish. (**D**) These diagrams show the predicated Pomca protein of *pomca* mutant compared with the WT Pomca protein. (**E**) Whole-mount ISH showing the expression of *pomc* transcripts in the pituitary in WT controls and *pomca* mutants larvae at 5 dpf. Scale bars = 200 µm (**F**) Expression patterns of α-MSH in 12 mpf Pomc neuron samples after Immunohistochemistry frozen section (IHC-F) staining. Scale bar = 50 µm. (**G**) Dorsal view of 2 dpf and 6 dpf *pomca* mutan larvae. Scale bars = 200 µm.

Three *pomca* mutants were generated here. Two independent α-MSH-depleted mu tant lines, α-MSH^{-7aa} and α-MSH^{-8aa} were generated with 21-bp and 24-bp bp in-fram. deletions, respectively. One independent *pomca* mutant line, POMCa^{141a}, was generated with a 13-bp frameshift deletion (Figure 1A,B). The α-MSH^{-7aa} and α-MSH^{-8aa} mutation. resulted in in-frame deletions that produced truncated proteins of 7 amino acids (aa) and 8 aa within the α-MSH regions, respectively (Figure 1C,D). No significantly decrease. expression pattern of *pomca* transcripts was observed in the pituitary gland or hypotha lamus of the α-MSH mutants, except for the POMCa^{141a} mutant compared with the W controls at 5 dpf (Figure 1E). The α-MSH mutants caused deficiencies of only the α-MSH peptide hormones (Figure 1F), whereas the POMCa^{141a} mutant resulted in premature stop codons that produced truncated proteins of 141 aa (Figure 1C,D), which caused a deficiency in the majority of melanocortin peptides derived from POMCa (Figure 1F). In additior three *pomca* mutants showed attenuated melanosome dispersion in dark conditions, and . significant decrease in melanosomes was observed in the POMCa^{141a} mutant (Figure 1G

3.2. Defective α-MSH Increased Food Intake in Zebrafish

To evaluate the pattern of food intake at the larval stage of the zebrafish mutants, w. used a previously described feeding protocol to prepare fluorescent food composed o larval fish food with lipophilic tracer 4-Di-10-ASP-labeled paramecia [35] (see Methods The larvae ingested this fluorescent food, which led to an accumulation of fluorescen signals within the intestine that could be visualized readily by fluorescence microscop. (Figure 2A).

The α-MSH^{-7aa} and α-MSH^{-8aa} larvae exhibited obvious hyperphagia at 7 dp (Figure 2B), when food intake significantly increased by 2.28-and 2.03-fold, respectively compared with control WT zebrafish (Figure 2C). However, no obvious differences be tween POMCa^{141a} and WT controls were observed in terms of food intake at 7 dpf (dat. not shown).

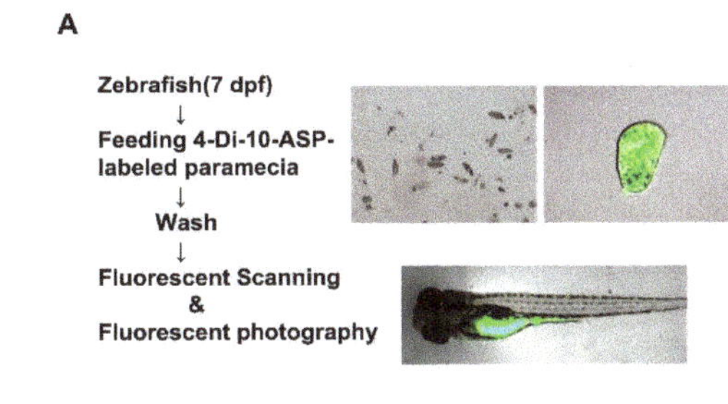

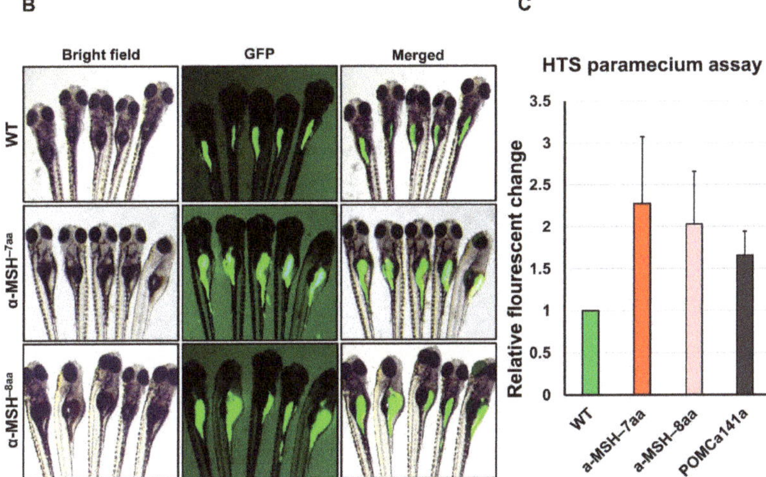

Figure 2. A qualitative food intake assay for zebrafish *pomca* mutant larvae. (**A**) Schematic representation of the feeding assay. Fluorescent intensities after free-feeding of 4-10-Di-ASP-labeled paramecia. Fluorescent intensities of ingested paramecia were at maximum levels 1.5 h after feeding (see Methods). (**B**) Side views of 7 dpf larvae examined under fluorescence illumination. The larvae were incubated with fluorescent microspheres coated with fish food for up to 1.5 h before visualizing the fluorescent contents in their gut. (**C**) Correlation between the relative amount of paramecia and fluorescent intensities of ingested paramecia in zebrafish. The fluorescent intensities were measured from the numbers of introduced paramecia thrice independently. All values are the mean ± SEM, *n* = 50.

3.3. Defective α-MSH Enhanced Somatic Growth and Decreased Energy Expenditure Concomitant with Liver Steatosis in Zebrafish Larvae

Next, we tested whether growth and energy budget were proportional to food intake in the POMC mutants, and simultaneous observation was necessary to estimate these parameters. Predictably, the α-MSH mutants significantly increased in BL at 8 dpf (Figure 3A). Somatic growth is predominantly controlled by regulating the GH/insulin-like growth factor (IGF) axis in fish [40,41]. Increased mRNA expression of genes controlling the GH/IGF axis accompanied by increased BL was detected in the *pomc* mutants compared with WT controls at 8 dpf (Figure 3B). Using whole-mount ISH analysis, a significantly increased expression pattern of gh was observed in *pomc* mutants compared with WT at 5 dpf (Figure 3C). Thus, these results indicate that the observed growth activation could be attributed to the upregulation of the GH/IGF axis.

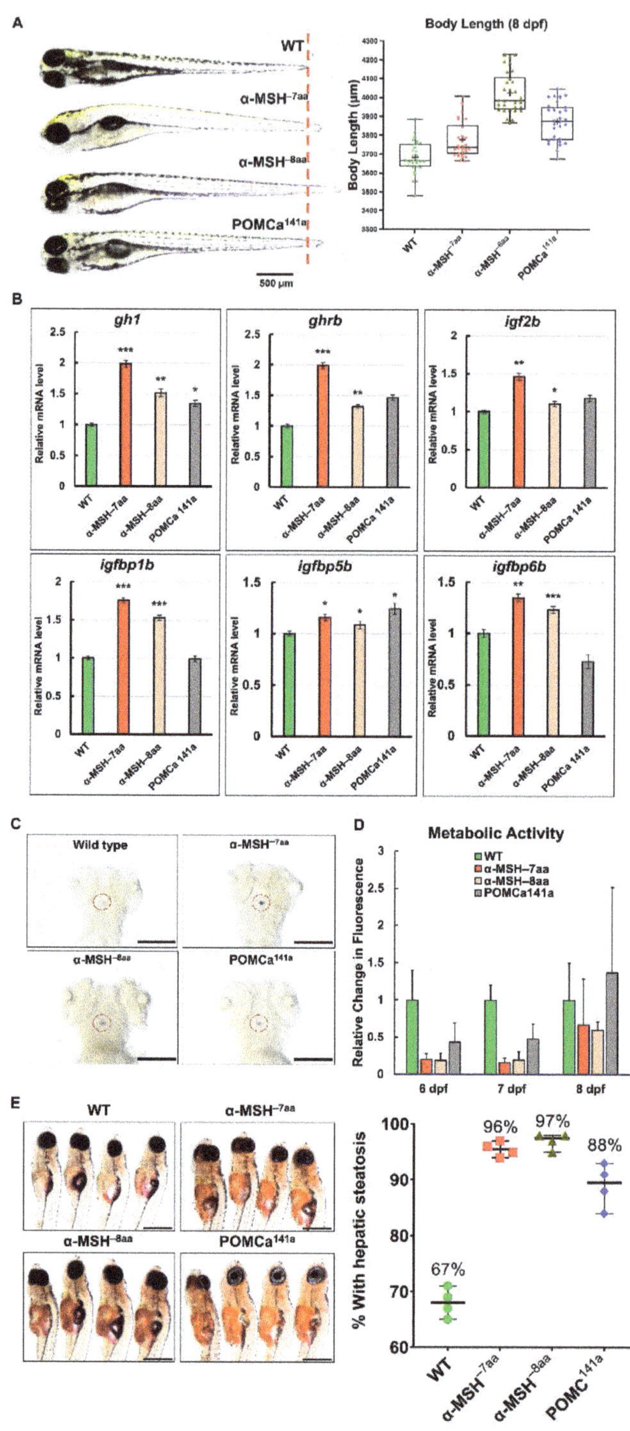

Figure 3. The level of α-MSH regulates normal somatic growth and energy balance in zebrafish embryos/larvae. (**A**) Left: Lateral view of wild-type and *pomca* mutant larvae at 8 dpf. Scale bar = 500 µm.

Right: Statistical analysis of BL (jaw to tail fin) in WT controls and *pomca* mutant larvae at 8 dpf. Data are shown as the mean $\pm$ SD ($n = 30$). (**B**) Expression of GH/IGF axis genes, *gh1*, *ghrb*, *igf2b*, *igfbp1b*, *igfbp15b*, and *igfbp6b*, in WT controls and *pomca* mutants larvae at 8 dpf. Values are the mean $\pm$ SEM. * $p < 0.05$, ** $p < 0.01$, and *** $p < 0.001$ compared with WT groups. (**C**) Whole-mount ISH showing the increased expression of *gh1* in the pituitary in *pomca* mutants larvae at 5 dpf. Scale bars = 200 µm. (**D**) Response to depletion of α-MSH in *pomca* mutant larvae by the Alamar Blue assay. Three larvae per well were incubated in 4 mM sodium bicarbonate with 1% Alamar Blue. The fluorescence of the solution was measured at different time points. Data are reported as the relative change in fluorescence intensity at least three times independently. (mean $\pm$ SEM, $n = 30$). (**E**) Left: Hepatic steatosis was observed by whole-body Oil Red O staining in *pomca* mutant larvae at 21dpf. Right: Percentages of WT and *pomca* mutant larvae with strong levels of hepatic steatosis at 21 dpf were calculated from at least 50 fish in each group. Data were representative of four independent experiments. Scale bar = 1 mm.

To assess the potential roles of α-MSH in the metabolism of the *pomc* mutants for further studies of energy expenditure that might influence obesity, BW, and appetite regulation, we performed the Alamar Blue metabolic rate assay in larvae at 6–8 dpf. The assay showed increased signals of the zebrafish mutants with incubation time compared with the WT controls (Figure 3D). All *pomc* mutants larvae exhibited significantly lower metabolic rates, with 53% to 84% between the 6 and 7 dpf stages. Both the α-MSH^{-7aa} and α-MSH^{-8aa} larvae exhibited lower metabolic rates, with decreases of 84% and 81%, respectively, compared with control WT zebrafish (Figure 3D). However, the POMCa^{141a} larvae exhibited a slightly higher metabolic rate, with an increase of 37% compared with control WT zebrafish at 8 dpf (Figure 3D). Furthermore, hepatic steatosis in the α-MSH mutants increased markedly compared with that in the WT larvae (Figure 3E), thus confirming that lipid accumulation as an energy reserve is reflected in hepatic steatosis.

3.4. α-MSH Mutant Adults Develop Characteristic Melanocortin-Related Obesity

Homozygous α-MSH mutant larvae developed normally and were slightly larger than the WT larvae (Figure 3A,E). Surprisingly, by measuring the growth rate of POMC mutants and WT zebrafish, a dramatic increase in BL (Figure 4A,B) and BW (Figure 4C) was detected in both α-MSH mutant adults compared with WT adults 12 months post-fertilization (mpf). The POMCa^{141a} adults showed the same BL as WT adults after 12 mpf and a slight increase in BW compared with WT adults until 9 mpf. (Figure 4C). There were also statistically significant increases in the BMI at 10 mpf for the α-MSH mutant adults (Figure 4D).

The BMI showed the same trend observed for BL and BW (Figure 4D). Examination of the BW data showed a significant increase in viscera and visceral fat contents in both the α-MSH^{-7aa} and α-MSH^{-8aa} groups at the 12 mpf stage (Figure 4E). Morphometric analysis of adipose tissues showed a dramatic difference in development between the α-MSH mutant adults and the WT controls. Both the α-MSH^{-7aa} and α-MSH^{-8aa} adult fish had larger amounts of adipose tissue, with increases of 8- and 12-fold, respectively, compared with control WT zebrafish (Figure 4F,G). Indeed, the average area of both visceral and subcutaneous adipose tissue showed a significant increase in α-MSH mutant adults compared with WT controls, indicating that an increase in food intake and decreased energy expenditure determined the growth of an abundant layer of adipose tissue, resulting in obesity.

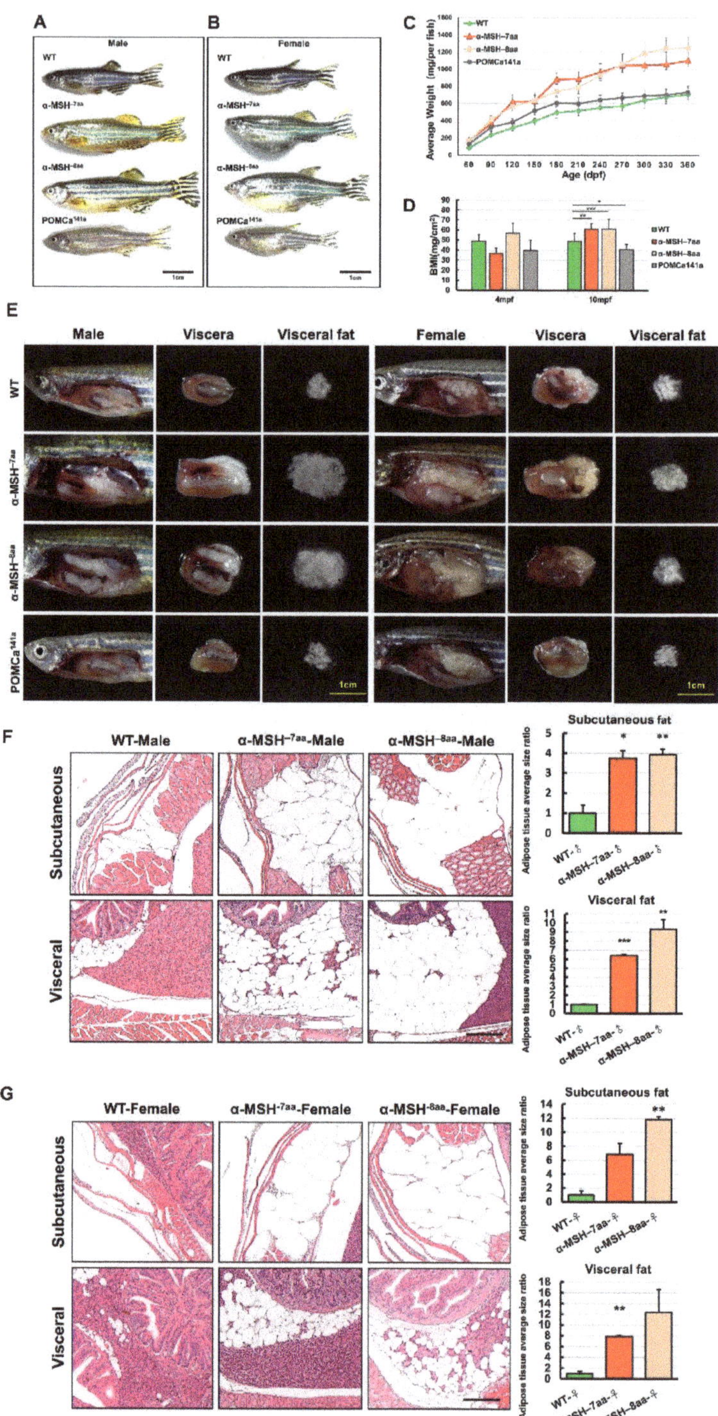

Figure 4. The level of α-MSH regulates somatic growth and sexual size dimorphism in *pomca* mutant adults. (**A**,**B**) Lateral view of F2 homozygous male (**A**) and female (**B**) mutants compared with WT

controls at 12 mpf. Scale bars = 1 cm. (**C**) BW curves of zebrafish from the juvenile to adult stages in the three experimental male groups (WT, α-MSH^{-7aa}, and α-MSH^{-8aa}; *n* = 20/group). (**D**) Bar graph showing the BWs in the three experimental male groups (*n* = 20/group) under normal feeding conditions at 4 and 10 mpf stages. (**E**) Whole mounts of viscera and visceral fat pads in the α-MSH^{-7aa} and α-MSH^{-8aa} fish groups at 12 mpf, showing increased fat pad size in the α-MSH. (**F,G**) Histological features of adipose tissue in hematoxylin and eosin (HE)-stained sagittal sections, showing (**F**) male and (**G**) female specimens. Left: HE-stained sagittal sections, showing visceral and subcutaneous adipose tissue contents in the three experimental male groups (*n* = 5/group). Right: Bar graph showing the body fat volume ratios calculated by morphometric analysis of fat on visceral and subcutaneous adipose tissue average size in each experimental group (*n* = 5 in each group). Values are the mean ± SEM. * $p < 0.05$, ** $p < 0.01$, and *** $p < 0.001$ compared with WT control groups.

3.5. Pre- and Postprandial Expression of Appetite-Related Genes in α-MSH Mutants

As several neuronal peptides are involved in regulating food intake and energy balance in zebrafish, as in mammals [42–44], appetite-related gene expression levels were investigated in whole α-MSH mutant brains using RT–qPCR. To compare the expression pattern of appetite-related genes between the WT and the *pomca* mutants, we examined several important factors, including four selected orexigenic genes: Promelanin-concentrating hormone (*pmch*), agouti-related protein 2 (*agrp2*), neuropeptide Y (*npy*), and hypothalamic hypocretin/orexin (*hcrt*) (Figure 5A). We also measured the expression levels of five selected anorexigenic genes: Brain-derived neurotrophic factor (*bdnf*), single-minded homolog 1-a (*sim1a*), corticotropin-releasing hormone b (*crhb*), thyrotropin-releasing hormone (*trh*), and prohormone convertase 2 (*pcsk2*) (Figure 5B). Although the analyses showed that the orexigenic mRNA levels were slightly downregulated in POMC mutants compared with those in WT controls at 7 dpf (Figure 5A), the anorexigenic mRNA levels were dramatically downregulated in *pomca* mutants compared with those in WT controls at 7 dpf (Figure 5B). Using whole-mount ISH analysis, significantly decreased expression patterns of *bdnf* and *sim1a* were observed in *pomca* mutants compared with WT at 5 dpf (Figure 5C). In addition, the intestinal orexigenic gene, *ghrelin* (*ghrl*), was dramatically upregulated in *pomca* mutants compared with that in WT adults (Figure 5D). Thus, functional α-MSH signaling is clearly required for transcriptional suppression to some extent of anorexigenic and orexigenic genes and interrupts the balance of hunger and satiety, which in turn possibly increases appetite under normal feeding conditions.

3.6. Administration of a Synthetic α-MSH Analog Rescued Hyperphagic Phenotypes in α-MSH Mutants

We investigated whether hindbrain ventricle injections of an α-MSH analog and MTII into α-MSH mutant larvae could reverse the hyperphagic phenotypes (Figure 6A). The feeding evaluation results after the administration of α-MSH analog (Figure 6B, left) and MTII (Figure 6C, left) were obvious in both the α-MSH^{-7aa} and α-MSH^{-8aa} specimens at 7dpf, respectively. Both α-MSH^{-7aa} and α-MSH^{-8aa} fish could be rescued by the α-MSH analog and MTII—similar to normal phenotypes—although their anorexigenic effects were stronger than in WT controls (Figure 6B,C).

To clarify whether the appetite-related gene expression levels were modulated directly by administration of the α-MSH analog in α-MSH mutant larvae, the mRNA levels of the orexigenic genes *npy*, *agrp2*, and *hcrt* were significantly lower than the mRNA levels of the anorexigenic genes *bdnf*, *sim1a*, *chrb*, and *trh* at 1.5 h after intracranial administration compared with the α-MSH mutant larvae (Figure 6D,E).

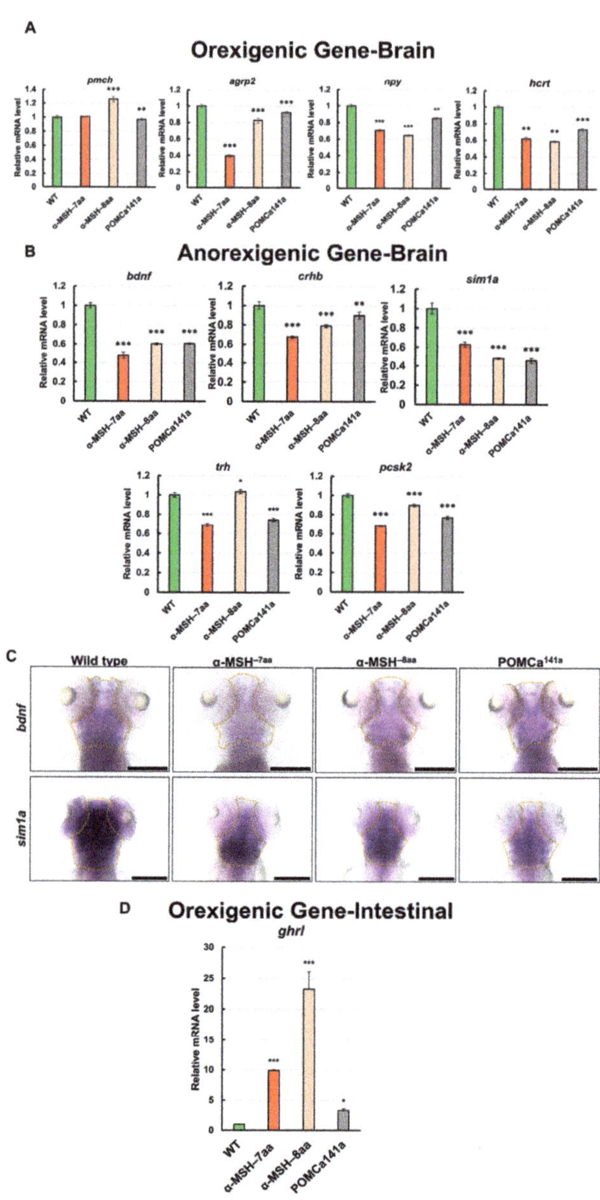

Figure 5. Effects of α-MSH on feeding regulation in the hypothalamus of zebrafish. (**A**,**B**) RT–qPCR analysis was used to measure the mRNA levels of four selected orexigenic genes, *pmch, agrp2, npy,* and *hcrt* (**A**), and five selected anorexigenic genes, *bdnf, sim1a, crhb, trh,* and *pcsk2* (**B**) in the WT and *pomca* mutants at 1.5 h after feeding. (**C**) Whole-mount ISH showing decreased expression of *bdnf* and *sim1a* in the pituitary in WT and *pomca* mutant larvae at 5 dpf. Scale bars, 200 μm. (**D**) RT–qPCR analysis was used to measure the mRNA level of the intestinal orexigenic gene, *ghrl*, in WT and *pomca* mutant adults. Values are means ± SEM. * $p < 0.05$, ** $p < 0.01$, and *** $p < 0.001$ compared with WT control groups.

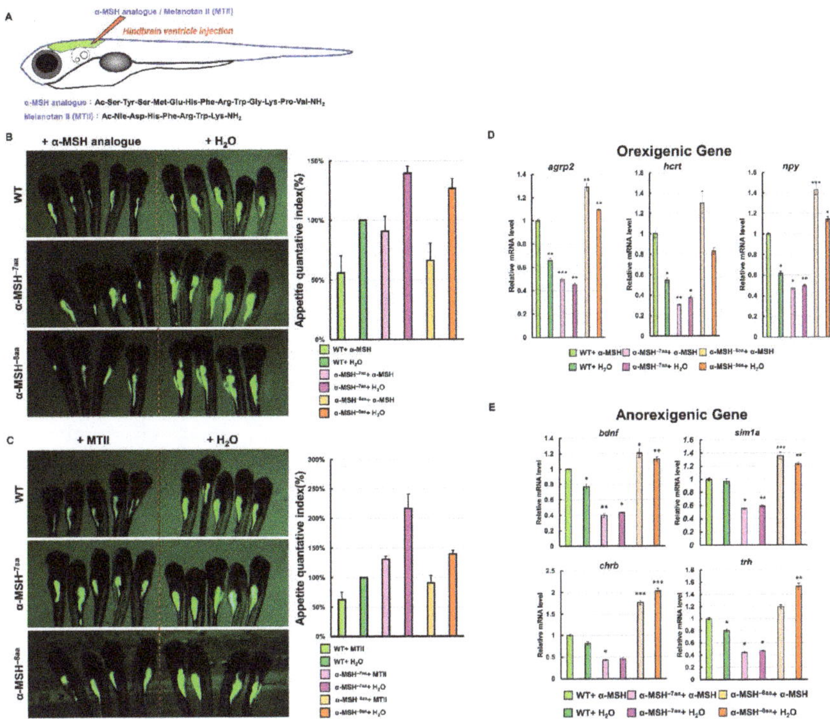

Figure 6. Effects of a synthetic α-MSH analog in rescuing hyperphagic phenotypes in α-MSH mutant larvae. (**A**) Schematic representation of hindbrain ventricle injection of an α-MSH analog/MTII into α-MSH mutant larvae. (**B**) α-MSH analog, and (**C**) MTII. Left: Administration decreased feeding volume in α-MSH mutant larvae at 7 dpf. Right: Bar graph showing the quantified appetite levels measured by morphometric analysis of fluorescent intensities in each experimental group (WT, α-MSH^{-7aa}, and α-MSH^{-8aa}) ($n = 50$/group). Values are means ± SEM. * $p < 0.05$, ** $p < 0.01$, and *** $p < 0.001$ compared with WT groups. (**D,E**) RT–qPCR analysis was used to measure the mRNA levels of three selected orexigenic genes, *agrp2*, *npy*, and *hcrt* (**D**), and four selected anorexigenic genes, *bdnf*, *sim1a*, *crhb*, and *trh* (**E**) in the WT and α-MSH mutant larvae at 1.5 h after feeding. We used relative fluorescent change and relative transcriptome expression for quantification three times ($n = 50$) independently.

4. Discussion

MSHs are well-known feeding and camouflage behavior-regulated hormones in mammals (Yaswen et al., 1999; Raffan et al., 2016). MC4R is a key gene controlling the α-MSH-mediated suppression of appetite in mammals [4,7,15,17,45]. Intracerebroventricular (icv) or intraperitoneal (ip) injection of MC4R agonists can suppress food intake in goldfish [46], zebrafish [47], spotted sea bass [48], and coho salmon [49], while transgenic zebrafish overexpressing *agrp* demonstrate adipocyte hypertrophy and increased linear growth, leading to an obese phenotype [50]. Additionally, in a zebrafish model, Mrap2a/Mrap2b can not only interact with Mc4r to regulate appetite, and growth patterns, but also help to modulate the sensitivity of Mc4r to α-MSH [47,51]. Increased food intake has also been observed in *mc4r*-deficient zebrafish [52].

The depletion of α-MSH in *Pomc*-null mice resulted in obesity, whereas the blockade of neuronal melanocortin signaling resulted in a decrease in response to centrally administered leptin [53,54]. Moreover, the central administration of α-MSH suppressed food intake and reduced BW gain in rodent models [55,56]. A deficiency of α-MSH in a *Pomc*$^{tm1/tm1}$ mouse produced exacerbated hyperphagia and obesity when feeding with a high-fat diet

(HFD) [12]. Our results—consistent with mouse models—showed significant increases in food intake in the α-MSH mutants (Figure 2B,C). Because no α-MSH peptide was produced in α-MSH mutant fish, a significant increase in food intake was found in their larval stages (Figure 2B,C). Increased BW and obesity in adult α-MSH mutants were also developed under regular feeding and overfeeding conditions (Figure 4A,C). However, a zebrafish *pomc* mutant, POMCa^{141a}, which was also α-MSH deficient, showed enhanced somatic growth without increasing food intake. This suggests that some aspects in addition to regulating appetite could be the key cause(s) for BW gain in our zebrafish model of POMCa^{141a}.

Hyperphagia alone did not promote fish growth. In medaka and zebrafish models, leptin-receptor deficiency also elicited an increased food intake pattern with a normal growth rate [57,58]. Because the energy homeostasis controlling system is highly complicated, simply increasing food intake might not explain the causal relationship with an acceleration of the growth rate or with weight gain [59–61]. In the Alamar Blue metabolic assay, we observed that the α-MSH mutant decreased energy expenditure by decreasing its metabolic rate (Figure 3D). Thus, the deletion of α-MSH in zebrafish induced hyperphagia and lower metabolism levels, which resulted in obesity. We found not only increases in food intake but also increases in the growth rate, including the standard BL, BW, and obesity, in the α-MSH mutant fish compared with the WT controls. The growth rate was determined by energy absorption and conversion efficiency, as well as the individual metabolic rate.

We selected appetite regulatory genes that have been demonstrated to take part in regulating food intake and to serve as indicators of controlling the anorexigenic and orexigenic functions and compared their expression patterns during preprandial stages between the pomc mutant fish and WT control. We observed that the anorexigenic genes (*sim1a*, *crhb*, *trh*, and *bdnf*) were significantly downregulated in the α-MSH mutant fish at 1.5 h after feeding. To further confirm the relationship between α-MSH and appetite regulation, we used intracranial administration of an α-MSH analog and MTII. During 1.5 h after administration, we discovered that fluorescence intensities were repressed significantly in the hypothalamus of both WT and α-MSH mutant fish. Moreover, the *bdnf* and *sim1a* expression levels in the α-MSH$^{-/-}$ and α-MSH analog/MTII administration groups were significantly enhanced compared with the WT and saline administration control groups. Lack of α-MSH leads to suppression of anorexigenic genes expression (*bdnf*, *sim1a*, *crhb*, and *trh*) results in a reversible imbalance between food intake and energy homeostasis in α-MSH mutant fish (Figure 7A). Most of these associated features might be regulated by α-MSH [19,45], suggesting their critical functions in controlling metabolism observed in our α-MSH mutant zebrafish (Figure 7B). In line with previous reports, most metabolic defects observed in fish [62], mammalian models [4,63,64], and humans [65–67] have been attributed to the impairment of functional MC4R signaling in various *pomc* mutants.

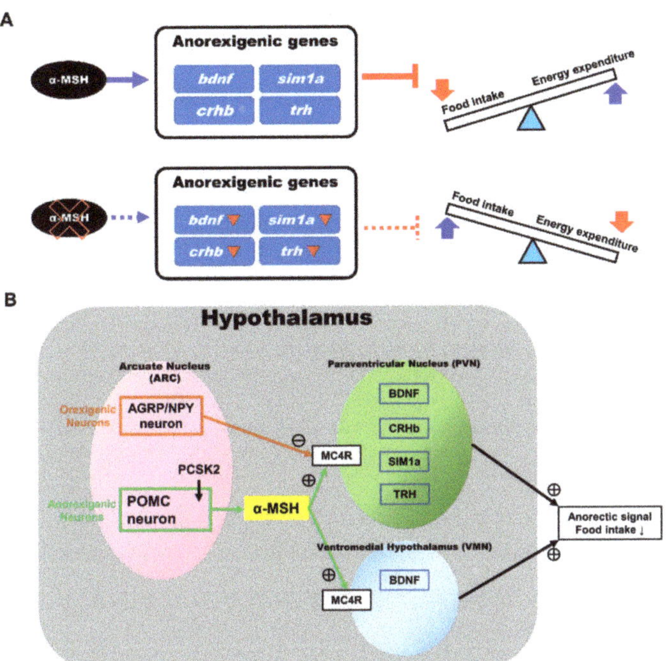

Figure 7. Anorexigenic signals in hypothalamic neural circuits promoted by α-MSH expression in zebrafish. (**A**) A proposed mechanism that α-MSH mediating food intake and energy expenditure via anorexigenic molecules manipulation which can reversed by α-MSH depletion. (**B**) Schematic presentation of anorexigenic factors (*bdnf*, *sim1a*, *crhb*, and *trh*) enhancement within the paraventricular nucleus (PVN) and the ventromedial hypothalamus (VMN) by α-MSH resulting in a dramatic outcome with the anorexigenic phenomenon.

5. Conclusions

Together with the above findings in our study, the BW gain (or obesity) and increase in body lipid contents were associated with increased appetite and decreased energy expenditure rates in this zebrafish model. In summary, we characterized a critical role for melanocortin signaling for appetite suppression. Therefore, our findings may help clarify underlying mechanisms between α-MSH signaling and obesity.

Supplementary Materials: The following are available online at https://www.mdpi.com/article/10.3390/biomedicines9080941/s1, Table S1: Primer sequences used for quantitative RT-PCR, Table S2: Primer sequences used for in situ hybridization probes.

Author Contributions: G.M.H. and Y.-W.T. designed the experiments, and Y.-W.H. wrote the manuscript. Y.-W.T. performed the majority of the experiments presented in this paper. Y.-W.H., H.-H.L., C.-Y.L. (Chi-Yu Lai), and C.-Y.L. (Chiu-Ya Lin) designed and performed the zebrafish experiments. All authors have read and agreed to the published version of the manuscript.

Funding: This study was supported by the Ministry of Science and Technology, Taiwan grants (MOST 108-2313-B-010-001-MY3) to Guor Mour Her. Chang Chung Memorial Hospital grants (CMRPG2I0071, CMRPG2I0072, and CMRPG2I0073) to Yi-Wen Tsai.

Institutional Review Board Statement: This study was carried out in animal experiments under approved guidelines (Approval ID: 1070708) by the Institutional Animal Care and Use Committee of National Yang Ming Chiao Tung University, Taipei 112, Taiwan.

Data Availability Statement: The data presented in this study are available upon request from the corresponding author.

Conflicts of Interest: The authors declare no conflict of interest.

References

1. Zhang, C.; Forlano, P.M.; Cone, R.D. AgRP and POMC neurons are hypophysiotropic and coordinately regulate multiple endocrine axes in a larval teleost. *Cell Metab.* **2012**, *15*, 256–264. [CrossRef] [PubMed]
2. Gonzalez-Nunez, V.; Gonzalez-Sarmiento, R.; Rodriguez, R.E. Identification of two proopiomelanocortin genes in zebrafish (*Danio rerio*). *Mol. Brain Res.* **2003**, *120*, 1–8. [CrossRef]
3. Wang, L.; Sui, L.; Panigrahi, S.K.; Meece, K.; Xin, Y.; Kim, J.; Gromada, J.; Doege, C.A.; Wardlaw, S.L.; Egli, D.; et al. PC1/3 Deficiency Impacts Pro-opiomelanocortin Processing in Human Embryonic Stem Cell-Derived Hypothalamic Neurons. *Stem Cell Rep.* **2017**, *8*, 264–277. [CrossRef] [PubMed]
4. Fisher, S.L.; Yagaloff, K.A.; Burn, P. Melanocortin-4 receptor: A novel signalling pathway involved in body weight regulation. *Int. J. Obes. Relat. Metab. Disord.* **1999**, *23*, 54–58. [CrossRef]
5. Kuhnen, P.; Krude, H.; Biebermann, H. Melanocortin-4 Receptor Signalling: Importance for Weight Regulation and Obesity Treatment. *Trends Mol. Med.* **2019**, *25*, 136–148. [CrossRef]
6. Yaswen, L.; Diehl, N.; Brennan, M.B.; Hochgeschwender, U. Obesity in the mouse model of pro-opiomelanocortin deficiency responds to peripheral melanocortin. *Nat. Med.* **1999**, *5*, 1066–1070. [CrossRef]
7. Huszar, D.; Lynch, C.A.; Fairchild-Huntress, V.; Dunmore, J.H.; Fang, Q.; Berkemeier, L.R.; Gu, W.; Kesterson, R.A.; Boston, B.A.; Cone, R.D.; et al. Targeted disruption of the melanocortin-4 receptor results in obesity in mice. *Cell* **1997**, *88*, 131–141. [CrossRef]
8. Krashes, M.J.; Lowell, B.B.; Garfield, A.S. Melanocortin-4 receptor-regulated energy homeostasis. *Nat. Neurosci.* **2016**, *19*, 206–219. [CrossRef]
9. Tung, Y.C.; Piper, S.J.; Yeung, D.; O'Rahilly, S.; Coll, A.P. A comparative study of the central effects of specific proopiomelanocortin (POMC)-derived melanocortin peptides on food intake and body weight in pomc null mice. *Endocrinology* **2006**, *147*, 5940–5947. [CrossRef]
10. Honda, K.; Saneyasu, T.; Hasegawa, S.; Kamisoyama, H. A comparative study of the central effects of melanocortin peptides on food intake in broiler and layer chicks. *Peptides* **2012**, *37*, 13–17. [CrossRef]
11. Lu, D.; Willard, D.; Patel, I.R.; Kadwell, S.; Overton, L.; Kost, T.; Luther, M.; Chen, W.; Woychik, R.P.; Wilkison, W.O.; et al. Agouti protein is an antagonist of the melanocyte-stimulating-hormone receptor. *Nature* **1994**, *371*, 799–802. [CrossRef] [PubMed]
12. Mountjoy, K.G.; Caron, A.; Hubbard, K.; Shome, A.; Grey, A.C.; Sun, B.; Bould, S.; Middleditch, M.; Pontre, B.; McGregor, A.; et al. Desacetyl-alpha-melanocyte stimulating hormone and alpha-melanocyte stimulating hormone are required to regulate energy balance. *Mol. Metab.* **2018**, *9*, 207–216. [CrossRef]
13. Hubbard, K.; Shome, A.; Sun, B.; Pontre, B.; McGregor, A.; Mountjoy, K.G. Chronic High-Fat Diet Exacerbates Sexually Dimorphic Pomctm1/tm1 Mouse Obesity. *Endocrinology* **2019**, *160*, 1081–1096. [CrossRef]
14. Mankowska, M.; Krzeminska, P.; Graczyk, M.; Switonski, M. Confirmation that a deletion in the POMC gene is associated with body weight of Labrador Retriever dogs. *Res. Vet. Sci.* **2017**, *112*, 116–118. [CrossRef]
15. Mayer, J.P.; Hsiung, H.M.; Flora, D.B.; Edwards, P.; Smith, D.P.; Zhang, X.Y.; Gadski, R.A.; Heiman, M.L.; Hertel, J.L.; Emmerson, P.J.; et al. Discovery of a beta-MSH-derived MC-4R selective agonist. *J. Med. Chem.* **2005**, *48*, 3095–3098. [CrossRef] [PubMed]
16. Harrold, J.A.; Williams, G. Melanocortin-4 receptors, beta-MSH and leptin: Key elements in the satiety pathway. *Peptides* **2006**, *27*, 365–371. [CrossRef]
17. Shome, A.; McGregor, A.; Cavadino, A.; Mountjoy, K.G. Central administration of beta-MSH reduces body weight in obese male Pomc(tm1/tm1) mice. *Biochim. Biophys. Acta Gen. Subj.* **2020**, *1864*, 129673. [CrossRef]
18. Appleyard, S.M.; Hayward, M.; Young, J.I.; Butler, A.A.; Cone, R.D.; Rubinstein, M.; Low, M.J. A role for the endogenous opioid beta-endorphin in energy homeostasis. *Endocrinology* **2003**, *144*, 1753–1760. [CrossRef] [PubMed]
19. Dutia, R.; Meece, K.; Dighe, S.; Kim, A.J.; Wardlaw, S.L. β-Endorphin antagonizes the effects of alpha-MSH on food intake and body weight. *Endocrinology* **2012**, *153*, 4246–4255. [CrossRef]
20. Raffan, E.; Dennis, R.J.; O'Donovan, C.J.; Becker, J.M.; Scott, R.A.; Smith, S.P.; Withers, D.J.; Wood, C.J.; Conci, E.; Clements, D.N.; et al. A Deletion in the Canine POMC Gene Is Associated with Weight and Appetite in Obesity-Prone Labrador Retriever Dogs. *Cell Metab.* **2016**, *23*, 893–900. [CrossRef] [PubMed]
21. Dubern, B.; Lubrano-Berthelier, C.; Mencarelli, M.; Ersoy, B.; Frelut, M.L.; Bougle, D.; Costes, B.; Simon, C.; Tounian, P.; Vaisse, C.; et al. Mutational analysis of the pro-opiomelanocortin gene in French obese children led to the identification of a novel deleterious heterozygous mutation located in the alpha-melanocyte stimulating hormone domain. *Pediatr. Res.* **2008**, *63*, 211–216. [CrossRef] [PubMed]
22. Samuels, M.E.; Gallo-Payet, N.; Pinard, S.; Hasselmann, C.; Magne, F.; Patry, L.; Chouinard, L.; Schwartzentruber, J.; Rene, P.; Sawyer, N.; et al. Bioactive ACTH causing glucocorticoid deficiency. *J. Clin. Endocrinol. Metab.* **2013**, *98*, 736–742. [CrossRef] [PubMed]
23. Santoro, N.; Cirillo, G.; Xiang, Z.; Tanas, R.; Greggio, N.; Morino, G.; Iughetti, L.; Vottero, A.; Salvatoni, A.; Di Pietro, M.; et al. Prevalence of pathogenetic MC4R mutations in Italian children with early onset obesity, tall stature and familial history of obesity. *BMC Med. Genet.* **2009**, *10*, 25. [CrossRef] [PubMed]
24. Farooqi, I.S.; Yeo, G.S.; Keogh, J.M.; Aminian, S.; Jebb, S.A.; Butler, G.; Cheetham, T.; O'Rahilly, S. Dominant and recessive inheritance of morbid obesity associated with melanocortin 4 receptor deficiency. *J. Clin. Investig.* **2000**, *106*, 271–279. [CrossRef]

5. Yeo, G.S.; Farooqi, I.S.; Challis, B.G.; Jackson, R.S.; O'Rahilly, S. The role of melanocortin signalling in the control of body weight: Evidence from human and murine genetic models. *QJM Int. J. Med.* **2000**, *93*, 7–14. [CrossRef] [PubMed]
6. Krude, H.; Gruters, A. Implications of proopiomelanocortin (POMC) mutations in humans: The POMC deficiency syndrome. *Trends Endocrinol. Metab.* **2000**, *11*, 15–22. [CrossRef]
7. Clark, A.J.; Weber, A. Adrenocorticotropin insensitivity syndromes. *Endocr. Rev.* **1998**, *19*, 828–843. [CrossRef] [PubMed]
8. Elias, L.L.; Huebner, A.; Metherell, L.A.; Canas, A.; Warne, G.L.; Bitti, M.L.; Cianfarani, S.; Clayton, P.E.; Savage, M.O.; Clark, A.J. Tall stature in familial glucocorticoid deficiency. *Clin. Endocrinol.* **2000**, *53*, 423–430. [CrossRef]
9. Imamine, H.; Mizuno, H.; Sugiyama, Y.; Ohro, Y.; Sugiura, T.; Togari, H. Possible relationship between elevated plasma ACTH and tall stature in familial glucocorticoid deficiency. *Tohoku J. Exp. Med.* **2005**, *205*, 123–131. [CrossRef]
10. Hansen, I.A.; To, T.T.; Wortmann, S.; Burmester, T.; Winkler, C.; Meyer, S.R.; Neuner, C.; Fassnacht, M.; Allolio, B. The proopiomelanocortin gene of the zebrafish (*Danio rerio*). *Biochem. Biophys. Res. Commun.* **2003**, *303*, 1121–1128. [CrossRef]
11. To, T.T.; Hahner, S.; Nica, G.; Rohr, K.B.; Hammerschmidt, M.; Winkler, C.; Allolio, B. Pituitary-interrenal interaction in zebrafish interrenal organ development. *Mol. Endocrinol.* **2007**, *21*, 472–485. [CrossRef] [PubMed]
12. Wagle, M.; Mathur, P.; Guo, S. Corticotropin-releasing factor critical for zebrafish camouflage behavior is regulated by light and sensitive to ethanol. *J. Neurosci.* **2011**, *31*, 214–224. [CrossRef] [PubMed]
13. Dang, Y.; Wang, F.E.; Liu, C. Real-time PCR array to study the effects of chemicals on the growth hormone/insulin-like growth factors (GH/IGFs) axis of zebrafish embryos/larvae. *Chemosphere* **2018**, *207*, 365–376. [CrossRef]
14. Jowett, T. Double in situ hybridization techniques in zebrafish. *Methods* **2001**, *23*, 345–358. [CrossRef] [PubMed]
15. Shimada, Y.; Hirano, M.; Nishimura, Y.; Tanaka, T. A high-throughput fluorescence-based assay system for appetite-regulating gene and drug screening. *PLoS ONE* **2012**, *7*, e52549. [CrossRef]
16. Lai, C.Y.; Yeh, K.Y.; Lin, C.Y.; Hsieh, Y.W.; Lai, H.H.; Chen, J.R.; Hsu, C.C.; Her, G.M. MicroRNA-21 Plays Multiple Oncometabolic Roles in the Process of NAFLD-Related Hepatocellular Carcinoma via PI3K/AKT, TGF-beta, and STAT3 Signaling. *Cancers* **2021**, *13*, 940. [CrossRef] [PubMed]
17. Lai, C.Y.; Lin, C.Y.; Hsu, C.C.; Yeh, K.Y.; Her, G.M. Liver-directed microRNA-7a depletion induces nonalcoholic fatty liver disease by stabilizing YY1-mediated lipogenic pathways in zebrafish. *Biochim. Biophys. Acta Mol. Cell Biol. Lipids* **2018**, *1863*, 844–856. [CrossRef]
18. Renquist, B.J.; Zhang, C.; Williams, S.Y.; Cone, R.D. Development of an assay for high-throughput energy expenditure monitoring in the zebrafish. *Zebrafish* **2013**, *10*, 343–352. [CrossRef]
19. Montalbano, G.; Mania, M.; Guerrera, M.C.; Abbate, F.; Laura, R.; Navarra, M.; Vega, J.A.; Ciriaco, E.; Germana, A. Morphological differences in adipose tissue and changes in BDNF/Trkb expression in brain and gut of a diet induced obese zebrafish model. *Ann. Anat.* **2016**, *204*, 36–44. [CrossRef] [PubMed]
20. Reinecke, M.; Bjornsson, B.T.; Dickhoff, W.W.; McCormick, S.D.; Navarro, I.; Power, D.M.; Gutierrez, J. Growth hormone and insulin-like growth factors in fish: Where we are and where to go. *Gen. Comp. Endocrinol.* **2005**, *142*, 20–24. [CrossRef] [PubMed]
21. Reinecke, M. Influences of the environment on the endocrine and paracrine fish growth hormone-insulin-like growth factor-I system. *J. Fish Biol.* **2010**, *76*, 1233–1254. [CrossRef]
22. Yokobori, E.; Azuma, M.; Nishiguchi, R.; Kang, K.S.; Kamijo, M.; Uchiyama, M.; Matsuda, K. Neuropeptide Y stimulates food intake in the Zebrafish, *Danio rerio*. *J. Neuroendocrinol.* **2012**, *24*, 766–773. [CrossRef]
23. Sundarrajan, L.; Unniappan, S. Small interfering RNA mediated knockdown of irisin suppresses food intake and modulates appetite regulatory peptides in zebrafish. *Gen. Comp. Endocrinol.* **2017**, *252*, 200–208. [CrossRef]
24. Zheng, B.; Li, S.; Liu, Y.; Li, Y.; Chen, H.; Tang, H.; Liu, X.; Lin, H.; Zhang, Y.; Cheng, C.H.K. Spexin Suppress Food Intake in Zebrafish: Evidence from Gene Knockout Study. *Sci. Rep.* **2017**, *7*, 14643. [CrossRef]
25. Kirwan, P.; Kay, R.G.; Brouwers, B.; Herranz-Perez, V.; Jura, M.; Larraufie, P.; Jerber, J.; Pembroke, J.; Bartels, T.; White, A.; et al. Quantitative mass spectrometry for human melanocortin peptides in vitro and in vivo suggests prominent roles for beta-MSH and desacetyl alpha-MSH in energy homeostasis. *Mol. Metab.* **2018**, *17*, 82–97. [CrossRef] [PubMed]
26. Cerda-Reverter, J.M.; Schioth, H.B.; Peter, R.E. The central melanocortin system regulates food intake in goldfish. *Regul. Pept.* **2003**, *115*, 101–113. [CrossRef]
27. Agulleiro, M.J.; Cortes, R.; Fernandez-Duran, B.; Navarro, S.; Guillot, R.; Meimaridou, E.; Clark, A.J.; Cerda-Reverter, J.M. Melanocortin 4 receptor becomes an ACTH receptor by coexpression of melanocortin receptor accessory protein 2. *Mol. Endocrinol.* **2013**, *27*, 1934–1945. [CrossRef] [PubMed]
28. Zhang, K.Q.; Hou, Z.S.; Wen, H.S.; Li, Y.; Qi, X.; Li, W.J.; Tao, Y.X. Melanocortin-4 Receptor in Spotted Sea Bass, Lateolabrax maculatus: Cloning, Tissue Distribution, Physiology, and Pharmacology. *Front. Endocrinol.* **2019**, *10*, 705. [CrossRef] [PubMed]
29. White, S.L.; Volkoff, H.; Devlin, R.H. Regulation of feeding behavior and food intake by appetite-regulating peptides in wild-type and growth hormone-transgenic coho salmon. *Horm. Behav.* **2016**, *84*, 18–28. [CrossRef]
30. Song, Y.; Cone, R.D. Creation of a genetic model of obesity in a teleost. *Fed. Am. Soc. Exp. Biol. J.* **2007**, *21*, 2042–2049. [CrossRef]
31. Sebag, J.A.; Zhang, C.; Hinkle, P.M.; Bradshaw, A.M.; Cone, R.D. Developmental control of the melanocortin-4 receptor by MRAP2 proteins in zebrafish. *Science* **2013**, *341*, 278–281. [CrossRef] [PubMed]
32. Fei, F.; Sun, S.Y.; Yao, Y.X.; Wang, X. Generation and phenotype analysis of zebrafish mutations of obesity-related genes lepr and mc4r. *Sheng Li Xue Bao* **2017**, *69*, 61–69. [PubMed]

53. Schwartz, M.W.; Woods, S.C.; Porte, D., Jr.; Seeley, R.J.; Baskin, D.G. Central nervous system control of food intake. *Nature* **2000**, *404*, 661–671. [CrossRef]

54. Morton, G.J.; Cummings, D.E.; Baskin, D.G.; Barsh, G.S.; Schwartz, M.W. Central nervous system control of food intake and body weight. *Nature* **2006**, *443*, 289–295. [CrossRef] [PubMed]

55. McMinn, J.E.; Wilkinson, C.W.; Havel, P.J.; Woods, S.C.; Schwartz, M.W. Effect of intracerebroventricular alpha-MSH on food intake, adiposity, c-Fos induction, and neuropeptide expression. *Am. J. Physiol. Regul. Integr. Comp. Physiol.* **2000**, *279*, R695–R703. [CrossRef]

56. Eerola, K.; Nordlund, W.; Virtanen, S.; Dickens, A.M.; Mattila, M.; Ruohonen, S.T.; Chua, S.C., Jr.; Wardlaw, S.L.; Savontaus, M.; Savontaus, E. Lentivirus-mediated alpha-melanocyte-stimulating hormone overexpression in the hypothalamus decreases diet induced obesity in mice. *J. Neuroendocrinol.* **2013**, *25*, 1298–1307. [CrossRef]

57. Chisada, S.; Kurokawa, T.; Murashita, K.; Ronnestad, I.; Taniguchi, Y.; Toyoda, A.; Sakaki, Y.; Takeda, S.; Yoshiura, Y. Leptin receptor-deficient (knockout) medaka, Oryzias latipes, show chronical up-regulated levels of orexigenic neuropeptides, elevated food intake and stage specific effects on growth and fat allocation. *Gen. Comp. Endocrinol.* **2014**, *195*, 9–20. [CrossRef]

58. Ahi, E.P.; Brunel, M.; Tsakoumis, E.; Schmitz, M. Transcriptional study of appetite regulating genes in the brain of zebrafish (*Danio rerio*) with impaired leptin signalling. *Sci. Rep.* **2019**, *9*, 20166. [CrossRef]

59. Hollis, J.H. The effect of mastication on food intake, satiety and body weight. *Physiol. Behav.* **2018**, *193*, 242–245. [CrossRef]

60. Luo, Y.; Zhang, X.; Tsauo, J.; Jung, H.Y.; Song, H.Y.; Zhao, H.; Li, J.; Gong, T.; Song, P.; Li, X. Intragastric satiety-inducing device reduces food intake and suppresses body weight gain in a rodent model. *Surg. Endosc.* **2020**, *35*, 1052–1057. [CrossRef]

61. Santiago-Garcia, P.A.; Lopez, M.G. Agavins from Agave angustifolia and Agave potatorum affect food intake, body weight gain and satiety-related hormones (GLP-1 and ghrelin) in mice. *Food Funct.* **2014**, *5*, 3311–3319. [CrossRef]

62. Liu, T.; Deng, Y.; Zhang, Z.; Cao, B.; Li, J.; Sun, C.; Hu, Z.; Zhang, J.; Li, J.; Wang, Y. Melanocortin Receptor 4 (MC4R) Signaling System in Nile Tilapia. *Int. J. Mol. Sci.* **2020**, *21*, 7036. [CrossRef]

63. Mastinu, A.; Premoli, M.; Maccarinelli, G.; Grilli, M.; Memo, M.; Bonini, S.A. Melanocortin 4 receptor stimulation improves social deficits in mice through oxytocin pathway. *Neuropharmacology* **2018**, *133*, 366–374. [CrossRef]

64. Cui, H.; Sohn, J.W.; Gautron, L.; Funahashi, H.; Williams, K.W.; Elmquist, J.K.; Lutter, M. Neuroanatomy of melanocortin-4 receptor pathway in the lateral hypothalamic area. *J. Comp. Neurol.* **2012**, *520*, 4168–4183. [CrossRef] [PubMed]

65. Lotta, L.A.; Mokrosinski, J.; de Oliveira, E.M.; Li, C.; Sharp, S.J.; Luan, J.; Brouwers, B.; Ayinampudi, V.; Bowker, N.; Kerrison, N.; et al. Human Gain-of-Function MC4R Variants Show Signaling Bias and Protect against Obesity. *Cell* **2019**, *177*, 597–607. [CrossRef] [PubMed]

66. Ayers, K.L.; Glicksberg, B.S.; Garfield, A.S.; Longerich, S.; White, J.A.; Yang, P.; Du, L.; Chittenden, T.W.; Gulcher, J.R.; Roy, S.; et al. Melanocortin 4 Receptor Pathway Dysfunction in Obesity: Patient Stratification Aimed at MC4R Agonist Treatment. *J. Clin. Endocrinol. Metab.* **2018**, *103*, 2601–2612. [CrossRef] [PubMed]

67. Eneli, I.; Xu, J.; Webster, M.; McCagg, A.; Van Der Ploeg, L.; Garfield, A.S.; Estrada, E. Tracing the effect of the melanocortin 4 receptor pathway in obesity: Study design and methodology of the TEMPO registry. *Appl. Clin. Genet.* **2019**, *12*, 87–93. [CrossRef] [PubMed]

 *biomedicines*

Article

Zebrafish Blunt-Force TBI Induces Heterogenous Injury Pathologies That Mimic Human TBI and Responds with Sonic Hedgehog-Dependent Cell Proliferation across the Neuroaxis

James Hentig [1,2,3] Kaylee Cloghessy [1,2,3], Manuela Lahne [1,2,3], Yoo Jin Jung [1,2,3], Rebecca A. Petersen [4], Ann C. Morris [4] and David R. Hyde [1,2,3,*]

1 Department of Biological Sciences, University of Notre Dame, South Bend, IN 46556, USA; jhentig@nd.edu (J.H.); kcloghes@nd.edu (K.C.); mlahne@nd.edu (M.L.); yjung1@nd.edu (Y.J.J.)
2 Center for Zebrafish Research, University of Notre Dame, South Bend, IN 46556, USA
3 Center for Stem Cells and Regenerative Medicine, Galvin Life Science Center, University of Notre Dame, South Bend, IN 46556, USA
4 Department of Biology, University of Kentucky, Lexington, KY 40506, USA; Rebecca.petersen@uky.edu (R.A.P.); ann.morris@uky.edu (A.C.M.)
* Correspondence: dhyde@nd.edu; Tel.: +1-574-631-8054

Citation: Hentig, J.; Cloghessy, K.; Lahne, M.; Jung, Y.J.; Petersen, R.A.; Morris, A.C.; Hyde, D.R. Zebrafish Blunt-Force TBI Induces Heterogenous Injury Pathologies That Mimic Human TBI and Responds with Sonic Hedgehog-Dependent Cell Proliferation across the Neuroaxis. *Biomedicines* **2021**, *9*, 861. https://doi.org/10.3390/biomedicines9080861

Academic Editors: James A. Marrs and Swapnalee Sarmah

Received: 16 June 2021
Accepted: 16 July 2021
Published: 22 July 2021

Publisher's Note: MDPI stays neutral with regard to jurisdictional claims in published maps and institutional affiliations.

Abstract: Blunt-force traumatic brain injury (TBI) affects an increasing number of people worldwide as the range of injury severity and heterogeneity of injury pathologies have been recognized. Most current damage models utilize non-regenerative organisms, less common TBI mechanisms (penetrating, chemical, blast), and are limited in scalability of injury severity. We describe a scalable blunt-force TBI model that exhibits a wide range of human clinical pathologies and allows for the study of both injury pathology/progression and mechanisms of regenerative recovery. We modified the Marmarou weight drop model for adult zebrafish, which delivers a scalable injury spanning mild, moderate, and severe phenotypes. Following injury, zebrafish display a wide range of severity-dependent, injury-induced pathologies, including seizures, blood–brain barrier disruption, neuroinflammation, edema, vascular injury, decreased recovery rate, neuronal cell death, sensorimotor difficulties, and cognitive deficits. Injury-induced pathologies rapidly dissipate 4–7 days post-injury as robust cell proliferation is observed across the neuroaxis. In the cerebellum, proliferating *nestin*:GFP-positive cells originated from the cerebellar crest by 60 h post-injury, which then infiltrated into the granule cell layer and differentiated into neurons. Shh pathway genes increased in expression shortly following injury. Injection of the Shh agonist purmorphamine in undamaged fish induced a significant proliferative response, while the proliferative response was inhibited in injured fish treated with cyclopamine, a Shh antagonist. Collectively, these data demonstrate that a scalable blunt-force TBI to adult zebrafish results in many pathologies similar to human TBI, followed by recovery, and neuronal regeneration in a Shh-dependent manner.

Keywords: traumatic brain injury; blunt-force TBI; regeneration; zebrafish; cerebellum; proliferation; learning; memory

1. Introduction

Traumatic brain injuries (TBIs) affect every age, ethnicity, and social class in society, with nearly 60 million people affected worldwide annually [1]. However, this number is likely an underestimation, as many people exposed to mild TBIs are often undiagnosed [2,3]. TBIs are categorized by the level of severity (mild, moderate, and severe), which is based on several metrics, including altered state of consciousness, hematoma and edema formation, and structural integrity. TBIs result from either a penetrating or blunt-force trauma. A penetrating trauma results from an object impaling the skull (stab or gunshot wounds) to cause focal brain damage. In contrast, blunt-force trauma, which accounts for over 90% of all TBIs [1,3], arises when the head is struck (motor vehicle accident, fall, sport incident, or combat) to produce a gradient of damage that is relative to the impact force and diffused across the entire brain, often inducing neurodegeneration [4–6].

Rodents serve as a powerful TBI model of human injury [7–9], with the Marmarou weight drop being a common method to generate a reproducible and scalable TBI [10,11]. This method allows for examining the pathophysiology of blunt-force trauma and the resultant sequelae and changes in cognitive function [10,12,13]. Based on these studies, the TBI-induced phenotypic deficits in rodents are similar to those observed in humans [8]. While mammalian brains possess resident neuronal progenitor cells, they are limited in number and migration potential [14–16]. Therefore, they are unable to sufficiently regenerate lost/damaged neurons and restore cognitive function.

Zebrafish, in contrast, possess region-specific populations of resident quiescent stem cells that are induced by injury to regenerate damaged and lost neurons across the nervous system: olfactory system, retina, spinal cord, and brain [17–23]. The regenerative capacity of the zebrafish brain has been studied following either penetrating TBI, due to either a stab wound or partial excision, by chemical toxins, or ultrasonic or pressure waves [24–29]. However, the damage in all these models is primarily focal and/or predominately categorized as severe. Thus, they do not adequately address the heterogeneity of blunt-force TBI severities or phenotypes [30,31]. A zebrafish blunt-force TBI model was recently described to examine changes in gene expression across the brain, but the investigation was limited to a mild injury and failed to address the extent of damage across the brain, the breadth of phenotypic pathologies produced, and the regenerative response [32].

In this study, we describe a scalable blunt-force TBI zebrafish model that recapitulates many of the features of human blunt-force TBI, examines recovery of cognitive function, and describes the extent of neuronal regeneration. We adapted the Marmarou weight drop model [10] to yield a reproducible TBI that is scalable (mild, moderate, and severe). Zebrafish exposed to this modified Marmarou weight drop (MMWD) experienced decreased recovery rate from anesthesia (loss of consciousness), seizures, subdural/intracerebral hematomas, blood–brain barrier disruption, neuroinflammation, cerebral edema, sensorimotor deficits, and learning/memory impairments, which recapitulate key diagnostic pathophysiological features of human TBI. Additionally, we report that the TBI results in widespread cell death across the brain, followed by a proliferative cell response in a severity-dependent manner. Within the cerebellum, injury induced progenitor cells proliferated and migrated over several days, ultimately differentiating into neurons. Upregulation of the Shh ligands, Shha and Shhb, was revealed by qRT-PCR, and we demonstrate that disrupting Shh signaling led to reduced cerebellar injury-induced cell proliferation and reduced production of neurons, while Shh activation induced increased proliferation in undamaged fish. Thus, our scalable zebrafish TBI model may be useful to study the effects of blunt-force trauma, as well as identify key pharmacological and genetic therapeutic targets that regulate injury-induced neuronal regeneration, cognitive recovery, and neuroprotection in an adult vertebrate.

2. Materials and Methods

2.1. Fish Lines and Maintenance

Adult wild-type AB, *Tg[nestin:GFP]^{tud100}* [19], *Tg[gfap:EGFP]^{nt11}* [33] transgenic lines, and *albino^{b4}* [34] and *roy^{a9};mitfa^{w2}* (referred to as *casper*) [35] mutant lines of zebrafish (*Danio rerio*) were maintained in the Center for Zebrafish Research at the University of Notre Dame Freimann Life Sciences Center. The study used approximately equal numbers of male and female adult zebrafish, 6–12 months old, 3 to 5 cm in length. All experimental protocols in this study were approved by the University of Notre Dame Animal Care and Use Committee protocol #18-03-4558 and adhered to the National Institutes of Health guide for the care and use of Laboratory animals (NIH Publications No. 8023, revised 1978).

2.2. TBI Induction via Modified Marmarou Weight Drop

Prior to inducing a blunt-force traumatic brain injury by a modified Marmarou weight drop [10,36], zebrafish were anesthetized in 1:1000 2-phenoxyethanol (2-PE, Sigma-Aldrich, St. Louis, MO, USA) until unresponsive to tail pinch. Anesthetized fish were secured onto a clay mold that stabilized the body and exposed the zebrafish head. To reduce cranial fracture and diffuse the impact, a 22-gauge steel disk was placed onto the skull over the analogous lambda and bregma cranial sutures that are centered over the intersection of the midbrain (mesencephalon and diencephalon; optic tectal lobes) and the hindbrain (rhombencephalon; cerebellum, Figure 1A). Either a 1.5 g or 3.3 g ball bearing weight was dropped down a shaft of either 7.6 or 12.7 cm length, which was placed approximately 1.5 cm above the skull of the fish to produce the desired force (Figure 1A). To induce mild traumatic brain injury (miTBI), a 1.5 g ball was dropped from 9.1 cm (v = 1.34 m/s) producing an energy of 1.33 mJ and an impact force of 1.47 N. A moderate traumatic brain injury (moTBI) was produced by dropping a 1.5 g ball from a height of 14.2 cm (v = 1.91 m/s) resulting in an energy of 2.08 mJ and an impact force of 2.45 N. To induce a severe traumatic brain injury (sTBI), a 3.3 g ball was dropped from 9.1 cm (v = 1.34 m/s) with an energy of 2.94 mJ and an impact force of 3.23 N. Following the TBI, the fish were placed into an aerated tank to recover, monitored for 1 h for seizure activity, and subsequently returned to Freimann Life Sciences Center until further investigation.

2.3. Mortality and Early/Latent Seizure

Zebrafish were assessed for survival and post-traumatic seizures within the first hour following blunt-force injury (miTBI, moTBI, sTBI), and at 12 h increments for 30 min at 1, 2, 3, 7, 14, and 28 dpi. Fish were scored for tonic-clonic seizure metrics defined in the zebrafish behavior catalog [37]: ataxia (ZBC 1.9), bending (ZBC 1.16), circling (ZBC 1.32), and corkscrew swimming (ZBC 1.37). At 1 h post-injury (hpi), fish that were unresponsive to tail pinch and displayed no operculum movement for 1 min were considered dead. Mortality and early seizure rates were calculated as an average of at least eight independent experiments (total of 200 fish).

2.4. Recovery Rate

Fish were anesthetized and subjected to either no, mild, moderate, or severe blunt-force injury. After 30 s, each fish was returned to a recovery tank and assessed for normal swimming behavior. Recovery rate was calculated as the time interval from entry into the recovery tank until the fish exhibited a complete lack of akinesia (ZBC 1.4), ataxia (ZBC 1.9), and motor incoordination (ZBC 1.99) as defined by Kalueff et al. [37]. The recovery rate of 15 individual fish from three independent trials was averaged. Comparisons were made using a One-way ANOVA followed by a Tukey post-hoc test.

2.5. Fluid Content Measurement

The extent of edema was measured using a modified Hoshi protocol [38]. Whole brains (undamaged, miTBI, moTBI, sTBI, n = 3, N = 3, 9 fish total) were extracted from control and injured fish at 1, 3, 5, 7, 14, 28 dpi, weighed wet in weigh boats, and placed in an oven at 60 °C for 12 h. Following drying, each brain was weighed again, and the percent fluid was calculated using the following formula:

$$\% \text{ Fluid} = \frac{\text{wet weight} - \text{dry weight}}{\text{wet weight}} \times 100\% \tag{1}$$

Comparisons were made using a Two-way ANOVA followed by a Tukey post-hoc test.

2.6. Tissue Processing

Undamaged control and blunt-force damaged fish were euthanized by exposure to 1:500 2-PE and brains were removed as described previously [36]. Briefly, fish were pinned dorsal side up on a clay dish, eyes were enucleated, and forceps were inserted behind the lambda suture, removing the right parietal calvarial bone along the sagittal suture. The remaining calvarial bones were removed and the olfactory nerves from the rosette to the olfactory bulbs were bluntly severed. The spinal cord was transected, and the brain was lifted by the caudal end of the spinal cord. Whole brains were extracted and fixed in 9:1 ethanolic formaldehyde (100% ethanol/37% formaldehyde) for 24 h at 4 °C. Brains were rehydrated in a 75%, 50%, and 25% ethanol series for 5 min each at room temperature, transferred into 5% sucrose/PBS for 1.5 h at room temperature, and cryoprotected in 30% sucrose/PBS overnight at 4 °C. Brains were immersed in two parts tissue freezing medium (TFM; VWR International, Radnor, PA, USA) to one part 30% sucrose/PBS for 24 h at 4 °C and finally 100% TFM for 24 h at 4 °C. Brains were mounted in TFM and stored at −80 °C until cryosectioned. Frozen serial coronal sections of 16 μm thickness were generated for immunohistochemistry/EdU labeling.

2.7. Vascular Injury

Undamaged and blunt-force injured (miTBI, moTBI, sTBI) roy^{a9};$mitfa^{w2}$ (casper) fish were qualitatively assessed at 4 hpi for vascular injury by the presence or absence of subdural pooling. Hematoma formation and resolution was qualitatively assessed in $albino^{b4}$ sTBI fish at 6 and 12 hpi, as well as 1, 1.5, 2, 2.5, and 3 dpi. Brains were prepared as described in the Immunohistochemistry section. Frozen serial coronal sections of 16 μm thickness from rostral tip of the olfactory bulb to caudal aspect of the cerebellum were generated, and intracerebral hematomas were assessed by the presence or absence of blood within serial sections during tissue collection and processing.

2.8. Sensorimotor Assay

Undamaged and sTBI fish were assessed at 1, 2, 6, and 12 hpi, as well as 1, 2, 3, 7, 14, and 28 dpi for sensorimotor activity, in regard to swimming orientation and response to a pain stimulus. At appropriate timepoints, individual fish were placed into a 7.5 × 15 cm tank with a water depth of 5 cm and given a 10 min acclimation period. Fish were assessed for swim orientation as defined in the zebrafish behavior catalog [37] (ZBC, 1.83, 1.164, 1.175). Using a 30 g needle, the lateral line was poked, and fish were assessed and scored for: pain response (ZBC 1.104), escape behaviors (ZBC 1.5, 1.52), and avoidance behavior (ZBC 1.12). Each fish was scored for each of the above tests as either responding (score of 1) or not responding (score of 0). The scores for each fish were then summed, such that a fish that exhibited normal swimming orientation and pain response scored a total of 4. A total of 10 fish were analyzed prior to, and following, injury and statistically compared using a repeated measures Kruskal-Wallis test followed by a Dunnett's multiple comparison post-hoc test.

2.9. Locomotion

Undamaged control and blunt-force injured fish (miTIB, moTBI, sTBI, n = 5, N = 3, 15 fish total) were observed at 4 hpi, 1, 2, 3, 4, 7, 14, and 28 dpi. Five fish of the same treatment group were acclimated in a circular observation tank that measured 15 cm in diameter with a water level depth of 10 cm. A digital video recorder, capturing images at a rate of 60 frames per second (fps), was placed above the testing area to monitor the swimming profile of each fish, which was used to quantify the distance that each fish traversed the tank over a 30 s period. Locomotion velocity for each group was calculated (v = swimming distance (m)/30 s) as the average of 15 fish from each treatment group. Comparisons were made using a repeated measures Two-way ANOVA followed by a Dunnett's multiple comparison post-hoc test.

2.10. Shuttle Box Assay Testing Apparatus

A shuttle box assay described by Truong et al. [39] was modified to assess associative learning and memory [40]. The testing apparatus was a modified electrophoresis gel box with a width of 19 cm, length of 30 cm, height of 7.5 cm, and 45-degree ramps on each side leading to an elevated center platform that measured 19 cm $\times$ 15 cm, and the water level was set at 5 cm. A standard gel electrophoresis power supply was joined to each tungsten wire to administer a 20 V, 1 mA electrical shock.

2.11. Learning

Learning was assessed in the shuttle box assay as previously described [40]. Briefly, undamaged control and blunt-force damaged fish (miTBI, moTBI, sTBI, n = 4, N = 3, 12 fish total) were individually placed into the shuttle box and examined at either 1, 2, 3, 4, 5, 6, 7, 14, or 28 dpi, with no fish being tested twice. After allowing the fish to acclimate in a dark and quiet room for 15 min, the fish was exposed to a red light visual stimulus that was placed against the plexiglass tank that contained the fish. The visual stimulus was applied alone for 15 s. If the fish swam to the other stall in response to the visual stimulus, the red light was turned off once it passed the half-way mark of the tank. However, if the fish failed to swim to the other stall, the red light was kept on and a pulsating electric current (20 V, 1 mA) was simultaneously applied until either the fish swam to the other stall or for 15 s (10 electrical shocks/15 s), whichever came first. The presentation of a light stimulus/electrical shock was repeated with 30 s rest intervals. Learning was defined as the fish successfully swimming across the tank within 15 s of light presentation on 5 consecutive trials. The number of trials that each fish required to learn were determined for each treatment group and averaged for each experiment. A Two-way ANOVA and Tukey's post-hoc test was performed to statistically compare the undamaged control fish and the different damage groups.

2.12. Immediate and Delayed Recall

Recall was assessed in the shuttle box assay as previously described [40]. Briefly, undamaged fish were individually placed into the shuttle box testing apparatus and tested as described under Learning. Fish were trained by 25 repetitions of exposure to red light and if they failed to swim to the other tank in response to the red light stimulus they were electrically shocked. Following this training period, fish were given 15 min to rest before they were tested 25 times by exposing the fish only to the red light stimulus (testing period 1). Each test involved applying the visual stimulus alone for 15 s. If the fish swam to the other stall within the initial 15 s, the red light was turned off after passing the half-way point of the tank and this was scored as a successful trial. However, if the fish failed to swim to the other stall, the pulsating electric current was simultaneously applied either until the fish swam to the other stall or for 15 s (10 shocks/15 s). The number of successful tests, defined as the fish crossing the tank without the electric shock, were counted to generate the initial testing baseline. Once all experimental fish were tested, they were randomly selected for either the undamaged control group or were administered a blunt-force injury (miTBI, moTBI, or sTBI). To assess immediate recall, the control group and TBI groups were retested either 4 h after the initial testing period or 4 hpi, respectively. All groups were subjected to a second testing period consisting of 25 iterations and scored for the number of successful tests. To test delayed recall, the undamaged fish were returned to Freimann Life Sciences Center after the initial testing period for four days. Fish were then randomly selected for either the undamaged control group or were administered a blunt-force injury (miTBI, moTBI, sTBI) and allowed to recover for 4 h. The fish were then subjected to a second testing period of 25 iterations and scored for the number of successful trials as described above. For both immediate and delayed recall, the percent difference in the number of successful secondary trials relative to the number of initial successful trials was calculated and averaged for each group (n = 3, N = 3, 9 fish total). Undamaged control and the different damaged groups were statistically assessed using a Two-way ANOVA and Tukey's post-hoc test.

2.13. Startle Response Habituation

Undamaged and sTBI fish (n = 3, N = 3, 9 fish total) were assayed for the startle response at 1, 4, and 7 dpi using a modified protocol from Chanin et al. [41]. Individual fish were placed in a testing tank (31 cm long × 19 cm wide × 15 cm high with water level depth of 10 cm) and allowed to acclimate for 15 min. A digital video recorder, capturing images at a rate of 60 fps, was placed above the testing area to monitor the swimming profile of each fish. A 100 g weight was dropped from 15 cm at the water level every 60 s for 10 iterations. The initial velocity following startle, velocity for each second, the time to recovery from the startle, and the total distance that each fish traversed the tank over 60 s following the startle was quantified. A One-way ANOVA followed by a Tukey's post-hoc test was used to statistically compare total distance swam by undamaged controls and sTBI fish.

2.14. Terminal Deoxynucleotidyl Transferase dUTP Nick-End Labeling (TUNEL) Assay

Undamaged control and blunt-force damaged fish at 16 hpi were anesthetized in 1:1000 2-PE until unresponsive to tail pinch. Fish were placed with the ventral side facing up, secured, and transcardially perfused with PBS (1 mL/1 min) for 5 min. Brains were removed from the euthanized fish and fixed in 9:1 ethanolic formaldehyde (100% ethanol/37% formaldehyde) for 24 h at 4 °C. For fluorescently labeled TUNEL: Brains were cryoprotected as described in the 'Tissue processing' section, and frozen serial coronal sections of 16 μm thickness of the entire brain were collected. Cell death was detected using ApopTag® Red In Situ Apoptosis Detection Kit (EMD Millipore, Burlington, MA, USA) following the manufacturer's instructions with the following deviations: Prior to labeling, cryosections were dried at 55 °C for 20 min then post-fixed for 30 min in 2% PFA, rehydrated in PBS, and permeabilized in PBS containing 0.5% Triton X-100 for 5 min at room temperature. The sections were then exposed to acetic acid:EtOH at −20 °C for 5 min, followed by an incubation in Proteinase K 10 mg/mL (1:200 dilution in PBS) for 20 min at room temperature, washed briefly in PBS for 5 min, and then incubated with equilibration buffer for 3 min at room temperature. Apoptotic cells were incubated with the manufacturer's TdT label mix at 37 °C for 1.5 h. To stop the enzymatic reaction, slides were incubated in 1× stop buffer for 10 min and then washed in PBS before TUNEL-positive cells were detected by a Dig-based system. Slides were incubated in a 1:1 ratio of anti-dig/blocker for 30 min, washed in PBS twice for 5 min each before slides were incubated in DAPI (1:1000) for 20 min. Slides were then repermeabilized in PBS-Tween 20 for 10 min, blocked in PBS containing 2% normal goat serum, 0.4% Triton X-100, and 1% DMSO, and incubated in primary rabbit anti-HuCD (1:100, Abcam, Cambridge, MA, USA) antibody diluted in blocking buffer in a humidity chamber at room temperature overnight. Slides were then rinsed three times in PBS-Tween-20 for 10 min each, and subsequently incubated in fluorescent-labeled secondary goat anti-rabbit antibody (diluted 1:500 in blocker) and DAPI (1:1000) in a humidity chamber at room temperature for 1.5 h. Slides were then briefly rinsed in PBS for 5 min before being mounted in ProLong Gold Antifade (Life Technologies, Carlsbad, CA, USA).

For DAB labeled TUNEL: Undamaged control and blunt-force damaged fish at 16 hpi were anesthetized in 1:1000 2-PE until unresponsive to tail pinch. Brains were perfused and fixed as described above for fluorescently labeled TUNEL. Whole heads were then decalcified in 100% filtered RDO Rapid Decalcifier (Electron Microscopy Sciences, Hatfield, PA, USA) for 1 h at room temperature. Brains were then dehydrated in a 70%, 80%, 90%, 95% ethanol series for 1 h each, followed by three 1 h exposures to 100% ethanol and xylene each. Subsequently, whole heads were submersed twice for 1 h each under vacuum in 60 °C paraffin. Whole heads were embedded in paraffin and 10 μm thick serial sections were prepared. Apoptotic cell death was visualized using NeuroTACS II In Situ Apop Detection Kit (Trevigen, Gaithersburg, MD, USA). Slides were deparaffinized in three xylene washes and rehydrated by incubating slides twice each in 100%, 95%, 70% ethanol for 5 min. Slides were washed for 10 min at room temperature in PBS and coated with the manufacturer's

NeuroPore® overnight. Slides were rinsed in PBS for 10 min and quenched in 9:1 methanol: 30% hydrogen peroxide for 5 min. Slides were washed in PBS for 10 min, coated with manufacturer's TdT label buffer for 5 min, followed by manufacturer's label reaction mix for 1 h at 37 °C. To stop the enzymatic reaction, $1\times$ TdT stop buffer was applied for 5 min and slides were then washed in PBS for 5 min. Slides were covered with Strep-HRP Solution at room temperature for 20 min, washed in PBS for 5 min, and then submerged in DAB solution for 4 min. Slides were exposed to 95% ethanol once, 100% ethanol twice, and xylene twice for 3 min each to dehydrate the tissue. Subsequently, slides were coverslipped with DPX (Sigma-Aldrich, St. Louis, MO, USA) to preserve DAB staining.

2.15. EdU Labeling

Fish were anesthetized in 1:1000 2-PE and intraperitoneally injected (IP) with ~40 µL of 10 mM 5-ethynyl-2′-deoxyuridine (EdU, Invitrogen, Carlsbad, CA, USA) using a 30-gauge needle. For cell migration experiments, undamaged control and blunt-force injured fish were injected with EdU at 48 hpi and collected at either 51, 60, 72, 84, or 96 hpi (n = 4, N = 3, 12 fish total). For recovery experiments, undamaged control and blunt-force injured fish were injected with EdU at 48 hpi and 60 hpi and collected at either 7 or 30 dpi (n = 3–5, N = 3, 10–15 fish total). For all other experiments utilizing EdU, undamaged control and blunt-force injured fish were injected at 48 hpi and collected at 60 hpi (N = 1–4 in triplicate, N total = 4–10). Tissue was collected, fixed, and sectioned as described in the 'Tissue processing' section. EdU detection was performed using the Click-iT® EdU Alexa Fluor® 594 Imaging Kit (Invitrogen, Carlsbad, CA, USA) as previously described [42,43], which was performed prior to immunohistochemistry.

2.16. Immunohistochemistry

Brain cryosections that were prepared as described in the 'Tissue processing' section (16 µm) were dried at 55 °C for 20 min before being rehydrated and permeabilized in PBS-Tween 20 (0.05%, Sigma-Aldrich, St. Louis, MO, USA) for 10 min. Sections were blocked in PBS containing 2% normal goat serum, 0.4% Triton X-100, and 1% DMSO for 1 h in a humidity chamber at room temperature. Slides were incubated in primary antibody diluted in blocking buffer in a humidity chamber at room temperature overnight. Primary antibodies (and their dilutions) included mouse anti-GFP monoclonal antibody (1:500, Sigma-Aldrich, St. Louis, MO, USA), chicken anti-GFP polyclonal antibody (1:500 for brain, 1:1000 for retina, Abcam, Cambridge, MA, USA), rabbit anti-PCNA polyclonal antibody (1:2000, Abcam, Cambridge, MA, USA), and rabbit anti-HuC/D polyclonal antibody (1:100, Abcam, Cambridge, MA, USA). Slides were then rinsed three times in PBS-Tween-20 for 10 min each, and subsequently incubated in fluorescent-labeled secondary antibody (diluted 1:500 in PBS-Tween-20) and DAPI (1:1000) in a humidity chamber at room temperature for 1.5 h. The secondary antibodies used in this study included either goat anti-mouse IgG or goat anti-rabbit IgG conjugated to Alexa Fluor 488, 594, or 647 and goat anti-chicken IgG conjugated to Alexa Fluor 488 (Jackson Immunoresearch Laboratories, West Grove, PA, USA). Sections were rinsed three times in PBS-Tween 20 for 10 min each and once in PBS for 5 min before being cover-slipped with ProLong Gold Antifade (Life Technologies, Carlsbad, CA, USA).

2.17. Image Acquisition

Confocal z-stacked images of 16 µm sections were taken every other section throughout the cerebellum. Z-stacks of 10 µm thickness were acquired in 0.5 µm steps, imaged at 1024 × 1024 on a Nikon A1R microscope using either a 20× (NA 0.75) or 40× (NA 1.3) oil-immersion objective. Channel crosstalk was minimized by acquiring images using the sequential channel series function (NIS-Elements 4.13.01, 5.20.02 software). Images across the entire brain were taken every 5th section and acquired using a 20× oil-immersion objective (NA 0.75) and employing the large-image acquisition function (15% overlap, NIS Elements). Brightfield images of DAB-TUNEL labeled sections were acquired using a

Nikon 90i microscope equipped with a 20× objective (NA 0.75) lens and a color camera (NIS-Elements AR 3.2 software). Lightsheet images were collected on a Z.1 Dual Illumination Lightsheet using a 5× objective with a refractive index of 1.45. Images were acquired in Multiview as 2 × 3 tiles. The image tiles were stitched together and rendered in the arivis Vision4D software to form the final images.

2.18. Tissue Clearing/EdU Labeling

We modified the brain clearing protocol described by Lindsey et al. [44]. EdU was intraperitoneally injected into undamaged control and sTBI fish (as described above) and transcardially perfused at 60 hpi with 2% paraformaldehyde (PFA, 1 mL/min) for 3 min. Brains were collected and fixed in 2% PFA at 4 °C overnight. Brains were washed four times in 0.3% Triton X-100/PBS for 30 min each on a shaker and then permeabilized in 1% Triton X-100/5% DMSO/PBS for 24 h on a shaker at 4 °C, and then washed four times for 30 min in 0.3% Triton X-100/PBS. EdU was labeled as previously described by Lindsey (3 days of labeling) [44]. Brains were washed four times in cold PBS, for 30 min each, to remove unbound fluorescently conjugated azide. Subsequently, brains were embedded in 1% low melting point agarose, dehydrated in 100% MeOH four times for 4 h each, and cleared during four washes in 2:1 benzyl benzoate and benzyl alcohol (Sigma-Aldrich, St. Louis, MO, USA) that lasted 4 h each.

2.19. Optical Density

Images of EdU-labeled undamaged and sTBI cleared brains at 60 hpi were taken at 40× using a Leica M205 FA epifluorescent microscope (Leica Application Suite 2.2.0 build 4765 software). Using ImageJ software (National Institutes of Health, Bethesda, MD, USA), images were converted to 8-bit gray scale and the gray area intensities were individually determined for each of the major brain regions (forebrain, midbrain, hindbrain). The background gray area intensity was measured within a region of interest (ROI) of each brain region that was placed at the outside edge, which did not contain a discernable focal point of fluorescence within its boundary of the background [22]. The optical density was calculated for each fish using the following formula:

$$OD = \log \frac{\text{Background intensity}}{\text{avg. intensity of ROI}} \tag{2}$$

Comparisons across groups were made using a Two-way ANOVA followed by a Tukey's post-hoc test.

2.20. Evans Blue Assay for Blood–Brain Barrier Disruption

Using modified Radu et al. and Eliceiri et al. protocols [45,46], undamaged and TBI injured fish (miTBI, moTBI, sTBI, n = 3, N = 3, 9 fish total) were intracardiacly injected immediately following sham/TBI, or at desired time point, with 50 µL of 1% Evans blue dye (Sigma-Aldrich, St. Louis, MO, USA) in PBS. Fish were allowed to survive for 2 h post-injection. Fish were then euthanized in 1:500 2-PE, brains were collected, weighed, and washed in PBS for 3 min to remove excess and superficial dye accumulated during dissection. For quantitative assays, 3 brains from a treatment group were pooled, placed in 300 µL of 100% formamide and incubated at 55 °C for 24 h. Brains were then centrifuged at 10× *g* for 10 min, the supernatant collected, 250 µL from each group was analyzed on a SpectraMax M5 plate reader (Molecular Devices, San Jose, CA, USA), and RFU value was normalized to mg of brain tissue. Comparisons were made using a One-way ANOVA followed by either a Tukey's or Dunnett's multiple comparison post-hoc test. For qualitative sections, brains were fixed in 4% PFA overnight. Brains were then embedded in 5% low melt agarose and a single coronal cut was made with a razor blade, and en face coronal images were taken.

2.21. Sonic Hedgehog Modulation

Sonic hedgehog signaling was modulated with either an agonist (purmorphamine, Sigma-Aldrich, St. Louis, MO, USA) or an antagonist (cyclopamine, Toronto Research Chemicals, Toronto, ON, Canada). Purmorphamine: Undamaged fish were IP-injected with 40 μL of 10 μM purmorphamine using a 30-gauge injection needle attached to a 1 mL syringe every 12 h for 48 h (0, 12, 24, 36, and 48 h). To assess cell proliferation in undamaged control and purmorphamine treated brains, EdU was coinjected (as described above) at the 48 h timepoint. Brains were collected 12 h later (60 h after the first purmorphamine injection) and prepared for EdU detection and immunohistochemistry as described above. Cyclopamine: Fish exposed to sTBI were IP-injected with 40 μL of 2 mM cyclopamine at 4, 12, 24, 36 hpi using a 30-gauge injection needle attached to a 1 mL syringe and coinjected with EdU at 48 hpi. Brains were collected at 60 hpi and prepared for EdU detection and immunohistochemistry as described above.

2.22. Quantitative Real-Time PCR (qRT-PCR)

Total RNA was isolated and purified from approximately the top 1/3 apical portion of cerebellums from five adult undamaged and sTBI fish. For neuroinflammation investigations, sTBI fish were collected at 12 hpi, 1, 2, 3, 7, 14, and 28 dpi using Trizol extraction and converted to cDNA from 1 ug of RNA using qScript cDNA SuperMix (Quantabio, Beverly, MA, USA) as described by Campbell et al. [47], and TaqMan probes were used according to manufacturer's instructions, with 10 ng of cDNA for amplification. TaqMan probes (Thermo Fisher, Waltham, MA, USA) for *il1β* (Dr03114367_g1), *tnfα* (Dr03126850_m1), and *il10* (Dr03103209_m1) were used to examine neuroinflammation. For sonic hedgehog investigations, sTBI fish were at 6, 12, 24, 36, 48, and 60 hpi, and TaqMan probes (Thermo Fisher, Waltham, MA, USA) for *shha* (Dr03432632_m1), *shhb* (Dr03112045_m1), *smo* (Dr03131349_m1), and *gli1* (Dr03093665_m1) were used. For quantitative real-time PCR (qRT-PCR) the data was normalized to 18s rRNA (Hs03003631_g1) in each well. Data was acquired using the ABI StepOnePlus Real-Time PCR System (Applied Biosystems, Foster City, CA, USA). Cycling conditions were as follows: 2 min at 50 °C, 10 min at 95 °C, 40 cycles of 15 s at 95 °C, and 1 min at 60 °C with data collection occurring after each extension cycle. The $\Delta\Delta$CT values were calculated and used to determine the $\log_2$-fold changes [33] of *il1β*, *tnfα*, *il10*, *shha*, *shhb*, *smo*, and *gli1*. Expression levels were examined in biological triplicate and technical replicates.

2.23. Statistical Analysis

All data within this study, with the exception of sTBI Blood-Brain Barrier disruption time course (n = 3 pooled brains, N = 2), whole brain fluorescent TUNEL (n = 1), whole brain proliferation (n = 2, N = 2, total of 4 fish), was obtained from at least three independent trials (N = 3) of at least 3 fish per independent trial (n = 3, total of 9 fish). The data are expressed as mean $\pm$ SE or as mean $\pm$ SD, each indicated within the figure legend, which was derived by averaging the data from the brains of individual fish from all combined trials. Data sets were analyzed in Prism 8 (GraphPad, San Diego, CA, USA) with a Student's *t*-test for single pairwise comparisons with control or One-way or Two-way ANOVA followed by either Tukey's, Bonferroni's, Dunnett's, or Sidik's post-hoc test for multiple comparisons. The statistical test used and significance indicator of # for $p < 0.05$ or ## for $p < 0.01$ are stated in each figure legend. In instances where comparisons were not statistically significant, the actual *p*-value was given in the figure.

3. Results

3.1. Modified Marmarou Weight Drop Results in a Reproducible and Scalable TBI

Blunt-force injuries, the most common form of TBI, range in severity and result in a heterogeneous set of injury-induced pathologies. While a blunt-force zebrafish TBI model has rarely been examined [32] it offers the unique opportunity to also examine the regenerative response. To develop a scalable blunt-force TBI model, we modified the commonly used Marmarou weight drop (Figure 1A, created with BioRender.com) [10,36] with the impact zone centered at the intersection of the midbrain (mesencephalon and diencephalon; optic tectal lobes) and hindbrain (rhombencephalon; cerebellum, Figure 1A).

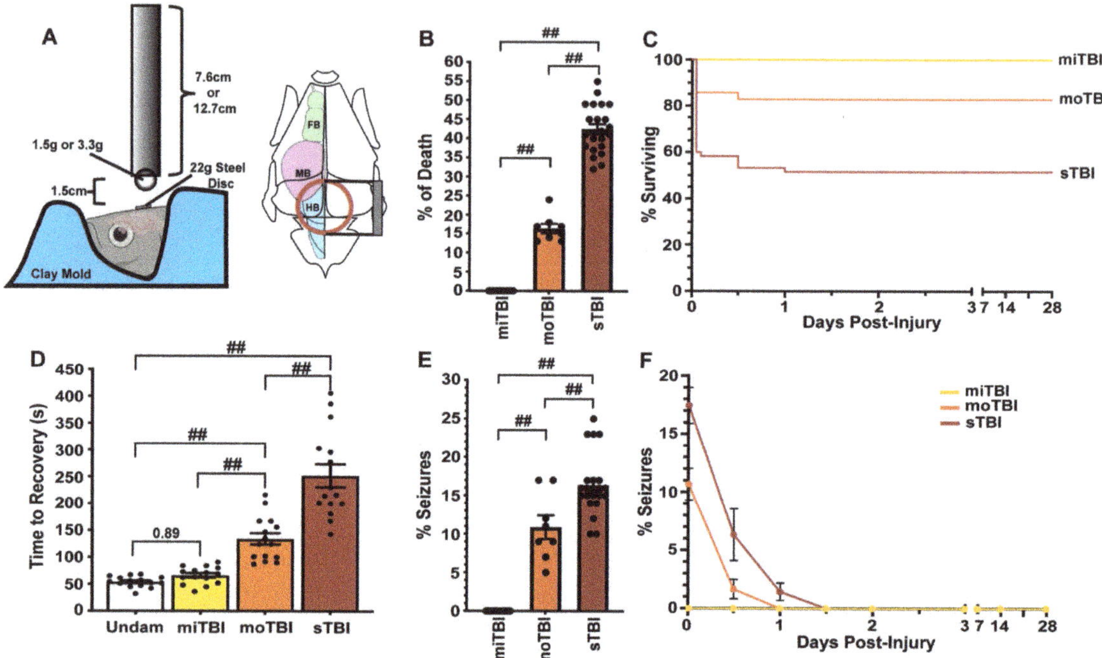

Figure 1. A reproducible and scalable blunt-force TBI in adult zebrafish. (**A**) Diagram depicting the conditions of the modified Marmarou weight drop scaled to the adult zebrafish. On the left, the tube through which the dropping ball falls is shown relative to the head and brain of the zebrafish. On the right, a red circle over the brain diagram shows the relative impact center of the dropping ball (forebrain in green, midbrain in purple, and hindbrain in blue). (**B**) Following damage, the percentage of fish dying significantly increased relative to the damage severity (n ≥ 200 fish). (**C**) Graph of the percentage of fish surviving following TBI out to 28 dpi. All mortality took place within 1 dpi (n = 30). (**D**) The recovery time for the fish to right themselves without exhibiting akinesia, ataxia, and motor incoordination significantly increased with damage severity (n = 15). (**E**) The percentage of fish that experienced intense tonic-clonic seizures significantly increased relative to the damage severity (n ≥ 200 fish). (**F**) The percentage of fish exhibiting post-traumatic seizures observed out to 28 dpi significantly increased relative to the damage severity. No seizures were observed after 1.5 dpi for any of the damage severities (n = 30). Forebrain, FB, hind-brain, HB, midbrain, MB All graph data points are Mean ± SEM. Statistical analyses were performed with either Two-way ANOVA or One-way ANOVA followed by a Tukey post-hoc test. ## $p < 0.01$.

To validate our model, we examined key pathophysiological features often used to categorize TBI, such as mortality, loss of consciousness, seizure, vascular injury, blood–brain barrier (BBB) disruption, edema, sensorimotor deficits, and neuroinflammation [3,48,49]. We determined the percentage of mortality for each level of damage. A mild traumatic brain injury (miTBI) never resulted in death (n = 225, Figure 1B). However, mortality significantly increased (16.37% ± 1.28%, $p < 0.01$, n = 143, Figure 1B) when the severity was increased to moderate TBI (moTBI) and was further significantly elevated (42.45% ± 1.33%, $p < 0.01$, n = 938, Figure 1B) following a severe TBI (sTBI). While we continued to monitor survival for 28 days post-injury (dpi), all mortality was observed within 1 dpi (Figure 1C). Thus, our model resulted in injuries with reproducible high, medium, and low survival rates, correlative to the prognostic outcomes in humans suffering from all three severity levels [1,49].

We next examined loss of consciousness, a commonly used diagnostic measure to rapidly categorize human TBI [50], by quantifying the amount of time required before returning to normal swimming behavior immediately following injury. Undamaged controls rapidly recovered from anesthesia in the recovery tank (Undam: 52 s $\pm$ 2.57 s) and the recovery rate following miTBI was not significantly different (61 s $\pm$ 4.1 s, $p = 0.89$, n = 15, Figure 1D). However, moTBI fish took significantly more time (133.93 s $\pm$ 10.56 s, $p < 0.01$, n = 15, Figure 1D) relative to undamaged controls before they regained consciousness and returned to normal swimming behavior. The sTBI fish took significantly longer than either undamaged control, miTBI, or moTBI fish, as they often remained motionless at the bottom of the tank for several minutes (252.2 s $\pm$ 21.19 s, $p < 0.01$, n = 15, Figure 1D). Thus, our model displayed increased times of lack of consciousness that were consistent with the level of TBI severity.

We also determined the percentage of zebrafish that displayed intense tonic-clonic seizure-like behaviors (akinesia, ataxia, bending, circling, and/or corkscrew swimming). This seizure-like behavior was never observed following a miTBI (n = 225 Figure 1E). However, the percentage of fish exhibiting tonic-clonic seizures significantly increased following moTBI (10.87% $\pm$ 1.54%, $p < 0.01$, n = 143 Figure 1E) relative to miTBI fish, and significantly increased further in sTBI fish (16.63% $\pm$ 0.84%, $p < 0.01$, n = 938, Figure 1E). Additionally, injured fish were observed for post-traumatic seizure activity from time of injury out to 28 dpi (Figure 1F). Following moTBI and sTBI, seizure activity was observed for 1 (moTBI) to 1.5 dpi (sTBI), after which all seizure behavior ceased and was not observed again through 28 dpi (Figure 1F). The increase in number of seizures relative to the injury severity observed in our model is in agreement with human blunt-force TBI populations [51,52].

3.2. Blunt-Force TBI Induces Severity-Dependent Vascular Injury with Blood–Brain Barrier Disruption, Neuroinflammation, and Edema

In human TBI, vascular injury, blood–brain barrier (BBB) disruption, edema, and neuroinflammation are critical metrics [53,54]. At 4 hpi, miTBI fish displayed minor bleeding (Figure 2C, arrowheads) that was not observed in undamaged controls (Figure 2A,B). As injury severity increased, we observed apparent pooling of blood following both moTBI and sTBI (Figure 2D,E), with intracerebral hematomas and blood-filled ventricles found in sTBI fish (Figure 2F). In sTBI fish, hematoma formation continued to expand between 6 and 12 hpi (Figure 2G,H), with hematomas visually resolving between 1–3 dpi (Figure 2I–M). We further assessed vascular injury in terms of BBB disruption using Evans blue dye [46] that was intracardiacly injected and permeated into the brain for 2 hpi. Undamaged and miTBI fish had low amounts of Evans blue diffuse across the BBB (91.18 RFU/mg $\pm$ 15.75 and 182.93 RFU/mg $\pm$ 20.68, $p = 0.15$, respectively, Figure 2N,P,Q). In contrast moTBI and sTBI had significantly more Evans blue dye diffuse across the BBB (moTBI: 227.16 RFU/mg $\pm$ 4.38, $p < 0.05$, sTBI: 574.26 RFU/mg $\pm$ 47.49, $p < 0.01$ Figure 2N,S,T). The significant increase of Evans blue dye in the sTBI fish was apparent in coronal brain sections (Figure 2U) relative to undamaged controls (Figure 2R). We also measured BBB disruption using Evans blue dye in sTBI fish from 2 hpi through 28 dpi. BBB disruption sharply peaked by 2hpi and remained significantly disrupted through 2 dpi (2 hpi: 524.01 RFU/mg, $p < 0.01$, 12 hpi: 285.06 RFU/mg, $p < 0.01$, 1 dpi: 163.49 RFU/mg, $p < 0.05$, 2 dpi: 140.72 RFU/mg, $p < 0.05$, Figure 2O). Extracted dye levels returned to near undamaged levels by 3 dpi and remained there through 28 dpi (undam: 70.73 RFU/mg, 3 dpi: 108.02 RFU/mg, $p = 0.44$, 7 dpi: 81.79 RFU/mg, $p = 0.99$, 28 dpi: 63.94 RFU/mg, $p = 0.99$, Figure 2O). The observed hematoma growth and BBB disruption led us to investigate edema and neuroinflammation.

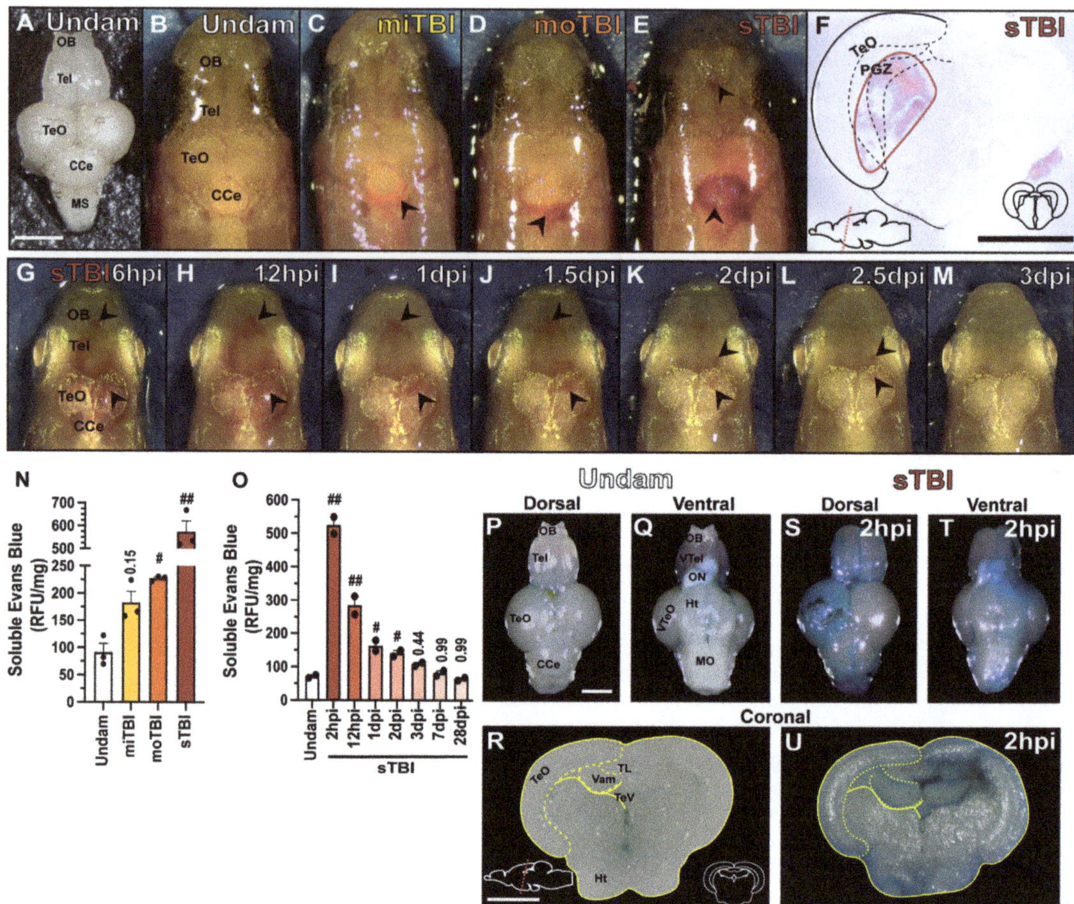

Figure 2. MMWD produces graded hematomas and blood–brain barrier disruption. (**A**) Isolated, undamaged whole brain with the major lobes labeled. Dorsal views of *roy^{a9};mitfa^{w2}* (*casper*) undamaged (**B**) and TBI fish displaying vascular injury 4 hpi (**C–E**). Compared to undamaged controls, vascular injury resulted in hemorrhaging (arrowheads) in all severity levels that increased in a severity-dependent manner. (**F**) Coronal section of sTBI brain with intracerebral hematoma (red boundary). (**G–M**) Repeated dorsal view of an individual sTBI *albino^{b4}* fish across time, in which hemorrhaging (arrowheads) qualitatively peaked at 12 hpi, and gradually resolved by 3 dpi. (**N**) Following injury, a significant increase in solubilized Evans Blue dye represented disruption of the BBB in a severity-dependent manner. (**O**) Following sTBI, a statistically significant increase in solubilized Evans Blue dye occurred by 2 hpi and then gradually decreased until it reached control levels at 3 dpi (3 pooled brains/group, n = 2–3 groups). (**P–U**) Dorsal, ventral, and coronal views of isolated undamaged (**P,Q,R**) and sTBI brains at 2 hpi (**S,T,U**) from fish injected with Evans Blue dye as a qualitative measure of BBB integrity. Solid lines in (**F,R,U**) denote tissue boundaries, while dotted lines denote internal anatomical boundaries. Corpus cerebelli, CCe, hypothalamus, Ht, medial valvula cerebelli, Vam, medulla oblongata, MO, medulla spinalis, MS, olfactory bulb, OB, optic tectum, TeO, periventricular grey zone, PGZ, tectal ventrical, TeV, telencephalon, Tel, torus longitudinalis, TL, ventral optic tectum, VTeO, ventral telencephalon, VTel. Scale bars, (**A–E,G–M,P–T**) = 500 μm, (**F,R–U**) = 250 μm. Mean ± SEM is depicted in (**N,O**). Statistical analyses were performed with a One-way ANOVA followed by a Tukey post-hoc test. # $p < 0.05$, ## $p < 0.01$.

Edema formation, which was measured as the percentage of fluid in the brain, was not significantly different between undamaged controls and miTBI fish between 1 through 28 dpi (Undam: 76.36% ± 0.98%, miTBI 1 dpi: 76.11% ± 0.93%, $p = 0.99$, n = 9, Figure 3A). However, as injury severity increased, so did the percentage of fluid content of dam aged brains, indicative of edemas. Relative to undamaged controls, both the moTBI (83.42% ± 0.83%, $p < 0.01$, n = 9) and sTBI (88.02% ± 0.64%, $p < 0.01$, n = 9, Figure 3A) brains contained significantly more fluid at 1 dpi. Increased edema remained significantly elevated in sTBI fish at 3 dpi (79.02% ± 0.76%, $p < 0.01$), while moTBI returned near undamaged levels (76.66% ± 0.99%, $p = 0.83$, Figure 3A). By 5 dpi, edema within sTBI brains returned to near undamaged levels (74.77 ± 0.95%, $p = 0.99$) and significantly elevated edema was not again observed in any severity out through 28 dpi (Figure 3A).

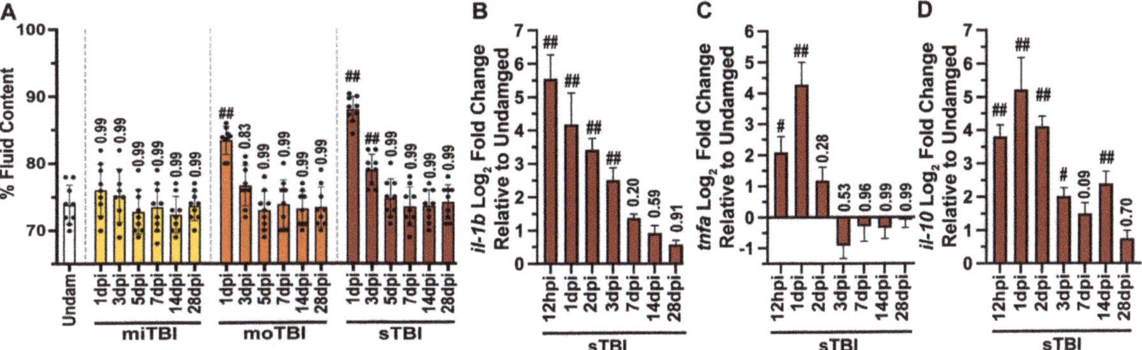

Figure 3. TBI inflicted zebrafish display edema and neuroinflammation. (**A**) Graph of level of edema, measured as the percentage of fluid in the brain, at all severities from 1–28 dpi, in which edema increased in a severity-dependent fashion, and remained elevated until 3 dpi and 5 dpi for moTBI and sTBI, respectively (n = 9). (**B,C**) Expression of pro-inflammatory cytokine genes *il-1b* (**B**) and *tnfa* (**C**) were significantly elevated in sTBI for 3 or 1 dpi, respectively, following injury before returning to near undamaged levels (5 pooled cerebellums/group, n = 3 groups). (**D**) Anti-inflammatory cytokine *il-10* gene expression was also elevated and remained highly upregulated for 14 dpi (5 pooled cerebellums/group, n = 3 groups). Mean ± SEM is depicted in all graphs. Statistical analyses were performed with a One-way ANOVA followed by a Tukey's or Dunnett's multiple comparison post-hoc test. # $p < 0.05$, ## $p < 0.01$.

We also assessed neuroinflammation by examining representative pro-inflammatory (*il1β, tnfα*) and anti-inflammatory (*il10*) cytokine expression using qRT-PCR of RNA collected from undamaged and sTBI isolated cerebellums across multiple timepoints following blunt force trauma (12 hpi, 1, 2, 3, 7, 14, and 28 dpi). The cytokines *il1β, tnfα*, and *il10* have been implicated as critical biomarkers in human TBI [55–57] and as playing a role in zebrafish tissue regeneration [42,58]. Relative to undamaged control cerebellums, *il1β* expression peaked at 12 hpi and remained highly upregulated through 3 dpi (Figure 3B), while *tnfα* expression peaked at 1 dpi, before rapidly decreasing and returning near undamaged levels (Figure 3C). The anti-inflammatory cytokine *il10* also began increasing by 12 hpi and remained elevated through 14 dpi before returning to near undamaged levels by 28 dpi (Figure 3D). Collectively, these data demonstrate our model recapitulates several injury-related pathologies consistent with key human diagnostic TBI measures and provides the sensitivity to reproducibly distinguish between mild, moderate, and severe TBI.

3.3. Blunt-Force TBI Results in Severity-Dependent Cell Death Spreading from the Impact Zone

We quantified the number of TUNEL-positive dying cells in brains following each of the three injury levels. In the cerebellum, the region primarily located within the impact zone, there was not a significant difference in the number of TUNEL-positive cells following miTBI (miTBI:103.62 ± 16.77, $p = 0.29$, n = 8, Figure 4B,E) relative to undamaged controls (7.25 ± 2.57 cells, n = 8, Figure 4A,E). This minor damage/cell death following miTBI is similar to human miTBI patients displaying negative CT/MRI scans [59,60]. However, there were significant increases in TUNEL-positive cells in moTBI (450 ± 54.65 cells, $p < 0.01$, n = 8, Figure 4C,E) and sTBI (705.5 ± 49.54 cells, $p < 0.01$, n = 8, Figure 4D,E) relative to undamaged controls. Additionally, there were significantly greater numbers of TUNEL positive cells between miTBI and moTBI ($p < 0.01$) and between moTBI and sTBI ($p < 0.01$) which is consistent with the scalable nature of the damage model.

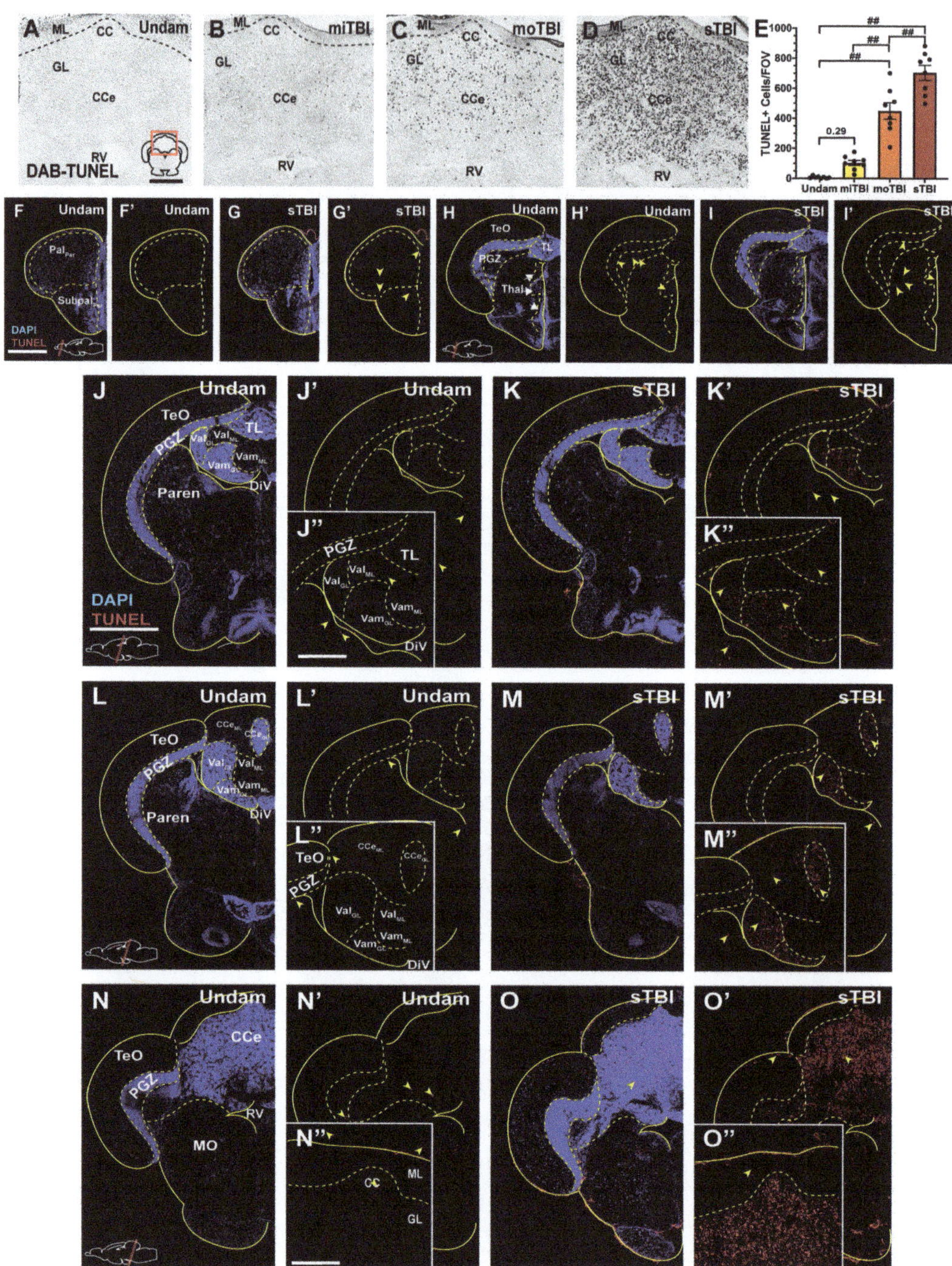

Figure 4. Cell death is severity-dependent and radiates out from the impact zone as a gradient. (**A–D**) Brightfield images of coronal cerebellar (CCe) sections stained with DAB-TUNEL 16 hpi. Prominent DAB TUNEL-positive cells were observed in

the CCe in a severity-dependent manner. (**E**) Quantification of the number of DAB-TUNEL-positive cells/section revealed significant increases in the number of TUNEL-positive cerebellar cells following moTBI and sTBI (n = 8). (**F–O″**) Confocal images of coronal sections, from rostral to caudal across the neuroaxis, stained with fluorescent TUNEL and DAPI-labeled at 16 hpi (**J–O**). Apoptotic cell death, emanating from the impact zone, was observed widely across the neuroaxis. The red box in panel A shows the relative position of the image on the appropriate brain cross-section. Solid lines denote tissue boundaries, while dotted lines denote internal anatomic boundaries. Cerebellar crest, CC, corpus cerebelli, CCe, diencephalic ventricle, DiV, granule cell layer, GL, molecular layer, ML, medulla oblongata, MO, parenchyma, Paren, periventricular grey zone, PGZ, optic tectum, TeO, torus longitudinalis, TL, rhombencephalic ventricle, RV, medial valvula cerebelli, Vam, lateral valvula cerebelli, Val. Scale bars: (**A**) = 100 µm, for panels (**A–D,F**) = 200 µm, for panels (**F–I′,J**) = 500 µm, for panels (**J–O′,J″**) = 200 µm, for panels (**J″–M″,N**) = 100 µm, for panels (**N″,O″**). Mean ± SEM is depicted in (**E**). Statistical analyses were performed with a One-way ANOVA followed by a Tukey's post-hoc test. ## $p < 0.01$.

We hypothesized a blunt-force trauma would result in a diffuse injury and cell death, unlike the focal injury associated with a penetrating TBI models. Therefore, we examined the entire neuroaxis of sTBI fish for apoptotic cell death following blunt-force TBI. We observed only a few TUNEL-positive cells in the most rostral parts of the brain (Figure 4F,F′,G,G′), though TUNEL-positive cells became more evident in the rostral parts of the midbrain approximately 0.5–1 mm outside of the impact zone (Figure 4I,I′). However, the most prominent labeling was observed in the granule cell layers of the medial and lateral valvula cerebelli (Vam and Val, Figure 4K,K′,M–M″), and the corpus cerebelli (CCe, Figure 4M–M″,O–O″), regions with high densities of cell bodies. Importantly, the number of TUNEL-positive cells increased in a gradient emanating from the epicenter of the impact zone. At a lower occurrence, we observed TUNEL-positive labeling in the Periventricular Gray Zone (PGZ, Figure 4K–K″,M–M″,O,O′) and other regions with high density of neuronal cell bodies located laterally away from the impact zone. We further examined apoptotic cell death in the cerebellum, the epicenter of the impact zone, and co-stained with the pan-neuronal marker HuCD. We observed double-positive TUNEL/HuCD labeling across the cerebellum of sTB fish (Figure 5A–A‴), to include large TUNEL/HuCD double positive cells within the Purkinje layer (Figure 5B–C″″) and smaller double-labeled cells within the granule layer (Figure 5B,D–E″″). These data suggest that blunt-force trauma results in a diffuse injury accompanied by apoptotic neuronal cell death that occurs in a gradient radiating outward from the impact zone.

3.4. TBI Results in Sensorimotor Impairments and Associative Learning and Memory Deficits with Rapid Recovery

A blunt-force injury often results in sensorimotor and cognitive impairment, with deficits increasing with injury-severity in human TBI populations [15,61,62]. Following TBI, we evaluated swim orientation, analogous to gait, and response to an adverse tactile stimulus to collectively assess sensorimotor coordination (Supplementary Table S1). Prior to injury, fish swam horizontally and parallel to the tank bottom/water surface, never breached the surface, and consistently displayed nocifensive, escape, and avoidance behaviors in response to an adverse tactile stimulus. At 1–2 hpi following sTBI, fish displayed disoriented swim profiles that were noticeably inclined or tilted and often breached the surface of the water with the most rostral portion of their head (Table 1). These abnormal swimming behaviors were absent by 6 hpi. Responses to tactile stimuli were impaired for longer durations following sTBI. All injured fish displayed and continued to display nocifensive behaviors at all timepoints, while at 1 hpi few fish displayed escape behaviors, and none avoided the stimulus (Table 1). Escape behaviors remained significantly impaired through 2 hpi ($p = 0.05$, Table 1) and returned to normal by 1 dpi, while avoidance behaviors remained significantly impaired through 12 hpi ($p = 0.05$, Table 1) and returned to normal by 2 dpi. Each behavior was individually scored and then collectively summed as a total sensorimotor score. Relative to preinjury, sTBI fish displayed significantly impaired sensorimotor scores at 1 hpi ($p < 0.01$) through 12 hpi ($p < 0.05$) and returned to near pre-damaged response by 1 dpi (*p* = 0.99, Figure 6A). Thus, following sTBI, our model induces orientation and sensorimotor coordination deficits similarly seen in human TBI patients.

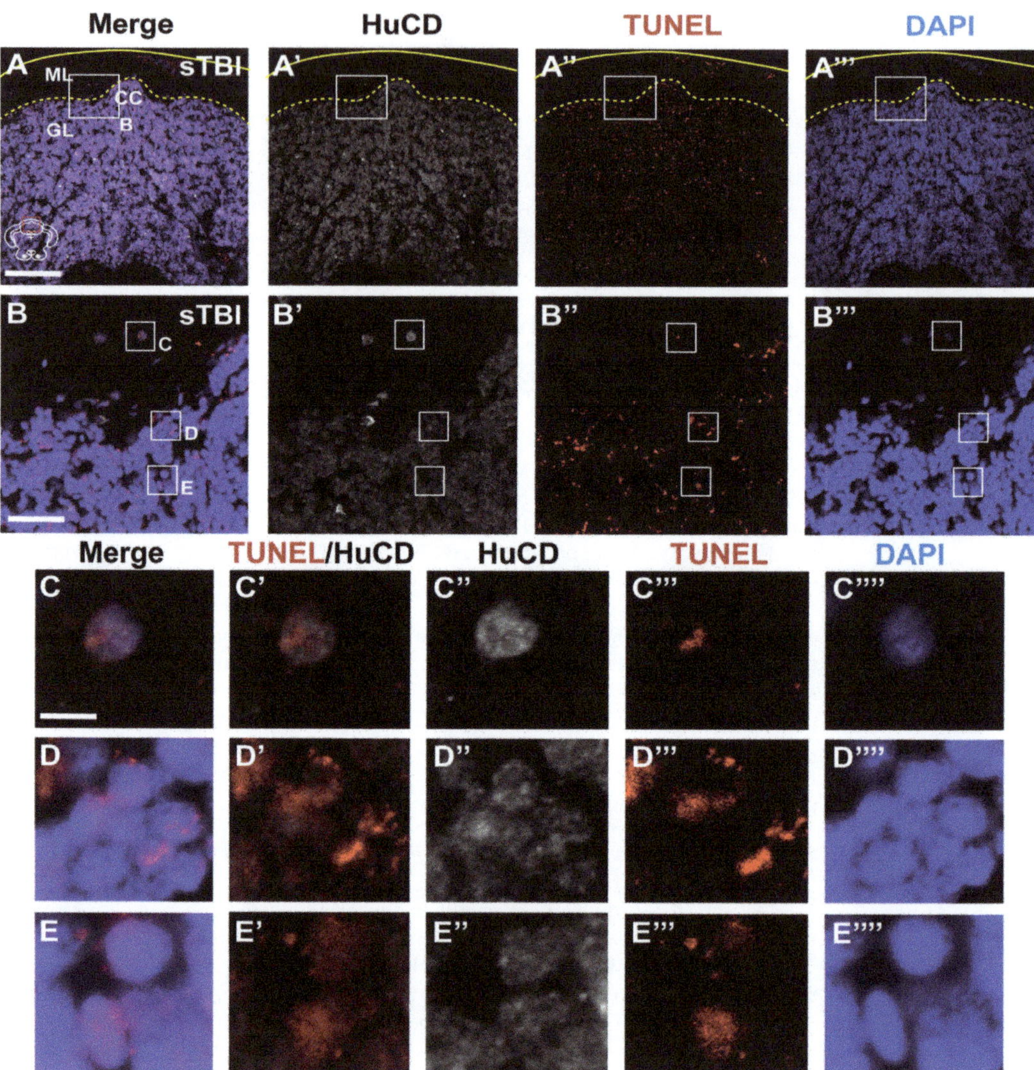

Figure 5. Apoptotic cell-death in cerebellar neurons. (**A–B‴**) Confocal images of coronal cerebellar sections stained with pan-neuronal marker HuCD (gray), fluorescent TUNEL (red), and DAPI (DAPI) at 16 h follow a sTBI. Apoptotic cell death was observed in the epicenter of the impact zone, in the densely packed granule layer of the cerebellum. (**C–E″″**) High magnification confocal images revealed colabelling of HuCD, TUNEL, and DAPI. The white box in panels (**A–A‴**) represent the region that is shown in corresponding panels (**B–B‴**). Lettered white boxes in panels (**B–B‴**) denote subsequent panels at higher magnification. A solid line denotes tissue boarder, dotted line denotes boarder between molecular and granule layer of the CCe. Cerebellar crest, CC, granule cell layer, GL, molecular layer, ML. Scale bars, (**A**) = 100 μm, for panels (**A–A‴**,**B**) = 50 μm, for panels (**B–B‴**,**C**) = 15 μm, for panels (**C–E″″**).

Table 1. Sensorimotor coordination is briefly impaired following blunt-force TBI. Quantification of swim orientation (Tilt) and behavioral responses to three adverse tactile stimulus responses (Pain Stimuli, Escape, and Avoidance) as defined in the zebrafish behavior catalog (Kalueff, et al., 2013) for 10 fish pre-injury and followed through several time points to 28 dpi (n = 10).

Time (n = 10)	Tilt Avg ZBC 1.83, 1.164, 1.175	Pain Stimuli Avg ZBC 1.104	Escape Avg ZBC 1.5, 1.52	Avoidance Avg ZBC 1.12	Total Score
Undam	1	1	1	1	4
sTBI 1 hpi	0.4	1	0.1	0	1.5
sTBI 2 hpi	0.7	1	0.4	0	2.1
sTBI 6 hpi	1	1	0.7	0.2	2.9
sTBI 12 hpi	1	1	0.7	0.4	3.1
sTBI 1 dpi	1	1	1	0.9	3.9
sTBI 2 dpi	1	1	1	1	4
sTBI 3 dpi	1	1	1	1	4
sTBI 7 dpi	1	1	1	1	4
sTBI 14 dpi	1	1	1	1	4
sTBI 28 dpi	1	1	1	1	4

We next evaluated habituation, a primitive non-associative learning response defined by a gradually decreased response over time to a continuous or repetitive stimulus [63]. In our model, the impact zone was centered over the intersection of the midbrain and hindbrain. The startle response is a well characterized behavioral assay that is predominantly initiated and executed by reticulospinal neurons in the hindbrain and also the granule neurons of the cerebellum, which have been shown to contribute to classical fear response [64]. Because the non-associative startle response quantifies total swim distance and velocities following the startle [41], we first assessed the general locomotion and swim profiles of undamaged and TBI fish for potential locomotor dysfunction. There was no significant difference in the swim velocity between undamaged and either miTBI, moTBI, or sTBI fish from 4 hpi to 28 days post-injury (dpi) (Figure 6B, n = 15). This suggests that locomotor function is not significantly affected in our TBI model.

To measure the startle response, a 100 g weight was dropped from water level next to the tank every 60 s for 10 iterations. During iteration 1, undamaged control fish responded with rapid bursts of swimming, with an initial velocity of 0.26 m/s $\pm$ 0.02 m/s and a total distance travelled of 2.48 m $\pm$ 0.11 m before returning back to relatively sedentary state within 22 s $\pm$ 1.59 s (Figure 6C, Table 2). However, undamaged fish quickly displayed signs of habituation, as their initial bursts of swimming shortened to 15 s $\pm$ 1.24 s and 10 s $\pm$ 0.6 s for iterations 5 and 10, respectively. Similarly, initial velocities of 0.19 m/s $\pm$ 0.01 m/s and 0.14 m/s $\pm$ 0.01 m/s and average swimming distances of 1.29 m $\pm$ 0.03 m and 0.69 m $\pm$ 0.03 m for were reduced for iterations 5 and 10, respectively (Figure 6C, Table 2).

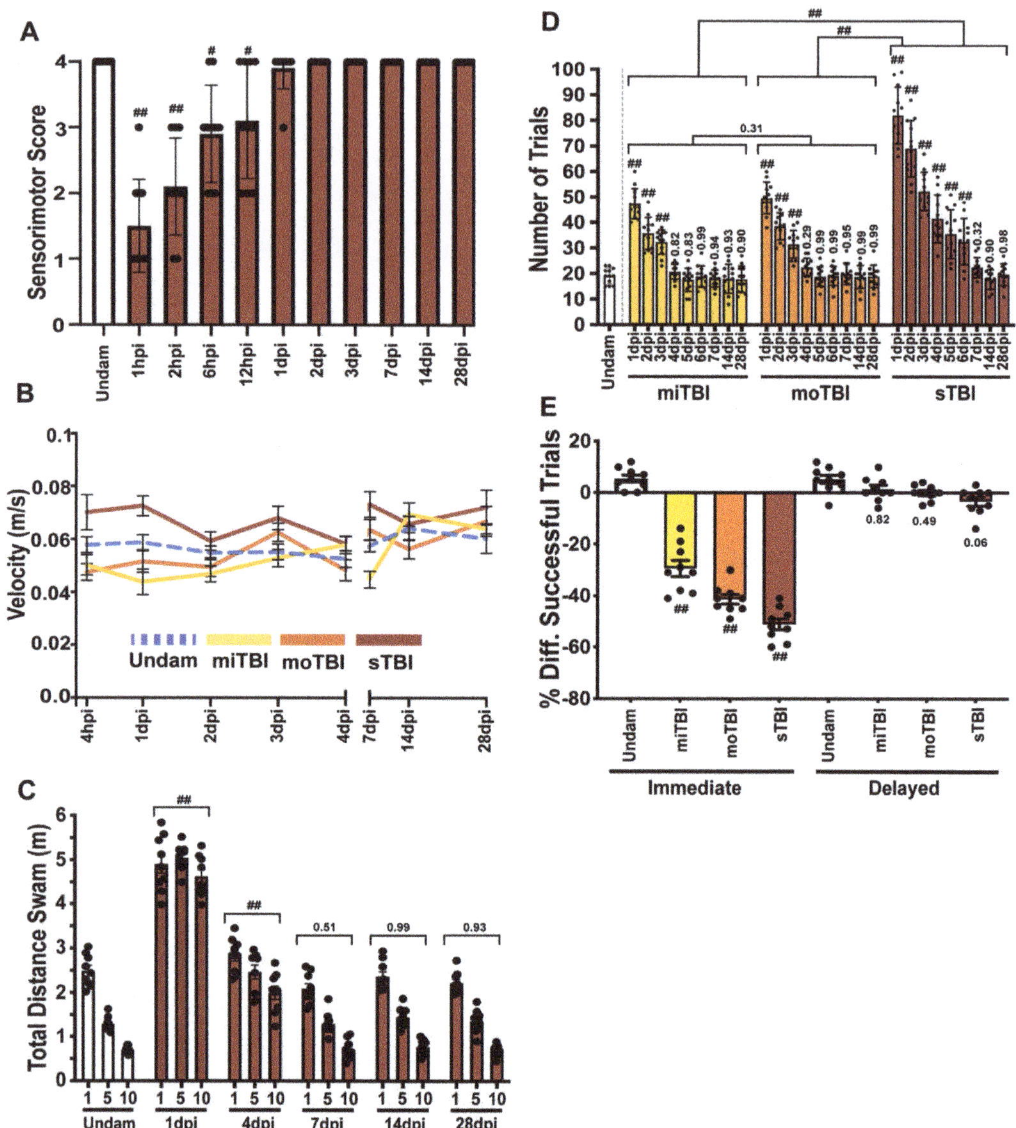

Figure 6. Injury-induced sensorimotor and cognitive deficits are severity-dependent with rapid recovery. (**A**) Sensorimotor coordination (plotted as a sum of four independent sensorimotor features involving swimming orientation and adverse tactile stimulus, Table 1) was significantly impaired following sTBI for up to 12 hpi compared to the pre-injury response (n = 10). (**B**) The unprovoked swim velocity over time was not significantly different between undamaged controls and all three severity models from 4 hpi to 28 dpi (n = 15). (**C**) Quantification of the swim distance of undamaged and sTBI fish following the startle response across iterations 1, 5, and 10 at 1, 4, and 7 dpi. TBI-damaged fish displayed cognitive deficits in this non-associative learning assay (n = 9). (**D**) Quantification of associative learning, using the shuttle box assay, of undamaged fish and all TBI severities from 1–28 dpi measuring the number of trials to master the assay (n = 12). All three damage severities resulted in a significant reduction in learning that returned to normal between 4–7 dpi. (**E**) Quantification of immediate- and delayed-recall of associative learning of undamaged and all three TBI severities using the shuttle box assay. All three TBI categories (miTBI, moTBI, sTBI) resulted in both learning and recall deficits (n = 9). Mean $\pm$ SEM is depicted in (**A–C,E**), while standard deviation is shown in (**D**). Statistical analyses were performed with either Two-way ANOVA, or a One-way ANOVA followed by a Tukey's or Dunnett's multiple comparison post-hoc test. # $p < 0.05$, ## $p < 0.01$.

Table 2. Blunt-force TBI induces non-associative learning impairments that rapidly recover. Quantification of initial swim velocities and recovery time following startle onset of undamaged and sTBI fish that were evaluated for startle response at iterations 1, 5, and 10 of the startle stimulus across multiple dpi (n = 9).

Group n = 9	Iteration		
	1	5	10
Undam Startle Velocity	0.256 ± 0.02 m/s	0.190 ± 0.01 m/s	0.138 ± 0.01 m/s
(Recovery Time)	$(22 \pm 1.59$ s)	$(15 \pm 1.24$ s)	$(10 \pm 0.60$ s)
sTBI 1 dpi Startle Velocity	0.253 ± 0.01 m/s	0.233 ± 0.02 m/s	0.246 ± 0.01 m/s
(Recovery Time)	(30 s+)	(30 s+)	(30 s+)
sTBI 4 dpi Startle Velocity	0.246 ± 0.01 m/s	0.231 ± 0.01 m/s	0.216 ± 0.01 m/s
(Recovery Time)	$(23 \pm 1.02$ s)	$(24 \pm 1.88$ s)	$(21 \pm 0.95$ s)
sTBI 7 dpi Startle Velocity	0.236 ± 0.02 m/s	0.153 ± 0.01 m/s	0.100 ± 0.01 m/s
(Recovery Time)	$(21 \pm 1.25$ s)	$(16 \pm 1.14$ s)	$(13 \pm 1.03$ s)
sTBI 14 dpi Startle Velocity	0.248 ± 0.02 m/s	0.171 ± 0.01 m/s	0.144 ± 0.02 m/s
(Recovery Time)	$(23 \pm 1.12$ s)	$(17 \pm 1.09$ s)	$(12 \pm 1.31$ s)
sTBI 28 dpi Startle Velocity	0.251 ± 0.01 m/s	0.182 ± 0.01 m/s	0.124 ± 0.01 m/s
(Recovery Time)	$(21 \pm 1.71$ s)	$(15 \pm 1.22$ s)	$(11 \pm 0.89$ s)

Following the startle, sTBI fish at 1 dpi responded with an initial velocity of 0.25 m/s $\pm$ 0.01 m/s during iteration 1, which was similar to undamaged fish (Table 2). However, sTBI fish at 1 dpi did not display any signs of habituation as the initial velocities for iterations 5 and 10 were 0.23 m/s $\pm$ 0.02 m/s, and 0.25 m/s $\pm$ 0.01 m/s, respectively, and exhibited persistent bursts of increased swimming activity that lasted longer than 30 s during all iterations (Table 2). Consequently, sTBI fish at 1 dpi swam a significantly greater distance than undamaged control fish for iterations 1, 5, and 10, averaging 4.90 m $\pm$ 0.26 m, 5.04 m $\pm$ 0.16 m, and 4.62 m $\pm$ 0.18 m, respectively (Figure 6C). The sTBI fish at 4 dpi swam a total distance that gradually decreased from 2.84 m $\pm$ 0.15 m during iteration 1 to 2.46 m $\pm$ 0.2 m and 1.99 m $\pm$ 0.19 m during iterations 5 and 10, respectively (Figure 6C), although there was no statistically significant difference between iterations 1, 5, and 10. While the sTBI fish began to habituate at 4 dpi, unlike at 1 dpi, they still displayed increased swimming velocities for 21–24 s (Table 2). In contrast, sTBI fish at 7 dpi displayed habituation of the startle response similar to undamaged controls in every metric, including decreased swim velocity, decreased swim distance, and returning to a sedentary state from iterations 1 to 10 (Figure 6C, Table 2). Startle responses resembling undamaged behaviors persisted through 14 and 28 dpi (Figure 6C, Table 2). These data demonstrate that following blunt-force injury, sTBI fish display an impairment in habituation, a non-associative form of learning, that rapidly recovers to near undamaged control levels by 7 dpi and persists through 28 dpi.

Because the sTBI resulted in cell death across the brain, we asked if this would result in broader cognitive dysfunction. Therefore, we tested associative learning and memory, higher-level cognitive tasks that are modulated in teleosts in part by the medial- and lateral-dorsal pallium of the telencephalon [65,66], a region adjacent to our impact zone. To assess associative learning, we used a modified shuttle box assay [39], with a visual stimulus and an electrical shock as an adverse stimulus prompting learning by negative reinforcement [40]. Because the shuttle box assay relies on fish recognizing a visual stimulus, we first assessed survival of retinal neurons in sTBI fish. The number of TUNEL-positive cells in each retinal layer were the same in both sTBI and undamaged fish (Supplementary Figure S1A). In addition, we did not detect a significant number of proliferating Müller glia in either undamaged or sTBI fish (Supplementary Figure S1B), which is a typical response following retinal damage [23].

Undamaged control fish required an average of 19 $\pm$ 1.09 trials (Figure 6D) to master the assay (completing 5 consecutive positive trials without the negative reinforcement). Following injury, miTBI (47 ± 1.88 trials, $p < 0.01$, n = 12, Figure 6D), moTBI (51 ± 3.93 trials, $p < 0.01$), and sTBI fish (82 ± 2.98 trials, $p < 0.01$) all required significantly more trials to master the shuttle box assay relative to undamaged control fish at 1 dpi. At 2 dpi and 3 dpi, significant learning deficits persisted for all three TBI severities (Figure 6D). By 4 dpi, the average number of trials for miTBI and moTBI to master the shuttle box assay decreased to a level similar to undamaged fish (Figure 6D). In contrast, sTBI fish continued to display significant, but declining, deficits through 6 dpi, and returned to the undamaged control levels by 7 dpi, where they remained through 28 dpi (Figure 6D). Thus, this learning deficit correlated with the level of damage severity, both in magnitude (number of trials to master the assay) and duration before returning to control levels.

While humans with miTBI display little physiological pathologies, difficulties with short-term memory are a common feature [67]. We assessed whether our model recapitulated both immediate- and delayed recall of a learned behavior in undamaged and TBI treated fish using the shuttle box assay [40]. For immediate recall, the fish learned a behavior and then 4 h later were either undamaged or subjected to a TBI and then retested for the behavior learned 4 h before the TBI. Undamaged fish displayed immediate recall with an increase of 5.44% ± 2.13% in successful trials when retested 4 h following testing period 1 (Figure 3E). In contrast, we observed a significant decrease in successful trials post-injury, suggesting TBI fish exhibited difficulty in immediate recall. Specifically, miTBI fish exhibited a $-29.44\% \pm 2.17\%$ decrease in immediate recall ($p < 0.01$), and the deficits were further decreased to $-41.33\% \pm 1.37\%$ ($p < 0.01$) and $-51.16\% \pm 1.66\%$ ($p < 0.01$) following moTBI and sTBI, respectively (Figure 6E, n = 9). For delayed recall, the fish learned a behavior and then 4 days later were either undamaged or subjected to a TBI and then retested for the behavior learned 4 days before the TBI. The miTBI and moTBI fish demonstrated delayed recall comparable to undamaged fish (Figure 6E), with sTBI fish displaying a slight, near significant decrease ($p = 0.06$). Similar to human TBI patients [15,68], zebrafish experience significant deficits in immediate recall following TBI with deficits increasing in a severity-dependent manner, while delayed recall retention was relatively unaffected.

3.5. TBI Induces Cell Proliferation across the Neuroaxis in a Severity-Dependent Manner

Previously, zebrafish were shown to have a robust injury-induced proliferative response that is localized to the injury site [22,25,27]. However, the sizable gradient of apoptotic cell death that radiated from the epicenter of the impact zone (Figure 4), coupled with the recovery of learning deficits over several days (Figure 6C,D), led us to examine the extent of cell proliferation across the neuroaxis in a severity-dependent manner. The timing of peak cell proliferation was assessed by intraperitoneally (IP) injecting undamaged and sTBI fish with EdU at 12 h intervals from 36 to 84 hpi (Figure 7A–F) and examining EdU incorporation in brains at 12 h post injection (48, 60, 72, 84, and 96 hpi). Compared to undamaged controls, we observed the largest significant increase in the number of EdU-positive cells in sTBI brains at 60 and 72 hpi (Figure 7C,D,G) in the molecular layer (ML) adjacent to the cerebellar crest (CC), the epicenter of the impact zone.

To initially obtain a broad overview of the extent of proliferation along the neuroaxis following TBI, we examined whole brains from undamaged and sTBI fish that were EdU injected at 48 hpi and collected 12 h later at the peak of cell proliferation (Figure 7G). Cleared brains were stained and EdU incorporation was assessed using an epifluorescent microscope and confirmed by lightsheet fluorescence microscopy. EdU incorporation in undamaged fish revealed constitutive pockets of basal levels of cell proliferation across the brain (Figure 8A). However, following injury, large increases of EdU incorporation were observed within the impact zone that expanded across the brain (Figure 8B,C). To compare the injury-induced proliferation across the neuroaxis, the optical density was determined for the forebrain, midbrain, and hindbrain of undamaged and sTBI fish. Relative to undamaged fish, we observed significant increases in optical density in sTBI fish across the forebrain (Undam: 0.11 ± 0.013, sTBI: 0.33 ± 0.038 A.U., $p < 0.01$, n = 9, Figure 8A,B,D), midbrain (Undam: 0.12 ± 0.017, sTBI: 0.39 ± 0.039 A.U., $p < 0.01$, n = 9, Figure 8A,B,D), and hindbrain (Undam: 0.19 AU ± 0.017, sTBI: 0.52 AU ± 0.017, $p < 0.01$, n = 9, Figure 8A,B,D).

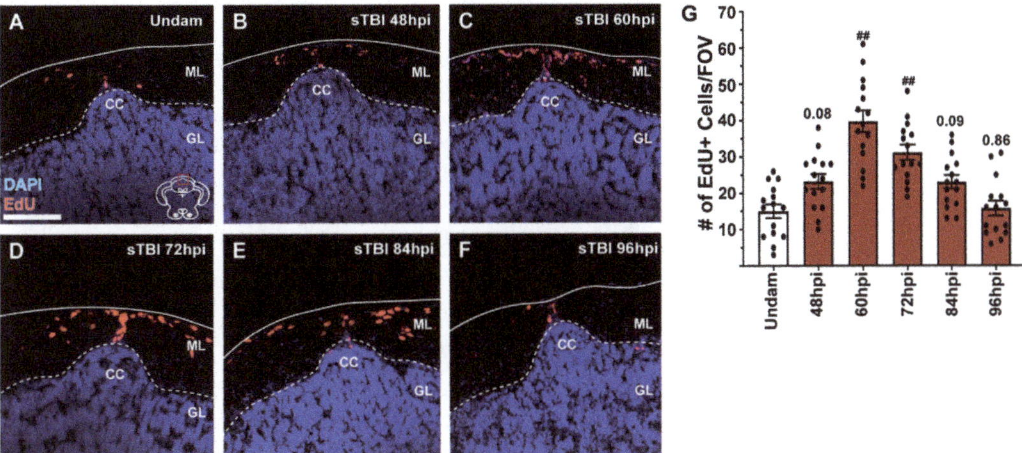

Figure 7. Blunt-force TBI induces cell proliferation at the cerebellar crest. (**A–F**) Confocal images of coronal cerebellar sections of undamaged and sTBI fish that were IP-injected with EdU 12 h prior to collection. EdU-labelled cells (red) are present radiating from the cerebellar crest in all panels. (**G**) Quantification of the number of EdU-positive cells at the CC at various timepoints following sTBI (n = 15). Increased EdU labeling, relative to undamaged fish, was observed as early as 48 hpi, with peak proliferation at 60 hpi. Solid lines in (**A–F**) denote tissue boundary, while dotted lines denote internal anatomical boundaries. Cerebellar crest, CC, granule cell layer, GL, molecular layer, ML. Scale bar = 100 μm for (**A–F**). Mean ± SEM is depicted in (**G**). Statistical analysis was performed using a One-way ANOVA followed by a Tukey's post-hoc test. # $p < 0.05$, ## $p < 0.01$.

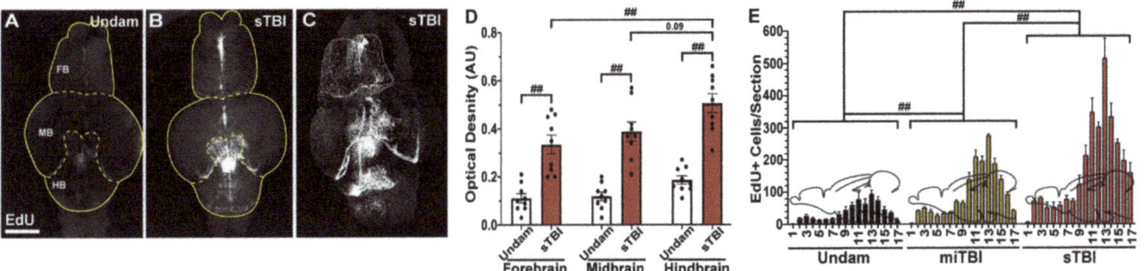

Figure 8. Blunt-force TBI results in increased proliferation in the brain. (**A,B**) Dorsal view of isolated and chemically-cleared undamaged (**A**) and sTBI (**B**) brains labeled with EdU and analyzed by fluorescent microscopy. The position of the forebrain (FB), midbrain (MB), and hindbrain (HB) are delineated. (**C**) Chemically-cleared sTBI brain imaged by lightsheet microscopy. (**D**) Quantification of optical density of EdU fluorescence by bulk brain region revealed significant increases in fluorescence in all sTBI regions relative to undamaged brains, with the largest increase seen in the hindbrain near the impact epicenter (n = 9). (**E**) Quantification of the total number of EdU-positive cells per section across the neuroaxis of undamaged, miTBI, and sTBI fish (n = 4). Significant increases in proliferation were observed in miTBI and sTBI brains relative to undamaged brains from the rostral tip of the olfactory bulb to caudal tip of the cerebellum. Scale bar = 500 μm. Mean ± SEM is depicted in D and E. Statistical analyses were performed with either a One-way ANOVA or Two-way ANOVA followed by a Tukey post-hoc test. ## $p < 0.01$.

To further investigate the distribution of proliferating cells within the different brain regions, EdU-injected brains (n = 4) of undamaged controls, miTBI, and sTBI fish were serially sectioned and the number of EdU-positive cells in each section was quantified (Figure 8E). EdU-positive cells were found throughout the entire undamaged brain, with the forebrain and rostral sections of the midbrain containing low numbers of EdU-positive cells, while larger numbers of EdU-positive cells were present in the caudal midbrain and hindbrain. In both miTBI and sTBI fish, proliferation was significantly increased across the entire neuroaxis compared to undamaged brains ($p < 0.01$, Figure 8E). Proliferation was also significantly further elevated across the entire sTBI brain relative to miTBI ($p < 0.01$). Thus, even following a miTBI, which exhibits minimal pathophysiological deficits, the damage induces a significant cell proliferation response.

A more detailed analysis of each section revealed specific subregions with increased EdU incorporation in TBI fish relative to undamaged controls. At the most rostral aspect of the brain, and furthest from the impact zone, the olfactory bulb displayed increased proliferation (Undam: 8.5 ± 2.05, miTBI: 16 ± 1.45, sTBI: 19 ± 2.89 cells). However these increases in the olfactory bulb were not significant relative to undamaged fish (sTBI $p < 0.12$, n = 4, Figure 9A,B,O), suggesting that the blunt-force injury did not reach the olfactory bulbs. More caudally, significantly greater numbers of EdU-positive cells were present in miTBI fish along the ventricular/subventricular zone (VZ) of the pallium (Pal_{vz}: miTBI: 79.5 ± 6.81 cells, $p < 0.01$, n = 4, Figure 9O) compared to undamaged controls (Pal_{vz}: Undam: 33 ± 2.85 cells). The number of EdU-positive cells along the pallium VZ was further elevated in sTBI brains (Pal_{vz}: sTBI: 144.75 ± 2.25 cells, $p < 0.01$, n = 4, Figure 9C,D) relative to undamaged controls and also between the miTBI and sTBI fish ($p < 0.01$, Figure 9O). Additionally, the VZ of the subpallium displayed significantly greater EdU incorporation in both miTBI ($Subpal_{vz}$: 40 ± 1.77 cells, $p < 0.01$, n = 4) and sTBI ($Subpal_{vz}$: 69.25 ± 9.56 cells, $p < 0.01$, n = 4) relative to undamaged controls ($Subpal_{vz}$: Undam: 22.75 ± 4.97 cells, Figure 9C–D′,O, Supplementary Figure S2), and between the miTBI and sTBI fish ($p < 0.01$, Figure 9O). In contrast, the telencephalon parenchyma possessed only a few EdU-positive cells in controls, with the miTBI and sTBI not showing a significant increase in EdU labeling, with the exception of the subpallium of sTBI fish relative to undamaged controls ($Subpal_{Par}$: sTBI: 29.5 ± 4.55 cells, $p < 0.05$, n = 4, Figure 9C–D′,O, Supplementary Figure S2).

Further caudally, approaching the impact zone, the midbrain was divided into six neuroanatomical regions, as defined by the neuroanatomical atlas of the adult zebrafish brain [69], including the Thalamus (Thal), Hypothalamus (Hypo), and the Periventricular grey zone (PGZ). Following miTBI and sTBI, there was a significantly greater number of EdU-positive cells within several regions relative to undamaged controls and across injury-severities. The Thal exhibited significantly more EdU-labeled cells in miTBI (60.5 ± 2.03, $p < 0.05$, n = 4) and sTBI (70 ± 2.28 cells, $p < 0.01$, n = 4, Figure 9E–F′,P) fish than in undamaged controls (22.75 ± 1.43 cells). Similarly, Hypo possessed significantly greater numbers of EdU-positive cells in miTBI fish (118 ± 9.1 cells, $p < 0.01$, n = 4) and sTBI fish (206.75 ± 18.89, $p < 0.01$, Figure 9G,G″,H,H″,P) than undamaged controls (52.25 ± 8.62 cells). Furthermore, in the midbrain, we observed the largest number EdU-labelled cells following TBI in the PGZ, which spans nearly the entirety of the midbrain. Relative to the undamaged PGZ (68.25 ± 13.59 cells), there were significantly greater numbers of EdU-labeled cells in the miTBI (185.5 ± 2.21 cells, $p < 0.01$, n = 4) and sTBI PGZ (315.75 ± 14.98, $p < 0.01$, n = 4, Figure 9G,H,P).

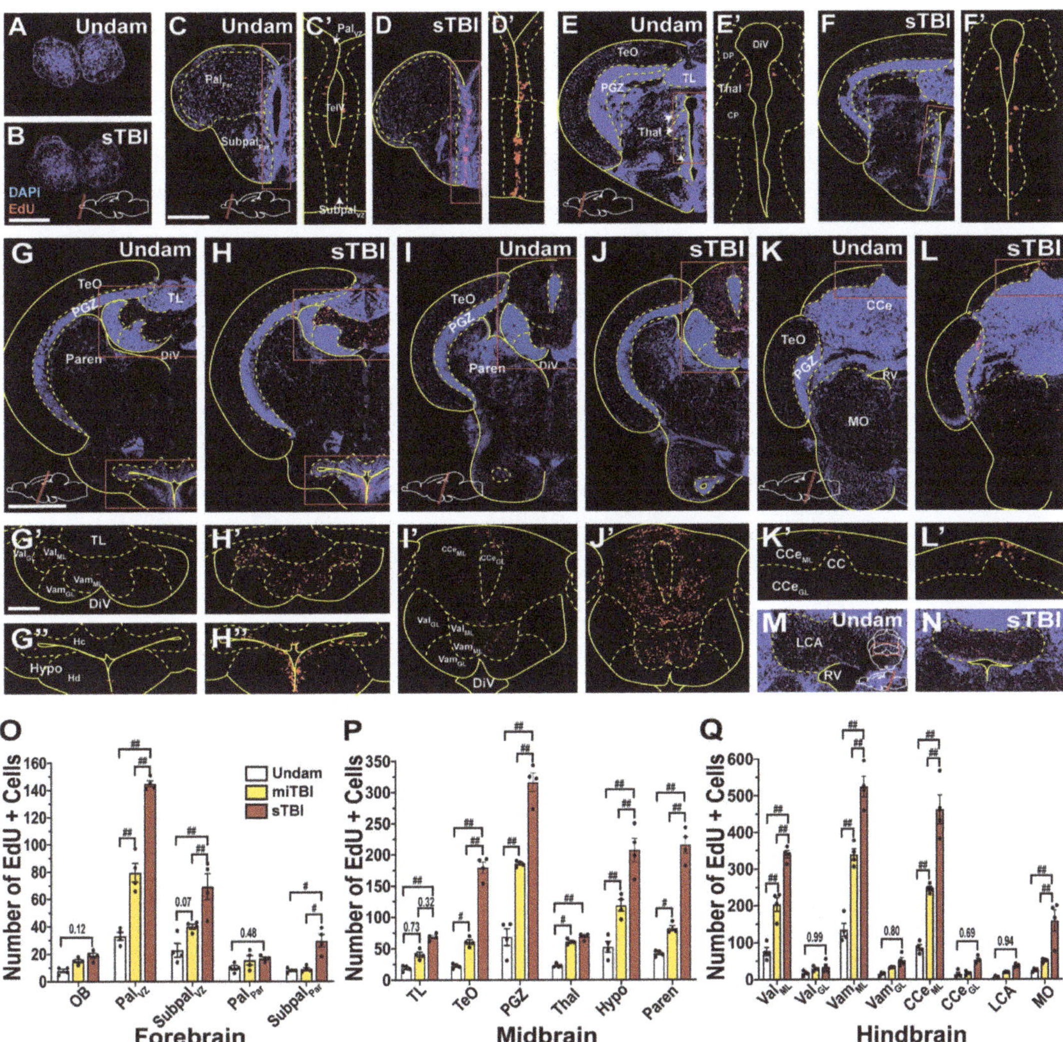

Figure 9. Injury-induced proliferation across the neuroaxis is region-specific and severity-dependent. (**A–N**) Coronal brain sections of undamaged and sTBI fish from the rostral aspect of the olfactory bulb to the caudal aspect of the lobus caudalis cerebelli. Red boxed regions in (**C–L**) are shown across the midline at a higher magnification in the corresponding prime and double prime panels. (**O–Q**) Quantification of the number of EdU-positive cells in brain subregions in undamaged, miTBI, and sTBI fish (n = 4). Solid lines in (**A–N**) denote tissue boundary, while dotted lines denote internal anatomical boundaries. Central posterior thalamic region, CP, corpus cerebelli, CCe, granule cell layer of corpus cerebelli, CCe$_{GL}$, molecular layer of corpus cerebelli, CCe$_{ML}$, diencephalic ventricle, DiV, dorsal posterior thalamic region, DP, lobus caudalis cerebelli, LCA, medulla oblongata, MO, olfactory bulbs, OB, parenchyma of pallium, Pal$_{Par}$, ventricular/subventricular zone of pallium, Pal$_{VZ}$, parenchyma of midbrain, Paren, periventricular grey zone of tectum optic, PGZ, rhombencephalic ventricle, RV, telencephalic ventricle, TeV, optic tectum, TeO, thalamus, Thal, torus longitudinalis, TL, parenchyma of subpallium, Subpal$_{Par}$, ventricular/subventricular zone of subpallium, Subpal$_{VZ}$, granule cell layer of lateral valvula cerebelli, Val$_{GL}$, molecular layer of lateral valvula cerebelli, Val$_{ML}$, granule cell layer of medial valvula cerebelli, Vam$_{GL}$, molecular layer of medial valvula cerebelli, Vam$_{ML}$. Scale bars: (**B**) = 100 μm, for panels (**A–C**) = 200 μm, for panels (**C–F'**,**G**) = 500 μm, for panels (**G–L**,**G'**) = 200 μm, for panels (**G'–L'**,**M,N**). Mean ± SEM is depicted in (**O–Q**). Statistical analyses were performed with a Two-way ANOVA followed by a Sidik's multiple comparison test. # $p < 0.05$, ## $p < 0.01$.

The hindbrain, which was the epicenter of the impact zone, was divided into the molecular and granular layers of the Lateral valvula cerebelli (Val$_{ML}$, Val$_{GL}$), Media valvula cerebelli (Vam$_{ML}$, Vam$_{GL}$), Corpus cerebelli (CCe$_{ML}$, CCe$_{GL}$), Lobus caudali cerebelli (LCA), and the Medulla oblongata (MO). The hindbrain possessed the larges basal levels of EdU-positive cells per neuroanatomical region in undamaged brains and the largest number of EdU-labeled cells following injury (Figure 8, Supplementary Figure S2 Following both miTBI and sTBI, significantly more EdU-positive cells were present in the molecular layers compared to undamaged controls and across severities (Figure 9G–L,C Supplementary Figure S2). Areas with prominent increases in EdU-positive cells following injury included the Val (miTBI: 203.25 ± 18.28, $p < 0.01$, n = 4, sTBI: 339.75 ± 9.97 cells $p < 0.01$, n = 4) and the Vam (miTBI:338.5 ± 15.21, $p < 0.01$, sTBI:523.75 ± 27.91, $p < 0.01$, n = 4 relative to undamaged brains (Val$_{ML}$: 76 ± 11, Vam$_{ML}$: 134 ± 17.7 cells Figure 9G–J′,Q One of the regions with the largest number of EdU-positive cells following injury wa the cerebellar crest (CC) of the CCe. The undamaged CC displayed 85.5 ± 7.57 EdU positive cells, while miTBI exhibited a significantly greater number (246.5 ± 5.73 cells $p < 0.01$, n = 4), with even greater numbers in sTBI CC (461.5 ± 39.08 cells, $p < 0.01$, n = 4 Figure 9K–L′,Q, Supplementary Figure S2). The most caudal and ventral sections of miTB and sTBI brains displayed significantly greater numbers of EdU-labeled cells in the MC (miTBI: 49.5 ± 5.26 cells, $p < 0.01$, sTBI: 157.75 ± 28.54 cells, $p < 0.01$, n = 4) relative tc undamaged controls (24 ± 3.91 cells, Figure 9K,L,Q). However, we observed no significan increases in EdU labeling in the LCA following either miTBI or sTBI relative to undamagec controls (Figure 9M,N,Q). Collectively, these data reveal that following injury, constitutive neurogenic regions significantly upregulated cell proliferation in a severity-dependen manner and the cell proliferation radiated beyond the impact zone.

3.6. TBI Results in Injury-Induced Cerebellar Proliferation, Progenitor Migration, and Differentiation

The Upper Rhombic Lip and the Cerebellar Recessuss, which collectively correspond: to the area (CC) in the adult hindbrain, has been heavily studied in development as a proliferative region that produces neuronal progenitors that migrate and differentiate intc most cerebellar cell types, including the most common, granule cell neurons [70]. As the fish age, the CC continues to generate basal levels of progenitor cells that migrate into the granule cell layer of the cerebellum and differentiate into granule cell neurons [71]. Because the CC exhibits significant increases in proliferation upon blunt-force trauma (Figure 7A–I and Figure 9K–L′), we sought to identify the migration and proliferative source of cerebella progenitors, as well as the fate of these cells.

To assess cell migration, undamaged and sTBI fish were IP-injected with EdU at 48 hp and collected at 51, 60, 72, 84, and 96 hpi. Following injury, the CC displayed increasec EdU incorporation relative to undamaged controls (Figure 10A–E′). These EdU-positive cells appeared to originate at the apical aspect of the CC (Figure 10A′,B′) and then migratec apically and laterally through the molecular layer at 60–72 hpi (Figure 10B′,C′). EdU positive cells then moved ventrally into the granule cell layer starting at 84 hpi and heavily infiltrated the granule cell layer by 96 hpi (Figure 10D′,E′).

During cerebellar development, the progenitors arose from populations of *nestin:EGFF* and *ptf1a:DsRed*-expressing cells [70,72]. However, following partial lateral excision of the adult zebrafish cerebellum, a significant increase of *nestin:EGFP*/PCNA double-positive cells was observed, while only limited numbers of *ptf1a:DsRed*/PCNA-positive cell: were present [27]. We assessed EdU incorporation in the Tg[*nestin*:GFP] line, during peak proliferation following blunt-force TBI. Undamaged and sTBI Tg[*nestin*:GFP] fisl were IP-injected with EdU at 48 hpi and collected 12 h later. Undamaged fish displayec low levels of EdU-incorporation (14.8 ± 1.72 cells, n = 10) and *nestin*:GFP-positive cells (14.6 ± 1.51 cells, n = 10), with most cells expressing both markers (11 ± 1.35 cells, n = 10 Figure 10F–F″,L). However, following injury, we observed a significant increase in the number of EdU-positive cells (43.9 ± 3.03 cells, $p < 0.01$, n = 10) and *nestin*:GFP-positive cells (36.4 ± 2.77 cells, $p < 0.01$, n = 10) at the CC relatively to the undamaged control. Fur thermore, there was a significant increase in the number of colabeled cells (30.6 ± 2.18 cells, Figure 10G–G″,L, $p < 0.01$, n = 10) at 60 hpi compared to undamaged controls.

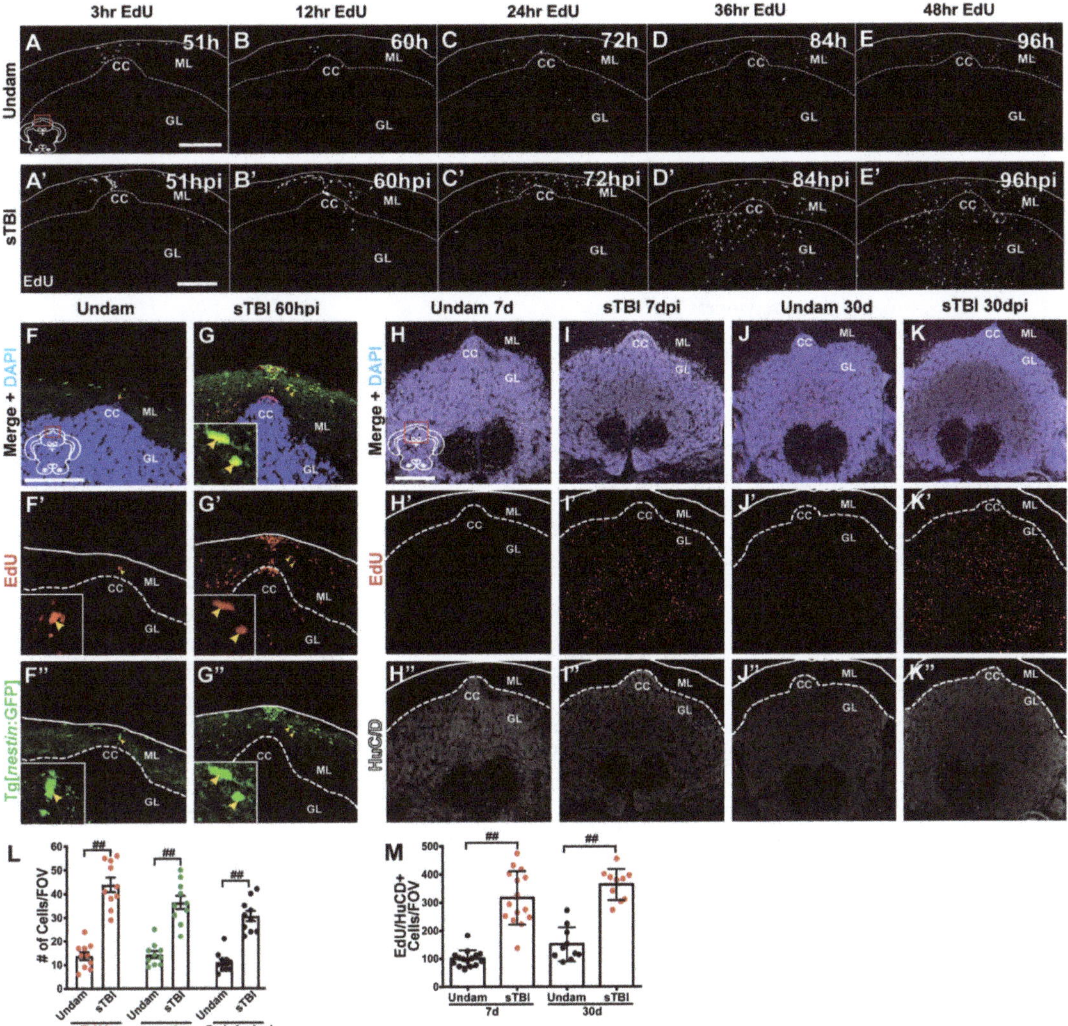

Figure 10. Blunt-force TBI induces cell proliferation, migration, and differentiation in the cerebellum. (**A–E′**) Coronal cerebellar sections of sTBI fish (**A′–E′**) that were IP-injected with EdU at 48 hpi and collected 51, 60, 72, 84, and 96 hpi to identify the migration pattern of injury-induced proliferative cells. Control undamaged fish (**A–E**) were also injected and brains assessed at similar intervals as sTBI fish. Coronal cerebellar sections of undamaged (**F**) and sTBI (**G**) Tg[*nestin*:GFP] fish with high magnification insets (**F′–F″,G′–G″**) that were IP-injected with EdU 12 h prior to collection at 60 hpi with colabeling of EdU and Tg[*nestin*:GFP] (yellow arrowheads). Coronal cerebellar sections of undamaged (**H–H″,J–J″**) and sTBI fish (**I–I″,K–K″**) that were IP-injected with EdU at 48 and 60 hpi to capture early onset and peak proliferative events and collected at either 7 (**H–I″**) or 30 dpi (**J–K″**) and costained with HuCD. (**L**) Quantification of the number of EdU-positive, *nestin*:GFP-positive, or colabeled cells for experiments in representative images (**F–G″**) (n = 10). (**M**) Quantification of the number of EdU/HuCD colabeled cells for experiments in representative images (**H–K″**) (n = 15). Solid lines in (**A–K″**) denote tissue boundary, while dotted lines denote internal anatomical boundaries. Cerebellar crest, CC, granule cell layer, GL, molecular layer, ML. All scale bars = 100 μm. Mean ± SEM is depicted in L and M. Statistical analyses were performed with either a One-way ANOVA or Two-way ANOVA followed by a Tukey's post-hoc test. ## $p < 0.01$.

We next asked if these EdU-positive cells that infiltrated the granule cell layer differentiated into neurons. Undamaged and sTBI fish were given EdU pulses at 48 and 60 hpi and we quantified the colocalization of EdU and the pan-neuronal marker HuCD at short (7 dpi) and long (30 dpi) recovery timepoints. Undamaged fish displayed basal levels of EdU/HuCD-colabeled cells in the granule cell layer of the cerebellum at 7 dpi (100.13 ± 7.6 cells, Figure 10H–H″,M) and the number of colabeled cells remained statistically unchanged at 30 d (151 ± 18.8 cells, $p = 0.24$, n $\geq$ 10, Figure 10J–J″,M). Following sTBI, we observed a significant increase in the number of EdU/HuCD-colabeled cells within the cerebellar granule cell layer relative to undamaged fish at both 7 dpi (316.36 ± 24.49 cells, $p < 0.01$, n = 15, Figure 10I–I″,M) and 30 dpi (356 ± 18.03, $p < 0.01$, n = 10, Figure 10K,K″,M). Similar to undamaged fish, we did not see a statistical difference in the number of EdU/HuCD-colabeled cells in sTBI fish between 7 dpi and 30 dpi ($p = 0.46$), suggesting that following injury, cells proliferate, migrate into the granule cell layer of the cerebellum, stably regenerate differentiated neurons, and then repress further regeneration.

3.7. Sonic Hedgehog Regulates Injury-Induced Proliferation in the Cerebellum

Sonic hedgehog (Shh) is a well characterized mitogen involved in development and regeneration and was demonstrated to play a critical role in regulating progenitors and neuronal regeneration following CNS trauma [73,74]. Therefore, we hypothesized that TBI-induced proliferation at the CC was regulated by Shh signaling and examined the temporal expression of Shh pathway components. We performed qRT-PCR to assess *shha*, *shhb*, *smo*, and *gli1* expression using RNA collected from the most dorsal third of undamaged and sTBI isolated cerebellums across early timepoints following blunt force trauma to the peak of cell proliferation (6, 12, 24, 36, 48, 60 hpi). Relative to undamaged control cerebellums, both *shha* and *shhb* RNAs (Shh ligands) were highly upregulated by 6 hpi, followed by increased expression of *smo* (Shh receptor) and *gli1* (downstream effector) by 12 hpi (Figure 11A).

To evaluate the influence of Shh signaling to stimulate TBI-induced cerebellar proliferation, undamaged controls were administered either 10 µM purmorphamine, a Smoothened (Smo) agonist, or vehicle control every 12 h for 48 h, then coinjected with EdU at 48 h, and collected at 60 h. Low basal levels of EdU-positive cells were observed at the CC in both untreated (9.75 ± 1.16 cells, Figure 11B) and vehicle-treated (Figure 11C) undamaged fish. In contrast, purmorphamine-treated undamaged fish exhibited a significant increase in the number of EdU-labeled cells at the CC (54.12 ± 5.51 cells, $p < 0.01$, n = 9, Figure 11D, L), relative to untreated controls. Interestingly, purmorphamine-induced proliferation in undamaged fish was similar to the proliferative response observed in untreated sTBI fish ($p = 0.95$, Figure 11L). Conversely, sTBI fish were administered either vehicle control or 2 mM cyclopamine, a Smo antagonist, at 4 hpi to account for early pathway activation and at 12, 24, 36, and 48 hpi, followed by coinjection with EdU at 48 hpi and collected 12 h later. Robust EdU-incorporation was observed in both untreated sTBI (56.5 ± 4.07 cells, Figure 11E) and vehicle-treated sTBI fish (Figure 11F). However, cyclopamine-treated sTBI fish had significantly fewer EdU-positive cells at the CC (4.25 ± 0.77 cells, $p = 0.01$, n = 9, Figure 11G,L) relative to the untreated control sTBI fish. Importantly, there was no significant difference in the number of EdU-positive cells between untreated undamaged controls and cyclopamine-treated sTBI fish ($p = 0.60$, Figure 11L), suggesting that Shh plays a role in TBI-induced cerebellar proliferation.

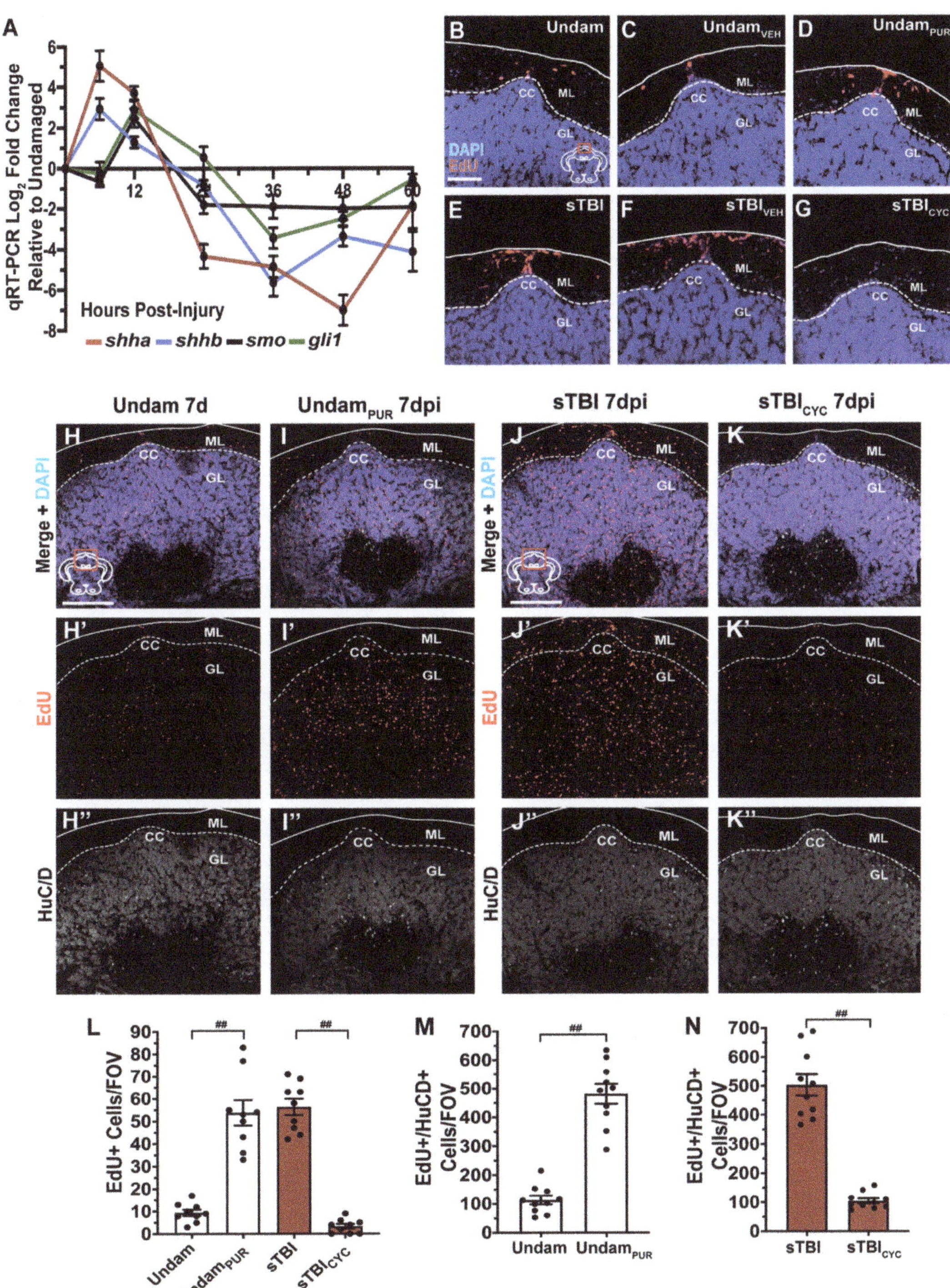

Figure 11. Shh regulates cerebellar proliferation and differentiation following injury. (**A**) Expression of Shh pathway genes by qRT-PCR reveals that the *shha* and *shhb* mRNAs are upregulated by 4 hpi in the top 1/3rd of the cerebellum, while *gli*

and *smo* expression increased by 12 hpi (5 pooled cerebellums/group, n = 3 groups). (**B–G**) Coronal cerebellar sections of the CC of undamaged (**B–D**) and sTBI fish (**E–G**) that were IP-injected with EdU and either vehicle (**C,F**), the Smo agonist purmorphamine (**D**), or the Smo antagonist cyclopamine (**G**). (**H,H″–I,I″**) Coronal cerebellar sections of undamaged fish that were either untreated (**H–H″**) or purmorphamine-treated (**I–I″**) that were IP-injected with EdU and collected at 7 dpi. Coronal cerebellar sections of sTBI fish that were either untreated (**J–J″**) or cyclopamine-treated (**K–K″**) that were IP-injected with EdU and collected at 7 dpi. (**L**) Quantification of the number of EdU-positive cells at the CC in undamaged and sTBI fish with Shh modulation (n = 9). (**M,N**) Quantification of the number of EdU/HuCD double-positive cells in the granule cell layer of the cerebellum in undamaged and sTBI fish with Shh modulation (n = 10). Solid lines in (**B–K″**) denote tissue boundary, while dotted lines denote internal anatomical boundaries. Cerebellar crest, CC, granule cell layer, GL, molecular layer, ML. All scale bars = 100 μm. Mean ± SEM is depicted in (**L–N**). Statistical analyses were performed with either a One-way ANOVA or Two-way ANOVA followed by a Tukey's post-hoc test. ## $p < 0.01$.

To further investigate the role of Shh in neuronal regeneration following a blunt force TBI, we examined the differentiation and production of new granule cell neurons in the cerebellum. Undamaged fish were IP-injected with purmorphamine every 12 h for 48 h, IP-injected with EdU at 48 and 60 h, and collected at 7 d to examine EdU and HuCD colabeling in the CC granule cell layer. Untreated, undamaged fish exhibited low numbers of differentiated HuCD neurons (113.3 ± 15.27 cells, n = 10, Figure 11H–H″,M), while purmorphamine-treated undamaged fish displayed a significantly greater number of EdU/HuCD double-positive cells in the CC granule cell layer (492 ± 32.52 cells, $p < 0.01$, n = 10, Figure 11I–I″,M). We also examined sTBI fish that were either uninjected or injected with cyclopamine at 4, 12, 24, 36, and 48 hpi. Both groups of sTBI fish were also IP-injected with EdU at 48 hpi and 60 hpi, collected at 7 dpi, and the number of EdU and HuCD double positive cells were quantified in the CC granule cell layer. Following injury, the uninjected sTBI fish possessed a large number of EdU/HuCD double-positive cells in the granule cell layer (508.2 ± 37.7 cells, n = 10, Figure 11J–J″,N). However, cyclopamine-treated sTBI fish exhibited significantly fewer double-positive cells (104.2 ± 8.7 cells, $p < 0.01$, n = 10, Figure 11K–K″,N) relative to the uninjected sTBI fish. Collectively, these data demonstrated that Shh signaling plays a role in regulating TBI-induced proliferation and the generation of HuCD-labeled neurons in the CC granule cell layer.

4. Discussion

Traumatic brain injuries produce a breadth of both acute and chronic pathologies [75] which largely correlate with injury severity. The current study describes a rapid and simple TBI model that utilizes the most common mechanism to induce human TBI: blunt force trauma [1,76]. We extensively characterized a scalable blunt-force injury model to examine mild, moderate, or severe TBIs in zebrafish, including the heterogeneity of severity-dependent injury-induced pathologies and the potential mechanisms underlying innate neuronal regeneration in the zebrafish brain. While the Marmarou weight drop is simple and rapid, it lacks the precision found in controlled cortical impact or lateral fluid percussive models, which are extensively studied in rodents [77]. Nevertheless, our results indicate that the zebrafish model induces many TBI sequelae analogous to those reported in the human population, including BBB disruption, neuroinflammatory response, and cognitive issues [3,48,49,78]. Consistency between human and zebrafish TBI-induced pathology makes zebrafish a useful tool to not only study the pathology and subsequent recovery post-TBI, but also provides the unique opportunity to study neuronal regeneration, which could reveal novel therapies for human TBI patients.

Previously, a blunt-force TBI model was described in adult zebrafish [32]. Our procedure was similar, except to dissipate the energy of the blunt-force trauma and to prevent cranial fractures in sTBI fish, we placed a small steel disc on the skull of the fish before dropping the weight, as is often done with rodents. The energy applied by Maheras et al. [32] was reported to be 35 mJ, which was over three-orders of magnitude greater than our calculated energies 1.33 mJ (miTBI), 2.08 mJ (moTBI), and 2.94 mJ (sTBI). The difference is that Maheras et al. [32] calculated their energy based on a fixed velocity of the falling ball, rather than an accelerating velocity, where the ball starts from a stationary position. If we used their parameters and calculated an accelerating velocity for the falling ball, then their energy is 0.35 mJ, which is 26% less than our mild blunt-force TBI (their steel ball had a mass of 0.33 g relative to our ball's mass of 1.5 g). Furthermore, the 35 mJ reported by Maheras et al. [32] is approaching the 40 mJ calculated from dropping a 20 g weight from a height of 20 cm onto a rat skull [79], which would likely crush the zebrafish skull. Thus, we feel that our miTBI is similar to the reported mild TBI energy reported by Maheras et al. [32].

Maheras et al. [32] did not report any significant phenotypic responses, outside of a learning and memory impairment using a T-box shoaling assay, or regenerative recovery in their mild blunt-force TBI. However, our study expanded the characterization to include several features of human TBI-induced pathologies across three levels of severity (mild, moderate, and severe) in adult zebrafish. Because classification of human TBI severity is often diagnosed with a collective, rather than a singular metric [3,48,49,78], we employed a variety of analogous tests to examine the breadth of zebrafish TBI pathologies. These diagnostic pathologies included injury-induced seizures and death, edema, BBB disruption, neuroinflammation, sensorimotor deficits, neuronal cell death, and cognitive impairments. While we did not exhaustively examine all known pathologies ascribed to human TBIs, our model validated a wide array of phenotypes that increased in severity with increasing levels of blunt force trauma. In humans, miTBI is the most reported injury [76], with many individuals experiencing little to no effects [59]. Similarly, our model demonstrated that following a miTBI, many injury-induced phenotypes (seizures, recovery rate, edema, vascular injury, and neuronal cell death) were not significantly different relative to undamaged controls, while there was a significant decrease in cognitive ability relative to undamaged controls.

Many of the injury-induced pathologies that we identified within 1 hpi could be early signs of the gradient of cell death we detected beginning at 16 h following sTBI, which emanated from the impact zone to more rostral portions of the brain. However, neuronal damage and possibly necrotic cell death likely occurred before 16 hpi and this could be associated with the early injury-induced pathologies we observed. While cell death was limited in the telencephalon, which is the analogous to the hippocampus and the location of many behaviors and cognitive ability [65,66], this region likely experienced severe disruption due to the significant cognitive deficits. While we did not measure learning within 16 hpi, we did examine the immediate recall of the fish at 4 hpi. In this case, we found that all three levels of TBI resulted in immediate memory deficits, suggesting that their cognitive function was negatively affected and supporting the idea that the telencephalon experienced sufficient damage. Alternatively, damage to the cerebellum, which has been implicated in fear learning and the escape response [64], may negatively affect a cerebellar brain circuit and the cognitive deficits. Additionally, while the neuronal cell death we observed is likely not the cause of the pathological and cognitive deficits, it is a significant outcome of the blunt-force trauma.

One of the major reasons to study TBI in zebrafish is its innate neuronal regenerative capacity across a wide range of tissues, which cannot be studied in mammalian TBI models [80,81]. While neuronal regeneration in zebrafish has been examined previously, most studies of injury-induced proliferation have focused on focal injuries and the surrounding injury site [24,25,27]. One of the few studies describing a proliferative response outside of the immediate injury site, Amamoto et al. [82] reported BrdU-positive cells in the rostral portion of the adult axolotl telencephalon after surgically removing a portion of the dorsal pallium. Similarly, Lindsey et al. [44] described increased proliferation beyond the stab wound site in the adult zebrafish telencephalon. However, they quantified the proliferative response as a measure of optical density and combined multiple regions of the brain into large bulk areas: the forebrain, midbrain, and hindbrain limiting the analysis of proliferation in subregions. Our bulk proliferative findings (Figure 8A–D) are largely in agreement with Lindsey et al. [44]. Additionally, we provide a comprehensive and quantitative comparative analysis of the proliferation across the neuroaxis from the rostral tip of the olfactory bulb to the caudal aspect of the rhombencephalon, including multiple subregions, following mild and severe blunt force TBI. This significant cell proliferation response, even following a miTBI, suggests that the blunt force trauma either induces widespread neuronal damage outside of the impact zone or generates a broad damage signal to initiate cell proliferation.

We focused on the cerebellum for the impact zone and our regeneration response because the zebrafish cerebellum is well characterized developmentally, with committed neuronal progenitors originating in the Upper rhombic lip and symmetrically dividing to produce the neuronal diversity in the cerebellum [83,84]. This developmental program persists into adulthood and mediates active neurogenesis throughout the zebrafish adult life [19,71]. Furthermore, the large quiescent neurogenic niche in the cerebellum (Figure 8) [71,85] has previously been studied in the context of localized injury-induced proliferation. Partial surgical excision of the cerebellum in adult zebrafish resulted in increased proliferation at the cerebellar crest and the proliferating cells subsequently migrated in a water fountain fashion to repopulate the cerebellar granule cell layer [19,27,71] We similarly identified the cerebellar crest as a source of proliferating cells that migrate to the cerebellar granule cell layer and differentiate into neurons. The identity of this proliferation source, the various mechanisms utilized to induce and regulate this regeneration response, the spectrum of neuronal types that can be regenerated, and the recovery of functional circuits are obvious questions to be explored further.

The Shh pathway is essential for early development and neurogenesis [86] Recently, Shh was shown to be critical for zebrafish neuronal regeneration following trauma in other parts of the CNS [73,74]. Similarly, we demonstrated that blunt-force trauma to the cerebellum induced the expression of *shh* pathway genes. However, the bulk RNA-Seq dataset reported by Maheras et al. [32] did not reveal an upregulation of genes associated with Shh signaling. This was likely due to Maheras et al. [32] performing their bulk RNA-Seq on RNA isolated from brains at 3 and 21 dpi. The increased expression of Shh signaling genes that we observed using qRT-PCR reached peak expression by 12 hpi. By 60 hpi, which is our latest timepoint and 12 h prior to the bulk RNA-Seq dataset, the expression of *shha*, *shhb*, and *smo* had all decreased below their baseline expression level and *gli1* had returned to its undamaged expression level. We also confirmed that Shh signaling is necessary for the subsequent proliferation response at the cerebellar crest as demonstrated by cyclopamine treatment eliminating nearly all injury-induced proliferation at the cerebellar crest and decreased numbers of EdU/HuCD double-positive cells 7dpi. Furthermore, Shh activation in undamaged fish, by purmorphamine exposure, provoked a proliferative response at the cerebellar crest similar to the amount of proliferation observed following blunt-force trauma, which differentiated into HuC/D-positive neurons in the cerebellar granule cell layer. It remains to be determined what role, if any, Shh signaling has on the other TBI-induced pathologies in zebrafish.

5. Conclusions

1. The modified TBI model for zebrafish is scalable for mild, moderate, and severe injury.
2. Zebrafish blunt-force TBI produces heterogeneous phenotypes replicating human injury.
3. Injury results in cognitive deficits that rapidly recover within 7 days.
4. Following injury, significant proliferation is observed across the entire brain.
5. Shh regulates injury-induced proliferation in the cerebellum.

Supplementary Materials: The following are available online at https://www.mdpi.com/article/10.3390/biomedicines9080861/s1, Figure S1: Blunt-force TBI does not induce retinal damage, Figure S2: sTBI induces increased cell proliferation across the neuroaxis, Table S1: Sensorimotor ethogram.

Author Contributions: Conceptualization, D.R.H. and J.H.; Methodology, D.R.H., J.H., K.C., M.L. and R.A.P.; Validation, D.R.H., J.H., K.C. and M.L.; Formal Analysis, J.H., K.C. and M.L.; Investigation, J.H., K.C., M.L., Y.J.J. and R.A.P.; Writing—Original Draft Preparation, D.R.H. and J.H.; Writing—Review & Editing, D.R.H., J.H., M.L., R.A.P. and A.C.M.; Visualization, D.R.H. and J.H.; Supervision, D.R.H. and A.C.M.; Project Administration, D.R.H.; Funding Acquisition, D.R.H., J.H. and A.C.M. All authors have read and agreed to the published version of the manuscript.

Funding: This project was funded, in part, with support from the Indiana Spinal Cord & Brain Injury Research Fund from the Indiana State Department of Health to DRH. Its contents are solely the responsibility of the providers and do not necessarily represent the official views of the Indiana State Department of Health. This work was also supported by grants from the National Science Foundation Graduate Research Fellowship Program #DGE-1841556 (JTH), LTC Neil Hyland Fellowship of Notre Dame (JTH), Sentinels of Freedom Fellowship (JTH), Pat Tillman Scholarship (JTH), the Office of The Director of the National Institutes of Health (NIH) under Award Number S10OD020067 (ACM), the Center for Zebrafish Research at the University of Notre Dame, and the Center for Stem Cells and Regenerative Medicine at the University of Notre Dame.

Institutional Review Board Statement: The study was approved and conducted according to the guidelines of the University of Notre Dame Institutional Animal Care and Use Committee (IACUC Protocol # 21-02-6445, most recent approval 22 March 2021).

Informed Consent Statement: Not applicable.

Data Availability Statement: Data are contained within this article and the Supplementary Materials.

Acknowledgments: The authors would like to thank the Hyde lab members for their thoughtful discussions, Laura Schaefer Hyde for suggesting the adverse tactile stimulus to assess sensorimotor activity, the Freimann Life Sciences Center technicians for zebrafish care and husbandry, and the University of Notre Dame Optical Microscopy Core/NDIIF for the use of instruments and their services. The graphical abstract was made with Biorender.com.

Conflicts of Interest: The authors declare no conflict of interest.

References

1. Korley, F.K.; Kelen, G.D.; Jones, C.M.; Diaz-Arrastia, R. Emergency department evaluation of traumatic brain injury in the United States, 2009-2010. *J. Head Trauma Rehabil.* **2016**, *31*, 379–387. [CrossRef] [PubMed]
2. Corrigan, J.D.; Selassie, A.W.; Orman, J.A. The epidemiology of traumatic brain injury. *J. Head Trauma Rehabil.* **2010**, *25*, 72–80. [CrossRef]
3. Levin, H.S.; Diaz-Arrastia, R.R. Diagnosis, prognosis, and clinical management of mild traumatic brain injury. *Lancet Neurol.* **2015**, *14*, 506–517. [CrossRef]
4. Deutsch, M.B.; Mendez, M.F.; Teng, E. Interactions between traumatic brain injury and frontotemporal degeneration. *Dement. Geriatr. Cogn. Disord.* **2015**, *39*, 143–153. [CrossRef] [PubMed]
5. Fleminger, S.; Oliver, D.L.; Lovestone, S.; Rabe-Hesketh, S.; Giora, A. Head injury as a risk factor for Alzheimer's disease: The evidence 10 years on; a partial replication. *J. Neurol. Neurosurg. Psychiatry* **2003**, *74*, 857–862. [CrossRef] [PubMed]
6. Gardner, R.C.; Burke, J.F.; Nettiksimmons, J.; Goldman, S.; Tanner, C.M.; Yaffe, K. Traumatic brain injury in later life increases risk for Parkinson disease. *Ann. Neurol.* **2015**, *77*, 987–995. [CrossRef]
7. Marklund, N. Rodent models of traumatic brain injury: Methods and challenges. *Methods Mol. Biol.* **2016**, *1462*, 29–46.
8. Xiong, Y.; Mahmood, A.; Chopp, M. Animal models of traumatic brain injuries. *Nat. Rev. Neurosci.* **2013**, *14*, 128–142. [CrossRef]
9. Morganti-Kossmann, M.C.; Yan, E.; Bye, N. Animal models of traumatic brain injury: Is there an optimal model to reproduce human brain injury in the laboratory? *Injury* **2010**, *41* (Suppl. 1), S10–S13. [CrossRef]
10. Marmarou, A.; Foda, M.A.; van den Brink, W.; Campbell, J.; Kita, H.; Demetriadou, K. A new model of diffuse brain injury in rats. Part I: Pathophysiology and biomechanics. *J. Neurosurg.* **1994**, *80*, 291–300. [CrossRef]
11. Flierl, M.A.; Stahel, P.F.; Beauchamp, K.M.; Morgan, S.J.; Smith, W.R.; Shohami, E. Mouse closed head injury model induced by a weight-drop device. *Nat. Protoc.* **2009**, *4*, 1328–1337. [CrossRef]
12. Foda, M.A.; Marmarou, A. A new model of diffuse brain injury in rats. Part II: Morphological characterization. *J. Neurosurg.* **1994**, *80*, 301–313. [CrossRef] [PubMed]
13. Zohar, O.; Schreiber, S.; Getslev, V.; Schwartz, J.P.; Mullins, P.G.; Pick, C.G. Closed-head minimal traumatic brain injury produces long-term cognitive deficits in mice. *Neuroscience* **2003**, *118*, 949–955. [CrossRef]
14. Babikian, T.; McArthur, D.; Asarnow, R.F. Predictors of 1-month and 1-year neurocognitive functioning from the UCLA longitudinal mild, uncomplicated, pediatric traumatic brain injury study. *J. Int. Neuropsychol. Soc.* **2013**, *19*, 145–154. [CrossRef] [PubMed]

15. Rabinowitz, A.R.; Levin, H.S. Cognitive sequelae of traumatic brain injury. *Psychiatr. Clin. N. Am.* **2014**, *37*, 1–11. [CrossRef] [PubMed]

16. McInnes, K.; Friesen, C.L.; MacKenzie, D.E.; Westwood, D.A.; Boe, S.G. Mild Traumatic Brain Injury (mTBI) and chronic cognitive impairment: A scoping review. *PLoS ONE* **2017**, *12*, e0174847. [CrossRef]

17. Vihtelic, T.S.; Hyde, D.R. Light-induced rod and cone cell death and regeneration in the adult albino zebrafish (Danio rerio) retina. *J. Neurobiol.* **2000**, *44*, 289–307. [CrossRef]

18. Becker, C.G.; Becker, T. Adult zebrafish as a model for successful central nervous system regeneration. *Restor. Neurol. Neurosci.* **2008**, *26*, 71–80. [PubMed]

19. Kaslin, J.; Ganz, J.; Geffarth, M.; Grandel, H.; Hans, S.; Brand, M. Stem cells in the adult zebrafish cerebellum: Initiation and maintenance of a novel stem cell niche. *J. Neurosci.* **2009**, *29*, 6142–6153. [CrossRef]

20. Ito, Y.; Tanaka, H.; Okamoto, H.; Ohshima, T. Characterization of neural stem cells and their progeny in the adult zebrafish optic tectum. *Dev. Biol.* **2010**, *342*, 26–38. [CrossRef]

21. Than-Trong, E.; Bally-Cuif, L. Radial glia and neural progenitors in the adult zebrafish central nervous system. *Glia* **2015**, *63*, 1406–1428. [CrossRef]

22. Hentig, J.T.; Byrd-Jacobs, C.A. Exposure to Zinc Sulfate Results in Differential Effects on Olfactory Sensory Neuron Subtypes in Adult Zebrafish. *Int. J. Mol. Sci.* **2016**, *17*, 1445. [CrossRef]

23. Lahne, M.; Brecker, M.; Jones, S.E.; Hyde, D.R. The Regenerating Adult Zebrafish Retina Recapitulates Developmental Fate Specification Programs. *Front. Cell Dev. Biol.* **2021**, *8*, 617923. [CrossRef] [PubMed]

24. Kroehne, V.; Freudenreich, D.; Hans, S.; Kaslin, J.; Brand, M. Regeneration of the adult zebrafish brain from neurogenic radial glia-type progenitors. *Development* **2011**, *138*, 4831–4841. [CrossRef] [PubMed]

25. Kishimoto, N.; Shimizu, K.; Sawamoto, K. Neuronal regeneration in a zebrafish model of adult brain injury. *Dis. Model Mech.* **2012**, *5*, 200–209. [CrossRef] [PubMed]

26. Skaggs, K.; Goldman, D.; Parent, J.M. Excitotoxic brain injury in adult zebrafish stimulates neurogenesis and long-distance neuronal integration. *Glia* **2014**, *62*, 2061–2079. [CrossRef] [PubMed]

27. Kaslin, J.; Kroehne, V.; Ganz, J.; Hans, S.; Brand, M. Distinct roles of neuroepithelial-like and radial glia-like progenitor cells in cerebellar regeneration. *Development* **2017**, *144*, 1462–1471. [CrossRef]

28. McCutcheon, V.; Park, E.; Liu, E.; Sobhebidari, P.; Tavakkoli, J.; Wen, X.Y.; Baker, A.J. A novel model of traumatic brain injury in adult zebrafish demonstrates response to injury and treatment comparable with mammalian models. *J. Neurotrauma.* **2017**, *34*, 1382–1393. [CrossRef]

29. Alyenbaawi, H.; Kanyo, R.; Locskai, L.F.; Kamali-Jamil, R.; DuVal, M.G.; Bai, Q.; Wille, H.; Burton, E.A.; Allison, W.T. Seizures are a druggable mechanistic link between TBI and subsequent tauopathy. *Elife* **2021**, *10*, e58744. [CrossRef]

30. Doppenberg, E.M.; Choi, S.C.; Bullock, R. Clinical trials in traumatic brain injury: Lessons for the future. *J. Neurosurg. Anesthesiol.* **2004**, *16*, 87–94. [CrossRef]

31. Marshall, L.F. Head injury: Recent past, present, and future. *Neurosurgery* **2000**, *47*, 546–561.

32. Maheras, A.L.; Dix, B.; Carmo, O.M.S.; Young, A.E.; Gill, V.N.; Sun, J.L.; Booker, A.R.; Thomason, H.A.; Ibrahim, A.E.; Stanislaw, L.; et al. Genetic Pathways of Neuroregeneration in a Novel Mild Traumatic Brain Injury Model in Adult Zebrafish. *eNeuro* **2018**, *5*, ENEURO.0208-17.2017. [CrossRef]

33. Kassen, S.C.; Ramanan, V.; Montgomery, J.E.; TBurket, C.; Liu, C.G.; Vihtelic, T.S.; Hyde, D.R. Time course analysis of gene expression during light-induced photoreceptor cell death and regeneration in albino zebrafish. *Dev. Neurobiol.* **2007**, *67*, 1009–1031. [CrossRef] [PubMed]

34. Tsetskhladze, Z.R.; Canfield, V.A.; Ang, K.C.; Wentzel, S.M.; Reid, K.P.; Berg, A.S.; Johnson, S.L.; Kawakami, K.; Cheng, K.C. Functional assessment of human coding mutations affecting skin pigmentation using zebrafish. *PLoS ONE* **2012**, *7*, e47398. [CrossRef] [PubMed]

35. White, R.M.; Sessa, A.; Burke, C.; Bowman, T.; LeBlanc, J.; Ceol, C.; Bourque, C.; Dovey, M.; Goessling, W.; Burns, C.E.; et al. Transparent adult zebrafish as a tool for in vivo transplantation analysis. *Cell Stem Cell* **2008**, *2*, 183–189. [CrossRef] [PubMed]

36. Hentig, J.; Cloghessy, K.; Dunseath, C.; Hyde, D.R. A scalable model to study the effects of blunt-force injury in adult zebrafish. *J. Vis. Exp.* **2021**, *171*. [CrossRef]

37. Kalueff, A.V.; Gebhardt, M.; Stewart, A.M.; Cachat, J.M.; Brimmer, M.; Chawla, J.S.; Craddock, C.; Kyzar, E.J.; Roth, A.; Landsman, S.; et al. Zebrafish neuroscience research consortium. Towards a comprehensive catalog of zebrafish behavior 1.0 and beyond. *Zebrafish* **2013**, *10*, 70–86. [CrossRef] [PubMed]

38. Hoshi, Y.; Okabe, K.; Shibasaki, K.; Funatsu, T.; Matsuki, N.; Ikegaya, Y.; Koyama, R. Ischemic brain injury leads to brain edema via hyperthermia-induced TRPV4 activation. *J. Neurosci.* **2018**, *38*, 5700–5709. [CrossRef] [PubMed]

39. Truong, L.; Mandrell, D.; Mandrell, R.; Simonich, M.; Tanguay, R.L. A rapid throughput approach identifies cognitive deficits in adult zebrafish from developmental exposure to polybrominated flame retardants. *Neurotoxicology* **2014**, *43*, 134–142. [CrossRef] [PubMed]

40. Hentig, J.; Cloghessy, K.; Hyde, D.R. Shuttle box assay as an associative learning tool for cognitive assessment in learning and memory studies using adult zebrafish. *J. Vis. Exp.* **2021**, *173*. [CrossRef]

41. Chanin, S.; Fryar, C.; Varga, D.; Raymond, J.; Kyzar, E.; Enriquez, J.; Bagawandoss, S.; Gaikwad, S.; Roth, A.; Pham, M. Assessing startle responses and their habituation in adult zebrafish. In *Zebrafish Protocols for Neurobehavioral Research*; Kalueff, A., Stewart, A., Eds.; Humana Press: Totowa, NJ, USA, 2012; Volume 66, pp. 287–300.

42. Conner, C.; Ackerman, K.M.; Lahne, M.; Hobgood, J.S.; Hyde, D.R. Repressing notch signaling and expressing TNFα are sufficient to mimic retinal regeneration by inducing Müller glial proliferation to generate committed progenitor cells. *J. Neurosci.* **2014**, *34*, 14403–14419. [CrossRef]

43. Lahne, M.; Li, J.; Marton, R.M.; Hyde, D.R. Actin-Cytoskeleton- and Rock-Mediated INM are required for photoreceptor regeneration in the adult zebrafish retina. *J. Neurosci.* **2015**, *35*, 15612–15634. [CrossRef]

44. Lindsey, B.W.; Douek, A.M.; Loosli, F.; Kaslin, J. A whole brain staining, embedding, and clearing pipeline for adult zebrafish to visualize cell proliferation and morphology in 3-Dimensions. *Front. Neurosci.* **2018**, *11*, 750. [CrossRef]

45. Radu, M.; Chernoff, J. An in vivo assay to test blood vessel permeability. *J. Vis. Exp.* **2013**, *73*, e50062.

46. Eliceiri, B.P.; Gonzalez, A.M.; Baird, A. Zebrafish model of the blood-brain barrier: Morphological and permeability studies. *Methods Mol. Biol.* **2011**, *686*, 371–378.

47. Campbell, L.J.; Hobgood, J.S.; Jia, M.; Boyd, P.; Hipp, R.I.; Hyde, D.R. Notch3 and DeltaB maintain Müller glia quiescence and act as negative regulators of regeneration in the light-damaged zebrafish retina. *Glia* **2021**, *69*, 546–566. [CrossRef]

48. Lund, S.B.; Gjeilo, K.H.; Moen, K.G.; Schirmer-Mikalsen, K.; Skandsen, T.; Vik, A. Moderate traumatic brain injury, acute phase course and deviations in physiological variables: An observational study. *Scand. J. Trauma Resusc. Emerg. Med.* **2016**, *24*, 77. [CrossRef] [PubMed]

49. Yamamoto, S.; Levin, H.S.; Prough, D.S. Mild, moderate and severe: Terminology implications for clinical and experimental traumatic brain injury. *Curr. Opin. Neurol.* **2018**, *31*, 672–680. [CrossRef] [PubMed]

50. Wasserman, E.B.; Shah, M.N.; Jones, C.M.; Cushman, J.T.; Caterino, J.M.; Bazarian, J.J.; Gillespie, S.M.; Cheng, J.D.; Dozier, A. Identification of a neurologic scale that optimizes EMS detection of older adult traumatic brain injury patients who require transport to a trauma center. *Prehosp. Emerg. Care.* **2015**, *19*, 202–212. [CrossRef]

51. Annegers, J.F.; Hauser, W.A.; Coan, S.P.; Rocca, W.A. A population-based study of seizures after traumatic brain injuries. *N. Engl. J. Med.* **1998**, *338*, 20–24. [CrossRef] [PubMed]

52. Frey, L.C. Epidemiology of posttraumatic epilepsy: A critical review. *Epilepsia* **2003**, *44*, 11–17. [CrossRef]

53. Servadei, F.; Nasi, M.T.; Cremonini, A.M.; Giuliani, G.; Cenni, P.; Nanni, A. Importance of a reliable admission Glasgow Coma Scale score for determining the need for evacuation of posttraumatic subdural hematomas: A prospective study of 65 patients. *J. Trauma* **1998**, *44*, 868–873. [CrossRef] [PubMed]

54. Mutch, C.A.; Talbott, J.F.; Gean, A. Imaging Evaluation of Acute Traumatic Brain Injury. *Neurosurg. Clin.* **2016**, *27*, 409–439. [CrossRef] [PubMed]

55. Murray, K.N.; Parry-Jones, A.R.; Allan, S.M. Interleukin-1 and acute brain injury. *Front. Cell Neurosci.* **2015**, *9*, 18. [CrossRef] [PubMed]

56. Gill, J.; Motamedi, V.; Osier, N.; Dell, K.; Arcurio, L.; Carr, W.; Walker, P.; Ahlers, S.; Lopresti, M.; Yarnell, A. Moderate blast exposure results in increased IL-6 and TNFα in peripheral blood. *Brain Behav. Immun.* **2017**, *65*, 90–94. [CrossRef] [PubMed]

57. Gill, J.; Mustapic, M.; Diaz-Arrastia, R.; Lange, R.; Gulyani, S.; Diehl, T.; Motamedi, V.; Osier, N.; Stern, R.A.; Kapogiannis, D. Higher exosomal tau, amyloid-beta 42 and IL-10 are associated with mild TBIs and chronic symptoms in military personnel. *Brain Inj.* **2018**, *32*, 1277–1284. [CrossRef] [PubMed]

58. Hasegawa, T.; Hall, C.J.; Crosier, P.S.; Abe, G.; Kawakami, K.; Kudo, A.; Kawakami, A. Transient inflammatory response mediated by interleukin-1β is required for proper regeneration in zebrafish fin fold. *Elife* **2017**, *6*, e22716. [CrossRef]

59. Eierud, C.; Craddock, R.C.; Fletcher, S.; Aulakh, M.; King-Casas, B.; Kuehl, D.; LaConte, S.M. Neuroimaging after mild traumatic brain injury: Review and meta-analysis. *Neuroimage Clin.* **2014**, *4*, 283–294. [CrossRef]

60. Gill, J.; Latour, L.; Diaz-Arrastia, R.; Motamedi, V.; Turtzo, C.; Shahim, P.; Mondello, S.; DeVoto, C.; Veras, E.; Hanlon, D.; et al. Glial fibrillary acidic protein elevations relate to neuroimaging abnormalities after mild TBI. *Neurology* **2018**, *91*, e1385–e1389. [CrossRef]

61. Dikmen, S.S.; Corrigan, J.D.; Levin, H.S.; Machamer, J.; Stiers, W.; Weisskopf, M.G. Cognitive outcome following traumatic brain injury. *J. Head Trauma Rehabil.* **2009**, *24*, 430–438. [CrossRef]

62. Vanderploeg, R.D.; Curtiss, G.; Belanger, H.G. Long-term neuropsychological outcomes following mild traumatic brain injury. *J. Int. Neuropsychol. Soc.* **2005**, *11*, 228–236. [CrossRef]

63. Thompson, R.F.; Spencer, W.A. Habituation: A model phenomenon for the study of neuronal substrates of behavior. *Psychol. Rev.* **1966**, *73*, 16–43. [CrossRef]

64. Matsuda, K.; Yoshida, M.; Kawakami, K.; Hibi, M.; Shimizu, T. Granule cells control recovery from classical conditioned fear responses in the zebrafish cerebellum. *Sci. Rep.* **2017**, *7*, 11865. [CrossRef] [PubMed]

65. Flood, N.C.; Overmier, J.B.; Savage, G.E. Teleost telencephalon and learning: An interpretive review of data and hypotheses. *Physiol. Behav.* **1976**, *16*, 783–788. [CrossRef]

66. Xu, X.; Bazner, J.; Qi, M.; Johnson, E.; Freidhoff, R. The role of telencephalic NMDA receptors in avoidance learning in goldfish (Carassius auratus). *Behav. Neurosci.* **2003**, *117*, 548–554. [CrossRef]

67. Greve, K.W.; Bianchini, K.J.; Mathias, C.W.; Houston, R.J.; Crouch, J.A. Detecting malingered performance on the Wechsler Adult Intelligence Scale. Validation of Mittenberg's approach in traumatic brain injury. *Arch. Clin. Neuropsychol.* **2003**, *18*, 245–260 [CrossRef]

68. Dunning, D.L.; Westgate, B.; Adlam, A.R. A meta-analysis of working memory impairments in survivors of moderate-to-severe traumatic brain injury. *Neuropsychology* **2016**, *30*, 811–819. [CrossRef] [PubMed]

69. Wullimann, M.; Rupp, B.; Reichert, H. *Neuroanatomy of the Zebrafish Brain. A Topological Atlas*, 1st ed.; Brikhäuser: Basel, Switzerland, 1996; pp. 1–160.

70. Kaslin, J.; Kroehne, V.; Benato, F.; Argenton, F.; Brand, M. Development and specification of cerebellar stem and progenitor cells in zebrafish: From embryo to adult. *Neural Dev.* **2013**, *8*, 9. [CrossRef]

71. Grandel, H.; Kaslin, J.; Ganz, J.; Wenzel, I.; Brand, M. Neural stem cells and neurogenesis in the adult zebrafish brain: Origin, proliferation dynamics, migration and cell fate. *Dev. Biol.* **2006**, *295*, 263–277. [CrossRef] [PubMed]

72. Kani, S.; Bae, Y.K.; Shimizu, T.; Tanabe, K.; Satou, C.; Parsons, M.J.; Scott, E.; Higashijima, S.; Hibi, M. Proneural gene-linked neurogenesis in zebrafish cerebellum. *Dev. Biol.* **2010**, *343*, 1–17. [CrossRef]

73. Ueda, Y.; Shimizu, Y.; Shimizu, N.; Ishitani, T.; Ohshima, T. Involvement of sonic hedgehog and notch signaling in regenerative neurogenesis in adult zebrafish optic tectum after stab injury. *J. Comp. Neurol.* **2018**, *526*, 2360–2372. [CrossRef]

74. Thomas, J.L.; Morgan, G.W.; Dolinski, K.M.; Thummel, R. Characterization of the pleiotropic roles of Sonic Hedgehog during retinal regeneration in adult zebrafish. *Exp. Eye Res.* **2018**, *166*, 106–115. [CrossRef]

75. Binder, S.; Corrigan, J.D.; Langlois, J.A. The public health approach to traumatic brain injury: An overview of CDC's research and programs. *J. Head Trauma Rehabil.* **2005**, *20*, 189–195. [CrossRef] [PubMed]

76. Cassidy, J.D.; Carroll, L.; Peloso, P.; Borg, J.; Von Holst, H.; Holm, L.; Kraus, J.; Coronado, V. Incidence, risk factors and prevention of mild traumatic brain injury: Results of the WHO Collaborating Centre Task Force on Mild Traumatic Brain Injury. *J. Rehabil. Med.* **2004**, *43*, 28–60. [CrossRef]

77. Marklund, N.; Hillered, L. Animal modelling of traumatic brain injury in preclinical drug development: Where do we go from here? *Br. J. Pharmacol.* **2011**, *164*, 1207–1229. [CrossRef]

78. Ruff, R.M.; Iverson, G.L.; Barth, J.T.; Bush, S.S.; Broshek, D.K.; NAN Policy and Planning Committee. Recommendations for diagnosing a mild traumatic brain injury: A National Academy of Neuropsychology education paper. *Arch. Clin. Neuropsychol.* **2009**, *24*, 3–10. [CrossRef]

79. Nasution, R.A.; Islam, A.A.; Hatta, M. Decreased neutrophil levels in mice with traumatic brain injury after cape administration. *Ann. Med. Surg.* **2020**, *54*, 89–92. [CrossRef]

80. Gemberling, M.; Bailey, T.J.; Hyde, D.R.; Poss, K.D. The zebrafish as a model for complex tissue regeneration. *Trends Genet.* **2013**, *29*, 611–620. [CrossRef]

81. Lahne, M.; Nagashima, M.; Hyde, D.R.; Hitchcock, P.F. Reprogramming Müller Glia to regenerate retinal neurons. *Annu. Rev. Vis. Sci.* **2020**, *6*, 171–193. [CrossRef] [PubMed]

82. Amamoto, R.; Huerta, V.G.; Takahashi, E.; Dai, G.; Grant, A.K.; Fu, Z.; Arlotta, P. Adult axolotls can regenerate original neuronal diversity in response to brain injury. *Elife* **2016**, *5*, e13998. [CrossRef] [PubMed]

83. McFarland, K.A.; Topczewska, J.M.; Weidinger, G.; Dorsky, R.I.; Appel, B. Hh and Wnt signaling regulate formation of olig2+ neurons in the zebrafish cerebellum. *Dev. Biol.* **2008**, *318*, 162–171. [CrossRef] [PubMed]

84. Chaplin, N.; Tendeng, C.; Wingate, R.J. Absence of an external germinal layer in zebrafish and shark reveals a distinct, anamniote ground plan of cerebellum development. *J. Neurosci.* **2010**, *30*, 3048–3057. [CrossRef] [PubMed]

85. Ganz, J.; Brand, M. Adult Neurogenesis in Fish. *Cold Spring Harb. Perspect. Biol.* **2016**, *8*, a019018. [CrossRef]

86. Vaillant, C.; Monard, D. SHH pathway and cerebellar development. *Cerebellum* **2009**, *8*, 291–301. [CrossRef] [PubMed]

Review

Ethanol Effects on Early Developmental Stages Studied Using the Zebrafish

Priyadharshini Manikandan, Swapnalee Sarmah and James A. Marrs *

Department of Biology, School of Science, Indiana University-Purdue University Indianapolis, Indianapolis, IN 46202, USA
* Correspondence: jmarrs@iupui.edu

Abstract: Fetal alcohol spectrum disorder (FASD) results from prenatal ethanol exposure. The zebrafish (*Danio rerio*) is an outstanding in vivo FASD model. Early development produced the three germ layers and embryonic axes patterning. A critical pluripotency transcriptional gene circuit of *sox2*, *pou5f1* (*oct4*; recently renamed *pou5f3*), and *nanog* maintain potency and self-renewal. Ethanol affects *sox2* expression, which functions with *pou5f1* to control target gene transcription. Various genes, like *elf3*, may interact and regulate *sox2*, and *elf3* knockdown affects early development. Downstream of the pluripotency transcriptional circuit, developmental signaling activities regulate morphogenetic cell movements and lineage specification. These activities are also affected by ethanol exposure. Hedgehog signaling is a critical developmental signaling pathway that controls numerous developmental events, including neural axis specification. Sonic hedgehog activities are affected by embryonic ethanol exposure. Activation of sonic hedgehog expression is controlled by TGF-ß family members, Nodal and Bmp, during dorsoventral (DV) embryonic axis establishment. Ethanol may perturb TGF-ß family receptors and signaling activities, including the sonic hedgehog pathway. Significantly, experiments show that activation of sonic hedgehog signaling rescues some embryonic ethanol exposure effects. More research is needed to understand how ethanol affects early developmental signaling and morphogenesis.

Keywords: fetal alcohol spectrum disorder; ethanol; zebrafish; development; gastrulation; *sox2*; *elf3*; *shh*

Citation: Manikandan, P.; Sarmah, S.; Marrs, J.A. Ethanol Effects on Early Developmental Stages Studied Using the Zebrafish. *Biomedicines* **2022**, *10*, 2555. https://doi.org/10.3390/biomedicines10102555

Academic Editor: Marc Ekker

Received: 23 September 2022
Accepted: 12 October 2022
Published: 13 October 2022

Publisher's Note: MDPI stays neutral with regard to jurisdictional claims in published maps and institutional affiliations.

1. Fetal Alcohol Spectrum Disorder

Alcohol is a common teratogen that causes adverse effects during pregnancy. Fetal alcohol spectrum disorder (FASD) covers a range of developmental defects and disorders of prenatal alcohol exposure (PAE), which occur when a woman consumes alcohol during their pregnancy [1]. The consequences of PAE are dependent on many factors, including, amount and duration of alcohol exposure, maternal and fetal age and genetics [1]. Premature death of the fetus also occurs with PAE [1]. Fetal alcohol syndrome (FAS) is the most severe form of the spectrum for babies born following PAE [1]. FAS is characterized by a set of craniofacial dysmorphology, neural defects, cardiac defects, sensory dysfunction, motor disabilities, and learning disabilities [2]. A recent study reported that globally an estimated 1700 babies are born every day with FASD [3]. There is a higher prevalence of FASD cases in higher-risk populations, such as those with a lower socioeconomic status [3]. A study conducted by the Center for Disease Control and Prevention (CDC), using data collected from pregnant women between 2015 and 2017, showed that one in nine women drank at least one alcoholic drink in the past month while pregnant, and around one third of these women reported binge drinking (drinking at least four alcoholic drinks in one sitting) [4]. Due to social biases against pregnant women consuming alcohol, there may be an underreporting of prenatal alcohol exposure incidences [4].

There is no cure for FASD, although treatments have been developed to help symptoms and aid in the development of a child with FASD [5]. Although folic acid has been shown to lessen the effects of early ethanol exposure in mouse, chicken and zebrafish embryos [5–7], it is not known whether folic acid protects the human baby from the deleterious effects of alcohol exposure. The only way to avoid FASD is through prevention, by abstaining from alcohol during pregnancy. Educating people about the consequences of FASD on a person's quality of life may help with prevention [5].

2. Use of Zebrafish as an FASD Model

The zebrafish (*Danio rerio*) is an established model for developmental studies of ethanol exposure effects, recapitulating FASD phenotypes [8]. Mammalian models, such as mice, are more similar to human development, but in utero development is difficult to study, particularly early developmental stages [8]. Zebrafish can produce hundreds of fertilized eggs per mating, allowing many embryos to be studied. Zebrafish development is very rapid. Early development, somitogenesis and establishment of the body plan occurs in 24 h. External fertilization eliminates ongoing parental influence during development and allows direct observation of embryos. Embryos and larvae are transparent, facilitating observation. Zebrafish also share extensive genetic evolutionary conservation with humans. The zebrafish genome has been completely sequenced enabling scientists to create mutations using reverse genetics and study the outcomes. Thus, the zebrafish can model human development and be used to study effects of teratogenic factors, like ethanol [5,9,10].

3. Early Zebrafish Development
3.1. Blastula Stage

Fertilized zebrafish embryos go through a series of rapid cleavages in the first 3 hours post fertilization (hpf) [11]. Initially, a blastodisc of 16 cells forms a syncytium with the yolk cells, and subsequent cleavages produce cells that are no longer connected, as well as cells that are cytoplasmically connected to the yolk cell (yolk syncytial layer; YSL). Early cleavage stages are directed by maternal transcripts deposited in the oocyte [12,13]. The zygotic genome is activated and midblastula transition occurs at the 1000-cell stage at 3 hpf. Afterwards, cell divisions are slower and asynchronous [11–14].

Zygotic gene expression activation regulates pluripotency and morphogenesis. The pluripotency gene circuit, its role in zygotic genome activation and pluripotency maintenance will be discussed below. The YSL is formed, which interacts with the overlying embryo and is a critical extraembryonic signaling center [15]. The YSL microtubule and actin cytoskeleton drives blastoderm spreading over the yolk cell. This spreading occurs by thinning and expansion of this cell layer in a process called epiboly [16]. Three processes combine to produce epiboly movements: (i) The blastoderm cells migrate toward the vegetal pole; (ii) Microtubules within the yolk cell pull on the membrane-actin junction with the enveloping layer at the germ ring, dragging this junction toward the vegetal pole and (iii) Radial intercalation thins the blastoderm cell layers, which expands the cell sheet spreading it over the yolk cell [16].

Epiboly starts around 4 hpf and is the first morphogenetic event in zebrafish development. As the epiboly process begins, the blastula is patterned by extraembryonic signals from the yolk cell [16]. This patterning establishes a pre-gastrulation fate map and is also a prelude to extensive morphogenesis that occurs at the onset of gastrulation [16,17].

The yolk cell is an extraembryonic tissue that lies beneath the blastoderm. The deep cells are a mass of cells that make the embryo proper. The enveloping layer is one cell thick sheet, enclosing the deep cells [13]. The YSL is a multinucleated syncytium within the yolk cell that forms during the blastula stage and matures by the 10th cell cycle [17]. It does not contribute cells or nuclei to the developing embryo, but the YSL secretes signaling factors that induce germ layer specification, embryo patterning, epiboly, and plays an important role in directing cell movements during gastrulation [13,16–18]. The yolk cell also provides critical nutrients during development. An array of genes control signaling from the YSL during embryogenesis, including Nodal and its related proteins, which are required for mesoderm induction and dorsal patterning of the blastoderm [13,18].

3.2. Gastrulation

When epiboly reaches the yolk cell equator (50% epiboly), mesoderm and endoderm precursors involute at the cell margins. The endoderm forms a ventral layer, and the mesoderm will populate the space between the endoderm and the ectoderm, which remains at the embryo surface. The embryo continues to elongate by epiboly progression to the vegetal pole, while the germ layer precursors converge on the midline and extend along the anterior-posterior axis, in a process called convergence and extension. These massive cell rearrangements, establishing the body axes, require a series of carefully orchestrated cell movements. Zygotic transcription drives morphogenesis, but the critical genes and activities are only partially understood. Axis specification is coupled with convergence and extension, and thus, the transcriptional mechanisms that specify the anterior-posterior and dorsal-ventral axes work in coordination with the morphogenetic movements that organize the body plan. These gastrulation events are preludes to full establishment of the body plan and organogenesis [16].

3.3. Pluripotency Circuit

Previously, zygotic genome activation was thought to be abrupt at midblastula transition. New evidence indicates that there is a progressive series of zygotic genome activation events. In addition to regulating morphogenesis, zygotic genome activation induces and maintains pluripotency [11,14]. One important result of zygotic genome activation is the expression of the pluripotency transcriptional gene circuit: sex-determining region Y-box containing gene 2 (*sox2*), octamer-binding protein 4 (*oct4*) also known as POU domain class 5 transcription factor 1 (*pou5f1*, recently renamed *pou5f3*) in zebrafish, and *nanog* q50 homeobox [14,19–22]. These transcription factors activate their own and each other's gene transcription, which produces a self-maintaining, feed-forward circuit that maintains pluripotent stem cell self-renewal and represses differentiation [17,23,24].

Pou5f1 and Sox2 proteins dimerize and work together to activate *nanog* and other pluripotency genes [17]. These genes also participate in the zygotic genome activation, while activating the pluripotency transcriptional program [14,25]. Sox2 has a high mobility group (HMG) DNA-binding domain and a transactivation domain [19]. Sox genes are grouped based on their homology within HMG domains. In the zebrafish, *sox2*, along with *sox1*, *sox3*, and *sox19a/b* are part of the SoxB1 group expressed in the early embryo that share a similarity in sequence and are functionally redundant to one another [24]. During early embryogenesis, maternally provided *sox19b* activates transcription, and *sox2* is one of the first zygotic genes to be transcribed [24,26]. Pou5f1 has two DNA binding domains, a low-affinity POU-specific domain and a higher affinitiy POU-homeodomain [19]. Nanog functions through its one homeodomain that binds to DNA [19]. These factors cooperate to accurately control a critical transcriptional program prior to gastrulation. At gastrulation, this transcriptional circuit is interupted, allowing specification of the 3 germ layers and the initation of appropriate differentiation programs [11,17,19].

4. Consequences of Ethanol Exposure during Early Zebrafish Development

Embryos treated with ethanol starting from 2 hpf display defects at early stages, including reduced epiboly progression, which is due to defects in cell adhesion, microtubule organization, and radial intercallation cell movements [6,16,25]. Effects of early ethanol exposure lead to signaling defects that persist and influence later embryogenesis stages [25], but mechanisms remain unclear. Epiboly cell movements are coupled to morphogenesis during gastrulation, and cell adhesion regulation orchestrates these morphogenetic events, particularly convergence and extension of the body axis [6,16]. Cadherin adhesion responds to developmental signaling during gastrulation [16,27]. When E-cadherin was measured, there was little or no change in its expression levels. However, cells in ethanol exposed embryos showed adhesion reduction and morphogenesis changes characteristic of reduced adhesion (radial intercallation and cell migration defects). Convergence and extension gastrulation cell movements also depend on cell intercallation and cell migration events [6].

Morphogenesis defects arise from ethanol exposure. Morphogenesis defects may be caused by ethanol effects on early zygotic gene expression, which regulates some of the earliest morphogenetic events. It is also possible that ethanol has direct biochemical effects on the cytoskeleton, developmental signaling machinery or other components. Ethanol exposure at 2 hpf leads to gastrulation defects due to cell adhesion and microtubule defects, which begin during the blastula stage [6,25]. Ethanol exposure also dysregulates genes that are evolutionarily conserved in the vertebrates and regulated during gastrulation like the reduction in *sox2* expression [6,25].

4.1. Pluripotency Gene Expression Defects
4.1.1. sox2

Gene expression analysis showed that expression of numerous genes are affected by ethanol exposure. A study on ethanol effects on mouse embryonic stem cell proteins reported that ethanol effects the stoichiometry of SOX2 and OCT4, and it skews the normal functional balance of the two [20]. Pluripotency regulator *sox2* was reduced in the pre-gastrulation zebrafish embryo, which then reduced *sox2* target gene expression [25]. Epiboly and gastrulation cell movements are reduced by ethanol. The pluripotency gene *pou5f1* works with *sox2* and also regulates epiboly and gastrulation cell movements. Injecting small amounts of *sox2* mRNA restores gene expression, epiboly and gastrulation cell movements. A gene-regulatory network affected by ethanol exposure was found that includes sox2 [25]. It is likely that ethanol produces defects through pleiotropic effects on this network, and restoration of normal developmental gene expression would require manipulation of several genes. It could be possible to identify a hierarchy of transcription regulators, making the manipulation easier by controling a small subset of genes that are at the top of the gene regulatory network hierarchy.

4.1.2. elf3

The Elf3 (E74 like ETS transcription factor 3) transcription factor is dysregulated by ethanol exposure in the early (4.5 hpf) zebrafish embryo [25]. Little is known about this factor's function in the early embryo. Elf3 is a member of the E26 transformation-specific family of transcription factors, which play a major role in the development and progression of various types of cancers. Elf3 is also involved during development. In humans, *ELF3* expression was detected in the mid-Carnegie stages [28]. In mice, the expression of *Elf3* was detected after fertilization, which remained high until the blastocyst stage [29]. *Elf3* knockout led to lethality of mice in utero [30], and the pups that survived had defects in small intestine epithelial tissue [30]. Studies showed that ELF3 plays role in terminal differentiation of skin epidermis, epithelia of the cornea, keratinocyte, and T cell differentiation. Our current work on understanding the role of Elf3 during zebrafish development indicates that it is critical for the development of epithelial, mesenchymal, and nervous tissues [31]. The *elf3* gene was among the most strongly dysregulated by ethanol in the early embryo, and the Elf3 transcription factor also targets many other genes, which may act as an important factor in a gene regulatory network dysregulated by ethanol exposure [25]. By dysregulating *sox2* and *elf3* and other ethanol sensitive transcription factors, ethanol exposure may disrupt the crucial balance between pluripotency and differentiation.

The interaction of human ELF3 with the pluripotency regulators SOX2, OCT4, and NANOG has been detected. ELF3 knockdown reduced SOX2 and POU5F1/OCT4 expression, whereas overexpression of ELF3 increased SOX2 and POU5F1 expression in high-grade serous ovarian cancer cells [32]. Human embryonic carcinoma NCCIT cell studies showed that ELF3 is a negative transcriptional regulator of OCT4 and NANOG. ELF3 controls the expression of those genes by directly binding to the promoters of OCT4 and NANOG [33]. The interaction of ELF3 with the pluripotency factors varies depending on the cell- and tissue types. A study manipulating *sox2* and *elf3* during development in the embryos with and without ethanol exposure may shed light on the roles of these genes in the FASD pathogenesis.

4.1.3. *pou5f1*

Maternal and zygotic *pou5f1* mutant (MZ*spg*) embryos showed defects in epiboly progression and in all three embryonic lineages [34,35]. Studies have shown, at this stage of development, *pou5f1* determines pluripotency, and then, *pou5f1* facilitates cellular reorganization, cytoskeletal reorganization, migration, and cell adhesion [27,34,35]. The *pou5f1* mutants have defects in the enveloping layer, deep layer cells, and the YSL [27]. Pou5f1 activates Rab5c-mediated endocytosis and recycling, which controls E-cadherin (Cdh1) dynamics during cell migration [36]. Cdh1 loss-of-function produces epiboly defects [37,38], and this phenotype resembles ethanol-treated embryos [6].

4.2. Epiboly Defects

The epiboly defect induced by ethanol exposure raised the hypothesis that ethanol reduced Cdh1 expression in the early embryo. However, measuring mRNA and protein levels showed no difference between control and ethanol-treated embryos [6]. We next examined known mechanisms of epiboly: yolk cell microtubule cytoskeleton; radial intercalation cell movements; and cell migration of deep cells [6].

Yolk cell microtubules connect with the leading edge of the embryo (germ band) where the enveloping layer and deep cells adhere to the yolk cell via cadherins. The yolk cell microtubules produce forces that drag this adhesive connection toward the vegetal pole during epiboly. Ethanol exposure from 2–3 hpf fragmented the yolk cell microtubule network, which may affect the forces pulling the embryo toward the vegetal pole as illustrated by the shapes of enveloping layer cells at the adhesive border [6].

Radial intercalation of deep cells occurs when cells at the interior move from the interior layers to the inner and outer surfaces of the deep cell layer, intercalating interior cells between the surface cell layers. These cell movements reduce the deep cell layer thickness and expand the dimensions of this cell sheet, spreading over the yolk cell during epiboly. Ethanol exposure reduced the frequency of radial intercalation events and increased the number of failed intercalation events, where cells move to the surface and then moved back to the interior, reducing epiboly [6].

Directed cell movements toward the vegetal pole and involution during gastrulation are coupled with epiboly and promoting normal convergent extension of the body axis. We tracked cell movements in time-lapse and measured their directionality. Calculating the meandering index showed that cells in ethanol-exposed gastrulating embryos had increased meandering. Furthermore, the shape of the embryonic axis (axial mesendoderm stained using *ntl* probe in situ hybridization) was shorter, wider, and wedge-shaped in ethanol exposed embryos at mid-gastrulation (8 hpf). These data indicate that directed cell movements and convergent-extension cell movements were affected by ethanol exposure [6].

These effects on epiboly (radial intercalation cell movements; and cell migration of deep cells) and gastrulation (convergent extension) phenocopy Cdh1 loss-of-function during early zebrafish development, which prompted us to examine Cdh1 expression levels and distribution. Indeed, we measured a reduced adhesion activity in blastomeres from ethanol exposed embryos in comparison to untreated embryos. Reduced adhesion occurred despite our results showing that there was no reduction in mRNA encoding Cdh1 and no reduction in the Cdh1 protein levels. The ratio of cell surface-to-cytoplasmic Cdh1 distribution was not different in the prechordal plate cells between control and ethanol treated embryos. However, there were cytoplasmic Cdh1 aggregates in the ethanol treated embryo prechordal plate cells, the significance of with remains unclear. The evidence indicated that there was no significant difference in Cdh1 levels or distribution [6].

Gene expression analysis comparing 8 hpf embryos treated with ethanol as compared with control embryos showed numerous ethanol dysregulated genes. One highly dysregulated gene was that encoding protocadherin-18a (Pcdh-18a), being reduced nearly 2-fold. We validated this gene expression change using quantitative PCR [6]. Protocadherins were shown to partner with classical cadherins to promote normal cell adhesion [39]. Perhaps reduced Pcdh-18a is responsible for aspects of the epiboly and gastrulation defects that resemble Cdh1 loss-of-function. This was tested by injecting synthetic mRNA encoding Pcdh-18a into embryos, which restored more normal epiboly and convergent extension phenotypes in ethanol and mRNA injected embryos that more closely resemble control embryos [6]. Together, our data showed that adhesion regulation was disrupted in ethanol exposed early embryos, producing gastrulation defects.

4.3. Sonic Hedgehog

Several studies have implicated sonic hedgehog (Shh) signaling defects in ethanol induced birth defects. The hedgehog (Hh) family of proteins are embryonic morphogens that mediate signal transduction pathways, regulating cell specification, differentiation and help maintain stem cells [17,40]. They form a spatial gradient in the tissue environment inducing differential gene expressions in a concentration dependent manner [17]. Shh, one of the three members of the Hh family in vertebrates, plays a critical role in embryonic cell proliferation, differentiation, and morphological patterning [40].

Shh processing regulates ligand secretion and, thus, signaling. Posttranslational lipid and cholesterol modification of Shh occurs in the Golgi. Modified Shh forms a protein complex with caveolin (Cav1), allowing for intracellular vesicular transport to lipid rafts in the plasma membrane, where it is then secreted [40,41]. In vertebrates, extracellular modified Shh binds to patched (Ptc), releasing smoothened (Smo) from the receptor complex. Smo signaling decouples suppressor of fused (SuFu), a negative regulator, from glioma-associated oncogene (Gli), allowing Gli to enter the nucleus and activate transcription [17,40].

Ethanol exposure during development can induce holoprosencephaly (HPE) [42,43]. HPE is characterized by defective rostroventral midline patterning of the forebrain with an array of other abnormalities, including failure of the forebrain to form hemispheres and cyclopia [44,45]. Disruption to various points in the Shh pathway, with or without exposure to ethanol, during development, can also lead to HPE [10,40,44,46].

Ethanol treated embryos have a similar phenotype to embryos deficient in *shh*, leading to the hypothesis that *shh* function is affected by ethanol exposure [10,40,46,47]. When exposed to alcohol, a defective posttranslational cholesterol modification on Shh may lead to reduced Shh signaling [40]. Studies have also shown that phenotypes produced by embryonic ethanol exposure such as cyclopia and other midfacial defects can arise from cell death of neural crest cells. Furthermore, *shh* developmental signaling was indirectly displaced by a synergistic interaction between ethanol and cyclopamine, a *shh* pathway inhibitor [43]. A rescue experiment using *shh* mRNA injection into ethanol treated zebrafish embryos reduced ethanol induced phenotypes, indicating that *shh* signaling is disrupted in FAS [47].

4.4. Cdon

In some patients with HPE, a loss of CDON (gene name for cell adhesion associated, oncogene regulated) function, a cell surface protein that facilitates the Shh pathway, was identified [45,46]. *Cdon* is a multi-functional co-receptor for Hh receptor and other receptor proteins [45,48]. Loss of *Cdon* in mice results in a mild HPE phenotype, but when coupled with ethanol exposure, more severe HPE phenotypes develop [49]. Similarly, a zebrafish study using *cdon* targeted morpholinos, *cdon* expression knockout produced mild craniofacial hypoplasia and did not produce and increase in cell death [50]. It was also shown that manipulation of the Shh pathway in zebrafish affected *cdon* expression during neural crest cell migration and epithelial mesenchymal transition, indicating that *cdon* responds to Shh signals [50]. Another study looked at the effect of Cdon in zebrafish eye development and found that Cdon functions as a negative regulator of Hh signaling in proximal-distal eye patterning [51].

Cdon also physically and genetically interacts with the Nodal pathway in mice, though the mechanism is still unclear [45]; Nodal signals prechordal plate (PCP) development from the anterior primitive streak, and the PCP produces Shh, which initiates forebrain patterning and rostroventral midline development [45]. Therefore, a defect in Nodal signaling during primitive streak formation, due to mutation or ethanol exposure, can exacerbate Hh signaling defects and HPE [45]. This study in mice suggests that Cdon plays an early role in development, prior its role as an Shh co-receptor.

Cdon may be redundant with other co-receptors like LRP2 [45]. Cdon and Lrp2 were shown to have similar functions in Nodal signaling. Mouse double mutants in Cdon and Lrp2 showed similar phenotypes as Nodal pathway mutations [45]. Additional studies are needed to understand ethanol effects on Nodal pathway and its downstream effects on Shh signaling during early development. Zebrafish may be useful to dissect the Nodal and Shh signaling pathway interactions and their interactions with ethanol.

5. Nodal and Bmp Gradients

Nodal and bone morphogenetic protein (Bmp) are transforming growth factor-*ß* (TGF-*ß)* superfamily members that regulate DV axis establishment [52–54]. Nodal and Bmp together, forming gradients in the DV axis of the zebrafish embryo consistent with the source/sink signal dispersal model hypothesized by Francis Crick in 1970 [55]. This model states that a signal is constantly produced at a localized source and diffuses through tissue where it is then destroyed or inhibited by a localized sink at specific distance away, forming a gradient signal that regulates morphogenesis [54,55]. Nodal and Bmp fit this model perfectly. Nodal signaling is concentrated on the dorsal end and Bmp signaling on the ventral end, both diffusing into the center of the embryo and inhibitors suppress further activation in distant areas [54,55].

Nodal and Bmp are first expressed from maternal transcripts in the YSL of a developing zebrafish embryo [13]. Nodal functions through two Nodal related genes, *nrd1* and *nrd2*, and its inhibitor, Antivin, to specify mesoderm and endoderm (mesendoderm) progenitors [13,52,56]. The ventral mesendoderm is formed receiving Bmp signals on the ventral-most side of the embryo. Chordin inhibits Bmp and is secreted by the dorsal organizer on the dorsal-most side of the embryo, which binds directly to Bmp to block its receptor interaction and signaling [52,54,57]. In zebrafish, the maternal Wnt/*ß*-catenin pathway activates the Nodal/Bmp cascade as well as Oct4, Drap1, and FoxH1 targets of Nodal signaling [56,58,59].

TGF-ß ligands, Nodal or Bmp, in their respective cellular locations, bind to and assemble the type I and type II activin transmembrane serine/threonine kinase receptor complex (Figure 1a) [60,61]. The heteromeric receptor complex, containing two type I and two type II receptors transduces the signal intracellularly by binding and phosphorylating receptor-regulated cytoplasmic Smad proteins (R-Smads) (Figure 1b,c). Common mediator Smad4 (Co-Smad) assembles with R-Smads in the cytoplasm forming heterotrimeric complexes which are then translocated into the nucleus to activate gene expression [54,62–65]. Nodal signaling leads to Smad2/3 phosphorylation, and Bmp signaling activates gene transcription through Smad 1/5/9 phosphorylation (Figure 1d) [54,62].

The TGF-ß pathway can be inhibited by Antivin or Chordin binding to EGF-CFC (epidermal growth factor- Crypto, FRL1, Cryptic) membrane linked coreceptor glycoproteins, blocking ligand signaling (Figure 1e) [54,57,66,67]. Inhibitory Smads (I-Smads), Smad6 and Smad7, can also inhibit this pathway by preventing intracellular Smad signaling by associating with the type I receptor (Figure 1f) [53,68–70].

A 2010 study using a 3% ethanol treatment for 3 h on mid-to late-blastula stage embryos showed a split axis phenotype starting at 24 hpf [71]. This phenotype resulted from cell movement disruption during the blastula and gastrulation stages [71]. This suggests that marginal tissue from the blastopore organize axis formation, but ethanol exposure delays epiboly progression, allowing a premature marginal axis to form and producing a split axis phenotype [71].

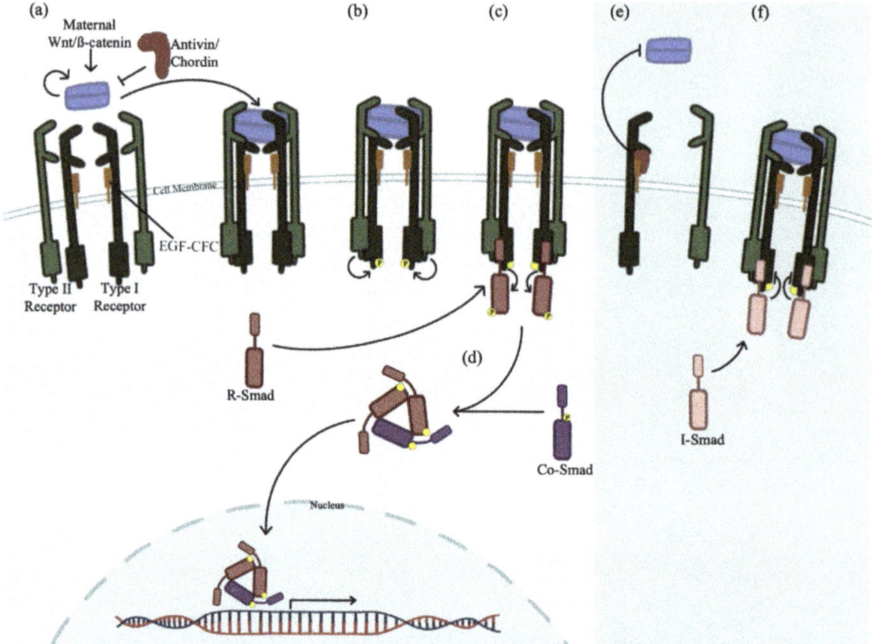

Figure 1. TGF-ß Signaling. TGF-ß ligands Nodal and Bmp activate Smad signaling. (**a**) Maternal Wnt/ß-catenin signals Nodal or Bmp ligands. These ligands bind and assemble the type I and type II heteromeric receptor complex. EGF-CFC co-receptors are bound to type I receptors. (**b**) Ligand binding transduces the signals intracellularly. Type II receptors phosphorylate type I receptors signaling Smad proteins. (**c**) R-Smad proteins are phosphorylated by type I receptors. Nodal specific R-Smads are Smad2 and 3. Bmp specific R-Smads are Smad1, 5, and 9. (**d**) Two R-Smads complex with one phosphorylated co-Smad4 and translocate into the nuclease and activate gene transcription. (**e**) Antivin or Chordin inhibits Nodal or Bmp, respectively, by binding to the EGF-CFC co-receptor on the type I receptor blocking ligand binding to the receptor. (**f**) After receptor phosphorylation, I-Smad6 or 7 can bind to the receptor complex blocking R-Smad binding and signaling.

6. Conclusions

Ethanol has detrimental effects on a developing embryo, and the zebrafish is a useful model for a developing human exposed to ethanol in utero (FASD) [8,25]. The range of defects depend on the concentration of ethanol and the timing of exposure, which produce developmental delays, brain defects, heart defects, craniofacial abnormalities, and potential lethality [1,2,9]. Ethanol affects transcriptional activity, but there may be independent effects within cells on proteins, protein complexes, lipid membrane structures and other effects. For example, our laboratory showed effects on the microtubule cytoskeleton in the yolk cell that occurred within 1 h of ethanol exposure at or prior to zygotic genome activation [6].

Our lab has previously studied the effects of embryonic ethanol exposure from the pre-gastrulation and mid-gastrulation stages, using Affymetrix GeneChip microarray gene expression analysis [6,25]. Ethanol exposure during embryogenesis alters the expression of important developmental genes. Sox2, Elf3, and their transcriptional targets produced potential ethanol dysregulated gene regulatory network changes [25]. Additional study is needed to understand the consequences of this gene regulatory network dysregulation.

A study by Hong et al., examined ethanol exposure effects on Nodal signaling using a Cdon mutation in a mouse model, showing ethanol effects on the interactions and trafficking of signaling proteins instead of directly disrupting early gene expression [49].

Many important early development activities are expressed from maternal transcripts in the early embryo [13]. If ethanol is altering these signaling activities, it could help explain the FASD phenotype that includes effects on the neural and body axes.

Pleiotropic effects of embryonic ethanol exposure make it difficult to sort out mechanisms. Furthermore, there are relatively few studies of ethanol exposure on the early embryo. The zebrafish is a particularly useful model for studying early development, like the experiments characterizing ethanol effects on cell adhesion and gene expression during zebrafish epiboly and gastrulation.

Author Contributions: Writing—original draft preparation, P.M., S.S. and J.A.M.; writing—review and editing, P.M., S.S. and J.A.M.; funding acquisition, S.S. and J.A.M. All authors have read and agreed to the published version of the manuscript.

Funding: This work was supported by NIH/NIAAA 1 R21 AA026711.

Institutional Review Board Statement: Not applicable.

Informed Consent Statement: Not applicable.

Data Availability Statement: Not applicable.

Acknowledgments: The authors thank members of the Marrs laboratory for helpful discussions.

Conflicts of Interest: The authors declare no conflict of interest.

References

1. Wallen, E.; Auvinen, P.; Kaminen-Ahola, N. The Effects of Early Prenatal Alcohol Exposure on Epigenome and Embryonic Development. *Genes* **2021**, *12*, 1095. [CrossRef]
2. Bilotta, J.; Barnett, J.A.; Hancock, L.; Saszik, S. Ethanol exposure alters zebrafish development: A novel model of fetal alcohol syndrome. *Neurotoxicol. Teratol.* **2004**, *26*, 737–743. [CrossRef] [PubMed]
3. Lange, S.; Probst, C.; Gmel, G.; Rehm, J.; Burd, L.; Popova, S. Global Prevalence of Fetal Alcohol Spectrum Disorder among Children and Youth: A Systematic Review and Meta-analysis. *JAMA Pediatr.* **2017**, *171*, 948–956. [CrossRef] [PubMed]
4. Denny, C.H.; Acero, C.S.; Naimi, T.S.; Kim, S.Y. Consumption of Alcohol Beverages and Binge Drinking among Pregnant Women Aged 18–44 Years-United States, 2015–2017. *Morbid. Mortal. Wkly. Rep.* **2019**, *68*, 365–368. [CrossRef] [PubMed]
5. Muralidharan, P.; Sarmah, S.; Marrs, J.A. Zebrafish retinal defects induced by ethanol exposure are rescued by retinoic acid and folic acid supplement. *Alcohol* **2015**, *49*, 149–163. [CrossRef]
6. Sarmah, S.; Muralidharan, P.; Curtis, C.L.; McClintick, J.N.; Buente, B.B.; Holdgrafer, D.J.; Ogbeifun, O.; Olorungbounmi, O.C.; Patino, L.; Lucas, R.; et al. Ethanol exposure disrupts extraembryonic microtubule cytoskeleton and embryonic blastomere cell adhesion, producing epiboly and gastrulation defects. *Biol. Open* **2013**, *2*, 1013–1021. [CrossRef]
7. Serrano, M.; Han, M.; Brinez, P.; Linask, K.K. Fetal alcohol syndrome: Cardiac birth defects in mice and prevention with folate. *Am. J. Obstet. Gynecol.* **2010**, *203*, 75.e7–75.e15. [CrossRef]
8. da Silva, J.P.; Luchiari, A.C. Embryonic ethanol exposure on zebrafish early development. *Brain Behav.* **2021**, *11*, e02062. [CrossRef]
9. Fernandes, Y.; Lovely, C.B. Zebrafish models of fetal alcohol spectrum disorders. *Genesis* **2021**, *59*, e23460. [CrossRef]
10. Lovely, C.B.; Fernandes, Y.; Eberhart, J.K. Fishing for Fetal Alcohol Spectrum Disorders: Zebrafish as a Model for Ethanol Teratogenesis. *Zebrafish* **2016**, *13*, 391–398. [CrossRef]
11. Paranjpe, S.S.; Veenstra, G.J. Establishing pluripotency in early development. *Biochim. Biophys. Acta* **2015**, *1849*, 626–636. [CrossRef] [PubMed]
12. Akdogan-Ozdilek, B.; Duval, K.L.; Goll, M.G. Chromatin dynamics at the maternal to zygotic transition: Recent advances from the zebrafish model. *F1000Res* **2020**, *9*, 299. [CrossRef] [PubMed]
13. Schauer, A.; Pinheiro, D.; Hauschild, R.; Heisenberg, C.P. Zebrafish embryonic explants undergo genetically encoded self-assembly. *Elife* **2020**, *9*, e55190. [CrossRef] [PubMed]
14. Leichsenring, M.; Maes, J.; Mossner, R.; Driever, W.; Onichtchouk, D. Pou5f1 transcription factor controls zygotic gene activation in vertebrates. *Science* **2013**, *341*, 1005–1009. [CrossRef] [PubMed]
15. Kimmel, C.B.; Ballard, W.W.; Kimmel, S.R.; Ullmann, B.; Schilling, T.F. Stages of embryonic development of the zebrafish. *Dev. Dyn.* **1995**, *203*, 253–310. [CrossRef]
16. Bruce, A.E.E.; Heisenberg, C.P. Mechanisms of zebrafish epiboly: A current view. *Curr. Top. Dev. Biol.* **2020**, *136*, 319–341. [CrossRef]
17. Gilbert, S.F.; Barresi, M.J.F. *Developmental Biology 11th Edition*; Sinauer Associates Inc.: Sunderland, MA, USA, 2016.
18. Chen, S.; Kimelman, D. The role of the yolk syncytial layer in germ layer patterning in zebrafish. *Development* **2000**, *127*, 4681–4689. [CrossRef]
19. Chambers, I.; Tomlinson, S.R. The transcriptional foundation of pluripotency. *Development* **2009**, *136*, 2311–2322. [CrossRef]

20. Ogony, J.W.; Malahias, E.; Vadigepalli, R.; Anni, H. Ethanol alters the balance of Sox2, Oct4, and Nanog expression in distinct subpopulations during differentiation of embryonic stem cells. *Stem Cells Dev.* **2013**, *22*, 2196–2210. [CrossRef]

21. Robles, V.; Marti, M.; Belmonte, J.C.I. Study of pluripotency markers in zebrafish embryos and transient embryonic stem cell cultures. *Zebrafish* **2011**, *8*, 57–63. [CrossRef]

22. Boyer, L.A.; Lee, T.I.; Cole, M.F.; Johnstone, S.E.; Levine, S.S.; Zucker, J.P.; Guenther, M.G.; Kumar, R.M.; Murray, H.L.; Jenner, R.G.; et al. Core transcriptional regulatory circuitry in human embryonic stem cells. *Cell* **2005**, *122*, 947–956. [CrossRef] [PubMed]

23. Rodda, D.J.; Chew, J.L.; Lim, L.H.; Loh, Y.H.; Wang, B.; Ng, H.H.; Robson, P. Transcriptional regulation of Nanog by Oct4 and Sox2. *J. Biol. Chem.* **2005**, *280*, 24731–24737. [CrossRef] [PubMed]

24. Zhang, S.; Cui, W. Sox2, a key factor in the regulation of pluripotency and neural differentiation. *World J. Stem Cells* **2014**, *6*, 305–311. [CrossRef] [PubMed]

25. Sarmah, S.; Srivastava, R.; McClintick, J.N.; Janga, S.C.; Edenberg, H.J.; Marrs, J.A. Embryonic ethanol exposure alters expression of sox2 and other early transcripts in zebrafish, producing gastrulation defects. *Sci. Rep.* **2020**, *10*, 3951. [CrossRef]

26. Okuda, Y.; Yoda, H.; Uchikawa, M.; Furutani-Seiki, M.; Takeda, H.; Kondoh, H.; Kamachi, Y. Comparative genomic and expression analysis of group B1 sox genes in zebrafish indicates their diversification during vertebrate evolution. *Dev. Dyn.* **2006**, *235*, 811–825. [CrossRef]

27. Lachnit, M.; Kur, E.; Driever, W. Alterations of the cytoskeleton in all three embryonic lineages contribute to the epiboly defect of Pou5f1/Oct4 deficient MZspg zebrafish embryos. *Dev. Biol.* **2008**, *315*, 1–17. [CrossRef]

28. E!Ensembl Gene: ELF3 ENSG00000163435. Available online: https://useast.ensembl.org/Homo_sapiens/Gene/ExpressionAtlas?db=core;g=ENSG00000163435;r=1:202007945-202017183 (accessed on 27 July 2022).

29. Kageyama, S.; Liu, H.; Nagata, M.; Aoki, F. The role of ETS transcription factors in transcription and development of mouse preimplantation embryos. *Biochem. Biophys. Res. Commun.* **2006**, *344*, 675–679. [CrossRef]

30. Oliver, J.R.; Kushwah, R.; Hu, J. Multiple roles of the epithelium-specific ETS transcription factor, ESE-1, in development and disease. *Lab. Investig.* **2012**, *92*, 320–330. [CrossRef]

31. Sarmah, S.; Hawkins, M.R.; Manikandan, P.; Farrell, M.; James, A. Marrs. Elf3 deficiency during zebrafish development alters extracellular matrix organization and disrupts tissue morphogenesis. *PLoS ONE*, 2022; *In Press*.

32. Wang, Z.; Yin, P.; Sun, Y.; Na, L.; Gao, J.; Wang, W.; Zhao, C. LGR4 maintains HGSOC cell epithelial phenotype and stem-like traits. *Gynecol. Oncol.* **2020**, *159*, 839–849. [CrossRef]

33. Park, S.W.; Do, H.J.; Choi, W.; Lim, D.S.; Park, K.H.; Kim, J.H. Epithelium-specific ETS transcription factor-1 regulates NANOG expression and inhibits NANOG-induced proliferation of human embryonic carcinoma cells. *Biochimie* **2021**, *186*, 33–42. [CrossRef]

34. Onichtchouk, D.; Geier, F.; Polok, B.; Messerschmidt, D.M.; Mossner, R.; Wendik, B.; Song, S.; Taylor, V.; Timmer, J.; Driever, W. Zebrafish Pou5f1-dependent transcriptional networks in temporal control of early development. *Mol. Syst. Biol.* **2010**, *6*, 354. [CrossRef] [PubMed]

35. Onichtchouk, D. Pou5f1/oct4 in pluripotency control: Insights from zebrafish. *Genesis* **2012**, *50*, 75–85. [CrossRef] [PubMed]

36. Song, S.; Eckerle, S.; Onichtchouk, D.; Marrs, J.A.; Nitschke, R.; Driever, W. Pou5f1-dependent EGF expression controls E-cadherin endocytosis, cell adhesion, and zebrafish epiboly movements. *Dev. Cell* **2013**, *24*, 486–501. [CrossRef] [PubMed]

37. Babb, S.G.; Marrs, J.A. E-cadherin regulates cell movements and tissue formation in early zebrafish embryos. *Dev. Dyn.* **2004**, *230*, 263–277. [CrossRef] [PubMed]

38. Kane, D.A.; McFarland, K.N.; Warga, R.M. Mutations in half baked/E-cadherin block cell behaviors that are necessary for teleost epiboly. *Development* **2005**, *132*, 1105–1116. [CrossRef]

39. Emond, M.R.; Biswas, S.; Blevins, C.J.; Jontes, J.D. A complex of Protocadherin-19 and N-cadherin mediates a novel mechanism of cell adhesion. *J. Cell Biol.* **2011**, *195*, 1115–1121. [CrossRef]

40. Mao, H.; Diehl, A.M.; Li, Y.X. Sonic hedgehog ligand partners with caveolin-1 for intracellular transport. *Lab. Investig.* **2009**, *89*, 290–300. [CrossRef]

41. Li, Y.X.; Yang, H.T.; Zdanowicz, M.; Sicklick, J.K.; Qi, Y.; Camp, T.J.; Diehl, A.M. Fetal alcohol exposure impairs Hedgehog cholesterol modification and signaling. *Lab. Investig.* **2007**, *87*, 231–240. [CrossRef]

42. Hong, M.; Krauss, R.S. Ethanol itself is a holoprosencephaly-inducing teratogen. *PLoS ONE* **2017**, *12*, e0176440. [CrossRef]

43. Sidik, A.; Dixon, G.B.; Kirby, H.G.; Eberhart, J.K. Gene-environment interactions characterized by single embryo transcriptomics. *BioRxiv* **2020**, 805556. [CrossRef]

44. Slavotinek, A.M. *Chapter 128 Dysmorphology-Nelson Textbook of Pediatrics*; Elsevier Inc.: Amsterdam, The Netherlands, 2020.

45. Hong, M.; Christ, A.; Christa, A.; Willnow, T.E.; Krauss, R.S. Cdon mutation and fetal alcohol converge on Nodal signaling in a mouse model of holoprosencephaly. *Elife* **2020**, *9*, e60351. [CrossRef] [PubMed]

46. Eberhart, J.K.; Parnell, S.E. The Genetics of Fetal Alcohol Spectrum Disorders. *Alcohol Clin. Exp. Res.* **2016**, *40*, 1154–1165. [CrossRef] [PubMed]

47. Loucks, E.J.; Ahlgren, S.C. Deciphering the role of Shh signaling in axial defects produced by ethanol exposure. *Birth Defects Res. A Clin. Mol. Teratol.* **2009**, *85*, 556–567. [CrossRef] [PubMed]

48. Echevarria-Andino, M.L.; Allen, B.L. The hedgehog co-receptor BOC differentially regulates SHH signaling during craniofacial development. *Development* **2020**, *147*, dev189076. [CrossRef] [PubMed]

49. Hong, M.I.; Krauss, R.S. Cdon Mutation and Fetal Ethanol Exposure Synergize to Produce Midline Signaling Defects and Holoprosencephaly Spectrum Disorders in Mice. *PLoS Genet.* **2012**, *8*, e1002999. [CrossRef]

50. Powell, D.R.; Williams, J.S.; Hernandez-Lagunas, L.; Salcedo, E.; O'Brien, J.H.; Artinger, K.B. Cdon promotes neural crest migration by regulating N-cadherin localization. *Dev. Biol.* **2015**, *407*, 289–299. [CrossRef]

51. Cardozo, M.J.; Sanchez-Arrones, L.; Sandonis, A.; Sanchez-Camacho, C.; Gestri, G.; Wilson, S.W.; Guerrero, I.; Bovolenta, P. Cdon acts as a Hedgehog decoy receptor during proximal-distal patterning of the optic vesicle. *Nat. Commun.* **2014**, *5*, 4272. [CrossRef]

52. Liu, Z.; Lin, X.; Cai, Z.; Zhang, Z.; Han, C.; Jia, S.; Meng, A.; Wang, Q. Global identification of SMAD2 target genes reveals a role for multiple co-regulatory factors in zebrafish early gastrulas. *J. Biol. Chem.* **2011**, *286*, 28520–28532. [CrossRef]

53. Wei, C.Y.; Wang, H.P.; Zhu, Z.Y.; Sun, Y.H. Transcriptional factors smad1 and smad9 act redundantly to mediate zebrafish ventral specification downstream of smad5. *J. Biol. Chem.* **2014**, *289*, 6604–6618. [CrossRef]

54. Rogers, K.W.; Muller, P. Nodal and BMP dispersal during early zebrafish development. *Dev. Biol.* **2019**, *447*, 14–23. [CrossRef]

55. Crick, F. Diffusion in embryogenesis. *Nature* **1970**, *225*, 420–422. [CrossRef] [PubMed]

56. Pinheiro, D.; Heisenberg, C.P. Zebrafish gastrulation: Putting fate in motion. *Curr. Top. Dev. Biol.* **2020**, *136*, 343–375. [CrossRef] [PubMed]

57. Troilo, H.; Zuk, A.V.; Tunnicliffe, R.B.; Wohl, A.P.; Berry, R.; Collins, R.F.; Jowitt, T.A.; Sengle, G.; Baldock, C. Nanoscale structure of the BMP antagonist chordin supports cooperative BMP binding. *Proc. Natl. Acad. Sci. USA* **2014**, *111*, 13063–13068. [CrossRef] [PubMed]

58. Marlow, F.L. Setting up for gastrulation in zebrafish. *Curr. Top. Dev. Biol.* **2020**, *136*, 33–83. [CrossRef] [PubMed]

59. Kumari, P.; Gilligan, P.C.; Lim, S.; Tran, L.D.; Winkler, S.; Philp, R.; Sampath, K. An essential role for maternal control of Nodal signaling. *Elife* **2013**, *2*, e00683. [CrossRef] [PubMed]

60. Massague, J. TGFbeta signalling in context. *Nat. Rev. Mol. Cell Biol.* **2012**, *13*, 616–630. [CrossRef] [PubMed]

61. Massague, J.; Chen, Y.G. Controlling TGF-beta signaling. *Genes. Dev.* **2000**, *14*, 627–644. [CrossRef] [PubMed]

62. Bennett, J.T.; Joubin, K.; Cheng, S.; Aanstad, P.; Herwig, R.; Clark, M.; Lehrach, H.; Schier, A.F. Nodal signaling activates differentiation genes during zebrafish gastrulation. *Dev. Biol.* **2007**, *304*, 525–540. [CrossRef]

63. Kumar, A.; Novoselov, V.; Celeste, A.J.; Wolfman, N.M.; ten Dijke, P.; Kuehn, M.R. Nodal signaling uses activin and transforming growth factor-beta receptor-regulated Smads. *J. Biol. Chem.* **2001**, *276*, 656–661. [CrossRef]

64. Guglielmo, G.M.D.; Roy, C.L.; Goodfellow, A.F.; Wrana, J.L. Distinct endocytic pathways regulate TGF-beta receptor signalling and turnover. *Nat. Cell Biol.* **2003**, *5*, 410–421. [CrossRef]

65. Dick, A.; Mayr, T.; Bauer, H.; Meier, A.; Hammerschmidt, M. Cloning and characterization of zebrafish smad2, smad3 and smad4. *Gene* **2000**, *246*, 69–80. [CrossRef]

66. Schier, A.F. Nodal morphogens. *Cold Spring Harb. Perspect. Biol.* **2009**, *1*, a003459. [CrossRef] [PubMed]

67. Meno, C.; Gritsman, K.; Ohishi, S.; Ohfuji, Y.; Heckscher, E.; Mochida, K.; Shimono, A.; Kondoh, H.; Talbot, W.S.; Robertson, E.J.; et al. Mouse Lefty2 and zebrafish antivin are feedback inhibitors of nodal signaling during vertebrate gastrulation. *Mol. Cell* **1999**, *4*, 287–298. [CrossRef]

68. Miyazawa, K.; Miyazono, K. Regulation of TGF-beta Family Signaling by Inhibitory Smads. *Cold Spring Harb. Perspect. Biol.* **2017**, *9*, a022095. [CrossRef]

69. Conidi, A.; Cazzola, S.; Beets, K.; Coddens, K.; Collart, C.; Cornelis, F.; Cox, L.; Joke, D.; Dobreva, M.P.; Dries, R.; et al. Few Smad proteins and many Smad-interacting proteins yield multiple functions and action modes in TGFbeta/BMP signaling in vivo. *Cytokine Growth Factor Rev.* **2011**, *22*, 287–300. [CrossRef]

70. Park, D.S.; Yoon, G.H.; Kim, E.Y.; Lee, T.; Kim, K.; Lee, P.C.; Chang, E.J.; Choi, S.C. Wip1 regulates Smad4 phosphorylation and inhibits TGF-beta signaling. *EMBO Rep.* **2020**, *21*, e48693. [CrossRef] [PubMed]

71. Zhang, Y.; Shao, M.; Wang, L.; Liu, Z.; Gao, M.; Liu, C.; Zhang, H. Ethanol exposure affects cell movement during gastrulation and induces split axes in zebrafish embryos. *Int. J. Dev. Neurosci.* **2010**, *28*, 283–288. [CrossRef] [PubMed]

MDPI AG
Grosspeteranlage 5
4052 Basel
Switzerland
Tel.: +41 61 683 77 34

Biomedicines Editorial Office
E-mail: biomedicines@mdpi.com
www.mdpi.com/journal/biomedicines